The Microscope in the Dutch Republic

Emphasizing the work of Jan Swammerdam and Antoni van Leeuwenhoek, *The Microscope in the Dutch Republic* dissects the social, cultural, and emotional circumstances that shaped early microscopic discovery. Arguing that the aspects of seventeenth-century Dutch culture widely assumed to have favored the lens actually impeded its serious use, Ruestow focuses on social contexts and on Swammerdam and Leeuwenhoek's social sensibilities as the key source of their commitment to the new instrument. He also analyzes how they drew upon their cultural background to vest microscopic images with meaning, though with strikingly different emphases. Having underscored how their influential contributions to the debates over generation also illustrated the problematic role of early microscopic observations, Ruestow concludes with reflections on the eighteenth-century decline and nineteenth-century resurgence of microscopic research and on the impact of institutionalization.

THE MICROSCOPE IN THE DUTCH REPUBLIC

The Shaping of Discovery

EDWARD G. RUESTOW

University of Colorado – Boulder

PUBLISHED BY THE PRESS SYNDICATE OF THE UNIVERSITY OF CAMBRIDGE
The Pitt Building, Trumpington Street, Cambridge, United Kingdom

CAMBRIDGE UNIVERSITY PRESS
The Edinburgh Building, Cambridge CB2 2RU, UK
40 West 20th Street, New York NY 10011–4211, USA
477 Williamstown Road, Port Melbourne, VIC 3207, Australia
Ruiz de Alarcón 13, 28014 Madrid, Spain
Dock House, The Waterfront, Cape Town 8001, South Africa

http://www.cambridge.org

First published 1996
First paperback edition 2004

A catalogue record for this book is available from the British Library

Library of Congress Cataloguing in Publication data
Ruestow, Edward G. (Edward Grant), 1937–
The microscope in the Dutch Republic: the shaping of discovery /
Edward G. Ruestow.
p. cm.
ISBN 0 521 47078 1 (hc)
1. Microscopes – Netherlands – History – 17th century.
2. Leeuwenhoek, Antoni van, 1632–1723.
3. Swammerdam, Jan, 1637–1680. I. Title.
QH204.R84 1996
502′.8′20949209032–dc20 95-45548 CIP

ISBN 0 521 47078 1 hardback
ISBN 0 521 52863 1 paperback

To Carmen
For abiding patience and so much more

Contents

List of Illustrations viii
Acknowledgments ix
List of Abbreviations xi

Introduction 1
1. Of Light, Lenses, and Glass Beads 6
2. Seeming Invitations 37
3. Obstacles 61
4. Discovery Preempted 81
5. Swammerdam 105
6. Leeuwenhoek I: A Clever Burgher 146
7. Leeuwenhoek II: Images and Ideas 175
8. Generation I: Turning against a Tradition 201
9. Generation II: The Search for First Beginnings 223
10. A New World 260
Conclusion 280

References 305
Index 339

Illustrations

1. A spectacle maker's storefront by Jan and Kaspar Luiken 9
2. Pierre Lyonet's dissecting microscope 10
3. Abraham Trembley's microscopic apparatus for observing hydras 11
4. Three typical Leeuwenhoek microscopes 13
5. Diatoms seen through a Leeuwenhoek lens 17
6. Christiaan Huygens's drawing of a droplet lens microscope 20
7. The lens in the Leeuwenhoek microscope in the Utrecht University Museum 21
8. Bouquet with insects by Johannes Goedaert 55
9. Hollow tubes in the nerves from the *Tractatus physico-medicus de homine* (1689), ascribed to Theodoor Craanen 67
10. A drawing of flies by Jacques de Gheyn II 75
11. A piece of the cortex of the brain prepared by Frederik Ruysch 99
12. Muscle fila by Wijer Willem Muys 103
13. The ovaries of a queen bee by Swammerdam 115
14. The viscera of a butterfly by Swammerdam 133
15. The capillaries of an eel drawn by one of Leeuwenhoek's draftsmen 177
16. Leeuwenhoek's depictions of muscle fiber striations 187
17. Rat erythrocytes 193
18. Fish corpuscles drawn by one of Leeuwenhoek's draftsmen 195
19. Leeuwenhoek's wax models of a yeast globule and a corpuscle composed of thirty-six smaller globules 197
20. The testicle of a flea by Leeuwenhoek 205
21. The ovaries of a louse by Swammerdam 207
22. A spermatozoon of a dragonfly by Leeuwenhoek 219
23. Vessels in the semen by Leeuwenhoek 254
24. A human spermatozoon by Leeuwenhoek 255
25. Spermatozoa drawn by one of Leeuwenhoek's draftsmen 256
26. "Vortices" of animalcules in the shipworm by Gottfried Sellius 266

Acknowledgments

A little over two decades ago, I launched this study of early microscopy in the Netherlands with a National Science Foundation Grant. That grant was followed four years later by a fellowship from the American Council of Learned Societies, itself supported in turn by a grant from the National Endowment for the Humanities. The assumptions about the dynamics of early microscopy with which I began the project proved ill-founded, however, and a more satisfying understanding took years to attain. Nonetheless, the initial period of supported study laid the essential foundations for what eventually emerged. After all the threads had ultimately come together, the Graduate Committee on the Arts and Humanities of the University of Colorado at Boulder and the Kayden Advisory Committee of the same institution contributed critically to the publication of the end result, this book. I remain deeply grateful to all of these sources of support.

I have profited immensely from the assistance of many individuals as well. Wittingly and unwittingly, numerous colleagues at the University of Colorado have contributed in diverse ways, be it by sharing their research experience, inviting me into their labs, trying their hand at a bit of Latin that resisted my best efforts, or helping to ease some of my own entangled prose. Among those colleagues to whom I am thus indebted I wish to mention in particular Todd Gleeson, Jane Bock, Shi-Kuei Wu, F. Marc Laforce, Url Lanham, Dick McIntosh, Anne Bekoff, Ernst Fredricksmeyer, Joop de Heer, Paul Levitt, Paul Winston, and Harold Walton. Their generosity in sharing their varied expertise represents the university at its best.

No less heartening has been the similar generosity of more distant scientists who responded to the rather odd inquiries of an historian in the Rocky Mountains. L. Lavenseau of the Laboratory of Neuroendocrinology at the University of Bordeaux enlightened me on what Jan Swammerdam may have seen in the ovaries of the caterpillar of *Orgyia antiqua;* Ruth D. Turner of Harvard University commented extensively on Gottfried Sellius's account of the shipworm *Teredo navalis* Linnaeus; M. Martoja of the Institut Océanographique in Paris shared her thoughts on the "crystalline globules" Swammerdam confronted in the snail *Viviparus viviparus* (Linnaeus), and J. van Zuylen of Zeist in the Netherlands deepened my understanding of seventeenth-century microscopes. In the realm of the arts as well, Ulysses Dietz of the Newark Museum allowed me to tap his knowledge of seventeenth-century domestic furnishings.

J. van Zuylen was kind enough indeed to read the manuscript of what was to become Chapter 1, and Albert Van Helden of Rice University likewise read Chapter 1 and the Introduction. I profited greatly from what both had to say. Christoph Lüthy of the graduate program at Harvard also gave the whole manuscript a very close and thoughtful critique, and Mordechai Feingold of the Virginia Polytechnic Institute and State University offered what proved a crucial suggestion for the manuscript's final reshaping.

Doubtless, few of those who have contributed in so many different ways will feel comfortable with all the arguments that follow in this book. Nonetheless, they have enriched those arguments immeasurably.

Abbreviations

Collections

AvL Letters — Antoni van Leeuwenhoek. Letters. Royal Society Library, London.

JS Papers — Jan Swammerdam. Papers. University of Göttingen Library, Göttingen.

Names

AvL	Antoni van Leeuwenhoek
ChrH	Christiaan Huygens
ConHj	Constantijn Huygens, Jr.
ConHs	Constantijn Huygens, Sr.
JS	Jan Swammerdam
NH	Nicolaas Hartsoeker
RS	The Royal Society, London

Works

AB — *Alle de brieven van Antoni van Leeuwenhoek*. 12 vols. to date. Amsterdam: Swets and Zeitlinger, 1939–.

Catalogus — *Catalogus van het vermaarde cabinet van vergrootglasen, met zeer veel moeite, en kosten in veele jaren geïnventeert, gemaakt, en nagelaten door wylen den heer Anthony van Leeuwenhoek*. Delft: Reinier Boitet, 1747.

DSB — *Dictionary of Scientific Biography*. Edited by Charles Coulston Gillispie. 16 vols. New York: Charles Scribner's Sons, 1970–80.

Gen. An., *Hist. An.*, *Parts An.* — *Generation of Animals, History of Animals,* and *Parts of Animals,* respectively, in *The Complete Works of Aristotle*. Edited by Jonathan Barnes. 2 vols. Princeton: Princeton University Press, 1984.

LJS — *The Letters of Jan Swammerdam to Melchisedec Thévenot*. Edited and translated by G. A. Lindeboom. Amsterdam: Swets and Zeitlinger, 1975.

Pliny *HN* Pliny. 1938–63. *Natural History [Historia naturalis]*. With an English translation by H. Rackham, W. H. S. Jones, and D. E. Eichholz. 10 vols. Cambridge: Harvard University Press / London: William Heinemann.

OCCH *Oeuvres complètes de Christiaan Huygens*. 22 vols. in 23. The Hague: Martinus Nijhoff, 1888–1950.

Introduction

The character of European science changed profoundly during the course of the seventeenth century. Aggressive experimentation established its place in scientific practice, and a new mathematical mechanics embracing both the terrestrial and celestial realms overthrew a divided Aristotelian cosmos and its qualitative, teleological physics. To prove no less distinctive of modern science, however, new scientific instruments created or discovered new phenomena and began, for the first time, to extend the reach of the human senses.[1]

The identification of science with discovery was accentuated, although it was discovery that assumed many forms. Architects of the new mechanics grasped new truths by exploring avenues of opportunity in the dematerialized realm of mathematics,[2] while anatomists cast about for previously unseen structures concealed in the gore of dissected bodies. A single, subtly designed pendulum experiment nudged the gradual unfolding of a deeper generality in an evolving system of ideas, and the air pump inspired a broad program of experimentation that explored the phenomena encountered in a vacuum.[3] But two other new instruments, the telescope and microscope, emerged as perhaps the most widely recognized and evocative new symbols of scientific discovery.

The beginnings of telescopic discovery dramatically impressed the age. In 1610 the English ambassador in Venice described Galileo's newly printed *Sidereus nuncius* as "the strangest piece of news (as I may justly call it) that he hath ever yet received from any part of the world." In it, Galileo recounted the discovery of mountains and craters on the Moon, new planets revolving around Jupiter, and the multitude of previously unrecognized stars that composed the Milky Way. In distant Wales, Sir William Lower, who himself had already observed the Moon through the telescope, exclaimed admiringly nonetheless that "my diligent Galileus hath done more in his three fold discoverie than Magellane in opening the streightes to the South sea or the dutch men that weare eaten by beares in Nova Zembla."[4]

Galileo soon published his account of the phases of Venus as well and reported several odd and inexplicably varying images of Saturn that Christiaan Huy-

1 Van Helden (1983a).
2 Yoder (1988), 26, 50, 59–63, 84–5, 95; see also I. B. Cohen (1980), 52–68.
3 Westfall (1971), 463; M. B. Hall (1965), 94–109.
4 Nicolson (1956), 33, 35, 37.

gens in the Netherlands finally, in 1656, grasped as manifestations of a planetary ring.[5] Huygens had also sighted the first satellite of Saturn the year before, and prompted by this young Dutchman's abilities, Henry Oldenburg, the future first secretary of the Royal Society of London, evoked again the older exemplar of discovery and the emerging expectation of more to come: With yet more perfect telescopes, he wrote, "we may by their means make navigations as well into ye Heavens and discover new Countries there" as Columbus had done in America.[6]

But discovery with the telescope proved halting. Huygens's discovery of the satellite of Saturn in 1655 was in fact the first major discovery in several decades, and after the sighting of the fourth and fifth satellites of Saturn in 1684, planetary discoveries ceased for nearly a century.[7] Shortly before his death in 1694, Huygens himself confided to his countryman Antoni van Leeuwenhoek that he believed astronomical observation had reached its limits, and that little more was to be seen.[8]

Delayed in its beginnings, the career of meaningful microscopic discovery also threatened to be a fleeting one. The pace of a quarter-century of successive discoveries in the late seventeenth century was not to be equaled until the nineteenth century, and by the early 1690s Robert Hooke was lamenting the waning ardor in the use of both the microscope and telescope. Most of those who had pursued these inquiries had now "gone off the Stage," he noted, and the opinion now prevailed "that the Subjects to be enquired into are exhausted, and no more is to be done."[9] Although the preeminent remaining microscopist, the Dutchman Leeuwenhoek, had over thirty years of persisting microscopic researches still ahead of him, the tempo of microscopic discovery had already slackened.

The telescope and microscope had made an enduring impact on the European imagination nonetheless and, together, left their distinctive impress on the new image of the world the seventeenth century had forged. The sweep from the astronomical imagery of the telescope to the newly discovered realm of the microscopic became a standard literary trope, embracing a symmetry of dimensions that stretched away endlessly in both directions.[10] The early English devotee of the microscope Henry Power proposed in 1661 "that the least Bodies we are able to see with our naked eyes, are but middle proportionals (as it were) 'twixt the greatest and smallest Bodies in nature, which two Extremes

5 On the struggle over the interpretation of Saturn's ring, see Van Helden (1974a,b).

6 Oldenburg (1965–), 1:277. Albert Van Helden points out that the expectation of a continuing improvement in scientific instruments emerged only during the course of the century (Van Helden 1983a, 49, 54–5, 65–6, 68–9).

7 Van Helden (1980), 147–9, 154; idem (1985), 129ff., esp. 154–5, 161, 163; idem (1983b), 137–8.

8 AvL to RS, 26 Feb. 1703, AvL Letters, fol. 240v. Huygens, however, had fallen well behind the cutting edge of telescopic observation and had made no new telescopic discoveries himself since the 1650s (Van Helden 1974a,167). Not everyone looked upon the progress of astronomy so pessimistically; see Basnage (1687–1709) 23:155, 159.

9 Hooke (1726), 261.

10 Hill (1752a), 105, 288–9; Schatzberg (1973), 69, 99, 258–9, 281; Jones (1966), 24–5, 128, 215–16; Saine (1976), 63–4.

lye equally beyond the reach of humane sensation." It was a prospect, however, whose testimony to the march of science was not without its ambiguity. Reflecting on Power's passage a century later, the German microscopist Wilhelm Friedrich von Gleichen-Russworm concluded that we could not but be shaken by the narrowness of our understanding; everything our dull senses perceive around us is pure mystery, he wrote, and still only the smallest part of God's boundless Creation. In less exalted tones, the Jesuit Athanasius Kircher had already remarked in 1646 that the microscope and telescope revealed that everything we see is very different from what it seems.[11]

Notable differences distinguished the usual contexts of microscopic and telescopic observation, however, and hence the nature of the discoveries to which they led. Despite the suggestion of limitless space, the revolutionary impact of telescopic discovery lay more in the recognition of what was familiar than in the encounter with what was not. The most consequential of Galileo's discoveries revealed that the Moon had mountains and valleys like those on Earth, that Jupiter had satellites like the Moon, and that Venus had phases like those of the Moon. Although Henry Oldenburg remarked that "ye vulgar opinion of ye unity of ye world" had now been "exploded," the new heavens in fact differed most fundamentally from the old in the new affinity of Earth and the other planets.[12] Apart from the vast distances and the surprise and uniqueness of Saturn's ring,[13] the telescope discovered little for which precedents and analogies did not readily come to mind. Not so in the realm of the microscope. Although many of its revelations echoed the familiar world as well, the microscope proved an increasing source of images and phenomena whose gripping impact lay rather in their strangeness.

Telescopic discoveries in the seventeenth century were also quickly incorporated into well-developed systems of astronomy. These discoveries helped overthrow a comfortable, traditional cosmos, to be sure, but they were most highly prized in scientifically progressive circles precisely because they simultaneously seemed to argue for a new alternative. They had confirmed the true system of the heavens, said Huygens, who perceived an argument for Copernican theory even in Saturn's ring.[14] In the case of notable microscopic discoveries, however, the immediate challenge was to make sense of them, and the conceptual resources at hand were usually crude, ill-fitting, or simply unpersuasive.

Adapted to one astronomical theory or another (Copernican, that is, or Tychonic), telescopic discoveries were immediately set within a framework of precise and measured order. In their slow progress along unchanging paths, the celestial motions had long been the epitome of regularity, while, even through the telescope, the appearance of the celestial bodies themselves conformed to an austere simplicity. Through better instruments, the stars still appeared as shin-

11 Power (1664), preface pp. [5]–[6]; Gleichen-Russworm (1764), [2]; Kircher (1646), 834–5.
12 Oldenburg (1965–) 1:277; Van Helden (1974c): 57.
13 Nothing had so surprised the astronomers of the century as had Saturn's ring, noted Huygens's countryman Nicolaas Hartsoeker (1694, 186); see also Van Helden (1974a), 158.
14 *OCCH* 13:434–5, 438–41, 740; see also ibid., 586; Van Helden (1974a), 163.

ing points, and the planets, apart from Saturn's anomalous ring, as starkly illuminated spheres. The backdrop of space was unrelieved blackness, against which even the colors of the planets appeared muted and pale. The microscopic observer, on the other hand, was confronted with the rich visual complexity of diverse and unusual textures, bizarre forms, and unexpected hues and light effects.[15] Dramatically accentuated by the lens,[16] the motion encountered through the microscope was often relentlessly irregular, frantic, and, as it was also remarked, inexpressible.[17]

Telescopic and microscopic observations in the seventeenth century tended to cater indeed to two contrasting aesthetics in modern science. Historically and emphatically a mathematical science, astronomy embraced the ideals of simplicity and demonstrable necessity in its conceptual construction. Most often applied to the realm of living things, on the other hand, the microscope accentuated rather the endless and often inexplicable diversity of natural forms and what seemed at times their superfluous and irrepressible abundance. By the end of the century, astronomy and the telescopic discoveries had been incorporated virtually in toto into a Newtonian "system of the world" derived from three universal laws and a framework of "mathematical principles"; but the microscope testified to an evermore intricate complexity in nature and a pervasive and continuing unexpectedness.

The difference between telescopic and microscopic discovery was not absolute. Huygens alluded to the novelty also encountered through the telescope,[18] and Saturn's ring in particular remained a unique and baffling phenomenon. The discovery of blood capillaries, on the other hand, was conceived from the start in terms of a broad, well-articulated theory, that of the circulation of the blood, and lent itself to the prospect, at least, of a mathematical rendering.[19] Microscopic researches into organic generation searched as well for a deeper order and unity underlying the diversity of life. Nonetheless, whereas the characteristic telescopic discoveries were quickly assimilated to a theoretical system and a simplifying mathematical order, early microscopic observations underscored nature's capacity for endless surprises and for images that challenged the limits of the imagination.

The microscope made the experience of discovery more widely accessible as well. Potential astronomical discoveries in the seventeenth century were narrowly limited in number, and, apart from the surfaces of the Moon and Sun,

15 One twentieth-century amateur naturalist – hence perhaps closer in spirit to the characteristic seventeenth- and eighteenth-century observer – with a reflective and lyrical bent describes the realm of microscopic life in particular as a "color-charged, glistening world" (Dillard 1974, 126). A contemporary microbiologist vividly conveys that sense as well in writing of the symbiotic relationship between *Paramecium bursaria* and the alga *Chlorella:* L. L. Larison Cudmore describes the paramecium as "golden, transparent and candescent, its body covered with thousands of hairlike cilia, beating in sensuous waves," and as having within that body "hundreds of tiny glowing roses, emerald green shining with refracted light" (Cudmore 1978, 43).

16 Lumsden (1980), 1: 132–3.

17 Müller (1786), xviii–xix.

18 *OCCH* 13: 586.

19 See Chapter 7, n. 42.

they presumed a technical knowledge of astronomy sufficient, at a minimum, for finding one's way through the night sky. Potential microscopic discovery, on the other hand, lay immediately at hand and all about, and everything was new. Moreover, microscopes of exceptional optical qualities (for the time, of course) could be made with minimal technical expertise; even a dexterous youngster toying with a bit of glass around a candle flame might contrive a simple but powerful instrument.[20]

For the first half-century, nonetheless, the opportunities for microscopic discovery were only hesitantly explored. To be sure, early microscopes were burdened with technical shortcomings, and those that in time proved optically the best – and simplest – proved the more troublesome to use. Nature's microscopic structures were difficult to manipulate, and to display them before an instrument could pose the greatest technical challenge of all. There were obstacles that pertained to the imagination as well. An indifference and perhaps blindness to the prospect of microscopic discovery were sustained in varied and subtle ways. So obstructive were such barriers that in the Netherlands, where Leeuwenhoek and Jan Swammerdam emerged among the principal pioneers of early microscopic discovery, the beginnings of that discovery required the stimulus of acute personal reactions to goading social circumstances. Early microscopic discovery unfolded hence as an intricate interplay of cultural traditions, social relations, and personal sensibilities. How that interplay forged the experience of discovery is indeed the subject of this book.

First, however, there had to be the instrument, and, though the source of so much that was unexpected, the microscope itself had required a preceding jolt to the imagination. That jolt was provided by the invention of the telescope.

20 See Chapter 1 regarding n. 90.

CHAPTER ONE

Of light, lenses, and glass beads

The news echoed the dreams of the magicians. At The Hague in late September 1608, a spectacle maker from Middelburg had presented Count Maurice, Stadholder and Captain-General in the Dutch Republic, with "glasses" (*brillen*) with which, through a tube, distant things could be seen as if nearby.[1] According to the account published shortly after the event, the clock in the city of Delft had been clearly seen from a tower in The Hague, as had the windows of the church at Leiden more than twice as far away.[2] The account was soon reprinted in France and reached Venice before the end of the year; by that time the Dutchman Cornelis Drebbel, living off his skills in natural magic, was also writing homefrom London to inquire into the new ability to see so far.[3] Conscious of the invention's military potential, Count Maurice and the States General of the Dutch Republic enjoined the Middelburg craftsman to keep its construction to himself, but its simplicity – a foot-long tube with a concave eyepiece at one end and a weaker convex lens at the other – could not be withheld from an inquisitive world.[4]

Indeed, within weeks of that initial scene in The Hague, two other Dutchmen, including yet another spectacle maker from Middelburg, came forward to testify that they also had made such instruments, and by the following spring the enemy in Brussels knew the secret too.[5] About the time of the demonstration in The Hague, in fact, a Dutch peddler was already hawking a telescope at the Frankfort fair, and it was perhaps the same peripatetic Dutchman who was advertising the instrument in northern Italy during the following spring and summer.[6] Galileo, who appears to have first heard of it at Venice in May, set his mind to constructing his own.[7]

1 Meteren (1614), fol. 661v.
2 Waard (1906), 205.
3 Ibid., 206, 231–2; Van Helden (1977), 26, 40–1nn. 9, 10; Bedini (1967), 273; Jaeger (1922), 35–7; Tierie (1932), 34.
4 Waard (1906), 202, 205, 290; Van Helden (1977), 11–12, 36.
5 Waard (1906), 23–4, 172–4, 199–200, 211, 219–20, 230.
6 Ibid., 162–8, 226, 233–6, 247–8; Van Helden (1977), 21 and n. 18.
7 Whitaker (1978), 157, 166; cf. S. Drake (1970), 144, 146.

From the telescope soon evolved the compound microscope.[8] Galileo early used what was in effect an extended telescope to peer at flies, but the Dutch diplomat Willem Boreel at midcentury claimed the invention of the microscope for Sacharias Janssen, a boyhood friend in Middelburg, and for Janssen's father.[9] Boreel described a microscope made by Sacharias that he himself had seen in 1619; constructed with a foot-and-a-half gilded brass tube rising vertically from three dolphin-shaped legs, it had fallen by then into the hands of their compatriot in London, Cornelis Drebbel.[10]

The compound microscope of the future, however, was to derive from a system of two convex lenses rather than the combined convex and concave lenses of the first telescopes. Paired convex lenses made the microscope a shorter and hence more manageable instrument while also offering a broader field of vision. The nature of the lenses in the Janssen instrument owned by Drebbel remains unknown, and the earliest sure description of a microscope with two convex lenses pertains to an instrument made by Drebbel himself.[11] Famous in his own time for such marvels as a fluid (barometric and thermometric) perpetual-motion machine and successful submarine trials on the Thames, Drebbel also prided himself on the "wonders and rare things" he had achieved through optics. For a time, at least, he turned to the making and marketing of microscopes.[12] In 1622, a member of the family was hawking Drebbel's instruments in courtly circles on the continent, and Nicolas-Claude Fabri de Peiresc described in a personal memorandum the microscope he had seen demonstrated before the Queen Mother of France.[13] An inch in diameter, the tube of the *lunette* was about the length of a traveler's quill case (*un canon d'escrittoire*), although so constructed of three pieces of gilded brass that its length could be adjusted. Both sides of the eyepiece lens at the top of the instrument had the curvature of "a rather small globe," while the single curved surface of the object lens had the curvature of a small cherry and peered through a hole no wider than the diameter of a pin.

8 In the nineteenth century, Pieter Harting argued that the microscope had been invented before the telescope (Harting 1848–54, 2[pt. 3]:30–4), but that contention has not remained convincing (see Waard 1906, 44, 160, 303–4; Van Helden 1977, 5n. 1; cf. Disney 1928, 94–5, 97).

9 Bedini (1967), 283–4; Borel (1655), 35–6.

10 Waard (1906), 162–8, 175–6, 234–6, 240.

11 In 1646 Francesco Fontana claimed that he himself had invented the microscope composed of only convex lenses in 1618 (Fontana 1646, 145–6), but the witness he cited referred only to the year 1625, when Drebbel's microscope was already known in Italy (Waard 1906, 297–8).

12 Drebbel (1732?), pt. 1:11; Tierie (1932), 41, 53–4, 59.

13 Humbert (1951), 155–8; Tierie (1932), 55–7. Drebbel's countryman and contemporary Isaac Beeckman included in his journal a section of a letter from Drebbel to James I describing the devices Drebbel could produce, to which Beeckman then added a very crude drawing of what seems to be one of Drebbel's compound microscopes (Beeckman 1939–53, 3:442; and G. L'E. Turner 1980, 3). J. van Zuylen believes that the microscope Christiaan Huygens describes in *OCCH* 13:674–5 was also very likely that which his father, Constantijn Huygens, Sr., had purchased from Drebbel and that Christiaan was studying thirty years later in anticipation of building his own instrument (letter to the author, 8 March 1990). If so, then we have a sketch and the proportions of a Drebbel compound microscope preserved for us by Huygens as well.

Microscope makers experimented in the following decades with additional lenses,[14] but this elaboration of lens systems contributed little if anything to the burgeoning of microscopic observation and discovery that began in the 1660s. Hooke, whose *Micrographia* in 1665 was the first of the classics of microscopic literature, reverted from three to the original two lenses when he wished to examine the small parts of an object more accurately,[15] and to no small extent the efflorescence of discovery rested on observations that were made in fact with a single lens. For the beginnings of serious microscopic exploration, after all, the most significant consequence of the encounter with the effects of lens systems was simply an awakened consciousness of the potential of magnification itself and, hence, of even a single lens.

Spectacle lenses had been known in Europe since the late Middle Ages and were being peddled in the streets when Count Maurice first looked through a telescope.[16] Nonetheless, scientific or literary references to lenses remained a rarity until the sixteenth century, when they were cited not only as an aid to troubled vision but also as a source of "strange sightes" and deceptive illusions, and with little apparent thought of the possible value of magnification to the investigation of nature.[17] The invention of the telescope, in contrast, evoked the immediate recognition of its potential usefulness to astronomy, and its application soon revealed what surprises might lie just beyond what the naked eye could see.[18] Likewise the early, two-lens microscope, despite its still very modest powers – Drebbel's microscope, according to Peiresc, enlarged a mite to the size of a large fly[19] – inspired anticipations of new reaches of nature to be found through magnification. Recalling his own experiences with one of Drebbel's instruments, Constantijn Huygens – the gifted servitor of the House of Orange

14 Rooseboom (1967), 269–70; Bedini (1963), 387–8; Griendel von Ach (1687), 3–4.
15 Hooke (1665), preface [xxii].
16 Hollstein (1949–), 15:36.
17 Fracastoro (1584), 13r–13v; see Van Helden (1977), 12–16. In 1523 the Florentine Giovanni Rucellai published in verse the observations he had made of the anatomy of a bee with a concave mirror, but no one followed up on this early use of magnification in the exploration of nature (Ronchi 1967, 203–4). On the other hand, one of the contributors to Thomas Moffett et al., *Insectorum sive minimorum animalium theatrum* (1634), wrote apparently in the late sixteenth century of the moving intestines and eggs within the flea without mentioning any instruments of magnification (p. 276). On the possible use of the lens in this work, see Singer (1915), 322.

Vasco Ronchi has argued that, in a "conspiracy of silence," scholars purposefully ignored lenses for three centuries because of a distrust of the images they formed and of vision itself – a bias, he asserts, finally overturned by Galileo and Kepler (Ronchi 1970, 71–4, 76, 97; idem 1967, 196–7, 203–4). David C. Lindberg and Nicholas H. Steneck have challenged Ronchi's arguments, however (Lindberg & Steneck 1972), and I have encountered nothing in my own research that supports his contention.

18 Although there is no evidence of astronomical observations by Galileo before 1610 (S. Drake 1970, 154), the initial account of the presentation of a telescope to Count Maurice published in The Hague had already recognized that the instrument would be of value in observing the stars (Waard 1906, 205). Regarding early telescopic observations of the heavens other than Galileo's, see Waard (1906), 141–2, 177–8, 188, 206, 259–60; G. L'E. Turner (1969), 75n. 17; Ploeg (1934), 21.
19 Humbert (1951), 155.

Figure 1. A spectacle maker's storefront with magnifying glasses and other optical instruments displayed among the wares (Luiken & Luiken 1694, 43). Photo courtesy of the Newberry Library, Chicago.

and father to Christiaan – wrote enthusiastically around 1630 of the "new theatre of nature," indeed "another world," discovered there.[20]

That prospect having now been grasped, the single lens soon also appeared as a magnifying glass or simple microscope, and in 1637 Descartes alluded to the

20 ConHs (1897), 120.

Figure 2. Pierre Lyonet's dissecting microscope. Courtesy of the Artis Library, University of Amsterdam.

flea glasses that had now become a commonplace.[21] Sold not only in the spectacle makers' shops (Fig. 1) but ultimately even door to door, such devices were only toys, but by the end of the century simple microscopes of a more serious nature were being marketed in a variety of forms. The simple microscope readily lent itself as well to special adaptations. In the following century, the Dutchman Pierre Lyonet described and illustrated the microscope apparatus (Fig. 2) he used in his insect dissections, the finest of the time; a lens was now suspended at the end of an arm composed of a succession of ball-and-socket joints above a small mahogany dissecting table, itself elevated above a wooden base with small drawers to hold Lyonet's other implements.[22] His Swiss friend Abraham Trembley also contrived a distinctive arrangement by which a similarly jointed arm held a lens before a vessel of water to facilitate his study of the freshwater hydra (Fig. 3), a study that provided the most unsettling scientific observations of the age.

The most celebrated microscopes of this period, however, were those made in the hundreds – though all for his personal use – by Leeuwenhoek.[23] As one

21 Descartes (1897–1910), 6:155.

22 Lyonet (1762), "Lettre à M. le Cat"; Star (1953), 49–52.

23 Leeuwenhoek himself testified in 1700 that he had hundreds and hundreds ("hondert en hondert") of ground microscopes (AvL 1702, 305), and Star estimates that there must have been no fewer than about 550 microscopes at Leeuwenhoek's death (Star 1953, 24). Leeuwenhoek also

Figure 3. Abraham Trembley's microscopic apparatus for observing hydras (*Philosophical Transactions,* no. 484 [1747], pl. 1). Photo courtesy of the Special Collections Department, University of Colorado at Boulder Libraries.

eighteenth-century aficionado remarked, they were as simple in construction as they could be.[24] The typical Leeuwenhoek instrument comprised a small lens clamped between two thin and roughly worked plates of brass, silver, or on occasion gold, and in the middle of one of the plates was fixed an arrangement of screws and a pin – or sometimes a clamp for a capillary tube – to hold and position the object before the lens (Fig. 4). The flat plates that hold the lenses of the few known surviving examples – only a half-dozen are complete and reasonably well authenticated – measure generally something over 4 cm in length and approximately 2 cm in width, so that they lie easily in the palm of the hand.[25]

What most distinguished the varied research instruments from the toys was the clarity, resolution, and power of their lenses, and the power of magnification was a function of the size of the lens itself. Indeed, an innovation more important to early microscopic discovery than the invention of new lens combinations – certainly, at least, in the Netherlands – was the ultimate recourse to lenses that themselves were very small.

Footnote 23 (cont.)
told interested visitors that he refused to either sell them or give them away (*OCCH* 9:38; Uffenbach 1753–4, 3:358).

24 H. Baker (1740), 505.

25 Ford (1991), 141–63. On their rough workmanship, see also Cittert (1934a), 14; Star (1953), 26–7; Uffenbach (1753–4), 3:360.

The first telescopes and microscopes were surely constructed with spectacle lenses, but the continuous effort henceforward to enhance their effectiveness as scientific instruments soon led to lenses of very different sizes and shapes. It was not obvious in the early seventeenth century that the smaller the lens – or more precisely, the smaller the radius of its surface curvature – the greater its power of magnification;[26] but smaller and more sharply curved lenses were soon being ground as microscope objectives, at first apparently because, with their shorter focal lengths, they allowed the instrument to be brought closer to the object being observed.[27] The curvature of a small cherry ascribed by Peiresc to the objective of Drebbel's microscope was already a considerable departure from a spectacle lens. Moreover, if Galileo's microscopes in 1624 were in fact inspired by Drebbel's, the difficulty acknowledged by the Pisan with the lenses of his own instruments might indeed reflect how truly unusual Drebbel's lenses were.[28]

Whatever the initial reason for resorting to smaller objective lenses, however, it was not such as to produce a continuing effort to reduce their size still further. (A lens, after all, could come too close to the object for convenience.) In 1654 a youthful Christiaan Huygens, already making his own first microscopes or preparing to do so, appears to have ordered a lens as sharply curved as a local lens maker could grind it, and it may indeed have been a planoconvex objective lens with which he worked that year whose curvature, with a radius of roughly 8 mm, was still similar to that of Drebbel's (i.e., to the curvature one might ascribe to a small cherry). Fourteen years later, however, Christiaan was inclined to lenses with a focal distance of roughly an inch, and he pointedly rejected small lenses as objectives – primarily, it seems, because of their shallow depth of focus.[29] To be sure, Hooke wrote in 1663 of attempting to make a microscope lens so small that he was "fain to use a magnifying glass to look upon it,"[30] but a fascination with the potential of truly minute lenses was aroused only by the observations Leeuwenhoek began to report in the 1670s. In 1680 members of the Royal Society were admiring a biconvex lens no more than one-twentieth of an inch in diameter,[31] and Christiaan Huygens, now with a very altered

26 Giovan Battista della Porta, who had cultivated an interest in lenses perhaps as long as anyone in the early seventeenth century, asserted the very opposite toward the end of his life: that smaller and more sharply curved surfaces magnified objects less than larger lenses and spheres (Porta 1962, 69, pr. 22; 93, pr. 26; 109, prs. 12,13; 115, pr. 2; 137, prs. 1,2).

27 See Descartes (1897–1910), 6:199–200, 206.

28 Bedini (1967), 285. Constantijn Huygens, Sr., also recorded that the objective lens of the Drebbel microscope he had seen was scarcely as large as the nail on the average little finger – "amplitudine auricularis digiti medium unguem vix adaequat" (ConHs 1897, 119).

29 *OCCH* 1:302, 13:675, 6:206; see also 13:cxxxvii–cxxxviii. Whereas J. van Zuylen believes it to be a Drebbel microscope (see n. 13 above), the editors of *OCCH* are inclined to believe that the microscope described in Christiaan's papers on 13:674–5 is in fact one of the early instruments constructed by Christiaan and his brother Constantijn (674n. 2). My own brief remarks earlier in the chapter are indebted in several ways to Dr. van Zuylen's discussion of the question in the letter cited earlier.

30 *OCCH* 4:382.

31 Birch (1756–7), 4:36–7; see also Singer (1914), 267–8; Frison (1965), 103.

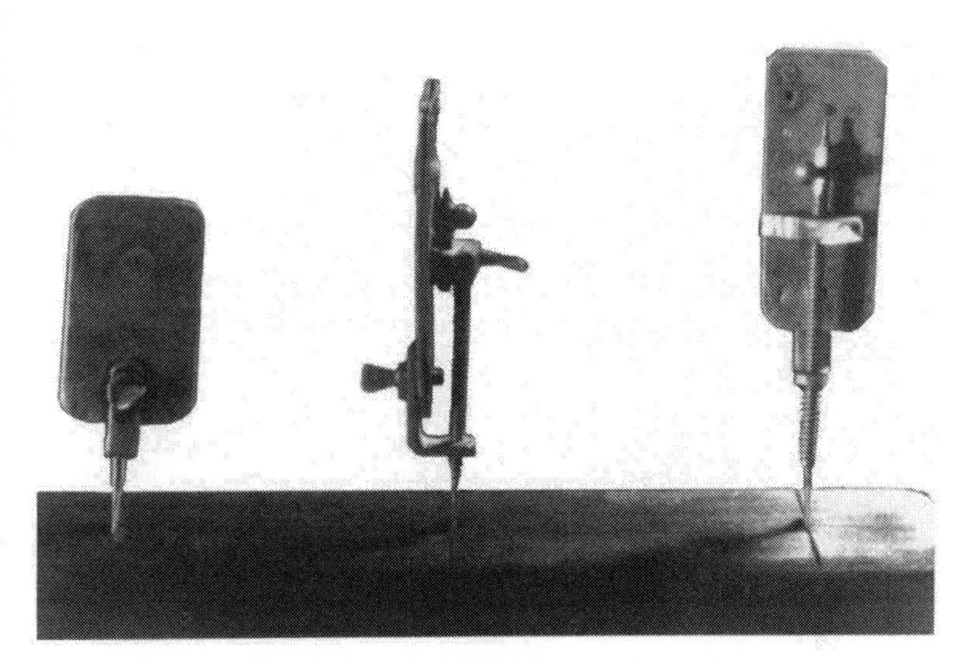

Figure 4. Three typical Leeuwenhoek microscopes from three different views. They stand a little over 4.5, 6.5, and 7 cm high, respectively. Photo courtesy of the Museum Boerhaave, Leiden.

outlook, would write that the perfection of the compound microscope (of two lenses) was to be sought in the smallness of its objective lens.[32] He claimed at the end of his life that the magnification such instruments could achieve was limited only by how small those lenses could be made and used.[33]

These tiny lenses achieved their most remarkable effects, however, as powerful simple microscopes. The lens of a typical flea glass had a curvature on both sides equivalent perhaps to a sphere an inch in radius,[34] but Huygens also now asserted that the best lens for a simple microscope (though he denied that its magnification could be increased indefinitely) had a curvature on both sides of a radius of no more than one-twelfth of an inch[35] – a lens no larger, that is, than one-sixth of an inch in any dimension. Hartsoeker maintained that the best microscopes consisted of a single lens with a focal length of from just under a centimeter at most to 0.3 mm at the least, which also meant radii of curvature that varied from perhaps 6.5 mm in the largest lenses to about 0.2 mm in the small-

32 *OCCH* 13:580–1.

33 Ibid., 13:530–5, 542–3, 568–9, 572–5. On the other hand, he also recognized that there was an absolute minimum for the size of any aperture, beyond which the image became confused (ibid., 13:695–6; see also pp. c–cii).

34 Hevelius (1647), 23; Borel (1655), 10; Zahn (1685–6), 3:111, 113 (cf. 109).

35 *OCCH* 13:514–17, 518–19, 530–1.

est.[36] The largest dimension of the latter lens was thus now less than half a millimeter.

The strongest and best of the surviving Leeuwenhoek lenses is 1.2 mm thick; it has a radius of curvature on both sides of roughly 0.7 mm, a focal length of slightly less than 1 mm, and a magnification of approximately ×270.[37] There are those who believe that this may perhaps be the best lens that Leeuwenhoek ever made,[38] but other scholars have found evidence in Leeuwenhoek's letters that he was familiar with magnifications of ×480, and contemporaries believed that he in fact had instruments capable of ×500.[39] Lenses of ×500 would certainly not have been unusual, and Hartsoeker did not fail to point out that the smallest lens he had recommended entailed a magnification of ×770.[40]

Without clarity and a greater resolution of detail, however, such magnification is meaningless, and what most struck some observers who looked through Leeuwenhoek's microscopes was indeed the distinctness (and the brightness) of their images.[41] In a report on twenty-six instruments Leeuwenhoek had left the Royal Society at his death, Henry Baker, the English popularizer of the microscope, noted the greater convenience of the mechanical apparatus of more recent microscopes; yet the lenses of the society's bequest were another matter, "for the Goodness of these before us gives just Reason to believe he might have others as excellent as can perhaps be ever made."[42] With a measured resolution now of 1.35 µm – an ability, that is, to distinguish a separation of 0.00135 mm between two points[43] – the surviving Utrecht lens is indeed, in terms of combined magnification and resolution, the best surviving simple mi-

36 NH (1694), 174. The smaller lenses of which both Huygens and Hartsoeker were speaking were in actuality small spherules of glass (see text below), and although the relationship between focal length and radius of curvature varies with the refractive index of the glass that is used, students of lenses have often taken the focal length of a spherule to be roughly equivalent to 1.5 times the radius; see also Kingma Boltjes (1941), 63.

37 Zuylen (1981), 178.

38 Kingma Boltjes (1941), 74.

39 Schierbeek (1950–1), 1:103–5; Boerhaave & Haller (1740–4), 3:357. Regarding the data pertaining to other surviving Leeuwenhoek lenses, see Ford (1991), 146–63, 166. Of the twenty-six microscopes Leeuwenhoek left the Royal Society, all of which are now lost, Henry Baker reported that the strongest of these had a focal distance of one-twentieth of an inch, which would mean a magnification of ×200 (H. Baker 1740, 506).

40 Hartsoeker in fact wrote in his *Essay de dioptrique* that a lens with a focal length of one-tenth of a *ligne* would magnify a diameter one thousand times larger than it was seen with the naked eye a foot away (NH 1694, 174). In Paris, where Hartsoeker had been working for some years before the *Essay de dioptrique* was published there, the standard foot was 324.839 mm, and Hartsoeker's calculation of the magnification (and see also NH 1722, 145) indicates that he conceived that foot as being divided into one hundred *lignes* (Zupko 1978, 134, 143). In terms of today's standard image distance of 250 mm, that would be equivalent to ×770. Given the potential power of bead lenses, which Hartsoeker indeed was using (see text below), that power is by no means out of the question.

41 Birch (1756–7), 4:365–6; Folkes (1723), 451; H. Baker (1740), 504.

42 H. Baker (1740), 515.

43 Zuylen (1981), 175, 178. Brian Ford reports having recently obtained clear photographic images with this lens of structures in fact only 0.75 µm in thickness (Ford 1985, 56–7, 129). The resolution of the other surviving Leeuwenhoek microscopes is decidedly inferior, however; see Ford (1991), 146–63, 166.

croscope from before the nineteenth century and far superior to the compound instruments of the day.[44]

With respect to such optical qualities, the simple microscope remained in general a better instrument than the compound well into the nineteenth century, and major observers, including all those in the Netherlands, long preferred it.[45] Hartsoeker had declared that the best microscopes were those with a powerful single lens, and Swammerdam agreed there was nothing better.[46] Leeuwenhoek may have combined lenses on occasion, but if he did so, these instruments remained rare and atypical, for none appears to have been found among the microscopes he left.[47] Converted from his initial attachment to the compound microscope, Christiaan Huygens came to believe that the simple microscope had been so improved in the final decades of the century (primarily, it would seem, by the making of very tiny lenses) that it surpassed all others in both distinctness and magnification.[48] It also remained the chosen instrument of Lyonet and Trembley in the following century.

A number of seventeenth-century observers, it is true, distinguished themselves with the compound instrument – among them Hooke, Nehemiah Grew, Francesco Redi, Georgius Baglivi, and perhaps the latter's great master, Marcello Malpighi[49] – but by the eighteenth century distrust of the compound microscope was widespread among practiced observers, who noted its particular susceptibility to illusory images, indistinctness, and distortion.[50] Even Henry Baker, who himself still used and recommended the compound instrument, acknowledged that the single lens displayed an object both larger and more distinctly, echoing Hooke's similar admission of the greater strength and clarity of the high-powered simple microscope – "to those," that is, "whose eyes can well endure it."[51]

44 Cittert (1934a), 6–8, 12–13; Frison (1948), 286; Bracegirdle (1978), 193.

45 Cittert & Cittert-Eymers (1951), 73; Cittert (1934a), 9, 12.

46 See above regarding n. 36. JS (1669), 81; idem (1737–8), 1:91.

47 After having visited Leeuwenhoek, the German diarist Zacharias von Uffenbach reported that Leeuwenhoek's microscopes included some with double lenses separated from one another at the proper distance presumably by a plate within; nonetheless, these particular instruments were not much thicker than the others (Uffenbach 1753–4, 3:359). In the catalogue of Leeuwenhoek instruments auctioned after the death of his daughter there are also a few items listed as having two or three lenses (*Catalogus,* items 4, 14, 128, 134); on the basis of this fact, Pieter Harting assumed that Leeuwenhoek had indeed used lens systems (i.e., doublets and triplets), but W. H. van Seters pointed out that the frontispiece of the catalogue and the contemporary engraved portrait of Leeuwenhoek by Johannes Verkolje depict Leeuwenhoek microscopes with three lenses side by side (Seters 1933, 4575–6). Seters also presumed that the double-lens combinations to which Uffenbach referred were composed in fact of two thin planoconvex lenses, which, with a larger aperture, would give a flatter and more corrected image (ibid.), but A. Schierbeek was inclined to believe that they were true doublets (Schierbeek 1950–1, 1:106).

48 *OCCH* 13:514–15, 697.

49 Hooke (1665), [22–3] of preface; Frison (1965), 98; Redi (1688), 125; Baglivi (1710), 399; Malpighi (1686–7), 2:142; see also Adelmann (1966), 2:828–32.

50 See Müller (1786), xviii, xxii–xxiii; P. Musschenbroek (1736), 543; idem (1762), 2:790; Hewson (1774), 304–5; Hill (1752a), 123–4; Gleichen-Russworm (1764), 9; Castellani (1973), 63–4.

51 Hooke (1678), 96–7; H. Baker (1753), 15–16; idem (1742), 7; Hooke (1665), preface [xxii].

Vigorous advocates of the simple microscope would in their turn readily grant that the single, high-powered lens was "difficult and disagreeable in the using."[52] It had to be brought very close to the object it was viewing; indeed, so close, noted Hartsoeker, that the object, necessarily transparent, could be lit only from behind.[53] (As Lyonet remarked, and as was relevant to much of the work of Swammerdam, Leeuwenhoek, and Trembley as well, such a short focal length also precluded the use of dissection tools beneath the lens.)[54] This restriction of the lighting aggravated the problem of diminishing illumination in more highly magnified images; thus, though extolling the smallest lens of an eighteenth-century English instrument as "the greatest single magnifying Power that Art has been able to contrive," the English naturalist John Hill also described it as dark and gloomy.[55] Moreover, the small single lens provided only a very constricted field of view.[56] Given such admitted inconveniences, many of the devotees of the simple microscope – including by their own testimony Swammerdam and Leeuwenhoek – appear themselves in fact to have only rarely used their smallest lenses.[57]

Nonetheless, "those who wish to ascertain the figure of minute bodies," said one accomplished eighteenth-century observer, should always prefer the simple microscope,[58] and many of his counterparts throughout Europe used the more comfortable compound instrument for observations at medium and lower magnification, turning to the simple microscope to explore smaller realms.[59] John Hill insisted in 1752 that all the great discoveries that had made the microscope famous had been achieved with the single lens, and he feared that neglecting it because of the difficulty of its use "will clip the Wings of all succeeding Discoveries."[60] The compound microscope was for those who sought entertainment, but for serious work "'tis the single Glass of the first [i.e., the greatest] Power that is to determine all"; "the View tho' dark is certain. . . ."[61]

All lenses of the period were burdened with difficulties (compounded by compound microscopes). In the first place, the imperfections of seventeenth- and

52 Hill (1752a), 123.
53 NH (1694), 176–7; see also Hooke (1665), preface [xxii]; idem (1678), 99; *OCCH* 13:526–7; J. Belkmeer (1719), 40, 58.
54 Lyonet (1757), 385.
55 Hill (1752a), 122–3, 124; see also *OCCH* 13:530–3; J. Belkmeer (1719), 40; P. Musschenbroek (1762), 2:788.
56 Hartsoeker (1694), 176; Hill (1752a), 122–3; P. Musschenbroek (1762), 2:788.
57 *LJS* 138; *AB* 1:292–3; 7:38–9; AvL (1702), 91; NH (1712), 58; Müller (1786), xviii; see also P. Musschenbroek (1762), 2:788.
58 Hewson (1774), 305.
59 Hill (1752a), 124; Torre (1776), 4; Cole (1930), 112n. 2 (cf. Castellani 1973, 64n. 70); Hughes (1955), 3; see also Müller (1786), xviii; Wrisberg (1765), 17 (cf. 8n).
60 Hill (1752a), 123.
61 Ibid., 124. See also Bradbury (1967b), 152; Harting (1846), 29. Not all the compound microscopes of the time deserved to be so summarily dismissed as research instruments; see Frison (1963), 3–4. In their examination of early compound microscopes in Utrecht and London, J. van Zuylen and J. C. Deiman have found that a good instrument can achieve a resolving power of 2 µm (letter to the author, March 1990).

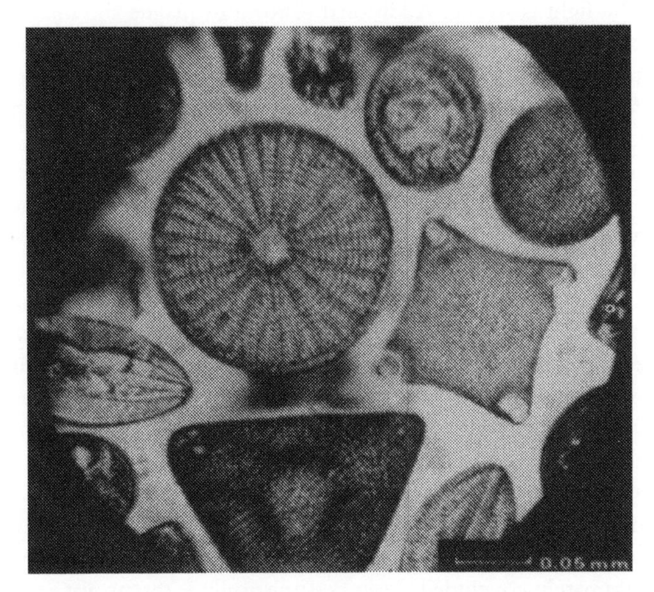

Figure 5. A photograph of diatoms made with the Leeuwenhoek lens in the University of Utrecht Museum. The dark blotches are caused by bubbles in the lens (Zuylen 1982, fig. 7[a]). Courtesy of the Utrecht University Museum.

eighteenth-century glass itself were a major frustration to those attempting to improve the new scientific instruments.[62] It tended to be colored with greenish tints, but more damaging in lenses were the bubbles, spots, and streaks that also often marred the glass of the age.[63] Indeed, bubbles and similar flaws could loom large as obscuring blotches in the image of a microscope lens (Fig. 5), and according to Georges-Louis Leclerc de Buffon, intendant in the mid-eighteenth century of the Jardin du Roi in Paris (where the local glass was particularly poor in quality), such flaws were always there.[64]

Early microscopic images were also disfigured by problems inherent in the shape and surfaces of the lens. A spherical surface, since it does not focus precisely, provides an image of uneven sharpness and distorted shapes toward the periphery. Moreover, the surfaces of a lens, like those of a prism, disperse the colors composing white light and produce colored fringes in the image, particu-

62 *OCCH* 4:249, 6:460; NH (1694), 91, 98.

63 Regarding the tinting of the glass in early lenses: Harting (1867), 265, 273–4; Bedini (1963), 390; Mills & Jones (1989), 179. With respect to other flaws within the glass: NH (1694), 91–8, 103; *OCCH* 6:148, 170, 206, 300, and 9:591; Mills & Jones (1989), 175 (fig. 2), 179; Buffon (1841–4), 3:57; Ledermüller (1756), 14; Harting (1867), 265, 274.

64 Buffon (1841–4), 3:57; see also Ledermüller (1756), 14. J. van Zuylen informed me by letter in March 1990 that good glass for lenses could be obtained in England, Holland, and Italy around 1700, Italian glass having a particularly fine reputation, but that Parisian glass simply did not measure up, being filled indeed with bubbles or black particles.

larly when the light is strong. Additional distortions plague the images of lenses and lens systems (such as astigmatism, uneven magnification, and the edges of the field curving up), but the consequences of color dispersion and spherical focusing – chromatic and spherical aberration, respectively – were the most prejudicial to early optical instruments.[65]

By the early seventeenth century it was understood that only certain aspherical surfaces could provide perfect focusing (and then, it was subsequently realized, only with monochromatic light).[66] Descartes in particular fired the hopes of a generation of scientists, dilettantes, and instrument makers that, with hyperbolic lenses, telescopes would be able to see objects in the stars as small as what one could see on Earth[67] – a prospect that also inspired talk of ultimately observing the "springy particles" of the air, the magnetic "effluviums" of the lodestone, and the corpuscles of light itself with the microscope. These aspirations were admittedly "vastly hyperbolical," punned Henry Power; yet he refused to rule out the possibility that the "darling Art" of dioptrics might yet realize the theoretical potential of conic sections.[68]

The nature of chromatic aberration, on the other hand, was not understood until the publication of Newton's experiments with prisms in 1672, from which he himself ultimately concluded – too pessimistically – that it was inescapable in transparent (or refractive) lenses. Although still more positive than Newton about glass lenses, Christiaan Huygens was also convinced by these experiments that aspherical lenses, at least, were not only too difficult to grind but had little to offer anyway, since, given the phenomenon of color dispersion, no glass lens would ever be able to focus perfectly.[69] Hartsoeker agreed that Descartes's high hopes for aspherical lenses had been dashed.[70]

To minimize spherical aberration in microscopes by other means, instrument makers resorted to a diaphragm with a tiny hole in its center; placed immediately behind the objective lens in compound instruments, it reduced the effect of spherical aberration by simply blocking out all the light except for that passing through the center of the lens. To be effective at higher magnifications, however, the hole had to be so small that it deprived the image of sufficient light, threatened the instrument's capacity to distinguish detail, and raised the possibility of further confusing the image with diffraction effects (such as small "halos" of light that may often have appeared as microscopic "globules").[71] As for

65 See Bracegirdle (1978), pl. 3, commentary; Bradbury (1967a), 26; idem (1967b), 151, 157, 160–5, 170; cf. Ford (1985), 125.

66 Ronchi (1970), 105–6; Kepler (1937–), 4:371–2.

67 Descartes (1897–1910), 6:165–96, 206, 210–11. Descartes also wrote more specifically of the prospect of seeing living creatures on the Moon (ibid., 1:69). For evidence of the hopes raised, see, for instance, *OCCH* 1:488, 5:542; Oldenburg (1965–) 1:327; Hooke (1665), preface [xvi].

68 Power (1664), preface [xvi–xvii]; see also Borel (1656), preface.

69 *OCCH* 7:302–3; 2:66; 7:3; 8:534; 7:350–1, 512.

70 NH (1706), 318–19; idem (1712), 72. See also idem (1694), 43, 101, 104, 122; idem (1710), 161.

71 Bradbury (1967b), 151–2, 170–1. Regarding this difficulty and the most powerful simple microscopes, tiny beads, see NH (1712), 58.

the problem of chromatic aberration, the possibility of significant improvement came only with the achromatic objectives, combining lenses of different kinds of glass, that were first successfully produced only in the late eighteenth century.

Lens-grinding techniques also left something to be desired. Hartsoeker declared near the turn of the century that no matter how well glass was polished, a thousand irregularities caused an infinite variety of refractions and reflections on its surface.[72] With one exception, even the surviving Leeuwenhoek lenses reveal under close examination the shallow, rounded pits often left by a brief polishing with a soft but resilient material.[73] Indeed, the tiny lenses for microscopic observing were particularly troublesome to work – our fingers are too thick, Hartsoeker lamented – and difficult even to shape.[74] Progress in grinding practices was, to be sure, a major source of improvement in seventeenth-century optical instruments,[75] but shortcomings remained, and they contributed to the cumulative deficiencies typical of early lenses.

Some of the most astonishing of the early microscope lenses, however, were not ground at all, as may in fact be the case with the best of Leeuwenhoek's surviving instruments. Leeuwenhoek was frustratingly secretive about his own techniques, but he did write in 1699 of his having ground (*geslepen*) his microscopes with increasing skill throughout the years, and he alluded in 1700 to the hundreds and hundreds of ground microscopes he had made.[76] In 1712 he characterized his "sharp-seeing" microscopes as "well-ground" as well and, when pressed by the German bibliophile Zacharias von Uffenbach two years earlier, he had remarked that all his microscopes were ground in the same grinding cup.[77] The instruments he left the Royal Society were also ground, as are all his surviving lenses with the apparent and significant exception of one.[78]

The lens now in the University of Utrecht Museum is not only by far the best of the remaining Leeuwenhoek microscopes, but differs from the others in other intriguing respects as well. It is the one lens whose smooth surface is free of pits, but it is also the one lens with minute bubbles trapped within it (Figs. 5 and 7). In addition, its curved surfaces seem to be aspherical, the sharpness of the curvature decreasing toward the rim.[79] In 1710 Leeuwenhoek had also told Uffenbach that, after ten years of reflection, he could now "blow" (*blasen*) lenses that were not actually round. Although he expressed at the same time a seemingly paradoxical contempt for "blown" lenses, the peculiarities of the Utrecht lens could all be explained if, as has been recently argued – and shown to be

72 NH (1706), 318; see also Hooke (1665), 4.
73 Zuylen (1981), 179; see also *OCCH* 6:208.
74 NH (1694), 102; see also *OCCH* 13:518–19; Rooseboom (1967), 272.
75 Van Helden (1974c), 45–6; Bedini (1966).
76 AvL (1702), 96, 305. As early as 1676 Leeuwenhoek also wrote of having a lathe in his study (*AB* 2:78–9).
77 AvL (1718), 21; Uffenbach (1753–4), 3:358.
78 Folkes (1723), 451; Zuylen (1981), 179, 182–5; see also Birch (1756–7), 4:365.
79 Zuylen (1981), 177, 182, 185.

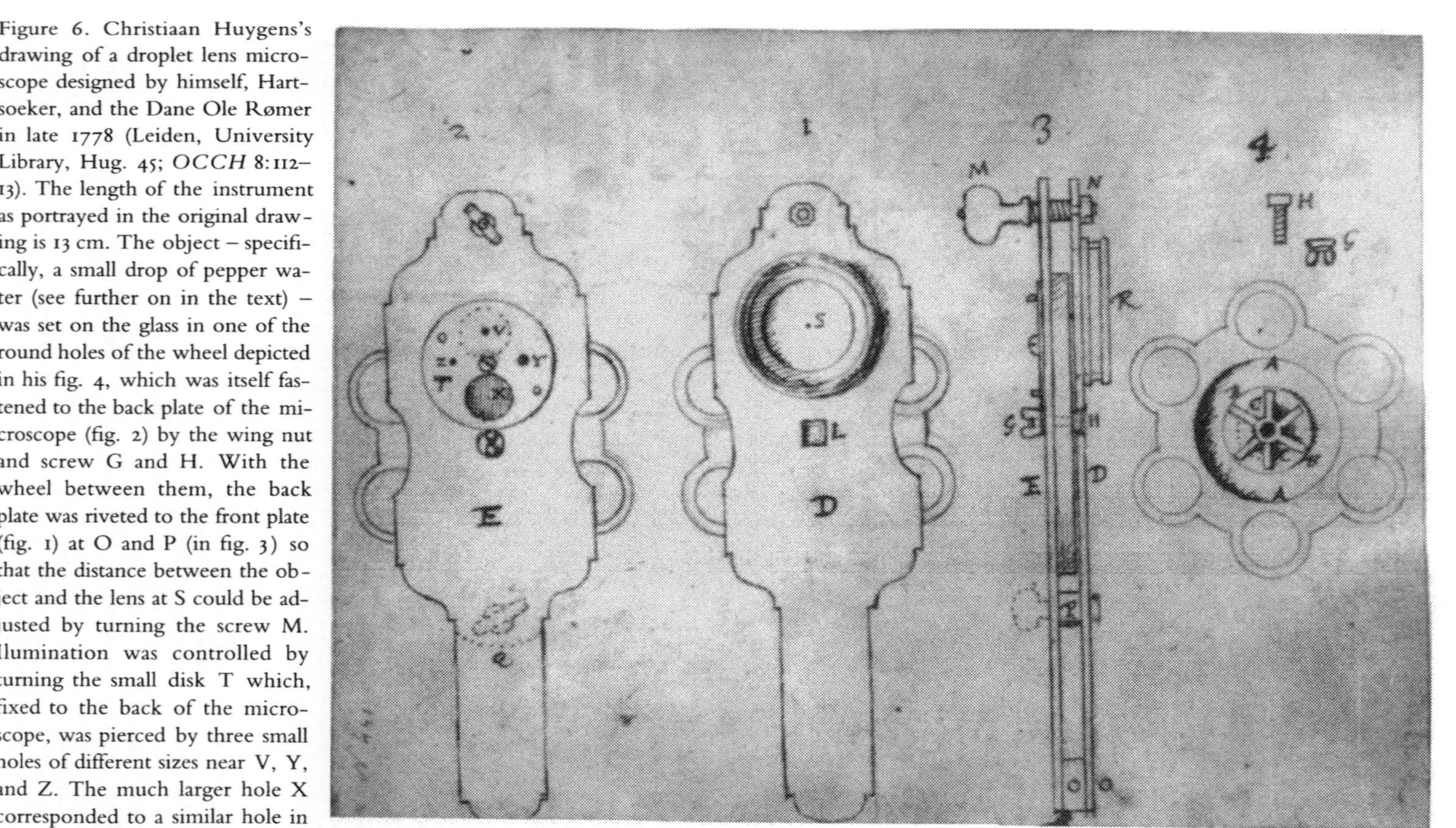

Figure 6. Christiaan Huygens's drawing of a droplet lens microscope designed by himself, Hartsoeker, and the Dane Ole Rømer in late 1778 (Leiden, University Library, Hug. 45; *OCCH* 8:112–13). The length of the instrument as portrayed in the original drawing is 13 cm. The object – specifically, a small drop of pepper water (see further on in the text) – was set on the glass in one of the round holes of the wheel depicted in his fig. 4, which was itself fastened to the back plate of the microscope (fig. 2) by the wing nut and screw G and H. With the wheel between them, the back plate was riveted to the front plate (fig. 1) at O and P (in fig. 3) so that the distance between the object and the lens at S could be adjusted by turning the screw M. Ilumination was controlled by turning the small disk T which, fixed to the back of the microscope, was pierced by three small holes of different sizes near V, Y, and Z. The much larger hole X corresponded to a similar hole in the back plate (shown by the dotted circle around V) and allowed the observer to align the object with the lens. Photo courtesy of the University Library, Leiden.

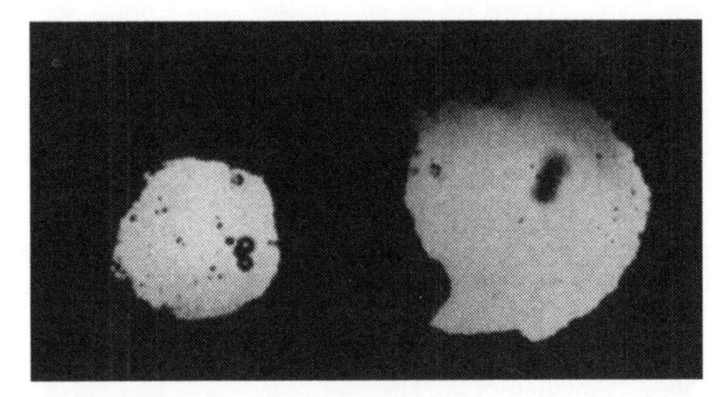

Figure 7. The lens in the Leeuwenhoek microscope in the Utrecht University Museum photographed from both sides in its original setting. Courtesy of the Utrecht University Museum.

technically possible – that lens had not been ground like the others but rather been formed naturally in the process of blowing and working molten glass.[80]

Uffenbach very likely understood Leeuwenhoek's talk of "blown" lenses as pertaining to small, spherical beads formed over a flame from molten glass, for the use of such beads was quite common. Now intrigued by magnification, the seventeenth century had discovered that such spherules of glass were themselves very powerful microscopes. Once Leeuwenhoek revealed how startlingly unexpected minute nature really was, these spherules became the most widely used high-powered microscopes of the age. Their power and clarity can still startle the modern microscopist.

In a brief treatise on the insect eye in 1644, the Sicilian mathematician Giovan Battista Odierna cited among the lenses he used a wonderfully magnifying crystal globule no bigger than a chickpea, and mention is made in the same year of small glass spherules fashioned by Galileo's former student Evangelista Torricelli.[81] Nothing is known directly about Torricelli's spherules, but he had succeeded his master at the Medici court in Florence, and it was Cardinal Gian Carlo de' Medici who, sometime shortly before 1646, presented Kircher in Rome with what the latter described as a new and very clever invention made with glass spheres no larger than the smallest pearls. When a flea's leg or a hair was placed against such a spherule at the end of a small tube, attested Kircher, the tiny globe produced huge, astonishing images.[82] The Italian Jesuit Mario Bettini wrote soon after that he had recently learned of a secret and new kind

80 Uffenbach (1753–4), 3:359; Zuylen (1981), 182, 184–5.

81 Pighetti (1961), 331; Pazzini (1959), 101.

82 Kircher (1646), 835. Such an arrangement, however, with a tube through which one peered toward the spherule at the other end, would in fact have greatly diminished the efficacy of the spherule as a lens.

of transparent body, a crystal sphere now smaller than a millet seed.[83] Mount such a sphere in a hole pierced through two pieces of paper, he explained, and though itself like a transparent atom, it would magnify other atomlike bodies to a prodigious size. So it was that in the smallest visible things, wrote Kircher's disciple Gaspar Schott, the smallest invisible things were seen.[84]

Apparently unaware of these previous magnifying beads, the Amsterdam mathematician and future *burgemeester* Johan Hudde by the early 1660s was also making simple microscopes with glass spherules – about the size of small peas, according to Christiaan Huygens.[85] Hudde picked up a particle of red-hot glass on the tip of a small, red-hot iron rod and melted the particle into a globule over a flame. As he revealed in his *Micrographia,* Hooke had in these same years developed another method of drawing out heated glass into very thin threads whose ends he then held over the flame until they also melted into droplets.[86] Whereas Hudde seems to have mounted his spherules in wood,[87] Hooke fixed his apparently much smaller beads against a needle hole pierced in a thin metal plate. Familiar with the *Micrographia* and very likely with Hudde's microscopes as well, the Dutch scholar Isaac Vossius declared in 1666 that, since the objects observed appeared larger as the spherules of glass became smaller, the power of vision could be almost infinitely enhanced.[88]

The English text of the *Micrographia* was in fact inaccessible to most Netherlanders;[89] but in Rotterdam in 1674, an eighteen-year-old Hartsoeker – while

83 Bettini (1659), 121. Bettini had died two years before the publication of this volume, and, although I have not located earlier editions, Gaspar Schott was already citing Bettini with respect to these small beads in the year of Bettini's death; see n. 84.

84 Schott (1657–77), pt. 1:472.

85 Monconys (1665–6), pt. 2:161–2; NH (1730), "Ext. crit.," 44–5; *OCCH* 8:90–1. I conclude that Hudde was unfamiliar with the previous mention of such magnifying glass beads in the Italian and Jesuit literature from the fact that the best-known of the authors who publicized the Italian innovations, Kircher and Schott, both represented the beads as being used by gazing through them at a light source with the object placed between that light source and the bead. (See n. 82 above, and Schott, 1657–77, pt. 1:533–4. Compared to Kircher's illustration, Schott's has turned the instrument around so that the lens is next to the eye. In the latter case it is difficult to conceive how, with a tube between the spherule and the object being viewed, the spherule was able to function as a lens at all, and even in Kircher's rendering, as noted previously, the tube would at best have seriously diminished the effectiveness of the spherule as a lens.) It apparently did not occur to Hudde, however, to use his glass spherules to look at objects directly against the light until a number of years had gone by; according to Hartsoeker, Hudde ascribed that ostensible innovation in usage to Leeuwenhoek (NH 1730, "Ext. crit.," 45; see also *OCCH* 8:91).

86 Hooke (1665), preface [xxii].

87 *OCCH* 8:91.

88 Vossius (1666), 102–3, 114. Isaac Vossius in fact spoke here of spherules of ice and water as well as glass, and, although ice is clearly impracticable, the use of drops of water as microscopes became quite common (Hooke 1935, 275; Gray 1696a,b, 1697; Massuet 1752, 670; Pluche 1732–50, 4:526; P. Musschenbroek 1762, 2:787).

89 Christiaan Huygens had to translate parts of the *Micrographia* for Hudde (*OCCH* 5:309, 318, 330), and Baruch Spinoza was also among those who were balked by the language in which Hooke's book was written (ibid., 359–60). Swammerdam was familiar with the *Micrographia,* valued it highly, and seemed to know some of the text in addition to the plates, which understandably would have impressed anyone who came into contact with the book (JS 1737–8, 2:376, 456, 501). But, otherwise, there is nothing to suggest that Swammerdam himself was

toying with a thread of glass in a candle flame, according to his own account – discovered what appears to have been virtually the same technique as Hooke's for making small beads of glass. Already aware of the magnifying power of a glass sphere, and recalling a visit with his father to Leeuwenhoek's home two years before, he immediately mounted a bead between two small plates of lead and "was as pleased as I could be to find myself the owner of a good microscope, and at such little cost." During the spring and much of the following summer, he later wrote, he used it to look at everything that he laid his hands upon.[90]

Over the course of thirty years, consequently, a number of European dilettantes had experienced the microscopic capacities of small glass spherules. But if we discount Hartsoeker's own doubtful claim years later that he had discovered spermatozoa during that same spring and summer of 1674,[91] the spherules had as yet made no obvious contribution to the early history of serious microscopic discovery, despite the astonishment their powers had more than once aroused. Whereas Hudde defended his own spherules against Christiaan's reservations,

Footnote 89 *(cont.)*
familiar with the English language, which proved a barrier for those interested in the *Micrographia* elsewhere in Europe as well (see Buonanni 1691, pt. 2:11).

90 NH (1730), "Ext. crit.," 43–5; see also p. 7. With respect to the possible influence of the *Micrographia,* Hartsoeker also betrays no personal familiarity with English, but it would appear to have been better known among the commercial circles of Rotterdam than elsewhere in the Netherlands (Osselton 1973, 18–19, 28–9). Brian Ford has argued that the *Micrographia* was in fact the origin of Leeuwenhoek's interest in the microscope as well (Ford 1985, 39–40, 59), but Leeuwenhoek himself would later suggest that he was indeed using droplet lenses as early as 1659 (see AvL 1702, 91).

In the following century, Martin Frobenius Ledermüller (1776, 3:47) would write that Otto von Guericke had also discovered a microscope in a small drop of glass by accident, but I have as yet encountered nothing to substantiate Ledermüller's assertion.

91 NH (1694), 227; see also NH (1730), "Ext. crit.," 45. Although Hartsoeker's claim won some influential adherents in the eighteenth century (see, e.g., Dionis 1729, 353–4; Fontenelle 1818, 1:363), it remains nonetheless very doubtful. Hartsoeker insisted that he had shared his discovery with Christiaan Huygens before Leeuwenhoek's November 1677 letter first describing his own observations of the spermatozoa to William Brouncker, president of the Royal Society, and Hartsoeker maintained that he would be able to verify this "if it were worth the trouble" by several letters written to him in Christiaan's hand (NH 1730, "Ext. crit.," 55). Hartsoeker never found it worth the trouble, and Christiaan himself had long been dead when Hartsoeker added this elaboration to his claim. Now published, Christiaan's earliest surviving correspondence with Hartsoeker deals indeed with spermatozoa, but Hartsoeker also asserted in his later account that, when he had first shown the spermatozoa to Christiaan, he had told the latter that what they were looking at was in fact saliva, of which there is no mention in the letters. Those letters date, in addition, from early 1678, after Leeuwenhoek's letter on the subject to William Brouncker, with which Christiaan was by then familiar, as was apparently Hartsoeker as well (*OCCH* 8:62–3, 65). In June of that same year, moreover, and without even mentioning Hartsoeker as a possible rival claimant, Christiaan wrote Nehemiah Grew that the first discoverer of the spermatozoa was said to be a student at Leiden named Ham (*OCCH* 8:77), which accords with Leeuwenhoek's account of a certain "Heer Ham"'s having brought the spermatozoa to his attention in 1677 (see Chapter 8, n. 74; see also F. Schrader 1681, 34–5; Halbertsma 1862). For Leeuwenhoek's indignant reaction to Hartsoeker's claim, indeed, see AvL (1702), 63–75. Nonetheless, those who defended Hartsoeker's claim and even abused Leeuwenhoek in the process were also to be found in the Dutch Republic in the following century (see Boddaert 1778, xxiii, 340–1).

Hooke shared Huygens's preference at this point for ground lenses and the compound instrument, declining to use the small bead lenses through most of his years of observation; "though exceeding easily made," he wrote in the *Micrographia,* they were "very troublesome to be us'd."[92] In Italy, meanwhile, they appear to have been fading from memory.[93]

It was Leeuwenhoek's letters in 1676 and 1677 on the observation of microscopic life in water and infusions and on the discovery of spermatozoa that first inspired a broad interest in tiny lenses in general and in bead lenses in particular. He revealed the existence of a new and very different realm of nature that was reached only through minute ground lenses or tiny droplets of glass. Apart from the ease with which these were made, it was also widely believed that Leeuwenhoek himself used the latter.[94]

The announcement of the discovery of spermatozoa in the fall of 1677 aroused the particular interest of Christiaan Huygens[95] and, through the young Hartsoeker, drew him belatedly to the bead microscope. Others in Holland were now also making such instruments; the Leiden instrument maker Samuel Musschenbroek had very likely been constructing microscopes with a single glass spherule for several years,[96] and Swammerdam wrote perhaps in January 1678 of a very easy method he had discovered for making his most powerful lenses (approximately 1 mm in diameter), a method he could teach to another in fifteen minutes.[97] But having apparently heard of a young man in Rotterdam whose microscopes could reveal the recently discovered spermatozoa, Christiaan got in touch with Hartsoeker.

92 *OCCH* 5:308–9, 318; Hooke (1665), preface [xxii]–[xxiii]; see also Hooke (1678), 99.

93 In 1686, in a letter describing a new Campani microscope, Emmanuel Schelstrate, Prefect of the Vatican Apostolic Library, wrote from Rome of "the spherical lenses already discovered by the Honorable Lewenhoek [*sic*]" (Bedini 1963, 405). The attribution suggests that, in Italian circles interested in the microscope, the earlier acquaintance with such beads in Italy was no longer widely remembered. In 1691, in a short history of microscopic literature, Filippo Buonanni cited the spherule instruments mentioned by Gaspar Schott, but did so also without any allusion to their Italian origins or, for that matter, to any subsequent use of them in Italy or elsewhere (Buonanni 1691, pt. 2:15).

94 Frison (1965), 99, 102; see also Hughes (1955), 2–3. With respect to the belief that Leeuwenhoek himself was using bead lenses, see Molyneux (1692), 281; Birch (1756–7), 4:365; Adams (1710), 24; Elsholt (1678–9), 280; Bedini (1963), 405; Pluche (1732–50), 4:525–6. Even in Holland there was uncertainty after Leeuwenhoek's death: Massuet (1752), 679; H. Baker [& M. Houttuyn] (1778), 417.

95 *OCCH* 8:77.

96 Ibid., 8:64; Elsholt (1678–9), 280. In November 1673, Hooke had already shown the members of the Royal Society "a microscope, with one only globule of glass, fastened to an instrument with many joints, to turn every way, and so to shew the object on every side with greater distinctness than other microscopes"; brought over from Holland, according to Hooke, the instrument as described is certainly suggestive of those characteristic of the Musschenbroek shop (Birch 1756–7, 3:110; Star 1953, 30–1). Although there was no mention now of the lens, Constantijn Huygens, Sr., also wrote a few months later of an impressive "machine microscopique" being used by Swammerdam and Arnold Syen, professor of Botany at Leiden, "qui tourne à tout sens sur des boulets proprement travaillez en cuivre" (ConHs 1911–17, 6:344).

97 JS to [MT], no. 41, undated, JS Papers, fol. 61r; *LJS* 80, 83. In the *LJS,* Lindeboom dates this letter Jan. 1678. In the letter itself, Swammerdam indicates the size of his strongest lenses with a small freehand circle that measures roughly 1 mm in diameter.

The essential account of their first contact, which is Hartsoeker's, is tainted by its entanglement with his later claim that he had in fact been the first to discover spermatozoa. The surviving correspondence begins with a reply from Hartsoeker in March 1678 in which he explained how he made the beads with which he observed the "animalcules" (*dierkens*) found in semen.[98] He presented Christiaan with a number of these spherules, as well as some wood and brass devices to hold them in place, and by the end of the month had himself come to The Hague to show Christiaan the spermatozoa of a dog.[99] Hartsoeker continued to correspond with Christiaan about the employment and improvement of these instruments, all of which Christiaan meanwhile shared with his brother Constantijn.[100] The following year Constantijn spoke of Hartsoeker as "the inventor of our microscopes," and years later Christiaan recalled Hartsoeker's having taught them to make the little spheres that served as lenses.[101]

Christiaan and Constantijn had soon begun making such beads themselves and, though still following Hartsoeker's lead, began experimenting with other methods, such as removing the molten globule from the thread of glass with a metal wire or, with one end of the wire moistened, picking up small fragments of glass directly to fuse them into globules over the flame.[102] The Huygens brothers mounted their own beads in small squares of thin, folded brass; with the bead trapped between opposing holes pierced with a needle through the two sides of the folded brass, those sides were clinched together with hammered pieces of wire.[103] The microscopes Christiaan used henceforth were constructed around such mounted droplets (see Fig. 6), which he now believed – as apparently did Swammerdam – had led to the perfection of the instrument.[104] He now doubted that the compound microscope would ever surpass them.[105]

Christiaan's eminence in European scientific circles gave his new enthusiasm an international forum. By the middle of July 1678 he had returned from The Hague to Paris to resume his role as the leading member of the Académie

98 NH (1730), "Ext. crit.," 46; *OCCH* 8:59–60.
99 See n. 98; also *OCCH* 8:62, 64–5, 67.
100 See *OCCH* 8:70, 74–5; ConHj (1881), 251, 255.
101 *OCCH* 8:206, 9:35. On Christiaan's own designs for such microscopes, see Fournier (1989), esp. 581–90.
102 *OCCH* 8:65, 67–8, 88–90, 90–1, and 13:520–1, as well as n. 2; NH [& ChrH] (1678).
103 *OCCH* 13:683–4, 520–1; see also 8:89.
104 Rooseboom (1959), 61; *OCCH* 13:749, 8:198. JS to [MT], no. 41, undated, JS Papers, fol. 61r; *LJS* 80, 83. Swammerdam wrote here of the "verfectie" of microscopes, but since there is no such word in Dutch, I assume he meant to write *perfectie,* "perfection," which would make sense in the context. Three words before (although it is not indicated in the published version of the letter) Swammerdam also misspelled a word by incorrectly repeating by anticipation a pairing of consonants, *verglootglasen* for *vergrootglasen;* that word, meaning "microscopes," begins itself with the syllable *ver-* and is the subject whose "verfectie" is at issue, suggesting a related pattern of thoughtless but loosely associative repetition underlying both of Swammerdam's slips here.
105 *OCCH* 9:125, 13:527n. 6, 514–15.

Royale des Sciences, and on July 16 he presented to the assembled company the "new microscope" he had brought back with him from Holland – one that, according to the academy minutes, was "extraordinarily small, like a grain of sand," and magnified incredibly.[106] Christiaan himself pointed out that the instrument in fact differed from others in use in England and Holland primarily only in the smallness of its spheres, some of which, he said, were scarcely visible.[107]

The impact of Christiaan's microscope was dramatically enhanced by association with Leeuwenhoek's recent and most astonishing observations. Before July was out, Christiaan used the instrument to show the members of the academy the microscopic life Leeuwenhoek had found in pepper water; soon after, publishing the first public announcement of their discovery in the *Journal des Sçavans,* Christiaan also identified it with the observation of spermatozoa; and in the following spring, he linked it to Leeuwenhoek's subsequent and startling account of their teeming multitudes as well.[108] In his own mind, Christiaan identified the new creatures and other microscopic spectacles of the late seventeenth century with the high-powered simple microscope, which he subsequently characterized as either a bead or a small ground lens.[109] In 1678 and 1679, however, he cultivated the impression that Leeuwenhoek's observations had been achieved with tiny droplets of glass similar to his own.

It was doubtless with some satisfaction that Christiaan wrote his brother in early August 1678 of the "great noise" occasioned in Paris by "my microscopes."[110] Instrument makers in the city had begun making "the new kind of microscopes that Monsieur Huygens has recently brought from Holland," and other members of the academy were soon offering their own observations.[111] In the first days of August, a wandering John Locke at Blois also heard from Paris of "the extraordinary goodness of a microscope Mr. Huygens has brought with him from Holland" and passed the news on to Robert Boyle in England.[112] Later in the fall, an account direct from Paris on the making of bead lenses "the bigness of a great Pins head and less" appeared in the *Philosophical Transactions;* "I Doubt not," wrote its author, "but you may be as busie at London as we are here in making of Microscopes of the manner lately brought out of Holland by Mr. Huigens. . . ."[113] Leeuwenhoek's letters on living microorganisms had in fact already provoked even Hooke – for whom, of course, Huygens's microscopes were not so "new" – to turn seriously to bead lenses for

106 *OCCH* 19:439.
107 ChrH (1678), 345–7.
108 *OCCH* 22:256, 267, 13:703; ChrH (1678), 347; Du Hamel (1701), 182.
109 *OCCH* 13:522–7.
110 *OCCH* 8:91.
111 *OCCH* 10:730n. 5, 730; Butterfield (1678), 1026; Fournier (1989), 591–3. With respect to demonstrations before the Académie by Ole Römer, Jean Picard, and Philippe de la Hire, see *OCCH* 19:439, 22:269.
112 *OCCH* 8:91n. 3.
113 Butterfield (1678), 1026.

a time, while the "noise" in Paris aroused an interest in "globular microscopes" in Germany as well.[114]

Hartsoeker managed to ensure himself a share of recognition too, a first step – though a premature one – in his own climb to eventual celebrity. With three years at the University of Leiden also now behind him, he had approached Christiaan when still in The Hague about some letters of introduction to the "sçavans" of Paris and had joined him on his journey south.[115] Once in Paris, however, and conscious of his own part in developing the microscope and observations that were causing such a stir, Hartsoeker lent himself to a scheme to bring Christiaan down a notch; but the latter quickly took his younger compatriot under tow and wrote a brief report for him, published in the influential *Journal des Sçavans,* that asserted Hartsoeker's active role in making the new bead-lens microscope.[116] The report appeared in late August, and for a time Hartsoeker continued to contribute to the further improvement of the instrument. A year later, however, he was back in Holland, apparently having laid aside for the moment whatever ambitions had taken him to Paris.[117]

It was from Rotterdam that Hartsoeker wrote Christiaan in September 1679 that he assumed the French interest in the microscope had passed by now, and the following year Christiaan indeed informed his brother Constantijn that he had written nothing of late about the microscope (or the making of telescopes, for that matter) because there was nothing new to report.[118] To be sure, the flurry of excitement in Paris had quickly passed. Nonetheless, the impact of Leeuwenhoek's discoveries endured, and the methods for making the high-powered glass droplets had been publicized across Europe, prompting others to experiment with their own techniques as well.[119]

Whatever the method used, however, few of the droplets made in fact proved serviceable as microscopes. Swammerdam said he could make forty lenses in an hour – hence, presumably the beads – but he also noted that they varied greatly among themselves, and we hear from others that only one in a hundred turned out perfectly.[120] To obtain a few usable lenses, consequently,

114 Indeed, Hooke had very recently reminded the members of the Royal Society and the English reading public of his own method of making bead lenses (Birch 1756–7, 3:351, 393; Hooke 1678, 97), which very likely pertained as well to what Hooke now spoke of as the "Lewenhooks microscope" he constructed in December 1677 (Hooke 1935, 332; see also Birch 1756–7, 3:358). Elsholt (1678–9), 280; F. Schrader (1681), 21.

115 NH (1730), "Ext. crit.," 46–7; *OCCH* 8:71 (see also pp. 73, 78).

116 NH (1730), "Ext. crit.," 47–78; NH [& ChrH] (1678), 371.

117 *OCCH* 8:112. Hartsoeker had returned to Rotterdam, married, and gone into the wine or tavern business (*OCCH* 8:213, 9:35).

118 *OCCH* 8:224, 297.

119 In addition to Hooke's earlier account in the *Micrographia,* see also Butterfield (1678), 1026; F. Schrader (1681), 21; NH [& ChrH], 371–2; Blanckaert (1680–8), 1:200–1. Regarding experimentation with other methods, see Gray (1696b), 280–1; Adams (1710), 24–6. See also Fournier (1989), 593–5.

120 *LJS* 138; J. Belkmeer (1733), 40; Ledermüller (1776), 3:47. See also Torre (1776), 34. William Hewson noted that the bead lenses made by della Torre had cloudy blotches from the smoke of the flame used in preparing them (Hewson 1774, 305–6). Cf. Kingma Boltjes (1941), 62.

large numbers had to be made in a process that thus still required patience.[121] (Time and trouble could be saved, to be sure, by buying them from Johan van Musschenbroek in Leiden forty for a guilder – roughly a day's wages for a skilled manual laborer in the Netherlands. Musschenbroek otherwise advertised his cheapest simple microscope for $7^1/_2$ guilders and his most elaborate, with nine separate and interchangeable lenses, for nearly ten times as much.)[122] Mounting the small beads was still the hardest part, however, for their very smallness made them difficult to handle. Christiaan's brother Constantijn often simply lost them, as did Giovanni Maria della Torre in Naples, who in the eighteenth century reported having made them 0.1 mm in diameter.[123]

A weightier inconvenience was that difficulty in the using that Hooke had earlier remarked. Henry Baker complained in the following century of the painful strain of looking through the tiny droplet lenses, and worried that he had actually harmed his eyes in examining some of della Torre's beads, the largest of which was said to be just under 0.5 mm in diameter.[124] Baker reported as well that he could see nothing through them satisfactorily and concluded that they were more objects of curiosity than useful instruments.[125] The prodigious power of such minute globules he readily admitted – della Torre's smallest lens purportedly achieved ×3,200 – but nothing could be more injudicious, he wrote, than the desire for such excessive magnification.[126]

Baker was an advocate of the compound microscope, but even the devotees of the single lens tended to avoid the very smallest beads, even though they were not only the most powerful but, according to Christiaan Huygens, the

121 *OCCH* 10:730, 13:520–1; Ledermüller (1776), 3:47; Stiles (1765), 248.

122 Johan van Musschenbroek advertised six beads – "Glaze dropjes, en bolletjes" – for three *stuivers,* which, there being twenty stuivers to the guilder, was a price equivalent to forty for a guilder (J. Musschenbroek 1736, 7); Vries (1976), x. Earlier in the century, Constantijn Huygens, Sr., had paid forty guilders for one of Drebbel's microscopes (Bachrach 1980, 52n. 34).

123 *OCCH* 8:89; Stiles (1765), 246–9; see also AvL (1702), 91. Della Torre reported that he had managed to make a spherule as small as half a "point" (Torre 1776, 38). He spoke in terms of a "Parisian inch," of the inch's being divided into twelve "lines," and of the line's being divided into ten points (ibid., 38, 86). If we assume the inch was also one-twelfth of the Paris foot (as was the case for the Parisian *pied de roi* as opposed to the *pied géométrique*), half a point, strictly speaking, would have been 0.11 mm (Zupko 1978, 134, 143). Although della Torre could hardly have adhered to measures of such precision, he did distinguish a next-larger bead of two-thirds of a point in diameter before that which was a whole point in diameter. Stiles assumed the point to be one-sixth of a line in speaking of these beads (Stiles 1765, 248), whereas Ronald Edward Zupko says that, in the French system of measures, it was normally one-twelfth of a line (Zupko 1978, 140).

Regarding della Torre's method of making beads, see Torre (1776), 5–6, 30–4.

124 H. Baker (1742), 36; idem (1766), 68, 69, 71; Stiles (1765), 248. Both Stiles and Baker inform us that the diameter of the largest of the della Torre beads Baker had at his disposal approached two "Paris Points," a determination that presumably reflected della Torre's initial assertion, which to della Torre would have meant about 0.45 mm (see previous note).

125 H. Baker (1766), 69, 71; see also idem (1742), 36.

126 H. Baker (1740), 514; idem (1766), 69. Della Torre reported that his smallest lens achieved ×2560 (Torre 1776, 43), but since he also assumed a standard image distance of eight inches (ibid., 38), I have adapted that magnification to today's standard image distance of ten inches. I have not adjusted, however, for the different length of della Torre's "Parisian" inch.

easiest to make.[127] Christiaan seemed to pride himself on the minuteness of the beads he took to Paris, and miffed Swammerdam by observing that the latter's lenses were not as small as his own;[128] but Swammerdam (while also rejecting Christiaan's claim) groused that Christiaan must have known that such small lenses were on the whole less useful than the large. Christiaan himself wrote that the "mediocre" beads of pinhead size were, after all, the most serviceable, for those that were smaller were less distinct and hence, though more powerful, no more effective.[129] (He was accustomed to those one-twelfth of an inch in diameter, he later added, for which at one point he assumed a hypothetical magnification of only ×160.)[130] Athough Hartsoeker had recommended lenses with focal lengths as close as one-thousandth of an inch – surely also droplets – he dismissed the tiniest droplets as simply useless.[131]

In addition to sooty stains from the flame and perhaps a greater susceptibility to bubbles, the difficulties surrounding the use of the tiny beads also included those common to all high-powered single lenses. Acknowledging their exceptional usefulness in observing an animal in its parts – presumably meaning the anatomy of insects and other small animals with which he was then preoccupied – Swammerdam otherwise minimized the general utility of the smallest beads apparently because of their narrow field of view. Moreover, while Christiaan noted that the observer's eye had to be as close as possible to the bead in order to see the broadest field, Swammerdam and others remarked that the object being viewed had to be so close that it almost touched the lens on the other side.[132] Again, the nearness of the object compounded the "deficiency of light" about which Baker also grumbled and limited the smallest beads to the observation of transparent bodies that could be illuminated from behind.[133]

As the beads diminished in size, moreover, not only did the greater magnification itself attenuate the luminosity of the image, but also the aperture through which the bead viewed the object and took in light necessarily decreased in size. To reduce the effects of aberrations (or "false" rays, as Christiaan said), Hartsoeker and Christiaan early decided that the diameter of the aperture should be no more than one-fifth the diameter of the bead; but Hartsoeker

127 H. Baker (1753), 15–16; *OCCH* 13:520–1; see also *OCCH* 8:124.

128 *LJS* 138.

129 Ibid.; *OCCH* 8:112, 123–4.

130 *OCCH* 13:518–19. Christiaan assumed in this instance a magnification of ×128 with respect to a standard image distance of eight inches, which would mean a magnification of ×160 when adjusted to the standard image distance of ten inches accepted today. At one point at least, Constantijn preferred smaller beads than did Christiaan, but still not as small as they could be (*OCCH* 8:129).

131 NH (1712), 58.

132 *LJS* 80, 138; *OCCH* 13:685; see also Hooke (1665), preface [xxii]; J. Belkmeer (1719), 58. With respect to the soot in bead lenses, see n. 120 above regarding William Hewson's comments on della Torre's beads. Zuylen's study of the surviving Leeuwenhoek lenses suggests that beads were also more liable to bubbles than were lenses ground directly from good glass; see n. 79 above.

133 H. Baker (1766), 69; see also P. Musschenbroek (1762), 2:788; NH (1694), 176–7; F. Schrader (1681), 22; J. Belkmeer (1719), 40.

subsequently reduced it to one-sixth, and their compatriot Jan Belkmeer later testified that an aperture made with the point of a fine needle was still sometimes too large.[134] According to Hartsoeker, the problem of lighting was the greatest drawback of the small glass beads as lenses and the reason for the uselessness of the smallest.[135]

Baker also complained of indistinct images in the minute spheres, and Martin Folkes, vice-president of the Royal Society, described the microscopes left it by Leeuwenhoek as less powerful but far more distinct than "those Drops, frequently us'd in other Microscopes."[136] Modern experience and theory suggest, however, that until glass beads become so small that diffraction becomes a problem, indistinctness is not among their inherent deficiencies. Quite to the contrary, beads made in recent times (with cleaner flames than in the past) have been found to surpass the optical performance not only of compound microscopes made before the middle of the nineteenth century but also, as theory predicts as well, of the Leeuwenhoek lens at Utrecht.[137]

Baker declared in 1766 that such spherules, though much used "many years ago," had long since been laid aside,[138] but they had continued for some time to play an important role in the high-powered microscopy of the eighteenth century. Wijer Willem Muys at the University of Utrecht relied on bead lenses in his extended study of muscle fiber in the opening decades of the century, and the greatest physiologist of the period, the Swiss Albrecht von Haller, still turned to them at midcentury to explore the inner structure of the glands, one of the central physiological mysteries of the age. Moreover, despite the troubles Baker had with della Torre's minuscule droplets, they offered vivid new images of the shape of human erythrocytes.[139]

At one point or another, Leeuwenhoek had also used such minute beads. Christiaan Huygens, whose father, at least, was close to Leeuwenhoek and his researches at the time, appears to have genuinely believed that the bead microscope he took to Paris in 1678 was similar to the instruments used by Leeuwenhoek, and Swammerdam, who ascribed what he himself had first learned about simple microscopes to Hudde, felt that a Leeuwenhoek microscope he

134 *OCCH* 8:68, 74, 13:678–9 (cf. 13:683–4); see also *OCCH* 13:690; NH [& ChrH] (1678), 372; J. Belkmeer (1719), 40.

135 NH (1712), 58.

136 H. Baker (1740), 514; idem (1742), 36; idem (1766), 69; Folkes (1723), 451. Christiaan Huygens agreed about the indistinctness of at least the smallest beads (*OCCH* 8:124).

137 Cittert (1954), 103, 108–10; Zuylen (1981), 195; Kingma Boltjes (1941), 71. See Cittert's caveat (Cittert 1954, 103n. 1).

138 H. Baker (1766), 69. More regretfully, John Hill had already noted nine years before that the use of powerful single lenses in general had fallen into "an unmerited Disregard" (Hill 1752a, 123).

139 Muys (1738), 2–3; Haller (1757–66), 2:376; Torre (1776), 83–8; Stiles (1765), 247, 249, 253–7; Hewson (1774), 305–6. In 1752, Pierre Massuet could still write that a good microscope required the smallest spheres of glass that could be found (Massuet 1752, 679), and among the lenses presently with Lyonet's dissecting microscope are four glass beads, one of which is the strongest and most precise of all the lenses (Star 1953, 50–2).

had seen had been inspired by Hudde's as well. Responding to word of a microscope "scarcely larger than a visible dot," Leeuwenhoek commented in 1699 that, forty years before – prior to Hooke's account in the *Micrographia,* be it noted, as well as to Hartsoeker's first visit and the latter's first trials over a candle flame – he had also made exceptionally tiny lenses, and his admission to Uffenbach that he had indeed made blown lenses in the past was explicit.[140]

Given the potential optical qualities of these droplet lenses, why had Leeuwenhoek ultimately rejected them?[141] Given his manual dexterity and his patience, it is unlikely that he simply made them badly. Although doubtless initially intrigued by their high powers of magnification, he would seem rather to have found them unsatisfactory for sustained observing. In 1699 he had added only that those exceptionally tiny lenses he once had made were not as good for making "first discoveries" as were lenses ground with larger diameters; but twelve years earlier he had also remarked that he used his stronger lenses only when he absolutely had to, for a body was magnified enough, he maintained, when its parts could be seen distinctly.[142] His reluctance to resort to higher-powered lenses doubtless reflected the familiar drawbacks they entailed. Leeuwenhoek's surviving instruments themselves suggest a concern with the easy manipulation of the object before the lens, which the greater focal distances of lenses of more modest power would have facilitated, and perhaps as well with the broadened field of view such lenses provided. Indeed, apart from the Utrecht lens, neither the magnification nor the resolution of the other surviving Leeuwenhoek lenses is exceptional, but they do offer an unusually broad area of relative clarity within the field of view, and perhaps his more mediocre lenses thus reflect what he valued more generally in his instruments.[143] If the "parts" were in fact seen clearly enough, a lower magnification would offer the eye a broader expanse of microscopic structures, and at lower powers ground lenses do perform better than beads.[144] As for even the smallest microscopic entities he observed to which some rough ascertainable size can be given, such as bacteria, they are within the visual capacity of at least the Utrecht lens,[145] whose own powers are still modest compared to the potential capacities of beads.

Whatever the advantages Leeuwenhoek found in his ground lenses, however, basic similarities with the experience of observing with beads remained.

140 See Chapter 6 regarding nn. 9, 10. JS (1669), 81; idem (1737–8), 1:91, 2:377; AvL (1702), 91 (regarding the reference to a lens scarcely larger than a dot, see [Plantade] 1699, 552); Uffenbach (1753–4), 3:359.

141 Having studied both bead lenses and Leeuwenhoek's surviving lenses, P. H. van Cittert was indeed still mystified in 1954 as to why Leeuwenhoek preferred ground lenses (Cittert 1954, 110). See also Star (1953), 19.

142 AvL (1702), 91; *AB* 7:38–9.

143 Star (1953), 20–1; also see nn. 39, 43 above; cf. Cittert (1954), 110–11. J. van Zuylen is of the opinion that it was indeed primarily the question of the working distance between the lens and the object that turned Leeuwenhoek against the high-powered beads. The leading student of Leeuwenhoek's surviving microscopes today, Zuylen emphasizes how the construction of those instruments reflects such a concern. (Letter to the author, 8 March 1990.)

144 Kingma Boltjes (1941), 71–2.

145 See above, n. 43, and Chapter 7, n. 24.

Peering through a small hole often less than a millimeter wide,[146] he encountered looming shapes that, as he turned some screws on his instrument ever so slightly, condensed out of a shifting haze. Now vague and indistinct, now startlingly clear, the shapes materialized as a spectacle of new forms and movements – indeed, in time, new forms of life – otherwise invisibly secreted away only a fraction of an inch before his face. What he stared at differed radically from what the naked eye had last seen on the other side of the lens and what his fingers had felt in placing it there. His memory of touch might recall the faint brush of insect parts, the stickiness of what had been smeared on a fragment of glass, or the smooth fragility of a capillary tube, but these remembered sensations no longer fit the images he saw. What he now felt with his hands was the sharply worked metal of the microscope, but all he now saw of the instrument itself was a ragged, circumscribing edge of darkness (see Fig. 7).[147] The oddness of the experience only heightened the strangeness of what he encountered staring into the tiny, concentrated nub of glass and light.

Leeuwenhoek's experience with the microscope was perhaps colored as well by the mystery that still hung about the instrument itself. Like most of his contemporaries, Leeuwenhoek surely "understood" how his microscopes created their remarkable images only in a crude and partial way at best. To be sure, the first appearance of the microscope coincided closely with the beginnings of the science of dioptrics; but even in many learned circles that developing new science appears to have done little to enhance the general understanding in the seventeenth century of how lenses did what they did.

Theoretical optics in the Middle Ages had taken an interest in the refractive effects of a transparent sphere, but not until the sixteenth century were the first awkward efforts made to provide a theoretical explanation of even the more obvious effects of lenses,[148] a meaningful science of which began only in 1611 with Johannes Kepler's *Dioptrice.* But the geometry of the new science proved from the very beginning too daunting or too alien not only for the layman but for many scholars as well. Even Galileo, whose telescopic observations had originally inspired Kepler's *Dioptrice,* found the work so obscure that he suggested Kepler did not understand it himself,[149] and its intensive geometry was sure to prove particularly daunting to those likely to be drawn more to the microscope than the telescope. Astronomers were mathematicians by definition, but physicians proved resistant to mathematical optics even as applied by Kepler to the eye and spectacles.[150] In his influential *Ophthalmographia,* published in three editions between 1632 and 1659, the Dutch physician (and subsequently professor at Louvain) Vopiscus Fortunatus Plemp did indeed embrace Kepler's understanding of vision and the effects of spectacles; but even Plemp was not fully

146 Star (1953), 27; Rooseboom (1939), 181.
147 See also Star (1953), figs. 6, 9, 10, 12, 13.
148 Lindberg (1976), 194, 279n. 82; Crombie (1967), 22.
149 Geymonat (1965), 37. On the response to the *Dioptrice,* see also Porta (1962), 18.
150 Koelbing (1968), 223; Crombie (1967), 48–9, 52, 54, 60–3.

comfortable with – indeed, was wary of – geometrical dioptrics. Meanwhile Franciscus dele Boë Sylvius, pride of the medical school at Leiden after mid-century, continued to ascribe presbyopia and myopia not to the geometry of light but to opacity or excess transparency (!) caused by differences in fluidity in the eye.[151]

Those aloof to the developing new dioptrics were not without their own versions of what was happening in lenses, of course. The traditional explanation of vision prevailing in the early seventeenth century revolved around the non-mathematical notion of "visible species" (*visibiles species*), conceived as replicas of the outward appearance of an object that emanated from the object and carried that appearance to the eye.[152] Plemp, who spoke of such species as stripping away from the object and then penetrating the medium through which the object was observed, defined them as a "quality" and denied that they were composed of fine corpuscles, as the ancient atomists had taught.[153] But Plemp's contemporary and compatriot David van Goorle, an atomist himself, maintained to the contrary that visible species were in fact corporeal, and he stressed how they filled the air, since a single given object could be seen at the same time by a thousand men and reflected in a thousand mirrors.[154] However visible species were explained – the physician Ysbrand van Diemerbroeck insisted they were imprinted on the air[155] – there was no denying that they were a puzzling notion. Their nature was as unknown to the intellect, acknowledged Goorle, as they themselves were familiar to the senses.[156] Nonetheless, easily conceived as refracting and expanding through transparent bodies, such species appear to have been easily accommodated to the new optical instruments.[157]

The theory of visible species was unlikely to provide a clear comprehension of how lenses actually worked, however, and it was in fact a widely held conviction, certainly in the Netherlands, that species were magnified simply by passing though a transparent medium – for everyone knew, remarked Plemp, that a coin appeared larger when submerged in water.[158] Plemp at least un-

151 Koelbing (1968), 223; Plemp (1659), 164, 275–6; Sylvius (1679), 403. Kepler had explained how concave and convex spectacle lenses alleviate myopia and presbyopia, respectively, as early as 1604 in his *Ad Vitellionem paralipomena* (Kepler 1937–, 2:6–391).

152 On the complex history behind this problematic concept, see Ronchi (1970), 30–1, 57–8, 63, 66–9, 83; Lindberg (1976), 98, 113–14, 130–1, 161–2.

153 Plemp (1659), 59, 61.

154 Goorle (1620), 109–11.

155 Diemerbroeck (1685), *Anatomes,* 421.

156 Goorle (1620), 108.

157 Goorle (1620), 111, 297; Kyper (1650), 231, 240, 248; idem (1654), 108; Plemp (1659), 48, 60, 67, 73, 82, 85, 184–5; Hortensius (1631), 22–5; Beeckman (1644), 3–4 (cf. pp. 10, 62, where it would appear that Beeckman did not really believe in the reality of the species, at least as traditionally conceived); Tulp (1641), 178.

158 Vossius (1662), 36–7, 39, 50–4, 75, 86–8; Plemp (1659), 184, 297, cf. p. 296. It was in these terms as well that Willebrord Snel, professor of mathematics at Leiden and one of the discoverers of the sine law of refraction, conceived the phenomenon of refraction (Waard 1935, 54–5; *OCCH* 10:405–6, 13:6), and Leeuwenhoek still echoed Plemp's point, at least, in 1680 (*AB* 3:206).

derstood, as not everyone did,[159] that magnification took place at the surface of the medium and not within it, but his conception of what was happening still reflected the poverty of the conceptual potential of the theory of visible species: More compactly gathered together in the denser medium of the water, he explained, the species of the coin was thinned out and expanded by refraction as it rose into the rarer medium of the air, thus enlarging the image of the coin.[160]

It is by no means clear, however, that the run-of-the-mill adherents of the major alternative doctrine at the time, the Cartesian, had a much better understanding of how magnification was brought about. In his own *La Dioptrique* in 1637, Descartes had attempted to make the new dioptrics more accessible, but Steven Blanckaert, a prominent Cartesian physician in Amsterdam toward the end of the century, explained for the benefit of his countrymen at large that lenses differed in their magnifying power "because they gather in more or fewer rays according to how their roundness differs, and the more rays that go through from all sides, the larger the objects are."[161] Descartes's espousal and elaboration of the new dioptrics had in this instance had little impact. Indeed, it is possible that by simultaneously emphasizing the inadequacy of older explanations while offering an alternative with its own baffling consequences, Cartesian doctrine, which was profoundly influential in the Netherlands, accentuated rather than relieved the puzzlement surrounding the effects of lenses.

Acutely conscious of the fact that the geometry of Keplerian dioptrics did not provide a sufficiently "physical" understanding of how lenses worked, Descartes sought to present that geometry as a necessary consequence of the physical nature of light, of its motion (or near motion, as it turns out to be) and corporeality. He conceived light, indeed, as an "action" or "inclination" in a sea of densely packed and imperceptibly small globules that filled the universe. For Descartes, this action or inclination was not yet actual motion itself, a paradox that exercised otherwise like-minded advocates of such a mechanical rendering of light; but whether in its pristine Cartesian form or as the outright motion of the subtle globules, the corpuscular conception of light acquired a pervasive influence in the Netherlands from the middle of the century on.[162] Yet this doctrine too had its obvious problems.

Both Plemp and Isaac Vossius (both also insisting on the incorporeality of light)[163] ridiculed this ocean of globules as pure fiction,[164] and Vossius in the

159 Vossius (1662), 56.

160 Plemp (1659), 73, 184–5, 297.

161 Blanckaert (1686), 235; see also idem (1701), 1:259.

162 Regius (1646), 51–3, 255; Craanen (1689), 593; Blanckaert (1701), 1:258; Senguerdius (1685), 109–13; Leclerc (1696), 399. Hartsoeker, however, conceived light very idiosyncratically as a perfect, *non*corpuscular fluid that streamed from the Sun, the stars, and fire through an infinite number of infinitely small tubes; see, for instance, NH (1696), 1–2; idem (1722), 104, 179–80; idem (1730), 63–4, 83, 86, and passim.

163 Plemp (1632), 71; idem (1659), 39; Vossius (1662), 13, 16, 29, 31. The scholastic instruction at the University of Leiden had also taught that light was incorporeal (Burgersdijck 1642, 107–8; Heereboord 1663, 202–3).

164 Plemp (1659), 40, 46; Vossius (1663), 12.

1660s underscored in particular the apparent absurdities in the Cartesian treatment of transparency and the focusing of light. According to the proponents of such a corpuscular light, transparent bodies were so riddled with rectilinear pores and interstices that light could pass straight through; but since light passed through every point in the glass and from all directions, protested Vossius, there seemed to be no glass at all![165] Moreover, if the rays of light were indeed composed of corpuscles, he argued, how could they pass through that mathematical point that was the focus of a parabolic mirror (and of the hyperbolic lenses Descartes had so earnestly proposed, he might have added)?[166]

Though rejecting the bulk of what Vossius had to say as worthless,[167] Christiaan Huygens indirectly echoed his objections. Christiaan also presumed light to be motion in a corpuscular ether,[168] but he acknowledged nonetheless that all that Descartes had said about light was beset with difficulties.[169] Indeed, no one, Christiaan noted, had as yet satisfactorily explained how light rays, coming from all directions, crossed without obstructing each other, and though he himself was also inclined to ascribe the transparency of glass to interstices between its particles, he recognized the problems that followed.[170]

The perplexities raised by both old and new doctrines did not prevent the recourse to the new optical instruments, however. Neither Vossius nor Blanckaert were deterred from using the microscope or magnifying glass, after all,[171] and nothing suggests that Swammerdam was more familiar or more at ease with dioptrics than was his master Sylvius. Leeuwenhoek had acquired a surveyor's knowledge of mathematics, yet he too betrays no intimate acquaintance with a science that was becoming only increasingly difficult.[172] Though frustrating the hopes for a clearer understanding of the workings of the lens (and lens systems), the recondite character of the new dioptrics and the difficulties that entangled the more accessible physical explanations appear to have raised no impediment to the use of the microscope.

165 Descartes (1897–1910), 6:197; Regius (1646), 82, 89–90, 106; Blanckaert (1701), 1:75; Bontekoe (1689), 2:400; Senguerdius (1685), 118; G. Vries (n.d.), 240, 256; Leclerc (1696), 406; Vossius (1662), 9–10, 66.

166 Vossius (1662), 14. Descartes had made a rather sad effort to deal with this problem (Descartes 1897–1910, 11:98, 101–2).

167 *OCCH* 4:149.

168 *OCCH* 19:461, 472–3, 482. In searching for a mechanical explanation of the sine law of refraction, Christiaan arrived at his own distinctive conception of light as a wave front moving through the corpuscular ether (see Ziggelaar 1980, 184, 186–7; *OCCH* 19:482–7).

169 *OCCH* 19:465–7.

170 *OCCH* 13:749–50, 19:459, 480–3.

171 Vossius (1662), 39; idem (1663), 75–7; see also Monconys (1665–6), pt. 2:153. Among the later Dutch followers of Descartes, Blanckaert indeed showed the greatest interest in the microscope, see, for instance, Blanckaert (1686), pl. 13 (figs. 2, 3), pl. 21 (fig. 8), pl. 24 (figs. 6, 7), pl. 25 (fig. 3), pl. 31 (figs. 1, 5), pl. 37 (figs. 1, 4), pl. 38 (figs 1, 2); also Blanckaert (1688b), 5, 149–51, 158–64, 167–70, and passim.

172 Schierbeek (1950–1), 1:22–3. For an example of what could only very generously be called a mathematical approach to dioptrical questions, see Leeuwenhoek's comparison of the eyes of fish and higher animals: AvL to RS, 22 July 1704, AvL Letters, fols. 297v–298r.

But impediments there were. As Hartsoeker had discovered, the microscope itself could require few resources and little effort to make, but the difficulties mounted with the seriousness of the observer's intent. To prepare a small and fragile subject for purposeful study did take time as well as ingenuity, adroitness, and experience; add the challenge of setting the subject before the lens so that the light could reveal what needed to be seen, and the effort could indeed become a trying one. Even with such preliminaries successfully negotiated, sustained and careful observing through tiny lenses could strain and tire the eyes, while the effort to make sense of the images thus encountered often entailed further frustrations as well. Persisting observers might well have questioned their perseverence. Hence, although a remarkable microscope was early and easily available in the form of a small glass bead, its application – or for that matter, the application of a strong ground lens – demanded not only considerable skill but a compelling motivation.

Renaissance magicians had earlier extolled the wonders that could be worked with special "glasses," but they had not included among those wonders the discovery of what because of smallness might yet remain unseen in nature. They fantasized rather about overcoming distances, about scouting a distant army or reading a letter that lay open in a distant room.[173] But the transformation of the telescope into the compound microscope jarred the imagination, evoking the first exhortations to use the microscope to explore an otherwise imperceptible realm. Perhaps inevitably turning at times to the transoceanic New World as a metaphor,[174] the exhortations echoed as well the rising hopes, no less indicative of the age, that a new and truer understanding of nature was in the offing.

Such exhortations bore fruit only slowly. For decades, the attempts to search for new knowledge with the microscope remained desultory and unadventurous. Galileo had startled Europe with the *Sidereus nuncius* within two years of the announced discovery of the telescope, but only at midcentury did developed accounts of microscopic observations begin to appear. Sustained efforts to apply the microscope in systematic research were even longer in coming, as was any systematic application of such powers of magnification as had been discovered by now in small glass beads. Even established cultural pursuits that might have been expected to seize upon the microscope with enthusiasm nurtured traditions of practice and thought that hindered the fuller exploitation of the new instrument. The idea of probing the imperceptibly small with a lens had emerged in the shadow of other aggressively promoted new programs for the radical advancement of learning, and, like them, it had to contend as well with older, entrenched commitments. The result was a concealed and at first glance paradoxical deflection of motivation that was nowhere more pronounced than in the Netherlands.

173 Van Helden (1977), 28–35.

174 ConHs (1897), 120.

CHAPTER TWO

Seeming invitations

Once grasped, the prospect of seeing with the lens what had remained until then invisibly small aroused vague but often heady expectations. Despite the more dramatic reach of the telescope, Descartes declared the usefulness of the microscope to be far greater, for it could probe the arrangements of the minute parts that composed and, in the case at least of inanimate things, determined the very essence of terrestrial bodies.[1] Although more ambivalent about the new instrument, Francis Bacon, like Descartes, attached the utmost importance to the search for aids to enhance the sense of vision, and his pointed emphasis on the investigation of the "hidden schematisms" and "true textures" of bodies later echoed in the works of early English microscopists.[2] Indeed, Bacon's English heirs dreamed at midcentury that the microscope might reach "the smallest Moleculae, or first collections of Atoms" and, detecting "Electrical Effluxions" and magnetical effluviums, finally decide between the Peripatetics and the atomists. If a lens of such powers were acquired, according to Henry Power, "we might hazard at last the discovery of the Spiritualities themselves."[3]

Like Power, who warned that otherwise "our best Philosophers will but prove empty Conjecturalists," Kircher also now doubted the worth of philosophizing about natural things without recourse to the microscope, and the early French microscopist (and collector of tales of microscopic observations) Pierre Borel wrote that the instrument was opening the gates of a new natural science.[4] "We live in a century," wrote Isaac Vossius, "wherein the senses of fools may gather in more than did the senses of all the wise men of the Greeks." The scholar Andreas Colvius, upon receiving one of the early microscopes of Christiaan Huygens, asked who among the ancient philosophers had indeed penetrated into the secrets of nature with the eyes of both the body and the mind as did the savants of the present age.[5]

Such oft-cited discoveries as hollows in hairs hardly threatened the ancient philosophers with obsolescence,[6] but other early observations did indeed suggest

1 Descartes (1897–1910), 6:226–7.
2 Bacon (1877–89), 1:168, 232–5, 307–8; Descartes (1897–1910), 6:81; Hooke (1665), 114, 204, etc.
3 Walter Charleton as cited in R. Frank (1980), 57; Power (1664), 57–8, see also preface [vii].
4 Power (1664), preface [xviii]; Kircher (1646), 834; Borel (1656), dedication.
5 Vossius (1666), 102; *OCCH* 1:322–3.
6 Kircher (1646), 834; Fontana (1646), 150–1; Borel (1656), 18.

an altered understanding of nature. Abundant but previously unseen or unrecognized living things were now reported in a variety of unexpected places. The "dust" on old cheese was found to be not dust at all but little animals, and swarms of minute worms were discovered tumbling about in vinegar.[7] Kircher announced that the blood of fever victims also teemed with worms, and there was talk that they infested sores and lurked in the pustules of smallpox and scabies.[8]

Borel, whose *Observationum microcospicarum centuria* in 1656 was the first book devoted exclusively to microscopic observations, noted the usefulness of the microscope to the study not only of disease but of the generation of insects. He was awestruck by the sight of spider eggs suspended like transparent pearls within which, in the candlelight, he thought at times he could see the beginnings of the spider's heart.[9] Francesco Fontana pressed the eggs out of the abdomen of a flea and observed the nits emerging prematurely from the damaged shells, but Kircher also claimed to have seen the spontaneous generation of diverse insects from putrefying fluids.[10] Others wrote more promisingly of having observed the early embryonic development of the chick within its egg and the rudiments of the plant within the seed.[11]

From the very first days of the microscope, however, the observation of insects themselves provided the most characteristic as well as the most astonishing microscopic images. In the first reported application of the principle of a compound microscope, Galileo had adapted his telescope to observing flies, and the earliest surviving microscopic illustrations, published in 1625 by Galileo's fellow academicians of the Accademia dei Lincei in Rome, depict the honeybee and its parts.[12] As recounted by Peiresc, gnats, fleas, mites, and lice served to demonstrate Drebbel's microscope to the Queen of France, and when Christiaan Huygens began constructing his own instruments, small insects again came most readily to mind as fitting subjects for observation.[13]

It was their unexpected complexity of construction that amazed observers throughout the century. Kircher wrote of the wondrous way the smallest insects were put together, and Borel marveled that the minute body of a mite contained feet, nerves, eyes (in which he sometimes noted a cheerful glint), and indeed all the parts of an animal.[14] Fontana had found that the spider in fact possessed six eyes (Borel later found eight), and Giovan Battista Odierna's remarkable study of the true insect eye suggested its deeper intricacy.[15] The

7 Fontana (1646), 148; Borel (1656), 19; Kircher (1646), 834; Singer (1915), 336; Belloni 1966a, 17–22.
8 Kircher (1646), 834; Borel (1656), 21.
9 Borel (1656), 38–9, 16.
10 Fontana (1646), 148–9; Kircher (1646), 834.
11 Borel (1656), 15; *OCCH* 1:321. See Chapter 9 as well.
12 Singer (1953), 198–201; idem (1915), 331–2.
13 Humbert (1951), 156–8; *OCCH* 1:321.
14 Kircher (1646), 834; Borel (1656), 16–17, 19.
15 Fontana (1646), 150; Borel (1656), 10–12; see Pighetti (1961).

transparency of some slighter insect bodies also revealed the movements of an internal anatomy that inspired vivid descriptions.[16] Both Robert Boyle in England and Pierre Gassendi in France saw the atomistic philosophy reflected in such images of minute anatomies, whereas to Blaise Pascal they suggested a fantastic descent through successive universes within the body of a mite.[17]

Suggestive though they may have been, however, these early ventures in microscopic observation offered little to advance the body of integrated scientific thought and understanding. With few exceptions,[18] the first half-century of microscopy was narrowly limited not only in its instruments but in its technical ingenuity and initiative, its standards of description and illustration, and its capacity to engender sustained and integrated research. The characteristic incentive had remained the delight of wonder, but a persisting application of small lenses, in particular to the systematic study of often dismayingly uncertain images, required a more compelling motivation.

Substantial obstacles had to be overcome. The technical difficulties remained crucial, for, even if the full potential of the instruments already at hand remained untapped, the lack of techniques for preparing and exposing microscopic structures for observation precluded literally endless possibilities. But the limited reach of the imagination and the set of the mind raised decisive barriers as well. They forestalled the search for techniques and discouraged the frustrating struggle with fragile objects difficult to handle, easy to mutilate, and quick to wither.

In the Dutch Republic, however, there flourished currents of thought and established pursuits that, at first glance, would seem to have offered encouraging contexts for the early and systematic use of the magnifying lens. They all implicitly or explicitly stressed the importance of the small in nature and, in some but not all cases, emphasized acute observation and cultivated an intense expectation of discovery. They drew upon both religious and philosophical commitments and flourished within the framework of both scientific and artistic practice. Prominent among them was the most influential scientific philosophy of the century, Cartesianism.

Descartes had withdrawn to the Netherlands from his native France in 1629, and there he remained for twenty years. The purpose of his retirement to the north was a fuller immersion in the philosophical tasks he had set himself, and one of the first fruits was the completion of a systematic philosophy of nature. Central to this new "physics" was the conviction that the underlying cause of the sensible effects of natural bodies were invisibly small mechanisms throughout the length and breadth of nature. All the properties perceived in the world hence derived from the motion of parts and structures so small that they were

16 Power (1664), 9; see also Hooke (1665), 212–13.
17 R. Frank (1980), 91, 94–5; Pascal (1977), 1:154–5.
18 Among the exceptions are the study of the bee by the members of the Accademia dei Lincei and of the insect eye by Odierna.

beyond the reach of the senses;[19] but the imaginary images of this unseen realm of diversely shaped moving particles and the hidden passageways through which they caromed captivated a rising generation of Dutch intellectuals. Shouldering aside a nascent atomism, Cartesianism became the dominant Dutch variant of the corpuscular mechanism that was exercising an increasing influence throughout Europe.[20]

Alarmed by the condemnation of Galileo, however, Descartes delayed the full publication of his developed system, choosing rather, in his *Discours de la méthode* in 1637 and the *Meditationes de prima philosophia* four years later, to argue first the philosophical and religious underpinnings of his "method." Except for his account of the physiology of the human body, his full system of nature finally appeared in his *Principia philosophiae* of 1644. Meanwhile, his presence in the Dutch Republic had sharpened the interest there in what he was about, and he soon became the subject of a storm of controversy that centered particularly in the Dutch universities.[21]

Cartesianism had begun to unsettle Dutch academic life shortly after Descartes's first prominent – if wayward – disciple Henri de Roy, better known as Regius, was in 1638 named extraordinary professor of theoretical medicine and botany at the recently founded university at Utrecht. There he waged a colorful but losing struggle that resulted in the early 1640s in an official reaffirmation of the university's commitment to Aristotle. The center of the growing commotion then shifted to the University of Leiden, the premier university of the Republic, where the confrontation between old and new philosophies both stimulated and embittered academic life for several turbulent decades. A whole generation was educated in the context of the persisting conflict over Cartesianism and, most alarming to the old guard, its theological ramifications. Except for a brief interlude at the end of the 1660s, however, a succession of influential professors at Leiden continued to propound a Cartesian view of nature until it was undermined at the turn of the century not by the persistence of old orthodoxies but by the influence of experimental physics and Isaac Newton.

In the interim, Descartes's imagery of an invisible but pervasive mechanism continued to fascinate those who led the Cartesian offensive in the universities.[22] Insisting that matter was everywhere divided into insensibly small parts that could alone account for the properties of natural bodies, they dwelt at length on ubiquitous networks of unseen interstices, gaps, and pores – "some wide, some narrow, some continuous, some closed, some straight, some twisting" – through which tumbled the corpuscles of fluids and of an ethereal "subtle matter."[23] The observable world was reduced to surfaces that masked the truly

19 See Descartes (1897–1910), 8(pt. 1):52, 100, 324–6.
20 Regarding atomism in the Dutch Republic, see Holwarda (1651), chaps. 3, 6; Beeckman (1939–53), 3:31, 43, 63, 86, 100, 218.
21 Verbeek (1992).
22 Ruestow (1973), chaps. 3–7; Hoog (1974), 3–71, 96ff.
23 Regius (1646), 3–4; Ruestow (1973), 51–4, 69–70; Raey (1654), 146–83; Craanen (1689), 45, 293–4; Geulincx (n.d.), 92–5, 103 (quotation from p. 93). See also Descartes (1897–1910), 6: 651–5.

critical workings beneath; "Or is it not with the bodies of God as it is with buildings made by man," asked Joannes de Raey, who taught the new philosophy at both Leiden and Amsterdam, "that the facade we see when we first arrive is surpassed by the skill and beauty concealed within?" Hence, a primary task of the natural philosopher, wrote de Raey, was to descend "into the interior, reveal the structure, nexus, and spaces of the tiniest parts, observe hidden movements, and fully expose at last the causes and ingenious art of nature."[24]

The fascination with the imagery of invisible mechanisms deeply affected medical thought as well, as the medical men prominent among Descartes's followers in the Netherlands embraced a physiology that was a natural extension of his physics.[25] The traditional physiology, with its Aristotelian and Galenic roots, understood the processes of the organs of the body in terms of the qualities of different tissues and the nature or "faculty" of the individual organ, a framework of explanation that denied significance to structure that might lie below the visible aspect of tissues and organs. Descartes, however – reviving, as it were, an alternative school of Greek physiological speculation – turned for explanation in physiology to invisibly small structures and moving parts. His own elaborate reconstruction of the human body, the *Traité de l'homme,* was published only posthumously, but Dutch friends and would-be disciples had gained early access to his manuscripts and had been teaching a Cartesian physiology for several decades before the 1670s, when they firmly entrenched themselves in the medical faculty at Leiden.[26] Along the way, they convinced an expanding circle of Dutch physicians and academics that the workings of the body were to be explained only in terms of the shape, arrangement, and movement of insensibly small particles interacting like the parts of a supremely intricate machine.[27]

The rise of the Cartesian physiology of invisible mechanisms coincided in the Netherlands, and particularly at Leiden, with an increasing Dutch commitment to "subtle anatomy" – so called, explained its foremost Italian practitioner, Marcello Malpighi, because of the "subtle" description of a thousand little glands, fibers, and tiny vessels.[28] Never had there been such rummaging through the human body in search of surprising structures as in the past century, noted Pierre Bayle in 1684;[29] indeed, the recent progress of anatomical exploration nurtured a keen expectation of discovery, particularly regarding small structures that only special techniques could render visible to the naked eye.

The most celebrated physiological discovery of the century, William Harvey's recognition and demonstration of the circulation of the blood, rested, it is true,

24 Raey (1654), 26, 32; NH (1730), "Ext. crit.," 45.

25 Regius (1646), 155, 206–7; idem (1668), 4; Craanen (1689), 35, 136, 208, 210–11, 274, 293–4, and passim; Terwen (1676); Fuhrmann (1676); see also Luyendijk-Elshout (1975), 294–307.

26 Descartes (1972), xxiv, xliii; idem (1897–1910), 4:566–7, 5:112.

27 See, for instance, Anton de Heide's updating of Cornelis vande Voorde: Voorde & Heide (1680), 219n., 231n. 1, 306n. 5.

28 Malpighi (1697b), p. 124 in the second series in the pagination.

29 Bayle (1686), 1:536.

on gross anatomical structures most of which had been long familiar to anatomists. But the year before the appearance of Harvey's classic *Exercitatio anatomica de motu cordis et sanguinis in animalibus* in 1628, another small volume had announced the beginning of the discovery of what would prove to be another vascular network that had remained hidden within the animal body. The Italian anatomist Gaspare Aselli, in his *De lactibus sive lacteis venis quarto vasorum mesaraicorum genere novo invento,* reported encountering a mass of white threads covering the intestine and the connecting membranes in the abdominal cavity of a dog he was vivisecting. When Aselli cut one of these threads, a white fluid oozed from its severed ends, identifying the threads as a new and unrecognized kind of vessel. That the sacrificed animal had just eaten was critical to Aselli's discovery, however, for it was only the white chyle they were removing from the intestine that had made these vessels – the lacteal vessels – visible.

In fact, the lacteals had already been observed by Galen and his predecessors,[30] but their rediscovery piqued the interest of anatomists throughout Europe nonetheless. The interest was only further heightened midcentury by new observations in France (also on vivisected animals still in the process of digestion) pertaining to the thoracic duct, a chyle-bearing canal that rises from the back of the abdominal cavity through the chest.[31] In the same years, the early 1650s, anatomists in Scandinavia and England discovered still other new vessels, the lymphatics, extending to every part of the body.[32] The beginnings of the lymphatics in small lymph spaces are invisible to the naked eye, and the main branches, filled now with the colorless lymph, also proved difficult to see, especially after the lymph escaped during dissection and the emptying vessels collapsed.[33] The thoracic duct was the largest vessel of the emerging network, but although first observed (in a horse) in the late sixteenth century, it remained difficult for even experienced anatomists to see because of its fineness and virtual transparency. Some very prominent anatomists never succeeded in finding it in the human body at all.[34]

The allure of these elusive vessels was greatly enhanced by the doubts they raised about the traditional physiology. Together with Harvey's demonstration of the circulation of the blood, the piecemeal discovery of a new vascular system challenged basic tenets of the Galenic doctrine that, at the beginning of the century, had still seemed unassailable in its coherence and authoritative con-

30 Cole (1944), 44; Longrigg (1981), 170. In England, Nathaniel Highmore also subsequently claimed to have seen the lacteals before 1622 (Eales 1974, 281).

31 Pecquet (1651). Bartolomeo Eustachi had observed the thoracic duct in the preceding century; Pecquet's major achievement was to have discovered the receptaculum chyli (*DSB,* s.v. "Pecquet, Jean").

32 Nordenskiöld (1928), 143–6; Eales (1974), 282; cf. R. Frank (1980), 28.

33 Davison (1923), 183. Davison also notes that the lymphatic vessels in general are so small as to be invisible unless injected (ibid., 184). Blanckaert (1686), 386; Diemerbroeck (1685), *Anatomes,* 58.

34 *DSB,* s.v. "Pecquet, Jean." I. Hahn (1728), 63; presumably this disputation reflects the experience of the distinguished German surgeon Lorenz Heister, who was presiding. Haller (1757–66), 7:204.

sensus. Harvey's demonstration denied the Galenic seepage of blood in separate arterial and venous systems, while the discovery that the contents of the lacteal vessels passed on ultimately to the thoracic duct and then emptied into the venous system near the vena cava denied the liver the central role it had formerly been granted in transforming food into blood.[35] Unlike the circulation of the blood, however, the theory of which had emerged from Harvey's mind with a completeness of its own, the function of much of the new vascular network and the nature of its contents remained uncertain, emphasizing that the disclosure of the true physiology of the body had only just begun. The discovery of further hidden structures that might reveal the purpose of this emerging network could also mean, consequently, a breakthrough to a new and truer understanding of the basic processes of the animal body. Johannes van Horne, who joined the medical faculty at Leiden in 1651 and himself independently discovered the thoracic duct (and first observed it in the human body), exulted that the discovery of the duct had drawn from nature's sanctuaries a "new and unheard-of doctrine" of nutrition, a doctrine "now first revealed."[36]

The sense of a new physiology looming on the horizon was encouraged as well by the discovery in the second half of the century of a number of glands – and supposed glands – and glandular ducts.[37] Given the impact of the recent discovery of both the circulation of the blood and the lymphatic network, it is hardly surprising that, in the wreckage of the Galenic system, the effort was now also made to incorporate the glands, with their own vessels and ducts, into a general system of fluid movements throughout the body.

While the anticipation of important discoveries to come stimulated further research into fine anatomical structures in the Netherlands as elsewhere, the crisis in physiological understanding oriented that research in specific directions. Even Harvey's account of the circulation of the blood focused attention on invisible structures that might yet be discovered, for while die-hard conservatives looked for the invisible pores Galen had posited in the septum of the heart, the rest of the Dutch medical community pondered whether the circulation of the blood was completed through invisible anastomoses, through the porosity of the flesh, or through some more complex connections between the veins and arteries.[38] The most gifted of a new generation of Dutch anatomists followed the paths of barely visible blood vessels in the substance of various organs and searched for

35 See R. Frank (1980), 104–5, 154; Foster (1901), 102–3; Nordenskiöld (1928), 143–7.

36 Horne (1652), dedication [iii]; Schierbeek (1947), 18.

37 Ruestow (1980), 265–6; Foster (1901), 101–18, 153–64.

38 See Elkana & Goodfield (1968), 61–73; Miller (1981), 258–9; Kyper (1650), 361; Wale (1647), 41, 54, 69–70; R. Drake (1647), 3, 8; Regius (1647), 151–2; Diemerbroeck (1685), *Anatomes*, 274–5; [Craanen] (1685), "Pars prima," 80; Back (1648), 183–4, 186–7; Sylvius (1679), 16, 891; Deusing (1655), 51. The accounts of those conservative anatomists who claimed to have demonstrated invisible pores in the septum of the heart by passing their stylus through it echoed through the middle of the century in the medical literature in the Netherlands: Wale (1647), 39; Kyper (1650), 68; Deusing (1655), 76; Diemerbroeck (1685), *Anatomes*, 278. Diemerbroeck cites such demonstrations by his own former masters at Leiden, Otto Heurnius and Adrianus van Valkenburg.

their interconnections.[39] Reports of success were rare,[40] but the prospect of ubiquitous invisible blood vessels was repeatedly raised, and was echoed by those who assumed similar invisible vessels in the systems they conceived for the lacteals and lymphatics.[41]

The debates surrounding the gradual discovery of the lymphatic network inspired a heated controversy in the Netherlands.[42] The center of the commotion was Lodewijk de Bils, a self-taught, amateur anatomist in Rotterdam who insisted that the digested food left the intestines in two different forms and in two different ways: the chyle, fit for conversion into blood, being transmitted through the mesenteric blood vessels to the liver, and a thinner fluid – not chyle properly speaking but a kind of "dew" – through the lacteals.[43] De Bils was not without supporters in the medical schools, and those who hoped to confirm some such system that would restore to the liver the function of converting chyle to blood would eventually include Jan Swammerdam.[44]

The more extended network of lymphatic vessels also provoked speculation and controversy. Wearying the spirits of the most dexterous anatomists, as one such anatomist remarked, the first beginnings of the lymphatics eluded their best efforts throughout the century.[45] Dispersed throughout the body, these vessels were variously said to arise in the tendons of the muscles, in various organs, in membranes, from the glands alone, from the blood vessels, or even from the nerves.[46] Claiming to have found an abundance of new lymphatic vessels spreading outward in the body in all directions, de Bils also contended that his "dew" was dispersed from the lacteals to various glands and from these to the "bedewing" of the entire body.[47] According to de Bils, the lymph so conceived was the source of a variety of excretions: tears, saliva, semen, milk, "&c."[48]

Such speculative elaboration reflected the general assumption of an integrated fluid system throughout the body, an expectation that itself sharpened the focus on the search for presumed but still invisible structures. The presumption, for instance, of vessels connecting the lacteals to the nipples of the breasts challenged the ingenuity of Dutch anatomists to the end of the century.[49] On occasion, anatomists believed that they had seen such vessels; but for the sake of the majority who understandably had not, the point was made that recent discoveries also revealed how much the body could conceal, so that the failure to have as yet observed these vessels did not preclude their existence. They may be in-

39 Horne (1668), [4–5]; Graaf (1677), 24–5, 59; JS (1667b), 104; Kerckring (1670), 22–5.
40 Deusing (1655), 81–2; Craanen (1689), 290; cf. Horne (1668), [4].
41 Kyper (1650), 361; Wale (1647), 69; Diemerbroeck (1672), 205, 824, 845; idem (1685), *Anatomes*, 58–9; Deusing (1661), 163; Horne (1652), preface [iv]. See also nn. 49, 50 below.
42 See Oudaen (1712) 2:20–2.
43 Bils (1658), 7–9.
44 Deusing (1662), 25, 171, 286–7, and passim; JS (1672), 29, 40; see also Deusing (1660), 275–368.
45 Nuck (1691a), 50–1; Deusing (1659), 511. Those first beginnings remained obscure, indeed, into the nineteenth century (see Eales 1974).
46 Deusing (1659), 506; Blanckaert (1686), 385; Nuck (1691a), 51–3.
47 Deusing (1660), 154, 187; Bils (1659b), 4.
48 Bils (1659b), 4; Graaf (1677), 55–6.
49 Nuck (1691a), 19–20.

accessible to our eyes, insisted Ysbrand van Diemerbroeck, professor of medicine at Utrecht, but reason convinces us they are there.[50]

Dutch anatomists and physicians debated as well whether the semen with its generative power came from the lymph, the blood, or the "animal spirits" of the nerves – or from the blood mixed either with the lymph or with the "animal spirits" of the nerves – a dispute that also brought into question the assumed but as yet unseen junctures of the respective vessels of these fluids with the seminiferous tubules of the testicles.[51] More widespread was the search for the ostensible hollow or pores within the nerves themselves and evidence of the spirit or fluid passing through them.[52] The nerves were generally believed, indeed, to carry the animal spirits from the brain to the rest of the body, and some Cartesians were inclined to endow them even with valves.[53] Linking the nerves to the larger fluid system, Horne's influential colleague at Leiden Franciscus dele Boë Sylvius (as well as Descartes) imagined the cerebral capillaries to have pores through which the spiritous portion of the blood passed into the brain to be elaborated into the animal spirits[54] There were those, to be sure, who were only amused by all this vascular elaboration, but such speculations about the "endless labyrinths" of the human body nurtured an acute awareness of a realm of pervasive but as yet undetected structure.[55]

The tradition of subtle anatomy embodied indeed the expectation of continuing discovery through ingenious techniques, skilled hands, and a practiced eye. Writing of his own discovery of the thoracic duct, Horne affirmed that the knowledge of such small parts of the body was far from complete and that another century would not suffice for the study of their endless diversity.[56] Roughly a decade later, Anton Deusing of the medical faculty at Groningen reemphasized that anatomical discovery had only just begun and that even a thousand centuries would not exhaust it.[57]

To the gratification of sharing in a common effort was added the prospect of personal fame, a powerful stimulus that, through the preoccupation with priority, surely contributed not only to intensified efforts but to premature claims and phantom structures. Almost the whole of Holland was shaken, we are told, by de Bils's ostensible discovery of an elaborate complex of rings that crowned the thoracic duct, which some confirmed and others more rightly denounced as fic-

50 Everaerts (1661), 282; Deusing (1659), 505; idem (1655), 551; Diemerbroeck (1672), 391.

51 Deusing (1661), 58ff; Graaf (1677), 54, 56–60; see also Ruestow (1983), 214n. 141.

52 Plemp (1632), 50–1; Kyper (1650), 129, 223; Diemerbroeck (1672), 143, 597, 607, 846–8; idem (1685), 503; Deusing (1659), 529–31; Bontekoe (1689), 1:62; Craanen (1689), 401; Blanckaert (1701), 1:206; idem (1686), 180. Regarding efforts to reveal and even trap the spirit or fluid flowing through the nerves, see in particular Graaf (1677), 527; Bidloo (1715a), 1, 121; see also E. Clarke (1968), 123–41.

53 Deusing (1659), 529–30; Diemerbroeck (1672), 143; Descartes (1972), pt. 1:26, pt. 2:18–19; idem (1897–1910), 11:280; Regius (1646), 232–7, 295–8; Craanen (1689), 459, 570.

54 Sylvius (1679), 20; Descartes (1972), pt. 1:19–21, pt. 2:10–12. According to Edwin Clarke, Regius first proposed that the animal spirits in the nerves also circulated (E. Clarke 1978, 294–5).

55 Needham (1667), 1–9; Deusing (1659), dedication.

56 Horne (1652), preface [i].

57 Deusing (1665), 12–13; idem (1660), dedication, 154.

tion.[58] Nonetheless, egoism and the competition between individuals added a powerful goad to the ethos of a persistent search for previously unseen structures in subtle anatomy.

In the 1660s, a brilliant circle of young anatomists linked to the medical school at Leiden gave full expression to that ethos and exemplified the incentives it brought into play. All destined for greatness as anatomists, Swammerdam, his compatriots Frederik Ruysch and Regnier de Graaf, and the Dane Nicolaus Steno were among those who gathered as students under Horne and Sylvius in the early years of the decade. Their interactions as both friends and competitors appears to have intensified their shared commitment to the aggressive exploration of the body and stimulated the development of their individual gifts.[59] Their common interest in a variety of problems made for fruitful periods of collaboration, but it also led in time to angry priority quarrels that betrayed the competitive drive and the play of personal ambition.

In 1665, taking issue with de Bils, Ruysch published a small but influential volume demonstrating the presence of valves throughout the lymphatics (to which de Bils responded by positing more unseen structure in the very walls of the veins, arteries, and lymphatics).[60] Ruysch himself pointed out that he was not the first to have seen these valves, and among those who had preceded him he included his professors Sylvius and Horne.[61] But Swammerdam, traveling in France the year before, had also shown the valves to a gathering in Saumur and, with the publication of Ruysch's book, suggested suspiciously that its author had managed an unacknowledged look at the drawings that he, Swammerdam, had sent to Steno in Amsterdam.[62] Indeed, Steno also took an early interest in the lymphatic network and at some point exposed such valves himself.[63]

As intriguing as the new vessels were, however, they by no means monopolized the interest of the Leiden group. Steno's career as an anatomist had begun just months before he enrolled at Leiden in 1660 with his discovery of the duct of the parotid gland; after three years of further research at Leiden, he published his studies of the texture of muscle tissue and revealed his discovery of numerous new glands and glandular ducts in the mouth and nose and elsewhere in the head.[64] In the following years, de Graaf settled at nearby Delft and there completed two celebrated treatises on the sex organs: The first, in 1668, showed that the testis was not composed, as was generally believed, of a spongy, pulpy, or

58 Deusing (1664), preface; Diemerbroeck (1685), *Anatomes*, 51–2; Horne & Stephanides (1660), thesis III, §§4–6.

59 The text above provides examples of their competitiveness, but their writings also bespeak their friendship: Graaf (1677), 35, 543–4; Ruysch (1725), 14; idem (1665), 24. Swammerdam would continue to think of Steno in particular as one of his very close friends; see Chapter 5, n. 82. See also Lindeboom (1973), 21–2.

60 Ruysch (1665), preface to the *candidus lector;* Diemerbroeck (1685), *Anatomes*, 58–9; Graaf (1677), 55–6.

61 Ruysch, loc. cit.

62 JS (1737–8), 1:3; idem (1667b), 90; *LJS*, 39–41; see also Nordström (1954–5), 41–50.

63 Graaf (1677), 35; Nordström (1954–5), 48–9; *DSB*, s.v. "Stensen, Niels."

64 Steno (1664).

"glandulous" substance but was a tangle of minute vessels.[65] The second, published four years later, was an extended and decisive demonstration of the true role of the ovaries in higher animals.[66] That former colleagues at Leiden had been following similar paths in the search for new discoveries became fully apparent in the bitter quarrel that followed.

Anticipating attempts to steal the glory of "this my discovery," de Graaf left no doubt that he considered himself the first to have recognized and revealed the remarkable construction of the testicle.[67] Having learned of de Graaf's treatise before it appeared, however, Horne had also rushed into print a brief and hasty affirmation of his own discovery of the thready character of the organ, in which research, as it turned out, he had closely collaborated with Swammerdam.[68] Horne acknowledged that independent research might lead indeed to the same results, but a more suspicious Swammerdam later recalled his frequent talks with de Graaf concerning Horne's researches.[69]

An open break between Swammerdam and de Graaf followed the publication of the latter's treatise on the female generative organs in 1672. Horne had died, and it was now a resentful Swammerdam who, frustrated as well by Horne's own failure to have mentioned their collaboration, undertook to make the case for their priority in exploring these organs. In doing so, he also called into question de Graaf's accuracy and integrity.[70] The critical contention in de Graaf's second treatise was that, in the female of the higher animals, organs that since the days of the Hellenistic anatomists had been called the "testicles" were in fact ovaries. De Graaf himself acknowledged that both Horne and Steno – whom Swammerdam also credited with the original insight – had reached that conclusion before him;[71] but the priority questions raised by Swammerdam now touched upon a variety of issues, and in the bitterness of the dispute both appealed to the Royal Society in London to decide between their claims. Before the society's effort to arbitrate reached the Netherlands, however, de Graaf had died,[72] and Leeuwenhoek, himself a resident of Delft, passed on the tale

65 *De virorum organis generationi inservientibus* (Graaf 1668); the version in de Graaf's *Opera omnia* (Graaf 1677) is cited throughout these notes. Graaf (1677), "Epistola ad Franciscum de le Boe Sylvium" and 38–41. Horne (1668), [5]; Deusing (1651), 8; Voorde & Heide (1680), 116.

66 *De mulierum organis generationi inservientibus tractatus novus* (Graaf 1672); the version in de Graaf's *Opera omnia* (Graaf 1677) is cited henceforward in these notes.

67 Graaf (1677), "Epistola ad Franciscum de le Boe Sylvium" and preface in *Tractatus de virorum organis*, 42.

68 Horne (1668), [1–2], [5]; JS (1672), 2, 41, 50–1.

69 Horne (1668), [2]; JS (1672), 51. Indeed, this particular discovery had been made elsewhere in Europe as well. A decade before de Graaf's treatise appeared, L. C. Auberius in Pisa had already published a fly sheet recounting his own discovery of the vascular structure of the testis (Belloni 1966b, 14–18).

70 JS (1672), 7–8, 15–16, 20, 47–8, 55–6, and passim.

71 Graaf (1677), 298, 300–3; JS (1672), 20, 54–5; see also Steno (1667), 117; Horne (1668), [8]. For other allusions by Swammerdam to the question of the recognition of the ovary as such see JS (1737–8), 1:305; idem (1675a), 9–10. Swammerdam's friend Justus Schrader soon testified, however, that as early as 1657 Willem Langly of Dordrecht had already recognized the female "testicles" as ovaries in the course of his observations in rabbits (J. Schrader 1674, preface [xiv]).

72 *LJS*, 15–16. Regarding Swammerdam's difficulties with de Graaf, see also Lindeboom (1973), 52–3, 111–13, 115–19; Oldenburg (1965–), 9:587–8.

that de Graaf's death had been a consequence of Swammerdam's attack. Even if the dispute lacked such fatal potency – and de Graaf gave as good as he got – its intensity bespeaks the personal ambition and competitiveness that spurred the field of subtle anatomy.[73] Together with the imaginative anticipation of still hidden but consequential structures, such drives forged an aggressive search for discovery in the intricate fabric of the animal body.

A very different preoccupation with nature's smaller forms had long been cultivated by Netherlandish painters. From master to apprentice, they had passed on a fascination with the depiction of small living things, and in the early seventeenth century, this craft tradition was unrivaled in its precise and detailed observation. The typical literature on insects, by contrast, still relied heavily on ancient learning and had in recent centuries focused on the significance of a chosen few – the bee, the ant, the silkworm, and the grasshopper – in terms of their significance to human society (be it as economic boon, moral or social exemplar, or supernatural punishment). Even where a more detached interest was shown in the description of their life and behavior, it was not accompanied by an equivalent concern for accurate depiction.[74] Representative of a reviving interest in insects among natural historians at the turn of the century, the University of Bologna's Ulisse Aldrovandi occasionally abandoned his literary sources for the fields, taking an artist along as well. The resulting woodcuts, however, were extremely crude, though still sufficing for another prominent survey of natural history a half-century later.[75] Such depictions were far removed from the detailed and sensitive renderings that had been offered by Netherlandish miniature painters for over a century.

Theirs was a tradition born in the margins of manuscript books, where late medieval illuminators had indulged their whimsy. In addition to bawdy irreverences and surrealistic grotesques, that whimsy had inclined as well to loving depictions of wildlife.[76] In small artists' notebooks as well as illuminated Psalters, breviaries, and Books of Hours, such motifs spread as models from the hands of innovators, initially in northern Italy, to the centers of the bookmaking trade across Europe.[77] By the late fifteenth century, the most vigorous center of that trade was the workshops of the southern Netherlands, where illuminators also explored a new illusionistic vision that found unique expression in the decorative motif of insects and other small naturalistic forms.[78]

73 AvL (1694), 670–1; *Partium genitalium defensio* in Graaf (1677), passim.

74 This was true, generally speaking, of natural history in the Middle Ages (see, e.g., Stannard 1978, 442, 450–1; Morge 1973, 38, 49), and the bias was still evident in the Dutch Republic in the seventeenth century. See Reneri (1634), [191–2]; D. Cluyt (1597); Nylant & Hextor (1672), 176–82. A striking example of a popular indifference to accurate depiction is provided by the fabulous form given the locust in sixteenth-century broadsheets (Ritterbush 1969, 563–6), a form that continued to reappear even among the more exacting collections of insect illustrations by Netherlandish painters that followed (ibid., 575–6; Jac. Hoefnagel 1592, pl. 267).

75 Aldrovandi (1602), preface "Ad lectorem"; Jonston (1657).

76 See Pächt (1950); Randall (1966); J. Evans (1931), 1:40–81; Hutchinson (1974).

77 Pächt (1950), 14–15, 18–19, 21; Hutchinson (1974), 162, 169; Randall (1966), 12.

78 Delaissé (1959), 12, 20, 182; Pächt (1948), 30.

The shaping of this vision was in large part the work of a single miniature painter known to us only as the Master of Mary of Burgundy.[79] Representing yet one more resurgence of naturalistic representation against the perpetual drift toward conventionalized decorativeness in manuscript margins, he also was gripped by a fascination with the powers of illusionism. Illustrated scenes from the text unfolded in the deep spaces of rural landscapes and urban thoroughfares, while panels bearing passages of the text might be suspended across such fictive space by painted ropes from the margins of the illustration.[80] The Master of Mary of Burgundy was reveling in the manipulation of illusion.

It was in the decorated borders, however, that the Master of Mary of Burgundy and the Netherlandish school of miniature illustration he inspired attempted the most complete illusionistic deceptions, and in the decorative motif of odds and ends of the field scattered about the edges of the page, insects now played a uniquely important role. In painted borders, artists could mimic even the scale and circumstances of commonplace reality, of an insect alighting on an open page. Capturing the intruder with careful fidelity and a painted light that cast shadows on the manuscript itself, the miniature painters delighted in toying with the borderline between the painted and the observer's world. Deceived apparently by painted flowers, a dragonfly settles on a page, the text still visible through its transparent wings – but the dragonfly is painted too.[81]

Such borders not only lent themselves to the new illusionism, but also preserved an older tradition of accumulated detail that, once the glory of Flemish panel painters, was now fast becoming an anachronism.[82] In his illustrations of textual themes, to be sure, the Master of Mary of Burgundy himself submerged detail in a simplified suggestion of form and light. But he accentuated the illusion of atmosphere and remoteness in these scenes by surrounding them with a different illusion of closeness and immediacy in the borders, an effect enhanced by contrasting the simplifying impressionism of the central picture with the continuing insistence on sharply delineated detail in the marginal illustrations.[83] Small nature studies committed to illusionism and a sense of fine detail hence became a characteristic of a major Netherlandish school of manuscript illumination.[84]

By the beginning of the seventeenth century, however, the social context of that tradition had been transformed. The Master of Mary of Burgundy had left a vigorous school of followers who sustained a high level of technical workmanship throughout much of the sixteenth century, but even in his own time, the rise of the printing press had foretold the imminent obsolescence of his craft in the bookmaking industry. As the sixteenth century progressed, indeed, miniature painters found themselves increasingly cast in the role of virtuosi clinging

79 Pächt (1948), 23–4, 26, and passim; Delaissé (1959), 184; Bergström (1956), 30.
80 Pächt (1948), 25, and see pls. 16 and 25.
81 Ibid., 28; Bol (1980), 372.
82 Pächt (1948), 31–2.
83 Ibid., 25, 27, 31–2, 37–8.
84 Delaissé (1959), 182, 190–1, 193.

to the skills of a craft tradition that had lost its original purpose,[85] and in time the workshops of illuminators passed away. Toward the end of the century, nonetheless, the skills and the naturalist motifs those workshops had preserved were revived as an artistic style prized by aristocratic collectors of art and curiosities.

The leading representative of the revival was Joris Hoefnagel of Antwerp.[86] Hoefnagel was no craftsman in the traditional mold, however, but a well-educated and widely traveled member of Antwerp's wealthy mercantile class. How he learned the art of miniature painting is something of a mystery. Soon after Hoefnagel's death, his first biographer implied that he had had no formal training, and Hoefnagel himself had pointedly acknowledged only nature as his master.[87] Having abandoned Antwerp, buffeted by the turmoil of the rebellion against Spain, he was enticed by the Duke of Bavaria into making a career of his artistic skills and ended his days in the service of the Habsburg emperor.

For his princely patrons, Hoefnagel filled the margins of manuscript volumes with a variety of motifs that included small nature studies and ventures into trompe-l'oeil illusionism (a hovering fly, for instance, and botanical specimens inserted through painted slits in the page).[88] His most extraordinary work, however, was a four-volume collection of such miniature nature studies he had apparently begun while still in Antwerp and which he continued in the years following his departure.[89] Painted in the traditional technique of miniature illumination,[90] watercolor on vellum, they are a distant descendant of the notebooks compiled by early illustrators to provide models for manuscript decorations. But Hoefnagel's four volumes now contained 277 leaves covered with more than 1,339 individual figures from natural history. Some seventy leaves in one volume are devoted to insects, usually several to a page, and in these insect studies any lingering trace of decorative flimsiness is overwhelmed by the sense of substantial presence, intense observation, and an absorption in the intrinsic beauty of natural things.

Often endowed with a firm solidity, Hoefnagel's insects capture as well the varying textures of insect bodies and the nuances of light and transparency. Throughout, there is an astonishing depiction of detail, from the structural tracery of wings to the minutiae of the color patterns on wings and bodies. In the finest examples from these leaves – such as an impressive dragonfly that alone

85 Kris (1927), 245; Pächt (1948), 42; Bergström (1963), 2; Verwey (1962), 15.

86 Bergström (1963), 2–4, 7–8; idem (1956), 33; Kris (1927), 245.

87 Mander (1604b), fol. 262v; Kris (1927), 249; *Biographie Nationale [Belgique]*, s.v. "Hoefnaeghel, Georges"; Hendrix (1984), 36 (cf. 39–42, 47).

88 Wilberg Vignau-Schuurman (1969), 2:pl. 33; Bergström (1963), 8–9, 66, and fig. on p. 7; Hoefnagel & Bocskay (1992), 99 and passim.

89 Now part of the Edith G. Rosenwald Collection, these volumes have recently been placed on deposit at the National Gallery of Art, Washington, D.C. (Hendrix 1984, 5). In contrast to the earlier belief that they had been executed for Emperor Rudolf II, Marjorie Lee Hendrix has made the case for their earlier origins (ibid., 15, 39–40, 47, 94–5). The volume devoted to insects is entitled *Ignis[:] Animalia rationalia et insecta.* See also Hendrix (1985).

90 Morge (1973), 72–4; Hutchinson (1974), 164–5; see also Delaissé (1968), 26, 30, 60, 88–9.

bestrides a single page – a high level of artistry merges with a precision of observation so insistent in its closeness that whoever peruses these pages cannot but remark it.[91] Nature seems to have been seized in all her diminishing intricacy – an essential ingredient, indeed, of the beauty of Hoefnagel's renderings.

Hoefnagel died in 1600 in Vienna, but his works in the Netherlands, where they were admittedly rare, appear to have spurred a revival of naturalist miniature painting in the Dutch Republic.[92] The influence of that revival persisted through the approaching golden age of Dutch painting and revealed itself most notably in the popular new genre of the flower piece. Having emerged as an autonomous genre only in the last decades of the sixteenth century, the self-sufficient flower still life in the Netherlands had its own roots in the border miniatures of earlier book illumination.[93] In their great fifteenth-century altarpieces, Flemish painters had introduced flowers essentially as religious symbols in larger pictorial contexts; but the illusionistic borders of the Master of Mary of Burgundy and his followers offer not only strewn flowers but isolated bouquets.[94] Though undated, the earliest known surviving example of an independent Netherlandish flower piece is also from the hand of Hoefnagel and is executed in the technique and manner of the miniature.[95] The earliest known specifically Dutch flower piece, dated 1600, appears in a small booklet of miniature studies of flowers, insects, and other small creatures also painted in watercolor on vellum by Jacques de Gheyn (II), a fashionable and important figure in the early, transitional period of Dutch seventeenth-century art.[96] The more

91 Joris Hoefnagel, *Ignis[:] Animalia rationalia et insecta,* fig. 53; see also figs. 30, 34–41, 43, 68.

92 Karel van Mander wrote that he knew of few of Hoefnagel's works done in the Netherlands, but he also noted that there was a fine little piece in the collection of Jaques Razet in Amsterdam (Mander 1604b, fol. 263r). Constantijn Huygens the Elder's mother was also Hoefnagel's sister, and consequently with Joris's death in 1600 part of his estate, including some artwork, went to the Huygens family; it is not unlikely that Joris's work was seen there by another friend of the family, Jacques de Gheyn, who was also a major figure in the revival of naturalist miniature painting in the Dutch Republic (Regteren Altena 1983, 1:66, 70, 177n. 56; see also Moes 1896, 179–80). Shortly after his arrival in Amsterdam in the early 1590s, moreover, de Gheyn had struck up an acquaintance with one Jacob Razet – presumably the Jaques Razet identified by Karel van Mander as also owning a work by Hoefnagel (Regteren Altena 1983, 1:27). Joris's son Jacob, who later returned to the Netherlands after also serving at the imperial court, continued his father's work and made his own and his father's nature studies more accessible through the publication of sets of engravings (Jac. Hoefnagel 1592, 1630). Copies reflect the influence of these engravings (see Hind 1915–32, 5:211–12, items 29, 33), as perhaps do similar sets of engravings that appeared in the Netherlands in the years following Jacob's first publication in 1592 (regarding Nicolaes de Bruyn and Marcus Gerards, see Hollstein 1949–, 4:24, 7:103). See also: Bol (n.d.), 15–17, 21–2, 24–6, figs. 1, 4, 7, 9, 11, 22, 24–26, 32, 34–38, 40; Gelder (1976), 347, 351.

93 Bergström (1956), 42, 13 (fig. 11); Rosenberg, Slive, & Kuile (1966), 334; see also Regteren Altena (1935), 25; Montias (1982), 55–6; Pächt (1948), 31–2; Sterling (1952), 27–8, 36, 41–3. (For a response to Sterling's emphasis on the importance of Italian influences to the emergence of Netherlandish still-life painting, see Bergström 1956, 292.)

94 Bergström (1956), 12–14, 29 (fig. 25), 31 (fig. 27); Sterling (1952), 25–6; Pächt (1948), 31, and pls. 12, 16.

95 Bergström (1956), 37–8; see also Wilberg Vignau-Schuurman (1969), 2:pls. 17, 28, 44; Chmelarz (1896), 286.

96 Bergström (1956), 44; Regteren Altena (1983), 2:141–2, items 909–30; vol. 3, "The Drawings by Jacques de Gheyn II," figs. 171–93, esp. fig. 173.

characteristic Dutch flower pieces, to be sure, would be painted with the oil-based pigments to which de Gheyn himself was more inclined, but Dutch flower painting, in its love of fine detail as well as in its format, continued to reflect its miniature background well into the middle of the century.[97]

Also preserved was the association of flowers and insects characteristic of the strewn borders of the illuminators. From virtually the first appearance of the new genre, the inclusion of insects was an established convention, and an early description of some of the earliest examples of the independent flower piece – paintings now lost – conveys the continuing preoccupation with nature's minutiae. It also testifies, however, to a continuing fascination with illusionism. The artist took so much time and worked with such patience and purity (*suyverheyt*) "that everything seemed natural"; so wrote Karel van Mander in 1604 of Lodewyck Jans van den Bosch and his paintings of flowers set in a glass of water and embellished with "heavenly dew" and a variety of insects.[98] Hoefnagel also enlivened his several flower pieces with insects painted with the same exacting detail as his isolated nature studies, and de Gheyn and other founders of the Dutch tradition of flower still lifes followed suit.[99] Vividly portrayed, the fine articulations and detailed patterns of butterflies, dragonflies, grasshoppers, caterpillars, bees, flies, and wasps served, in effect, as nature's own ornamental miniatures.

But the practitioners of this revived craft at the turn of the century were now also aware of the service they could render natural history. Hoefnagel's son Jacob, also a miniature painter who later served at the imperial court before returning to the Netherlands, published in 1592 a set of over fifty engraved plates filled with insects, flowers, and other creatures and natural objects from his father's studies.[100] In 1630 he issued another sixteen plates of insects and their ilk now done from life.[101] Together with his father, he painted or otherwise

97 Mander (1604b), fol. 294v; Bergström (1956), 42, 44–9, 52; Regteren Altena (1935), 31–2; idem (1983), 1:73, 112; Bol (1960), 21–2, 58, 73; Rosenberg et al. (1966), 355.

98 Mander (1604b), fol. 217r.

99 Bergström (1956), 37–8, 50–1, 58, 73; Wilberg Vignau-Schuurman (1969), 2:pls. 17, 28, 44; Regteren Altena (1983), 2:20–2, items 31, 39, 41–3; vol. 3, "The Paintings of Jacques de Gheyn II," pls. 1, 8, 9, 10, 20, 22; Bol (1960), 22, 28, 37, 44, pl. 45.

100 Jac. Hoefnagel (1592).

101 Jac. Hoefnagel (1630). This work has sometimes been credited to a hypothetical second son of Joris Hoefnagel named Jan or Jean (Fétis 1857–65, 1:117; R. Evans 1973, 172). The existence of this second son seems doubtful, however. The only reason for assuming his existence would seem to be that Jacob appears to have died in 1629 or 1630, the latter being the year the work in question, the *Diversae insectarum volatilium icones,* appeared under the name of simply "I." Hoefnagel. As Eduard Chmelarz observes, however, Jacob's death on the eve of the publication hardly offers a compelling reason for assuming an otherwise unknown Jan (Chmelarz 1896, 287n. 3), and Swammerdam, for one, explicitly cites only Jacob as the artist responsible for the many Hoefnagel insect illustrations. Swammerdam also characterizes these illustrations as having been done from life, echoing the title page of the *Diversae . . . icones* in particular and thus suggesting that Swammerdam had this work specifically in mind (JS 1669, pt. 1:70, 89–90, 96, 120). (The earlier *Archetypa studiaque patris Georgii Hoefnagelii* [Hoefnagel 1592], explicitly by "Jacobus," derived no less explicitly from his father's prior studies.) That Jan is never mentioned in contemporary sources also hardly accords with the description of "I. Hoefnagel" on the title page of the *Diversae . . . icones* as "celeberrimus pictor."

depicted such an extensive variety of "bloodless little animals" – over three hundred different kinds, Swammerdam later testified[102] – that the artist now seems to have consciously become a natural historian as well. Indeed, Jacob appears to have also studied aspects of insect life cycles, and Auger Cluyt, a Dutch botanist with whom Jacob had shared such an effort, cited him as *diligentissimus naturae perscrutator.*[103] Beyond their occasional collections of dead insects to paint,[104] subsequent Dutch painters might also prove to be not inconsequential naturalists. It was the painter "Otto Marsilius" – Otto Marseus van Schriek, presumably, who kept a small menagerie – who first informed Swammerdam of having observed a parasitic wasp or fly depositing its egg within a caterpillar. Of great relevance to his intense concern with spontaneous generation and the nature of metamorphosis, it was an observation that eluded Swammerdam himself.[105]

The supreme example of the miniature painter turned naturalist was Johannes Goedaert of Middelburg. In the early seventeenth century, the city sheltered not only the primary claimants to the invention of the telescope (and perhaps the compound microscope) but an influential family of early flower painters whose characteristic style was marked by its affinity with miniatures.[106] As a flower painter, Goedaert represented the prolonged survival of that early style, decidedly old-fashioned by his death in 1667 or 1668 (Fig. 8); his watercolors of birds and insects on parchment, though much admired, harked back no less to older artistic practices.[107]

In the observation of insects, however, Goedaert considered himself a pioneer, and in the decade before he died he published a three-volume study of insect development, his *Metamorphosis naturalis,* the product of over thirty years of continuous research.[108] Although familiar with the classical as well as more recent literature on natural history, he recited the empiricist emphasis on direct observation and insisted that he offered only what he himself had seen.[109] Critical as well to the *Metamorphosis naturalis* were Goedaert's roots in the miniaturist tradition. The work included more than one hundred and fifty engravings of varied insects and their larvae that derived from watercolor studies done from life (still available for purchase, he did not neglect to note).[110]

Subsequent naturalists varied in their assessment of Goedaert's achievement. Citing him often, Swammerdam found his depictions liable to error, but to err was human, Swammerdam allowed. Their compatriot in the following century,

102 JS (1669), pt. 1:70, 89–90; idem (1737–8), 1:54, 280.
103 A. Cluyt (1634), [66], 100.
104 Bredius (1915–22), 2:705, 3:873.
105 JS (1737–8), 2:709; Bergström (1974), 24.
106 Bol (1960), 21 and passim.
107 Bol (1960), 56; idem (1959), 12–13, 16; Ruë (1741), 64.
108 Goedaert (1662?–9), dedications for vols. 1–3; Ruë (1741), 63. See also Bol (1984–5), no.3, pp. 64–9; no. 4, pp. 48–52.
109 Bol (1959), 3; Goedaert (1662?–9), vol. 1, dedication and preface "Aen den goedtwilligen leser"; vol. 2, dedication. Indeed, he makes the point in the extended title of this work that his account comes not from books but from his own experience.
110 Bol (1959), 1–2, 4; Goedaert (1662?–9), vol. 1, preface "Aen den goedtwilligen leser."

Pierre Lyonet, was harsher. Many of Goedaert's insects were wholly unrecognizable, he wrote, and other depictions were useless until clarified by still other illustrations or by a text that itself left something to be desired. Presiding over the burgeoning enterprise of natural history in Lyonet's time, René-Antoine Ferchault de Réaumur concluded that Goedaert was more talented as a painter than observer;[111] but Réaumur also recognized the *Metamorphosis naturalis* as one of the most extensive works of its kind and its author as one of the first serious observers of metamorphosis. As Swammerdam had already noted, Goedaert, despite all his errors, had observed in a few years more about caterpillars than had centuries of scholars.[112]

Goedaert provided a striking example, indeed, of how the practice of naturalistic miniature painting could lend itself to a more scientific bent. Hence, given its absorption in minute forms, that practice might also have seemed likely to encourage an early recourse to the lens. Goedaert himself resorted to the magnifying glass, and in 1691 Filippo Buonanni cited Jacob Hoefnagel's 1592 engravings of his father's insect studies as the first published instance of the use of the microscope.[113] Buonanni's claim remains problematic, but the tradition of naturalist miniature painting was nonetheless another thread in the fabric of Dutch culture that accentuated an awareness of nature's smaller constructions.

So likewise did a more exalted quest for things transcendental. The contemplation of small creatures could also entail a searching that strained to reach beyond both science and aesthetics. All across Europe, intimations that the appearances of the world were laden with a higher, supernatural import nourished a passion for symbols, allegories, and religious allusions. The conviction that, as "mysticall letters," "Symbolic Images," "Hieroglyphical Characters," or "marks" of God,[114] nature's varied forms declared higher truths was widely shared, and among the "hieroglyphs" identified in emblem books were creatures found creeping about in Dutch painted bouquets.[115]

It is by no means clear, however, that such emblematic allusions as tenacity, insolence, understanding, or the intellectual self-sufficiency of philosophers commanded wide recognition as attributes of these creatures, much less attributes of compelling significance. Indeed, the original Renaissance tradition of hieroglyphs stressed a very arcane wisdom acquired through recondite scholarship, and the example of Joris Hoefnagel, who filled his decorative work with emblematic images and styled himself *inventor hieroglyphicus et allegoricus*, suggests how that tradition could degenerate into a display of abstruse learning and

111 JS (1669), pt. 1:21, 45, 49, 53; JS (1737–8), 1:5, 13, 33, 220, 279–80, 2:398, 557; Lesser (1745) 1:31n; Réaumur (1734), 1:11.

112 Rèaumur (1734–1929), 1:11–12; JS (1669), pt. 1:45; idem (1737–8), 1:30.

113 Goedaert (1662?–9), 2:101–2, 234, 236, 272; Buonanni (1691), pt. 2:7.

114 Browne (1964) 1:25. See the citation from Christoforo Giarda in Gombrich (1948), 190. Walter Raleigh is also cited by R. Evans (1973), 246; Vondel (1820–4), 14:140.

115 Wilberg Vignau-Schuurman (1969), 1:11–16, 172, 176–7; R. Evans (1973), 246; Giehlow (1915), 180, 202, 207, 218, 219, 224.

Figure 8. Bouquet with insects by Johannes Goedaert (28 × 22 cm). Courtesy of the Rijksbureau voor Kunsthistorische Documentatie, The Hague.

ingenuity.[116] Karel van Mander, a Dutch painter and poet not unfamiliar himself with the atmosphere at the imperial court, recalled Hoefnagel not as a wise but a "clever" (*geestigen*) man, "witty, learned and very inventive" (*gheestigh, gheleert, en seer vindigh*).[117] The extent to which subsequent Dutch artists regarded the flowers and insects they depicted as burdened with symbolic meaning is still disputed, but in the opening years of the eighteenth century, the influential painter's guide Gerard de Lairesse characterized Dutch still-life paintings in general as devoid of meaning and urged flower painters in particular to abandon their beetles, butterflies, and spider webs.[118]

Nonetheless, there is abundant evidence of an inclination in the sixteenth and seventeenth centuries to grant even the slightest living things their spiritual significance. Notably, metamorphosis was often presented as a symbol of personal resurrection, and Goedaert and Swammerdam were among those who elaborated the theme, the latter intending an entire treatise to demonstrate the resurrection of the dead "tangibly and visibly by means of nature herself."[119] The perception of insects was often also colored by the taste in early seventeenth-century Dutch culture for allusions to the brevity of life and the ultimate futility of human aspiration.[120] Certain creatures in particular, such as the spider and the fly, acquired a *vanitas* or *memento mori* significance,[121] but the observation of insect life required no promptings from emblem books to evoke somber reflections on mortality. The writhings of a caterpillar moved Goedaert to a long sermon on the fear of death, and Swammerdam offered in his *Ephemeri vita* a remarkable study of the mayfly *Palingenia longicauda* (Olivier) as an extended allegory of the abruptness and misery of human life.[122]

116 Wilberg Vignau-Schuurman (1969), 1:258, and passim; Kris (1927), 245, 249; R. Evans (1973), 172; Chmelarz (1896), 283, 285.

117 Mander (1604b), fols. 262v, 263r.

118 Lairesse (1707), 2:268, 357–8. L. J. Bol has pointed out that, a century before Lairesse, Karel van Mander, who painted flowerpieces himself, had also made no mention of any hidden emblematic meanings when he spoke in his *schilder-boeck* of the insects that often appeared in such paintings (Bol 1960, 46–7). See also Wilberg Vignau-Schuurman (1969), 1:72. Contention over the extent to which symbolic meaning should be read into Dutch seventeenth-century painting has recently been sharpened by Svetlana Alpers's *The Art of Describing: Dutch Art in the Seventeenth Century* (1983); see in particular her appendix "On the Emblematic Interpretation of Dutch Art," pp. 229–33. Cf., for instance, Schama (1984), 25–31.

119 Goedaert (1662?–9), 1:7–15; JS (1737–8), 1:207–8, 333, 346–7; 2:571–2, 666–7, 684, 829 (quotation from 1:333), cf. 1:20.

120 Apart from the literary elaborations of the theme, the city of Leiden became the center of the one unmistakably symbolic still-life genre, the *vanitas,* which flourished with particular vigor during the decades preceding midcentury (Bergström 1956, 154–5, 158–9, 188; idem 1970a,b). The increasing – and often quite public – practice of anatomy also offered irresistible opportunities to exploit the symbolism of *vanitas* and *memento mori,* and not only in the pages of anatomical atlases but in elaborate displays of anatomical preparations and in the ornamentation of anatomical amphitheaters (Bidloo 1685, frontis. and pls. 87, 89; Ruysch 1739a, 106, 113, 123, 169, etc.; idem 1739b, 8, 9, 11, etc.; Judson 1973, 37; Bleyswijck 1667, 573–4; Möller 1959).

121 Wilberg Vignau-Schuurman (1969), 1:240; Bergström (1955), 346; idem (1956), 158, and see p. 213 (fig. 179).

122 Goedaert (1662?–9), 2:171–227; JS (1675a), "Ernstige aanspraak" [1–2], 3, 420; see also the letter dated 15 Jan. 1675 from Antoinette Bourignon that Swammerdam includes at the beginning. Schierbeek (1947), 266.

It was for their spiritual significance, indeed, that the closer study of such small and frequently repugnant forms of life was most often urged, but it was a significance, after all, that exceeded symbolism and allegory. That each created thing was a direct revelation of the attributes of God was proclaimed insistently and almost universally throughout the century. God's creatures were his reflection, declared Joost van den Vondel, the most majestic of Dutch poets, and the pastor Johannes Feylingius taught a more popular audience that God had created the whole of nature as his image, so that the world was nothing other than God discovered and revealed.[123] The metaphor of nature as God's second book – this wondrous book of God's six days of work, this book of books, wrote the elder Constantijn Huygens – also echoed through the century in both elite and popular literature. The one book, the Scripture, taught his will; the second taught his power, declared Jacob Cats, long the most popular poet in the Netherlands.[124] It was a theme intimately familiar to all members of the Dutch Reformed Church. God was known not only through Scripture, proclaimed their profession of faith, but also through the creation, preservation, and government of the universe, "which is before our eyes as a most elegant book, wherein all creatures great and small, are as so many characters leading us to contemplate the invisible things of God, namely, his eternal power and Godhead. . . ." Having cited the same passage from the Apostle Paul, Goedaert also emphasized at length that the knowledge of natural things, the works of God, was a noble and godly business.[125]

To venerate the revelation of the Deity in nature was an ancient injunction still cherished by Catholics and Protestants alike and embraced by scientists and naturalists across the breadth of Europe. Among the religious leaders of the sixteenth century, however, none had given it greater emphasis than Calvin. Although incomprehensible, insisted the founder of the Reformed movement, God made himself known to and communicated with humanity through his works, on each of which he had engraved unmistakable signs of his glory. Every spot in the universe hence harbored sparks of that glory, through which we, absorbed in wonder, were to contemplate and adore the Creator.[126]

Indeed, Calvin taught, God would have us perpetually occupied in the contemplation of his attributes revealed in the fashioning of the universe. Nor was that contemplation to be hurried; rather, it was to be thoughtful and intent, for the knowledge of God was the final goal of the blessed life.[127] So Goedaert, reaffirming that humanity's foremost duty was to honor God, recounted how he had sought some field of study that might serve that end. Since so much had

123 Vondel (1820–4), 14:140; Feylingius (1665), pt. 1:3.

124 ConHs (1892–9), 4:307, 309; Cats (1726), 2:598–9; Nylant & Hextor (1672), foreword; J.H.S.M.F. (1694), foreword; Verwey (1962), 31–2.

125 Cochrane (1966), 189–90; Goedaert (1662?–9), vol. 1, preface "Aen den goedtwilligen Leser"; vol. 2, dedication.

126 Calvin (1559), 1:51–3, 62, 69n. 44; Dowey (1952), 72–8, 135–7, 139–42.

127 Calvin (1559), 1:51, 179–81; Hooykaas (1972), 106.

already been discovered and written about the heavens, the larger animals, plants, and the form of the world, he had chosen for himself, he wrote, a subject virtually ignored because of its difficulty or seeming triviality.[128]

Difficult and persistent research could have its own further spiritual overtones. It was a paradox in Calvin's teaching that he also insisted the Fall had virtually blinded Adam's lineage to that very evidence of God in his works that Calvin enjoined his followers to contemplate: Man's perversion increasingly dulled whatever perception he still had of the Creator in his creation.[129] Through his decline into sin, remarked Goedaert, man subjected himself not only to misery but to an ignorance of natural things. Nonetheless, Goedaert at least allowed, there remained a light in man by which, through diligent research and observation, he might still grasp the invisible aspects of God in visible things.[130] The endeavor to recognize the glory of God in his works was consequently transformed into a personal struggle to break the grip of spiritual alienation and moral corruption, and discovery emerged as a sign of grace.[131]

The study of nature's living minutiae thus acquired a religious aura. Luther, to be sure, had voiced an unsympathetic view of vermin, among which he included butterflies as well as flies and bedbugs, as themselves by-products of the Fall,[132] but naturalists across late-sixteenth and seventeenth-century Europe emphasized rather the exceptional evidence of God's power and wisdom to be found in his smaller creations. Aldrovandi declared that God and his handmaid Nature had wrought greater miracles in the smaller than in the larger animals, and apart from man himself, wrote the Englishman Thomas Moffett, nothing in the universe was more divine than insects.[133]

From the very beginning, microscopic images spoke to such religious sensibilities. Peiresc had perceived the work of divine providence in the body of the cheese mite, and Francesco Stelluti and Federico Cesi of the Accademia dei Lincei declared that, through their microscopic study of the bee, the eye had learned to magnify its faith. Kircher struggled to convey a sense of the omnipotence, wisdom, and goodness of God inspired by microscopic observations,

128 Goedaert (1662?–9), vol. 1, preface "Aen den goedtwilligen Leser."

129 Calvin (1559), 1:liii, 40, 40n. 2, 43n. 1, 51n. 2, 63, 68, 69, 341; Dowey (1952), 73, 81–2; Niesel (1956), 43–4.

130 Goedaert (1662?–9), vol. 1, preface "Aen den goedtwilligen Leser."

131 In Calvin's thought, indeed, it was the acquisition of faith that enabled one to see God's revelation in Nature (Dowey 1952, 135–9). With respect to the exhilaration occasioned by the belief that every discovery was a gift from God, see Chapter 4, n. 46, and Chapter 5, nn. 164–9.

132 Pelikan (1961), 467.

133 Aldrovandi (1602), 1–2; Moffett et al. (1634), preface; see also Pighetti (1961), 320–1. Augustine and the Bible were sometimes cited in such passages (Aldrovandi 1602, 2; Goedaert 1700, 1:166; Goedaert (1662?–9), vol. 1, preface "Aen den goedtwilligen Leser"; Mey 1742, 1:558–63), but it was the statement of the pagan Pliny that nature was never more complete than in the smallest things (Pliny *HN* 11.1.4) that echoed most frequently, even into the following century (A. Cluyt 1634, [62]; Moffett et al., preface; *OCCH* 1:323; JS 1737–8, 1: 300; H. Baker 1742, title page; idem 1753, title page; Rösel von Rosenhof 1746–61, vol. 1, frontispiece).

and Borel proclaimed that the structure of minute insects, "those animated atoms," would inspire even the most committed atheist with a proper reverence for God. Henry Power, whose English account of microscopical observations had preceded Hooke's, lamented that Antiquity, lacking the microscope, was ignorant of the "eminent signatures of Divine Providence" with which the minutest things were enriched and embellished. "This [the microscope] to our mind the a'theriall wisdome brings," he had earlier rhymed,

> how God is greatest in ye Least of things
> And in the smallest print wee gather hence
> the world may Best reade his omnipotence.[134]

Such evocations flourished in the succeeding century as well, but the emotion, we may surmise, was most intense when familiarity – which cooled our admiration for nature's other wonders, the elder Huygens noted – had not yet dulled the impact of these unexpected images.[135]

That God was to be discovered in the least of his works was emphatic in Dutch thought as well. The confession of faith of the Dutch Reformed Church deviated from its French prototype (though not from Calvin) in explicitly affirming that small creatures as well as large reflected the invisible attributes of God,[136] a theme that was rehearsed by both naturalists and poets. Inspired by a mosquito that lit upon his hand (and apparently by Pliny), Cats marveled at the parts and senses that had been fashioned in such a narrow frame and concluded that the mosquito and its buzz cried out the wisdom of God.[137] Goedaert explained that the very purpose of his years of study had been to show that God could be admired in all his creatures and that the wonders of nature were most pronounced in the smallest, indeed, and least esteemed.[138] It was to be expected, hence, that Netherlanders would be quick to invest microscopic images with religious import. Recalling what he had once observed through Drebbel's microscope, the elder Huygens proclaimed that nothing so compelled us to worship the wisdom and power of God as did the discovery of the care that had been bestowed upon the smallest and most despised creatures and the encounter with the ineffable majesty confronted there.[139]

Old as well as new traditions of both belief and practice in philosophy, anatomy, art, and religion thus shaped a persisting and, in some instances, insistent awareness of the significance of minute forms. It might well be argued, there-

134 Singer (1915), 328; idem (1953), 201; Kircher (1646), 834; Borel (1656), dedication; Power (1664), preface [ii]; idem (1934), 73.

135 Buonanni (1691), pt. 1:1–6; Pluche (1732–50) 4:526; Hoogvliet (1738), 1:54; Muys (1738), preface [xxxiv]; H. Baker (1742), 310; Gleichen-Russworm (1764), 3 ff.; ConHs (1897), 120.

136 Cochrane (1966), 189–90; Calvin (1559), 1:181.

137 Cats (1726), 2:356; Pliny *HN*, 11.1.2–3; see also Cats (1726) 2:353; Westerbaen (1672), 173; Oudaen (1712) 1:4. Regarding Philibert van Borselen, see Veen (1960), 16–17.

138 Goedaert [1662?–9], vol. 1, dedication.

139 ConHs (1897), 120.

fore, that these traditions, though by no means unique to the Netherland, were cultivated there with a distinctive interest and intensity, so that Dutch society in the seventeenth century could have been expected to have proved singularly responsive to the potential of the microscope.

To the contrary, however, that potential was only seized upon in the Netherlands belatedly. The first half-century of microscopic literature came from the pens and researches of Italian and English virtuosi, German Jesuits, and a French royal physician. For nearly two generations, the Dutch, having played such a prominent role in the early development of the microscope, contributed little of significance to what is known of its early application. The most notable contribution by a Dutchman reaches us through secondhand accounts of the sales ventures and self-advertisement of that typically hybrid inventor, engineer, and magician Cornelis Drebbel, by then resident abroad. The apparent inefficacy of seemingly obvious incentives to take up the instrument in the Netherlands underscores the complexity of the interaction between early scientific innovation and social and cultural contexts.

CHAPTER THREE

Obstacles

It was not that no one in the Netherlands entertained any early hopes for the microscope. Inspired by Drebbel and his instrument, the elder Huygens, when still a young man in the 1620s, had urged de Gheyn to undertake an atlas of a "new world" of tiny creatures as seen now through the microscope, a project toward which Constantijn believed the older artist was not ill-disposed. Henricus Reneri, a friend of Descartes and first professor of philosophy at the newly chartered University of Utrecht in the 1630s, hoped by means of the microscope to discover what had until then remained unknown in seeds, flowers, and the other parts of plants, and Johan Hudde wrote in 1657 of his intention to explore the processes of generation with the microscope.[1] As far as we know, however, these and presumably similar aspirations bore no fruit, and although Huygens believed it was only the death of de Gheyn that frustrated his hopes, the dearth, for decades, of any signs of sustained microscopic observation in the Netherlands suggests some hidden dissuasion.

Differing in character and conspicuity – now explicitly revealed in conscious discourse, now buried in unarticulated structures of thought that underlay intent and anticipation – the limitations imposed by established ways of doing and thinking, though perhaps never demonstrated conclusively, can constrain us nonetheless. Our conceptions of science and knowledge limit the methods and techniques we recognize, and the accomplishments of our predecessors shape what we aspire to do. Culturally validated aims and purposes entail less conspicuous habits of ingrained indifference and exclusion. Even our acquired understanding of nature determines what we are capable of seeking. Such limiting predispositions can indeed assume diverse, elusive, and unexpected forms, and even those contexts of thought and practice in seventeenth-century Dutch culture that would appear at first glance to have encouraged a recourse to the microscope – Cartesian corpuscularism, subtle anatomy, the traditions of miniature painting, and the exaltation of God in his smallest creatures – nurtured outlooks and attitudes that may well have thwarted that encouragement and, in some cases, rendered these contexts more barriers than bridges.

1 ConHs (1897), 120; Descartes (1897–1910), 2:102; Hudde to Lambert van Velthuysen, 13 Oct. 1657, University of Amsterdam Library, fols. 1v–2r.

At the very core of Cartesianism, indeed, resided an absorbing commitment that implicitly denigrated the value of the microscope. Descartes had not only offered the most elaborate and systematic version of the mechanical philosophy, but also had sharply accentuated the epistemological basis of its appeal. Mechanism promised to make the operations of nature comprehensible, a promise Descartes embraced and amplified in his broader philosophical quest for certain and unsullied understanding. The possibility of achieving such understanding was an exhilarating prospect that remained a preoccupation of Descartes's more doctrinaire followers in the Netherlands throughout the century; but it was a preoccupation that, despite the Cartesian fascination with minute mechanisms and Descartes's anticipations of the value of the microscope, worked against the pursuit of microscopic research within the framework of the Cartesian program. The vanguard of the Cartesian movement in the Dutch universities consisted in large part of medical men who showed little appreciation of the potential of the microscope and more interest in the pursuit of philosophy than, for instance, the practice of subtle anatomy.[2] Their distinguishing purpose as Cartesians was rather to depict the living body and its processes in terms of mechanistic images that the mind could fully grasp.

Descartes's "method" lent itself to an extreme rationalist reading according to which the physical world could be comprehended only through deduction from ideas discovered within the mind itself, and his prominent followers in the Dutch universities echoed his distrust of the senses and his reliance rather on the independent capacities of the intellect as the only source of knowledge that was not only true but absolutely clear.[3] Teaching at Leiden in the 1650s, de Raey argued emphatically that, although sense experience might suffice for the needs of the multitude, the philosopher had to seek more certain knowledge deriving from "the internal light of the mind alone." The criterion of truth was not to be sought in things existing outside our thoughts, asserted a Cartesian successor toward the end of the century, but in our thoughts themselves.[4] The Cartesian preoccupation with comprehensibility and certainty, consequently, meant a preoccupation with the mind.

2 Joannes de Raey was explicit that he had undertaken the study of medicine for the sake of philosophy (Raey 1654, 2–3). There was, to be sure, another less distinctive stream of Cartesian influence that simply encouraged the desire to investigate nature without imposing Descartes's specific teachings. Henricus Reneri, for example, is to be associated with this brand of "Cartesianism" (Sassen 1941, 882–5, 891), which continued to surface through the middle of the century. Swammerdam would even cite Descartes against the overemphasis on reason that was characteristic of the latter's more doctrinaire followers (JS 1669, pt. 1:154; idem 1737, 2:868–73), and modern scholars still endeavor to affirm and clarify the place of empirical research in Descartes's conception of method (e.g., see D. Clarke 1982). Nonetheless, his most prominent disciples in the Dutch universities tended to simplify his doctrine rather than to cultivate its subtleties and complexities, and the brand of Cartesianism that loomed largest as an identifiable movement in Dutch intellectual life in the middle of the seventeenth century – and that, indeed, was most given to the elaboration of the imagery of invisible mechanism – was characterized by its commitment to dogmatic speculation.

3 Heereboord (1654), 170; Raey (1654), 21ff., 36, 41–2, 45ff.; Geulincx (n.d.), 1–2, 120; Craanen (1689), 517; Volder (1695), 48, 68–9.

4 Raey (1654), 1ff., 41–2; Volder (1695), 48; see also n. 25 below.

The realm of unseen particles and pores in particular was to be explored by reason alone. Descartes's early disciple Regius did not fully accept his master's rationalist theory of knowledge, but he declared nonetheless that it was the "keenness of the mind" and not the senses that perceived the network of fissures and the flux of corpuscles within even the most subtle bodies. In all natural things including the diverse tissues of the living body, asserted Regius, the "intellect alone" observed the varied particles – such as the branched particles of oil and the oblong, pliant particles of water – without which no clear explanation of the qualities observed in bodies was possible.[5] Cartesian thinking had posited the minute, invisible mechanisms throughout nature in the pursuit of intelligibility, but the instrument by which this intelligibility was won was explicitly the mind alone. The Cartesian fascination with invisibly small structures thus offered little incentive to resort to the awkward new microscopes and their problematic images or to search for new techniques that might open up new realms for microscopic exploration.

More consciously concerned than Regius with the philosophy of knowledge, de Raey revealed even more vividly how the commitment to Cartesian rationalism worked against the microscope. The assumption that, in addition to the subtle matter itself, all kinds of particles and structures in the interior of bodies were beyond the senses was critical to his own demonstration that the foundations (*praecognita*) of the *scientia* of natural bodies were provided by reason, the intellect, the internal light of the mind alone.[6] It was in no small part in order to make the case for a rationalist epistemology that de Raey dwelt at such length on the minute mechanisms throughout nature, and their essential significance to de Raey was hence not only that they rendered phenomena comprehensible but that they were also beyond the reach of the senses – even, implicitly, when aided by instruments. The accessibility of those minute structures to the mind but not the senses was dictated by de Raey's philosophy, and the lens was thus a priori denied any prospect of contributing in any meaningful way to the exploration of this realm of the small.

When the age of early microscopic discovery began in the Netherlands, the most notorious representative of militant Cartesianism in the Dutch universities was Theodoor Craanen. He joined the faculty of philosophy at Leiden in 1670 but, in the context of a final crisis over Cartesianism in the Republic, was transferred to the faculty of medicine three years later, where he continued to teach until 1687. The books ascribed to him, including an anonymous *Oeconomia animalis* published during his lifetime and a *Tractatus physico-medicus de homine* in 1689, were published by his students and admirers,[7] and specific thoughts and

5 Regius (1646), 3–4; idem (1668), 4. Regius, to be sure, did not himself speak of "tissues" but of the different "parts," *partes*, of the body. He differed from Descartes in his rejection of the need for innate ideas (Regius 1646, 251) and in his less exalted understanding of "method" (ibid., 288).

6 Raey (1654), 23, 26, 28, 41–2.

7 Luyendijk-Elshout (1975), 299.

phrasings cannot be attributed to Craanen without caution; but their Cartesian emphasis corresponds to his reputation. (Hartsoeker recalled his dogmatic instruction in Cartesian physiology.)[8] Irrespective of the actual hand that wrote them, their problematic allusions to the microscope – even though its prestige was now greatly enhanced by the work of Swammerdam, Leeuwenhoek, and Malpighi – still reflect the reservation of the Cartesian tradition that Craanen represented.

He was acquainted with both Swammerdam and Leeuwenhoek[9] and, unlike Regius and de Raey, often spoke of the microscope in his attributed works; yet even during those decades of exciting new microscopic discoveries, his attitude toward the instrument remained ambivalent at best. He cited it on occasion to challenge the conventional beliefs in anatomy,[10] but more often than not, when he brought it up, the real point seemed to be that the minute structures of real significance were beyond any that the microscope could reach.

When he intimated, for instance, that the origin of fevers lay in a disturbance of the particles of the blood, not only did Craanen disregard Leeuwenhoek's discovery of the red corpuscles, he explicitly insisted that the particles in question were inaccessible to the microscope. On the other hand, he was surely alluding to Leeuwenhoek's early accounts when he noted that the microscope found everything to be composed of globules, yet the microscopic globules themselves were apparently of very little potential significance to Craanen. He mentioned them, rather, to focus on the spaces in between and the infinitely diminishing particles those spaces perhaps contained. He also cited the myriad animalcules that the microscope – again, it would be Leeuwenhoek's – had discovered in a drop of fluid, but he did so only in order to emphasize once more what he presumed to be still unseen: the parts, fluids, and "spirits" within their bodies. Hence his conclusion that "the subtlety of nature surpasses our powers of thought."[11] It was not his intention to denigrate these powers, however, but to spur them to reach far beyond those microscopic images. Undaunted, he himself characteristically forged ahead to consider the fine structure of real interest to him, the particles and pores too small to be detected by any lens. Taking up the subject of nutrition, he again alluded to the microscope only in order to note what it could not see: in this instance, the particles of water and, by implication, all those particles of nutriment that, passing through the pores of the arteries, dispelled the obscurity of past explanations of the nourishment of the body.[12] Comprehensibility still lay far, perhaps infinitely far, beyond the discoveries of the microscope.

It did so because the Cartesians expected comprehensibility to rest on an understanding of the most basic level of natural processes, which they insisted no

8 NH (1730), "Ext. crit.," 45.

9 Craanen (1689), 290, 731. Leeuwenhoek wrote that Craanen had often visited him (*AB* 2: 280).

10 Craanen (1689), 22, 68, 363, 647.

11 Ibid., 234, 211; see also p. 512. Regarding Leeuwenhoek's early reports on microscopic globules in a wide variety of different substances, see Chapter 7, n. 63.

12 Ibid., 208–9; [Craanen] (1685), pt. 1:65–6.

observations could reach. Citing microscopic studies of the embryonic chick that ostensibly described the beginnings of the heart, Craanen pointed to the failure of those studies not only to reveal the size, figure, and motion of the particles involved – "since vision's keenness does not reach these subtleties" – but to explain why the heart began to beat. Nor, he noted, had microscopic studies of the animal fetus been able to explain how the parts of the fetus were produced, for such things also eluded the dullness of the senses. Craanen had made his own very Cartesian outlook on such issues very clear: the whole *ratio* of generation or conception lay in the insensible motions of the diverse particles in the male and female semens when they mixed. Hence, the kind and level of explanation he was ultimately demanding inevitably relegated microscopic observations to what was necessarily superficial. Although granting that it would be no trivial observation if, as some had claimed, the microscope could discover the rudiments of the plant within the seed, he concluded by doubting that the microscope could do it.[13]

Craanen did recognize some positive contributions made by the microscope to the knowledge of the body – it had discovered that hair (or some hair) was hollow, challenged the distinction between the "conglomerate" and "conglobate" (essentially, the true and lymphatic) glands, revealed the complex structure of the walls of the stomach and esophagus, and, through Swammerdam, confirmed the anastomoses of veins and arteries[14] – but "reason" remained a no less powerful instrument for discovering the minute structure of the body, and an instrument, to all appearances, much more to Craanen's taste. (It is not clear that he ever used the microscope himself.) Just as reason could demonstrate the necessity of unseen muscles, blood, and nerves within the microscopic animalcules, so in the human body it revealed an infinite number of invisibly small lymphatic vessels and even muscles in the walls of the veins and arteries. To those who might protest that no anatomist could demonstrate that there were valves within the nerves, Craanen replied that the acuteness of the mind revealed how such valves were necessarily formed, and "what we can clearly and distinctly conceive cannot be denied. . . ."[15]

Not surprisingly, what reason dictated often conformed to the teachings of Descartes (as indeed in the case of the valves in the nerves).[16] Nevertheless, the typical Dutch Cartesians – even when deviating from the master on particulars or, in the later decades of the century, granting the microscope what they thought its due – still looked to reason as the instrument of ultimate authority in determining the minute construction of the body. Although also deeply committed to the Cartesian outlook, Craanen's influential former student Cornelis Bontekoe attempted to combine that loyalty with an aggressive advocacy of in-

13 Craanen (1689), 712–13, 747.

14 Ibid., 380, 363, 647, 68, 22, 731. Regarding the anastomoses of the capillaries, compare the treatment of the subject in [Craanen] (1685), pt. 1:80, where the microscope remains unmentioned.

15 Craanen (1689), 512, 251–2, 459.

16 Descartes (1897–1910), 11:135–6, 200–1.

tellectual independence. While urging other physicians to turn to Descartes's philosophy if they wished to see medicine perfected, he also rebuked the followers of Descartes for having, in their loose talk of particles, pores, and subtle matter, succumbed to the same useless theorizing as had the followers of Aristotle. Anatomists must turn from useless speculation to the knife, the injection syringe, and the microscope, wrote Bontekoe at one point;[17] yet his own personal deference to speculative reason, was not, after all, so easily overcome.

Among the specific doctrines Bontekoe rejected was Descartes's description of the nerves, which (though Bontekoe's own brief synopsis is somewhat puzzling) posited bundles of threads (*filets*) for the sensory function within tubular nerves through which, for the motor function, also flowed animal spirits from the brain. Although Descartes remained "our great philosopher," Bontekoe still asserted that no anatomy provided the least proof of this construction.[18] Shortly thereafter, however – and without so much as a blink of an eye, as it were – Bontekoe continued that to deny that the nerves were hollow and conveyed a fluid was to trust one's eyes more than reason, "which you must always believe, nonetheless, as our only and unfailing teacher." To conclude that there was no hollow in the nerves because none could be seen even with the microscope was against all reason, he now insisted.[19]

The text of Craanen's *Tractatus* adhered more or less to Descartes's original doctrine of the nerves, but the plate that accompanied the text (provided, presumably, by the editor of the posthumous work) depicted the threads themselves within the nerve as hollow tubes (Fig. 9) – a bit of updated "reason" doubtless surreptitiously slipped in. Moreover, instead of citing Descartes, as did the text, the legend of the plate – which depicted the threads as a bundle of cylindrical pipes neatly revealed by laying back the surrounding membrane of the nerve – declared that it derived from the microscope![20] Let us ascribe this less to purposeful deceit than to the conviction that the microscope could ultimately only confirm what reason taught.

The Cartesians were not alone in exalting the power of the "mind's eye" to discover the minute construction of the body,[21] and the doctrine of the hollow nerve was after all an ancient one with many non-Cartesian adherents as well in the seventeenth century.[22] Still, the Cartesians as a sect were notorious for their preoccupation with doctrine they believed reason had revealed. Despite his own debt to Descartes, an older Christiaan Huygens rebuked the Cartesians for their delusion that they knew the causes of everything and their failure,

17 Bontekoe (1689), vol. 1, "Voor-reden" and pp. 35, 45. He maintained that the microscope did offer support, however, for the assumption of pores in all solid bodies (ibid. 2:377).

18 Ibid. 1:61–2; Descartes (1897–1910), 11:132–3, 141, 143, 6:110–11.

19 Bontekoe (1689), 1:63.

20 Craanen (1689), 400, and pl. 17. Craanen introduced an element of ambiguity in his own account, to be sure, by speaking of the spaces between the threads – *filamenta* and *fibrillae* – within the nerve as *tubuli*.

21 See, in particular, Deusing (1662), 1–28 and passim.

22 See E. Clarke (1968).

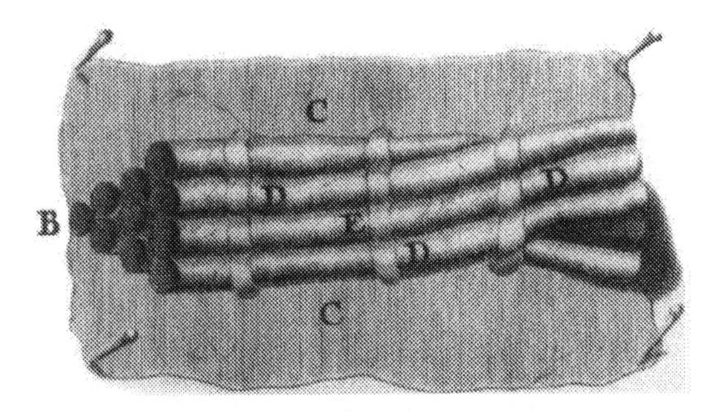

Figure 9. The hollow tubes ostensibly observed in the nerves with the microscope, an illustration included in the *Tractatus physico-medicus de homine* (1689) posthumously ascribed to Theodoor Craanen. Photo courtesy of the National Library of Medicine, Bethesda, Md.

hence, to seek for what truly lay behind natural phenomena. So likewise Leibniz, unable to recall anyone besides Craanen who had recently dealt with medicine like a philosopher, regretfully added that the Cartesians were too committed to their hypotheses: "I care more for a Leeuwenhoek who tells me what he sees than a Cartesian who tells me what he thinks."[23]

Doubtless, this overindulgence in speculation often reflected little more than a distaste for the exertions of research;[24] but a longing for clear and certain understanding remained a driving force behind Cartesian rationalism as well,[25] and it was a force that predisposed Cartesian thought against the microscope. Divorced from this epistemological obsession, the mechanistic imagination could prove stimulating indeed to microscopic exploration, and while Cartesians speculated about the sharp shapes of salt and acid particles, Leeuwenhoek may well have been searching for the analogous shape of pepper particles when he stumbled upon a memorable encounter with microscopic life.[26] Descartes and his

23 *OCCH* 10:405, 52.

24 It is perhaps worth noting that clinical instruction at Leiden as well as the encouragement of research declined during Craanen's tenure there (Luyendijk-Elshout 1975, 298, 305).

25 Burchardus de Volder, who taught philosophy at Leiden during 1670–1705, was also remembered by Hartsoeker as no less a Cartesian than Craanen (NH 1730, "Ext. crit.," 45), even though he became disillusioned with Cartesianism in the closing years of the century and disavowed the publication in 1695, presumably by former students, of a series of Cartesian lectures ascribed to him (Volder, 1695, *Exercitationes academicae*); it was a disavowal, however, that must also be understood in the light of the storm of controversy and the prohibitions that had marked the career of Cartesianism at Leiden (Verbeek 1992, 34ff; Ruestow 1973, 44–8, 61–3, 74–8, 111–12). In these lectures, although he was a pioneer in introducing experimental physics at Leiden as well, de Volder had still emphasized the ideal of a rationalist basis for certain knowledge (Ruestow 1973, 91–2, 96–8).

26 Descartes (1897–1910), 6:249–64, 11:146; Bontekoe (1689), 1:271; Overkamp (1686), 299; *AB* 2:90–1, see also 1:264–5; see in addition Beeckman (1939–53),1:216. Some scholars have indeed stressed the linkage between mechanism and the beginnings of microscopic research (Belloni 1971, 100; Grmek 1970, 303–4; Fournier 1981b, 82–4), but the Dutch Cartesians had other priorities. Regarding Leeuwenhoek's use of pepper, see Chapter 7.

followers, however, had developed their imagery of invisible mechanisms in a quest for a comprehensibility that could rest only on foundations laid in the mind, isolated from the uncertainty and deceptiveness of the senses. What that mind fabricated in order to render the sensible world intelligible was a vision of the invisible innards of nature reduced to no more than structures and moving shapes whose collisions and interactions left nothing obscure. The microscope only clouded this prospect by refocusing attention on sensible appearances in the realm now of the microscopically small. In an age of such vibrant scientific activity, the Cartesians could not be indifferent to the remarkable advances in empirical research; but as Cartesians – and it was in this guise that they articulated their elaborate spectacle of hidden mechanisms – their philosophical aspirations posed a formidable barrier to the appreciation of the potential of the lens.

The craft tradition of naturalist miniature painting – a very different kind of endeavor than the philosophical search for certainty – nurtured attitudes and aspirations that also inhibited the meaningful use of the microscope. Although insects were universally perceived as the most obvious subjects for microscopic scrutiny, the masters of their depiction in the miniature tradition in the Netherlands gave little if any heed to the new instrument. To be sure, Filippo Buonanni subsequently stated that Jacob Hoefnagel had turned to the microscope in his engravings of insects in 1592, and the finely detailed depiction of insect life by Netherlandish miniature and still-life painters of the period has also persuaded recent students of the genre that a magnifying glass had indeed been used.[27] But if so – and the point is open to dispute – what is most striking then in the light of their familiarity with magnification is how little use they had for it. Both creatures and details are rendered as they would appear in their true scale to the naked eye and often smaller. Even in the few exceptions that have been found,[28] the enlargement, if it is granted, is so minimal that it hardly constitutes compelling testimony to the employment of a lens: The draftsman's unaided eye and hand can do as much even by unconscious error, and can intentionally do so more easily without having to manipulate a needless instrument. After all, even Goedaert, although self-consciously now a naturalist, turned only rarely to the microscope, and even more rarely with respect to observations in any way similar to the motifs of the miniature tradition.[29]

27 Buonanni (1691), pt. 2: 7; Bergström (1956), 40; Judson (1973), 16. P. van der Star points out, however, that there are no details in Hoefnagel's plates that cannot be seen with the naked eye, especially if one were slightly myopic (Star 1953, 7–8).

28 Singer (1915), 319–20; Regteren Altena (1983), 1:137.

29 Goedaert refers to the microscope in affirming that, contrary to what some said, caterpillars do have nerves (although it is highly unlikely that he observed other than external features) and other organs by which they feel and move (Goedaert 1662?–9, 2:101–2), in observing and depicting the emergence of beetle larvae "no thicker than a horse's hair" from their eggs (ibid. 2:234, 236), and in reporting what he took to be the spontaneous generation of worms in a fluid from a mushroom (ibid. 2:272). Only the first of these instances would seem to have any bearing on traditional miniature motifs.

The lens as such, in fact, was nothing new to the atelier of the miniature painter. In 1558 Simon Bening, one of the more renowned of the Master of Mary of Burgundy's successors (and perhaps his son), portrayed himself before his work with spectacles in hand.[30] Given the close nature of that work, his advanced age, and the relative likelihood of near- as opposed to farsightedness, the lenses in his spectacles were most likely convex.[31] His exposure to the phenomenon of magnification had hence very probably been recurrent as, perhaps toying with his glasses, he labored at his bench. Like Bening's spectacles, the isolated lens may well have been used to clarify detail for miniaturists; but even when, by the early seventeenth century, they surely recognized the effect of magnification in the flux of optical distortions, they failed to find it relevant to what they were about.

Although precisely what they were about is perhaps beyond our grasp – historical and cultural distance being compounded by the notorious difficulty of explicating artistic intent – recent efforts at such explication offer suggestive insights.[32] The tradition of miniature painting in the Netherlands was in many ways the antithesis of the program of the Cartesian rationalists. While the latter tried to escape the distracting turmoil of sensations and to grasp, rather, an unseen substratum of uniform, colorless matter, the former fixated on aspects of surfaces – colors, patterns, textures, and the tricks of light – that the Cartesians purposefully thrust aside.[33] The Cartesians struggled for ideas purged of all obscurity, ambiguity, and deceit, whereas the miniature painter dealt ultimately in illusory evocation and, if recent analyses of the artistic endeavor are right, with the expression of "feelings" that by their very nature may exceed the competence of language.[34] While discouraging hopes of fully capturing or conveying the artist's intent in words, however, such analyses do suggest what may after all have been the root of the miniature painter's indifference to the magnifying lens.

Naturalist miniature painting was not without its own scientific affinities, to be sure, exemplified particularly in what was in fact a long tradition of fruitful interaction with natural history.[35] However, despite Joris Hoefnagel's motto

30 Durrieu (1921), pl. 87; Pächt (1948), 43–5.

31 In most places, myopia was not as common as presbyopia, writes Albert Van Helden, who also notes that concave lenses were probably only just becoming common in northern Europe about the middle of the sixteenth century (Van Helden 1977, 10–11).

32 General assumptions about the nature of art and aesthetics in the following discussion have been influenced particularly by the thoughts of Susanne K. Langer (as developed in Langer 1967–82, 1:chaps. 3–7), bolstered by Ellen Winner. E. H. Gombrich (1961) has proven very stimulative as well.

33 On the interaction of light and surface textures as a traditional and distinguishing fascination of northern painters, see Gombrich (1976), 20, 30–3; see also Sterling (1952), 36.

34 Langer (1967–82), 1:64–5, 87, 102–4, 155, 157, 166–9, 202, 213, 219, 223; Winner (1982), 7–8. Although the context is narrower, see also Gombrich (1961), 49.

35 Pächt (1950), 22, 25, 31; idem (1948), 53n. 22; Arber (1953), 1:321–5; J. Evans (1931), 1:76, 78.

that claimed nature as his only teacher – *natura sola magistra*[36] – his typical heirs among Dutch seventeenth-century still-life painters were not instilled with a compelling ethic that called for the direct, much less original, observation of nature. The repetition of the same blossoms in successive painted bouquets and the presence side by side of flowers that bloom in different seasons betray a widespread reliance on floral patterns, which reappear even in the work of different painters.[37] Insects recur similarly. Goedaert as a naturalist prided himself on his original observations, and indeed other seventeenth-century painters kept their own collections of insects to paint; perhaps as often as not, however, they turned for models to the work of other artists, including such plates as Jacob Hoefnagel had engraved.[38]

By 1600 naturalist miniature painting itself was something of a stylistic convention perpetuated by repetition. A half-dead tradition resuscitated by Hoefnagel, it remained stylistically arrested even in his gifted hands.[39] For de Gheyn and the artistic circle from which he emerged, it was one of a number of styles from which to choose,[40] and if de Gheyn himself was stimulated as well by the heightened interest in natural history at Leiden, where he did his first naturalist miniature studies,[41] a more immediate incentive would appear to have been the advancement of his artistic career through the mastery of new techniques. Having concentrated for years on engraving, he had turned to the minute rendering of small flower pieces in order to practice the handling of color

36 Kris (1927), 249. Hoefnagel's motto, however, must also be weighed in the light of E. H. Gombrich's central contention in *Art and Illusion* that "art" – and he is speaking of representational art – "is born of art, not of nature" (Gombrich 1961, 24).

37 Rosenberg, Slive, & Kuile (1966), 335; Regteren Altena (1935), 32; idem (1983), 1:110; Bergström (1956), 50–1, 62–3, 69, 76, 78–9, 84; idem (1974), 28–9; E. Jongh (1982), 153–4, 164, cf. 171. See also Bredius (1915–22), 1:115.

38 Rosenberg et al. (1966), 335; Stuldreher-Nienhuis (1944), 46–7; Ritterbush (1969), 574. For various insects copied from Jacob Hoefnagel's *Archetypa studiaque patris,* see Hind (1915–32), 5:211–12, items 29, 33. Jacob, in turn, copied from the album of flowers and insects done by de Gheyn (Regteren Altena 1983, 2:142), just as his father, Joris Hoefnagel, had also copied from others before him. Marjorie Lee Hendrix underscores the extensive copying Joris's natural history illustrations indeed entailed, but she also points out that this seems to have been least true in the case of his insect illustrations (Hendrix 1984, 17, 41–2, 65–70, 121, 157; cf. 70, 73, 235–6).

Judging from Arthur M. Hind's *Catalogue of Drawings* and Leo van Puyvelde's *Dutch Drawings,* insects do not in fact appear to have been a very common or popular subject for the kind of casual observation by artists that found expression in drawing (Hind 1915–32; Puyvelde 1944). To be sure, biases in subject matter may well reflect the tastes of the collectors rather than the artists, but one of the major contributors to the collection of prints and drawings in the British Museum, to which Hind's catalog pertains, was Hans Sloane, who was not without an interest in natural history.

39 Kris (1927), 243, 245; Pächt (1948), 42.

40 De Gheyn's friend and fellow-artist Karel van Mander touted de Gheyn's master Hendrik Goltzius precisely for his mastery of a number of distinct styles (Mander 1604b, fol. 284r–v), and Arnold Hauser associates "mannerism," an artistic movement to which de Gheyn was clearly linked, with the emergence of a coexistence of different styles and with an unprecedented consciousness of style as such (Hauser 1986, 21).

41 Regteren Altena (1983), 1:66, 112; Judson (1973), 11, 14–16.

and oil-based pigments; his most memorable watercolor studies of insects and flowers followed soon thereafter. According to his friend and fellow painter Karel van Mander, however, de Gheyn really desired to undertake larger subjects and to paint more broadly, and he soon turned with delight to a commission for a life-sized painting of a horse![42] Flower paintings (with their accompanying insects) remained a part of his repertoire, but as a distinct genre and style alongside other very different styles, including the fashionable "mannerism" that, in contrast to the miniaturist's detailed fidelity to nature, reveled in stylized faces and artificially postured and proportioned nudes.[43] Of these styles, moreover, de Gheyn's miniature work remained the most subservient to the practice of past masters,[44] and it is perhaps not without significance that, near the end of de Gheyn's life, his friend Constantijn Huygens the elder felt it was now time to distinguish that style with a name of its own: *miniatura,* a name Huygens implies he had newly coined.[45]

More characteristic of the seventeenth century than the cultivation of such a repertoire of styles was the tendency to specialize in genres set apart by their own iconographic if not always stylistic rules. Still-life painting was thus divided into several distinct and persisting types, among them the fruit and flower pieces with their typical insects that seldom appeared in other paintings.[46] During the greatest age of Dutch painting, the naturalist miniature tradition thus survived most visibly within the framework of this genre, but the small animals that embellished its fruit platters and bouquets also testified to the persisting grip of iconographic convention. Within the framework of the painter's profession, consequently, the Dutch artist's approach to insect life was narrowly limited by the craft conventions to be mastered.

Moreover, the stylistic changes that affected the genre during the course of the century also bespoke broader artistic priorities that not only differed from those of the observational sciences but discouraged any recourse to the microscope. The beginnings of the seventeenth-century Golden Age of Dutch paint-

42 Mander (1604b), fol. 294v.

43 See, for instance, Judson (1973), fig. 11 (and compare it with "The Triumph of Wisdom" by Bartholomaeus Spranger, the source of the mannerism cultivated by Mander and Goltzius: R. Evans 1973, pl. 3; Rosenberg et al. 1966, 23). See also Regteren Altena (1983), 1:95, 110.

44 Regteren Altena (1935), 25.

45 ConHs (1897), 65, 67. According to Constant van de Wall's English translation of Karel van Mander's *Schilder-boeck,* Mander had already written in 1604 of "miniatures," "miniature painting[s]," and even of a "miniaturist" (Mander 1604a, 282, 284, 401, 422), but Mander's original text employs no such cognates of "miniature" (or, hence, of Huygens's *miniatura*) in these particular instances, and no terms more indicative of a distinct style than *verlichterije,* "illuminations" (Mander 1604b, fols. 262r–263v, 294r, 299v).

46 Gombrich (1982), 296; Bergström (1956), 3, 65, 68; Bol (1960), 21–2. The conventions that marked the earlier miniature paintings of insects as a distinct art form persisted as well. Arnold Houbraken informs us at the beginning of the next century that, like his father Mathias, Pieter Withoos, who died in Amsterdam in 1693, also painted small animals in watercolor one to a page, and that those pages were then gathered into a book that was still prized by those who loved such things (Houbraken 1753, 2:189).

ing are commonly associated with a new "realism" that, unconcerned with theories of art or aesthetics, seized upon the everyday, unembellished aspect of the world. Yet this was a realism shaped by a continuing – indeed intensified – engrossment with illusionism and experimentation with aesthetic effects. The subdued tonalities imposed upon "reality" in the pioneering new genres of landscape, domestic still lifes, and genre scenes amply testify to this aesthetic preoccupation,[47] and, in the genre of flower painting in particular, new strategies of illusionism challenged the lingering imprint of an earlier deference to more narrowly horticultural concerns.

A burgeoning interest in horticulture and new flowers from the East had provided the immediate backdrop to the emergence of flower painting as an autonomous genre in the Netherlands. Joris Hoefnagel's manuscript illuminations had already reflected the sixteenth-century fascination with the new plant and animal life encountered overseas, and, in Holland, the founding of a botanical garden at the University of Leiden in 1593 added an academic focus to a widespread amateur enthusiasm that, at such cities as Middelburg, was ensuring a warm welcome for Flemish flower painters fleeing the civil turmoil at home. The first supervisor of the Leiden garden, the eminent French botanist Charles de L'Écluse, encouraged that enthusiasm and fanned in particular the rising passion for the exotic flowers that also loomed large in early flower paintings.[48]

But the potentially deadening influence (for art, that is) of botanical didacticism could already be detected even in Hoefnagel's illustrations, and the effort to display the structure and distinctive hue of every flower with equal clarity and attention gave early still-life bouquets an almost primitive quality. Displayed in a uniform light that played no favorites, the blossoms pressed forward all together in a common but shallow front.[49] By midcentury, however, yielding to a tide that had swept through European painting and crested in the Netherlands in the work of Rembrandt van Rijn, Dutch flower painters were learning to construct their paintings through the manipulation of a rich chiaroscuro. The illumination now focused sharply on chosen blossoms disposed more irregularly – and artfully – across the canvas, while others, their form only partly revealed and their half-lit color merging into the general tonality, slipped into an enveloping background of shadow. The effect was now of more freely and deeply sculpted forms and of a more dramatic, real, and penetrating light. Insects still appeared, but they were subordinated as well to the orchestrated flux of illumination, their role now more clearly to accentuate light's revelation of a

47 Arthur Wheelock, Jr., who offers some interesting observations on the problematic nature of the "realism" of Dutch seventeenth-century paintings (Wheelock 1981, 28ff.), points out another instructive example of the manipulation of reality for aesthetic effect in the alteration of the positions and proportions of buildings in Jan Vermeer's *View of Delft*, which gives, nonetheless, such a striking impression of reality accurately and faithfully transcribed (ibid., 32–3, 94).

48 Kris (1927), 246–7; Bol (1960), 15–18; Bergström (1956), 48–50, 62; R. Gibson (1976), 4.

49 Kris (1927), 245; Regteren Altena (1935), 31–2; Bergström (1956), 50, 58, 66; Bol (1960), 20 and passim.

diversity of textures – a hard exoskeleton glistening against the soft darkness of shadow, or a luminous tracery suggesting the half-seen intricacy of a locust's wing.[50] Thus the artist's goal emerges more clearly as less an undistorted and comprehensive depiction of flowers and insects than emotional arousal through the dramatic manipulation of light and color and illusionistic effects. This was the power artists wielded as the unique magic of their craft.[51]

Historically, such play with light effects had been a characteristic emphasis in the illusionism practiced by northern painters,[52] but it was an illusionism with many levels. The Dutch painter Samuel van Hoogstraten wrote in 1678 that painting was a science of fooling the eye, for a perfect painting was like a mirror of nature in which what was not really there appeared to be there nonetheless.[53] The painter deceived, however, not by copying nature directly but by depicting "ideas" nature provided, and if Hoogstraten's aesthetic theory thus betrayed Italian roots,[54] the illusionism characteristic of Dutch painting indeed reached well beyond those qualities that, strictly speaking, pertained to the eye alone. A twentieth-century devotee of Dutch art alludes, for instance, to the fluttering restlessness of flowers painted by de Gheyn,[55] and we may speak in a similar vein of a sense conveyed of the brittleness of an exoskeleton, the crisp thinness of a locust's wing, or, for that matter, the lightness and fragility of a small wasp that has alighted – the evocation of interrupted flight as well – on the broad expanse of a flower petal. Hence, Dutch artists cultivated an illusionism that not only counterfeits the thing depicted but evokes a variety of subtly shaded feelings akin to those that twentieth-century commentaries on the nature of art have also placed at the very heart of the aesthetic experience.[56] If anachronistic aesthetic sensibilities do not mislead us, then, it is likely that seventeenth-

50 Martin (1977), 41; Rosenberg et al. (1966), 304–16; Bergström (1956), 74; R. Gibson (1976), 4 and pls. 14, 15, 19, 20, 21.

51 Speaking before the St. Lucas Guild of Leiden about the handling of light and shadow, the painter Philips Angel referred indeed in 1641 to the "magical power," *tooverachtighe kracht,* by which a painter made things seem real (Angel 1642, 39).

52 See n. 33 above.

53 For examples, including works by Hoogstraten himself, of a fascination with illusionism pushed to the point of trompe l'oeil in Dutch painting, see E. Jongh (1982), 16, 144–9, 193, 199. See also Wheelock (1981), 28–30, where mention is made of an instrument of even more pronounced illusionism – and a "unique Dutch art form" – the perspective box (the example reproduced is also by Hoogstraten). In what was a commonplace allusion to Pliny's story concerning the grapes painted by Zeuxis (see Brenninkmeyer-de Rooij 1984, 67), Joost van den Vondel rhymed about a bee deceived by the flowers painted by Daniel Seghers (E. Jongh 1982, 185). Samuel Pepys wrote of the drops of dew in a small flowerpiece – "the finest thing that ever, I think, I saw in my life" – by Simon Pieterszoon Verelst, a Dutch painter who had settled in England, that he was "forced, again and again, to put my finger to it, to feel whether my eyes were deceived or no" (quoted by R. Gibson 1976, 10).

54 Hoogstraten (1678), 24–5; see also 286–7. Regarding the Italian origins of such aesthetic theorizing, see Blunt (1940), 36, 100, 141.

55 Bergström (1956), 48.

56 Langer (1967–82), 1:64, 87, 166–9; see also idem (1957), 95–6; Winner (1982), 104, 110–11. It is true that Langer, at least, would not be sympathetic to all the implications of the verb "evokes" (see, e.g., Langer 1957, 95), but she does conceive of illusion, very broadly understood, as pertaining to the very essence of art in general (Langer 1967–82, 1:chap. 7 passim).

century Dutch artists prized their power to create such illusions as the hallmark of their craft.[57]

Indeed, modern devotees of the art of the period remark still subtler and more complex levels in the illusionism of naturalist miniature painting. Having captured the characteristic "attitudes" and "movements" of the insects he portrayed, Hoefnagel succeeds, we are told, in conveying a "striking impression" of their being "alive." Ambrosius Bosschaert the Elder, the founder of the school of painters at Middelburg, infused his butterflies and dragonflies with "life and expression," and de Gheyn captured not only the "curious evanescence" of flies (Fig. 10) but also, in his sketches of other small animals, the very "essence of their being."[58] Such commentaries bespeak the evocation of clusters of feelings that are evoked by the real presence of the living creatures as well – clusters that, in a primitive if enduring and perhaps most fundamental way of experiencing the world, constitute the recognizable identities of which that world is perceived to be composed.[59]

57 To be sure, the Dutch painters of the period who were most inclined to leave some literary expression of their thoughts about art tended to be those who also felt that art should aspire to more noble subjects than insects; hence, those artist-authors were unlikely to elaborate on the subtleties of the illusionism of insect painting (but see Hendrix 1984, 77). Nonetheless, these artists clearly indicated a consciousness of attempting to convey more than merely visual qualities through illusionism. As cited by Karel van Mander, the sixteenth-century painter Lucas de Heere commented (in verse) not only on the illusionistic impact of the jewelry in the Ghent altarpiece by the van Eyck brothers but remarked as well on the veneration, *innigheyt,* one seemed to be able to "read" on the mouth of the Virgin Mary (Mander 1604b, fol. 201v). In a passage stressing indeed that art derived from the artist's ideas could far surpass that which simply copied nature, Samuel van Hoogstraten wrote now of classical artists who captured the majesty and glory of the Olympian gods (Hoogstraten 1678, 286–7). Likewise, at the beginning of the next century, and also in a section stressing the insufficiency of merely copying nature, Arnold Houbraken (d. 1719) called attention to the ability to display the diverse passions of the soul (Houbraken 1753, 1:264, 270). Vivid references to the evocation of qualities surpassing the purely visual occur in the discussions of portraiture as well (see n. 58 below). With respect to the earlier founders of the Netherlandish tradition, James H. Marrow has recently argued that the essential innovation with which northern painters of the fifteenth century were most absorbed was the effort indeed to manipulate the emotional experience of those who looked at their paintings, an effort intimately interwoven with the exploration of illusionistic techniques (Marrow 1986).

58 Bergström (1956), 34, 38, 67; Judson (1973), 14; Regteren Altena (1935), 64. Marjorie Lee Hendrix insists that among Joris Hoefnagel's chief concerns was indeed the imparting of life and "character" to his naturalist illustrations (Hendrix 1984, 75–9, 182, 201).

Again, those Dutch painters who elaborated in writing about art were not inclined to make much of insect painting, but they expressed similar sensibilities to those alluded to above in the text when talking in particular of portraiture. Karel van Mander wrote of Hans Holbein's life-sized portrait of Henry VIII as being so thoroughly alive, *soo gheheel levendigh,* that its head and whole body appeared to move, frightening those who saw it (Mander 1604b, fol. 222v). A century later, Arnold Houbraken admired a Rembrandt portrait in which the head, he said, seemed to thrust out of the picture and speak to you (Houbraken 1753, 1:269). On the passion for portraiture during the period, see n. 59 below.

59 Langer (1967–82), 1:59; Winner (1982), 106. E. H. Gombrich has stressed how readily portraits or other images of faces also evoke a sense of a "presence" (Gombrich 1961, 113), and there is perhaps some significance to the fact that the intensification of illusionism in miniature manuscript illumination in the Netherlands in the late fifteenth century was accompanied by a further individualization of depicted figures as well, so that they often had an air of portraiture

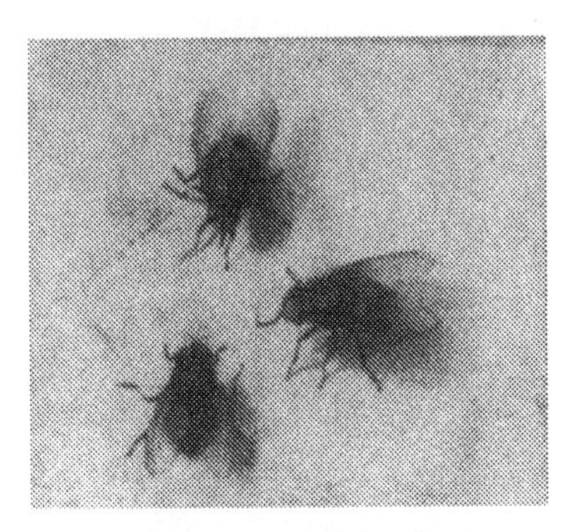

Figure 10. Pen, ink, and wash drawing of flies by Jacques de Gheyn II (4.2 × 4.5 cm). De Gheyn depicted single flies as well on squares of paper as small as 1.8 × 1.9 cm (Regteren Altena 1983, 3[cat. 2]:pls. 480, 479, 2:140). Courtesy of the Städelsches Kunstinstitut, Frankfurt am Main.

If such illusionism was indeed the core of the artist's purpose, however, even the naturalist miniature painter would have felt little attraction to the microscope: For observers of a painting to experience such illusionistic feelings, they must also initially recognize what the artist is intending to represent. The evocation of the crispness of an insect's wing depends in the first place on the recognition of a reference to a wing in the artist's strokes and colors, a recognition that entails a readiness to perceive the light effects that are also mimicked there. The efficacy of illusionism therefore rests on an unproblematic familiarity with the subject painted – whence follows the particular aptness of the "everyday world." The corollary, however, was the inappropriateness of new microscopic images, whose very strangeness was their most compelling attribute.[60] To be sure, the full illusionary effect is a circular one, for familiarity with the thing depicted undergirds an evocation of feelings that, in turn, enhance the illusion of the presence of the thing. Unfamiliar images break the circle and thus blunt the immediacy and power of the impact of the illusion.

In the case of insect life, moreover, its very smallness was one of its distinguishing qualities in the familiar "visible world" it was the artist's purpose to

Footnote 59 *(cont.)*
about them (Delaissé 1959, 182–3). The oeuvre of Hendrik Goltzius also includes among his several styles a continuing series of often very vivid and striking miniature portraits (see, in particular, Reznicek 1961, figs. 3–10, 17–26, 30–5, 37–44, etc.), a genre that his student de Gheyn continued to cultivate (see Regteren Altena 1983, 3:cat. 2, pls. 8, 19–23, 38–43, 45, 60–1, etc.). Indeed, portraiture flourished in seventeenth-century Dutch painting and drawing as nowhere else (Haverkamp-Begemann 1976, 356).

60 Gombrich has pointed out that the evocations of illusionism rest on familiarity (Gombrich 1961, 211, 243, 260–1), and he has noted the disorienting effect of the absence of a familiar context specifically with reference to Constantijn Huygens's project for an atlas of creatures seen through the microscope, which he had hoped Jacques de Gheyn would undertake (Gombrich 1983, 13).

represent, and the evocation of associated feelings rested in no small part on the very minuteness of suggested detail.[61] To the extent, hence, that illusionism is in fact enhanced by a distortion of the appearance of reality, by the subtle accentuation of some aspects of that appearance and a selective suppression of others,[62] the artist conjuring the illusion of insects would have been more inclined to emphasize that quality of smallness and disappearing detail than to destroy that quality through magnification. De Gheyn, indeed, would appear to have been toying with just such an accentuation in his sketches of flies on pieces of paper or parchment as small as two centimeters square (see Fig. 10). Whatever his intent, however, it differed profoundly from Hooke's in the latter's celebrated fold-out representation of a flea – as big as a cat, noted an admiring Christiaan Huygens – in the *Micrographia* more than a half-century later.[63] Magnified images of insects exercised their own gripping fascination, to be sure, but they were inimical to the illusionistic commitment that Dutch painters of the Golden Age shared with the Master of Mary of Burgundy and his sixteenth-century successors.

As it persisted in the seventeenth century, the legacy of naturalist miniature painting consequently remained narrowly constrained in its approach to minute life by craft traditions that entailed not only iconographic conventions but an accentuated commitment to illusionism. Reinforced by the growing power of illusionistic techniques, that commitment was now assimilated as well into the movement in seventeenth-century Dutch painting to pay homage to the everyday life of the painters and their patrons. Thus absorbed in creating the illusion of the small crawling, scurrying, and flitting forms of an intimately familiar world, the artists shaped in their masters' ateliers had no use for the powers of the microscope. To Constantijn Huygens, an atlas of microscopic images of small creatures seemed a natural extension of the miniaturist's skills, but the

61 What are referred to as "qualities" here correspond to what Winner speaks of as "expressive properties" (Winner 1982, 106). Langer uses "quality" to mean an aspect, indeed a "projected feeling," in the work of art itself (Langer 1967–82, 1:106), but see her remarks on "images" (ibid. 1: 59, 63–4).

62 Experiments seem to indicate that appropriately exaggerated drawings – indeed, cartoons – prompt recognition more readily than more accurate representations that include photographs and, in at least one experiment, the objects themselves (E. Gibson 1969, 102–4). Hence, since the illusions of art presuppose recognition (see n. 60 above), it is not improbable that illusionism plays upon analogous distortions, so that the most vivid illusions are evoked by a selective exaggeration of the kinds of identifying qualities alluded to in the text above. Of the "subjectively created images" in terms of which the mind perceives things in the world (see Langer 1967–82, 1: 63–4), Susanne Langer writes: "An image does not exemplify the same principles of construction as the object it symbolizes but abstracts its phenomenal character, its immediate effect on our sensibility or the way it presents itself as something of importance, magnitude, strength or fragility, permanence or transience, etc. . . . Suffice it now to point out that we apprehend everything which comes to us as impact from the world by imposing some image on it that stresses its salient features and shapes it for recognition and memory" (ibid. 1: 59).

63 Regteren Altena (1983), 2: 140, items 898–900, and 3:226, pls. 479–81; Hooke (1665), pl. 34; *OCCH* 5: 304.

seventeenth-century Dutch heirs of Hoefnagel and the Master of Mary of Burgundy declined to make the endeavor.[64]

Hidden obstacles to an innovative application of the lens lay concealed also within the theological doctrine of God's self-revelation in his creation. Although even the particular significance of God's smaller creatures was sometimes emphasized – as it was in the Netherlands – the doctrine harbored implicit assumptions adverse to the beginnings of microscopic discovery.

As shaped particularly in Calvin's thought, the theology surrounding the study of nature wove together a complex of ideas colored by ambivalence. While insisting that the contemplation of God's works was indeed an obligatory form of worship, Calvin also warned (echoing earlier movements of Netherlandish pietism) of the spiritual dangers of being distracted by the works themselves from reflecting on their Creator. A happy intellectual life, successive classes of students were taught at Leiden, consisted in knowing the wisdom, power, and goodness of God and in loving and worshiping him; but to wander in one's thoughts from his glorification was inexcusable.[65] Although a pious obligation incumbent upon all, the study of nature was also potentially a dangerous seduction.

It was in adapting a revelatory creation to his central tenet regarding the moral state of humanity, however, that Calvin's thought impinged most prejudicially on the prospect of microscopic research, for the implications of his linkage of those two elements of his theology worked against the very conception of a microscopic realm. Central to that theology was his insistence on the intrinsic spiritual wretchedness of man, and as both a manifestation and further accentuation of that wretchedness, Calvin stressed as well the congenital inadequacy of man's capacity even to see the revelation in nature he was nonetheless obliged to con-

64 See ConHs (1897), 120. For that matter, Constantijn Huygens himself also declined to perform the task, though he had taken lessons in miniature painting from his cousin Jacob Hoefnagel and, indeed, did not think poorly of his own accomplishments therein (ibid., 65–6). The most celebrated illustrator of insects in the Netherlands in the late seventeenth and early eighteenth century was Maria Sibylla Merian; born and raised in Germany, she lived her later, productive years in the Netherlands, and her training in the genre of flower and insect painting had been much influenced by the Dutch tradition. Although she did occasionally resort to the microscope in her researches, she almost never depicted magnification in her final paintings or illustrations of insects (see Merian n.d., pp. and pls. 2, 6, 8, 9, 10; idem 1730, 22 [pl. 42], 52 [pl. 101]; only in the last plate cited is magnification evident in the depiction of an ant, but even in this instance it is hardly very revealing). The most prominent Dutch successors in the following century showed a similar lack of use for microscopic images; see l'Admiral (1774) and Sepp & Sepp (1762-1860). (Jacob l'Admiral's case is all the more interesting in that his brother Jan did microscopic anatomical illustrations for Frederik Ruysch and Bernhard Siegfried Albinus.) Huygens's hopes would be most fully realized in the early modern period by the eighteenth-century German miniaturist August Johann Rösel von Rosenhof.

I should note that the argument developed in the text above is very much at odds with that set forth by Svetlana Alpers, who stresses a basic compatibility between microscopic observation and the way of looking at the world expressed in Dutch seventeenth-century painting (Alpers 1983, 25, 32–3, 83–5). On the other hand, see Ackerman (1961), 73–5, for an approach analogous to that I take here vis-à-vis the microscope.

65 Calvin (1559), 1: 63, 69; Thomas à Kempis (1892), bk. 3, chap. 31; Burgersdijck (1642), 351.

template.[66] Wherever they turned their eyes, Calvin declared, even the most unlearned and dimwitted of men inescapably confronted the glory and wisdom of God; but since it still remained unseen, the very ubiquity of the revelation only further bespoke man's blindness and moral bankruptcy. Thus underscoring humanity's reprobate blindness, that ubiquity and the immediate accessibility of the revelation in nature constituted a tenet upon which Calvin insisted.[67] That we were consequently "without excuse" for our failure to see the divinity thus revealed "before our eyes" was reiterated in the confession of the Dutch Reformed Church and echoed – though far less frequently than the happier theme of the revelation itself – in the works of both poets and naturalists.[68] To expose these failings so incontestably, the revelation of divinity in God's creations had to be openly emblazoned before mankind; hence, that it might be so deeply concealed as to require an instrument such as the microscope ran contrary to the profound moral coloring of Calvinist doctrine. The mentality that sustained this doctrine was not concerned, after all, with a philosophical understanding of natural processes such as the Cartesians sought; rather, it was striving to recognize a revelation of God hidden not by the secretiveness of nature but by the moral depravity of man.[69]

Moreover, several fully visible aspects of nature already seemed the obvious vehicles of the revelation to which the theologians pointed. The boundless taxonomical diversity of visible creatures and their remarkably ordered and purposeful existence appeared a fitting reflection of both the wonder and the wisdom of God's creative power. The practical as well as morally didactic usefulness of things found in the world seemed also to express that divine beneficence that Calvin exhorted his flock to learn to appreciate more fully.[70] Tantalizing in its greater elusiveness, the intuition of symbolic meanings in the multitude of creatures conceived as hieroglyphs focused the attention of would-be adepts on the suggestive behavior and shapes of natural things and on the patterns and colors blazoned across their surfaces. These were patterns, indeed, that seemed intended to be observed and, hence, to be scrutinized for deeper, embedded messages.[71] There was thus little cause to suspect that God's revelation in nature was not already fully displayed for all to see.

66 Calvin (1559), 1:liii, 40n. 2, 63, 68–9, 341; Niesel (1956), 46; Dowey (1952), 72–3.

67 Calvin (1559), 1:52–3, 68–9; Niesel (1956), 49–50; Dowey (1952), 72–3, 82–3, 131, 135, 143. Although Calvin granted that the learned sciences did attain to deeper insights, his emphasis on the immediacy and conspicuousness of God's revelation was more germane to the essential thrust of his teachings (Calvin 1559, 1:53; see also Dowey 1952, 142).

68 Cochrane (1966), 189–90; Vondel (1820–4), 14:140; JS (1669), pt. 2:48.

69 According to Calvin, however, nature was also affected by man's fall (Calvin 1559, 1:341).

70 Ibid., 1:181.

71 "Nature is trying to tell us something," writes a twentieth-century pioneer in electron microscopy; "the investigator's goal is to get the message" (Anderson 1975, 16). Such an intuition can become profoundly compelling, but it can also assume forms very alien to the modern scientific mentality, as in the conviction that one was confronted everywhere with semiconcealed hieroglyphs, whose message, again, it was the researcher's goal to "get." For a twentieth-century expression of a sense of there being an as-yet hidden significance in the outward forms and surface displays of organisms, see Portmann (1952), 18, 25, 35–6, 83, 86, 206–7, 214, 216, 218, 220.

Authoritative religious sources also taught that God intended his creation for the benefit of humanity – to serve man that he might serve God, declared the Dutch Reformed confession[72] – and if the creation had been thus designed with man in mind, did that not imply as well that the religious instruction embodied there for his edification had been accommodated to the compass of his senses? As late as 1705, Leeuwenhoek wrote of a visitor who, having been shown some microscopic structure, still puzzled over what its purpose might be if no eye could see it.[73] It was the microscope itself that now elicited the expression of the thought, but father to the thought was a prior presumption that precluded the idea of a divine workmanship in nature that was too small for the God-given senses.

Contexts that would seem otherwise to have invited a recourse to the microscope thus nurtured biases that contravened any such incentive. The very purpose of the Cartesians in fabricating a world of hidden mechanism discouraged an appeal to sense experience, and seventeenth-century Dutch painters who took up the legacy of naturalist miniatures were intent on capturing, not overcoming, the suggestiveness of intricate smallness and diminishing detail. In time, to be sure, the conception of nature as a divine revelation would interact with microscopic research, once begun, to expand vastly the imagined potential for discovery; but in the early seventeenth century that conception still worked, rather, to discount the possibility that God's creation continued into a realm of the invisibly small. That such biases, moreover, were rooted in those very traditions and schools of thought that, in Dutch seventeenth-century culture, were most concerned with nature's smaller forms could only enhance the impact of those prejudicial restraints in delaying the beginnings of microscopic discovery in the Netherlands.

Nonetheless, one vigorous tradition of research was already identified with the discovery of hidden aspects of the revelation in nature and was increasingly searching for that revelation in nature's smaller dimensions. Echoing Galen, Calvin himself had cited the human body as testimony to God's wonder-working powers,[74] and it was indeed a commonplace in the Dutch Republic as elsewhere that anatomy vividly displayed the attributes of God.[75] Plemp's prediction in 1630 that the microscope would reveal the power and wisdom of God even in the internal anatomy of insects would be borne out only after the passing of several decades,[76] but anatomical research in human bodies and those of

72 Cochrane (1966), 196; Calvin (1559), 1:181; Burgersdijck (1642), 351; idem (1627), 41.

73 Leeuwenhoek to the Royal Society, 24 April 1705, AvL Letters, fol. 369v. The idea, voiced by the English physician Thomas Sydenham, that the very effort to overstep the natural limits of the sense was impious appears not to have found expression in the Netherlands, however (Wolfe 1961, 210; see also p. 215 with respect to Locke).

74 Calvin (1559), 1:53–4; Lloyd (1973), 151.

75 Vondel (1820–4), 14:229; Graaf (1677), 2. See Swammerdam's poem "Godts almoogende wijsheid in't formeeren van den mensch" in the opening pages of Blaes (1675); Deusing (1659), dedication; Drelincourt (1680), 17–19, 25–6.

76 Plemp (1630), 16.

the larger animals was already grappling with ever-finer structures. Here indeed appeared the prospect of a persistent, systematic, and aggressive program of microscopic observation.

But here also, after all, lay veiled impediments to a fuller exploitation of the microscope. In the subtle anatomy practiced in the Dutch Republic, the potential for the application of the lens was paradoxically to be narrowly circumscribed by the very success of new techniques and by the vivid anticipations of great things soon to be discovered in the fine structure of the body. The potential of the microscope was now obscured by the brighter promise of other avenues to discovery.

CHAPTER FOUR

Discovery preempted

In 1656, Pierre Borel wrote of the microscope's contribution to anatomy, including the observation of white particles in blood serum and chyle, valves in pores, and the rough scaliness of the skin. It would reveal as well, he noted, that the "parenchymatous" organs of the body – those whose origins Hellenistic anatomists had ascribed to congealing blood – were composed of finer structures that might include even sieves with shaped pores for correspondingly shaped particles.[1] In fact, however, the lens contributed little to the impressive sequence of anatomical discoveries during the first sixty years of the century and aroused no excitement in those medical circles responsible for the progress of subtle anatomy. It was doubtless used more often than it was mentioned, which was only rarely, but that silence itself argues the absence of any general feeling among anatomists that the lens offered any extraordinary promise. As the seventeenth century moved toward its close, the microscope did assume a more prominent place in anatomical research, but, in the Netherlands, it was another new technique that inspired hopes of a final, great age of discovery in subtle anatomy – an expectation, ironically, that also worked in its own way to limit the further realization of the potential of the microscope.

Microscopic anatomy in any real sense had to await Marcello Malpighi. In 1661, as a member of the medical faculty at Bologna, he published two epochal letters on lung tissue, followed in the middle years of the decade by studies of other major viscera and of the structures in the skin and tongue.[2] Dutch medical literature soon reflected Malpighi's impact, particularly on the conception of what lay within the substance of the major organs. One Middelburg physician, Anton de Heide, explained for the sake of his less learned countrymen that, for centuries, most anatomists had held such parts of the body to be coagulated lumps of fluid, especially blood, and had hence called them *parenchyma* (or *stolsel,* in effect "coagulum," in Dutch), conceived as a "confused lump" lacking any fine, organized structure within.[3]

1 Borel (1656), 36, 43.
2 Malpighi (1661a,b, 1665a,b, 1666). Subsequent citations of these works will refer to idem (1686–7).
3 Voorde & Heide (1680), 496n; see also 98, 98–9n. Cf. Horne (1660a), 68, 74; Deusing (1659), 73, 83, 88–9, 192–3.

But Malpighi had dramatically transformed the understanding of the organs, and his descriptions of various tissues echoed vividly in the Netherlands. According to the accounts in the Dutch medical literature, Malpighi had with his microscope completely overturned the notion of the spleen as coagulated blood, for he had discovered that the organ was rather a mass of membranes and little chambers – "membranous pockets" or "cells" – among which were strewn an abundance of small white globules or glands.[4] He had also found the supposed parenchyma of the lung to be similarly composed of fine membranes that formed infinite small vesicles embedded in a "most cunning" network of microscopic blood vessels.[5] While the liver was described now as an aggregation of small glandular *acinos* – Malpighi's word, literally meaning "grapes" – also enclosed in membranes and each the terminus of a variety of vessels,[6] the kidney had become an accumulation of tubules and finely branching blood vessels proceeding, again, to small glands at their ends.[7] The impact of these discoveries on the general perception of the viscera was profound and carried over into an eager anticipation of analogous microscopic structures everywhere.[8]

De Graaf's demonstration that the parenchyma of the testicles were in fact densely tangled gatherings of vessels also added to the evidence, dramatically accumulating in the 1660s, for the ubiquity of minute structure throughout even the seemingly solid masses of the body, and so likewise did Steno's emphasis on the fibrous nature of muscle tissue.[9] It was primarily Malpighi, however, who lay behind de Heide's insistence in 1680 that, with the aid of the microscope, modern anatomists had discredited the idea of parenchyma by revealing the astonishingly intricate construction in all the body's solid parts.[10]

Inevitably, Malpighi's influence heightened the awareness of the potential of the lens in anatomy, and allusions to the microscope, rare in Dutch medical literature before the 1660s,[11] became commonplace in the succeeding decades. Gerard Blaes, who was to play a role in the careers of both Steno and Swammerdam, had inaugurated his instruction in medicine at the Athenaeum in Amsterdam in 1660 with an address on what humanity owed to nature and what to "art." When he had turned – briefly – to the microscope, he spoke of fleas, lice,

4 Voorde & Heide (1680), 104n. 2; Diemerbroeck (1672), 145–6; Blanckaert (1686), 411.

5 Diemerbroeck (1672), 508–9, 517; Voorde & Heide (1680), 151n. 1; Blanckaert (1686), 72; Bontekoe (1689), 1:72.

6 Diemerbroeck (1672), 113; Voorde & Heide (1680), 99n; Blanckaert (1686), 397. Reflecting the rapid incorporation of Malpighi's account into the instruction at Leiden, see Alsem (1671), §22.

7 Diemerbroeck (1672), 168–70; Blanckaert (1683a), 63.

8 JS (1669), pt. 1:[105]–6. With respect to a readiness now to find microscopic glands everywhere, see Bidloo (1685), pl. 10 (fig. 2), pl. 13, pl. 14 (fig. 4), pl. 23 (fig. 2), pl. 38 (fig. 1), pl. 43 (fig. 5), pl. 44 (fig. 1), pl. 46 (fig. 4).

9 See, for instance, Voorde & Heide (1680), 116n, 10n. 5, 69n, 70n. 5. See below, n. 114.

10 Ibid., 98n, 337n. 16, 495–6n; see also Bontekoe (1689), 1:57–8.

11 Moreover, those few instances usually pertained to various "worms" or the shape of the hair (Tulp 1641, 178 and pl. 7 [fig. 2]; Plemp 1630, 224; Linden 1653, 735; see also A. Cluyt 1634, 79).

and spider webs, and, echoing Kircher, of the worms observed in buboes, in the blood of fever victims, and in pieces of flesh exposed to the moon throughout the night; but he made no mention of anatomy.[12] By the 1680s, however, microscopical descriptions and illustrations figured prominently in major anatomical publications by leading Amsterdam physicians.[13] Leeuwenhoek's observations were now also finding their way into these works,[14] and his growing celebrity during these same decades contributed further to the greater regard for the microscope in Dutch medical circles. Whereas Leeuwenhoek's impact on those circles was enhanced by proximity and national affinity, however, Malpighi shared with these northern physicians, as Leeuwenhoek did not, the cultural and professional affinity of the world of learned, academic medicine. Consequently, it was in Malpighi's works rather than Leeuwenhoek's that these Netherlandish physicians more readily recognized a systematic and coherent new orientation in anatomy associated with the microscope.

Nonetheless, it was still possible in 1702 for Bernard Albinus to avoid any explicit reference to the instrument when he launched his professorship in theoretical and practical medicine at Leiden with an address on the rise and progress of medicine.[15] The microscope, it is true, was now to be found in the hands of professors and students alike;[16] but, despite the respect paid both Malpighi and Leeuwenhoek, the Dutch medical faculties still did not perceive the lens as the key to a new age of anatomical and medical research.[17] After the appear-

12 Blaes (1660), 6.

13 Bidloo (1685), passim in the plates; Blanckaert (1686), 175 and pl. 11 (fig. 3), 176–7 and pl. 13 (figs. 2, 3), 239 and pl. 18 (figs. 2–5), 278 and pl. 21 (fig. 8), 314 and pl. 24 (figs. 6, 7), 339 and pl. 25 (fig. 3), 432–3 and pl. 31 (figs. 1, 5), 499 and pl. 37 (figs. 1–4), 506–7 and pl. 38 (figs. 1, 2), 589 and pl. 50 (fig. 11). See also Fournier (1985).

14 See, for instance, Blanckaert (1686), 136–46, 227–30, 239, 594–5, and pl. 18 (figs. 2–5).

15 Albinus did indeed allude to what was by then a very well-known microscopic observation first made known by Leeuwenhoek, the movement of the blood in the capillaries (B. Albinus 1702, 25), but the microscope itself still remained unmentioned even when Albinus bemoaned how little in medicine was accessible to the senses (ibid., 27).

16 Rooseboom (1967), 273, 290n. 26.

17 In an address similar to that he gave in 1702 but delivered almost a decade later, Albinus did now mention the microscope as well as Leeuwenhoek's observations of the spermatozoa (B. Albinus 1711, 49–50), just as another decade later his son, Bernhard Siegfried Albinus, in his own inaugural oration on the "true way" to explore the structure of the human body, also mentioned the microscope and Leeuwenhoek's studies of muscle fibers in insects (B. S. Albinus 1721, 22–4). But having mentioned the instrument only briefly, neither offered any comment on its general significance for medicine or, as will be developed more fully below, evidenced anything approaching the enthusiasm voiced for the technique of injection. Other professorial orations in the medical school at Leiden in the early decades of the eighteenth century may mention Malpighi and Leeuwenhoek with respect, but they are also curiously silent nonetheless about the microscope itself; see Rau (1713), 27–9, 33; and Schacht (1723), 13. We know that Rau had personally used the microscope in his own research as a student at Leiden (Rau 1694, §§3, 10, 15) and, in the classes in anatomy he had offered in Amsterdam before coming to Leiden, had provided his own students with microscopes as well (Rooseboom 1967, 290n. 26); his silence about the instrument in an address specifically on teaching and learning anatomy is hence particularly striking. Concerning the continuing use of the microscope despite such reticence, see also n. 105 below regarding Herman Boerhaave.

ance of Malpighi's first works, de Heide had argued that, without microscopic studies in anatomy, the field of medicine would be plagued by continuing uncertainty, yet even some of those who now made use of the microscope remained ambivalent about its place in the arsenal of the anatomist. Such was not the case, however, with another new technology of research, the injection syringe, about which even those who were silent or reserved about the lens were usually more forthcoming.[18]

The technical difficulties of preparation remained the major obstacle to the fruitful application of the microscope in anatomy. Those structures of the body that lay just beyond visibility were concealed not only by their dimensions; they were veiled as well by an absence of distinguishing color, firmness, and texture, which otherwise would have often revealed them even to the naked eye. In discovering the lacteals, Aselli had indeed been fortunate to be dissecting an animal that had recently eaten, so that the vessels were revealed by the white chyle from the gut, and Ruysch (before he developed his injection techniques) affirmed that, even though anatomists now knew that the lacteals existed, they and their valves still remained scarcely visible in animals that had not just been fed.[19] It was the passing effect of distinctive color that momentarily stripped them of their invisibility. Anatomists lamented in particular what was lost to them by the fading of the more delicate parts of the body in death,[20] so that the eager quest for techniques of preservation proved of great consequence to the search for otherwise invisible structures as well. Consequently, it was the development of injection techniques, not the microscope, that was perceived by the end of the seventeenth century as the great technical breakthrough in anatomy.

Seventeenth-century anatomists had cast about unceasingly for techniques other than magnification to reveal hidden structures, from simple strangulation (to make certain vessels stand out in dissection) to the recourse to ants to eat away the soft flesh around the vessels of the liver.[21] Turning to the ancient technique of inflation, Ruysch had initially explored the lacteal system by blowing into it through very fine tubes, and it was by means of inflation that he had ultimately displayed the valves of the lymphatics.[22] Of more recent origin, however, the

18 Voorde & Heide (1680), "Voorreden." In 1680 in his new journal of medicine and natural science, Steven Blanckaert proposed to discard the "idle speculations" of previous systems of medicine for the discoveries of the microscope as well as for other advances in anatomy and chemistry (Blanckaert 1680–8, vol. 1, "Toe-eigening en voorreden"); nonetheless, he did not include the microscope among the necessary instruments of anatomy in his account of the "new, reformed anatomy," although he did include the syringe (Blanckaert 1686, 5). In those addresses in which they declined to mention the microscope explicitly (see nn. 15, 17 above), Bernard Albinus and Johannes Jacobus Rau cited injection with a clear note of admiration (B. Albinus 1702, 25; Rau 1713, 30). Herman Oosterdijk Schacht, on the other hand, spoke explicitly of neither injection nor the microscope in his 1723 addresses.

19 Ruysch (1665), 9.

20 Horne (1660b), 20; Deusing (1655), 81; Ruysch in Vater (1727), 14; Ruysch (1744b), 6; B. S. Albinus (1721), 25.

21 Nuck (1685), 16–17; M. Frank (1916), 303.

22 Ruysch (1726), 10; idem (1665), 1, 10.

related technique of injecting the vessels of the body with fluids was to prove of far greater moment.

Introduced as a practice in the sixteenth century, injection was ultimately to lead to the most exciting new technique of seventeenth- and eighteenth-century anatomy. Soon after the middle of the seventeenth century, anatomists across Europe were indeed busily injecting bodies with mercury, milk, ink, and a variety of other variously tinted liquids.[23] De Graaf made wide use of such colored fluids in the exploration of a number of organs and vascular structures, and Leiden's Anton Nuck was particularly renowned for his use of what he either fondly or possessively called "my mercury." Whereas Craanen had reported Swammerdam's confirmation of capillary anastomoses by means of the microscope, Swammerdam himself wrote, rather, of having used mercury to make such anastomoses visible to the naked eye.[24]

Such practices, and not the lens, were touted for making possible the major discoveries in subtle anatomy in the Netherlands. De Graaf wrote at one point of having employed a microscope in looking for the cavity and fluid within the nerve, and Leeuwenhoek relates that de Graaf made use of a microscope in his historic search for the mammalian egg in rabbits as well. The latter instance, however, remains unmentioned by de Graaf himself in his classic treatise on the female generative organs, and there is no mention of the instrument in his companion treatise on the organs of the male. Swammerdam's references to the microscope in his early anatomical works are likewise not only rare but, in the context of the key anatomical discoveries of the time, inconsequential, and it is not without significance that they usually pertain to injected vessels.[25]

Ruysch's celebrated work on the valves of the lymphatics (like the works, for that matter, of de Bils before him) also lacked any references to the microscope, even though those valves are depicted larger than life at one point for the sake, we are told, of clarity. Steno had occasion to repeat Malpighi's observation of the tissue of the lung, but he seems to have had reservations even about that experience (he confides that he was still not fully convinced) and in his published works was also silent about any role the microscope may have played in

23 J. Baker (1945), 10; Cole (1921), 288–9; idem (1944), 274; Belloni (1966b), 23; idem (1966c), 109; idem (1967), 289; Scheltema (1886), 40; R. Frank (1980), 174–5, 184, 205.

24 Graaf (1677), 42, 65, 67, 79–80, 633, 706–8, 712, 714; Nuck (1691a), 13, 32–4, 40, 51, 54, 141; Craanen (1689), 731; JS (1672), dedication; idem (1737–8), 2:832.

As he claimed (Graaf 1677, 458, 705), de Graaf also seems to have played a major role in introducing the use of the injection syringe at Leiden. Frederik Ruysch later acknowledged that it was indeed de Graaf's use of the instrument that induced him to take it up as well (Ruysch 1725–51, 3). Although de Graaf described the syringe in a small treatise in 1668 (Graaf 1677, 703–17), however, he was not the first to have dealt with it in print (see Glisson 1654, 2, 5, 210–16). The year before de Graaf's treatise, indeed, Swammerdam had also published an illustration and account of the glass tubes and attachments he used for inflating small vessels, the finest of which tubes were as thin as a hair, he asserted (JS 1667b, 91–3).

25 Graaf (1677), 527; AvL to RS, 30 March 1685, *AB* 5:170; see also Lindeboom (1973), 55. The only reference to the microscope in de Graaf's treatise on the female reproductive organs pertains to an observation of the fimbriae of the oviduct (Graaf 1677, 337). JS (1672), 28, 33, 55–6.

his research.[26] Again, silence about the microscope is evidence not of it remaining unused, but of how little excitement it aroused among the leading Dutch practitioners of subtle anatomy. The contrast with the response to new techniques of injection is striking.

These techniques were forged by this same circle of Dutch anatomists, and their search for such technical innovations was spurred on by the same mixed stimulus of common endeavor and competitiveness that goaded their search for anatomical discoveries. Swammerdam and de Graaf also contended over the authorship and comparative usefulness of such innovations, and Ruysch, noting that others had indeed observed the valves of the lymphatics before him, prided himself rather on his having found a way to display them. De Bils conjured up a particularly tantalizing prospect when he boasted of a method of preparing bodies that made possible bloodless dissections, to which supporters already attributed a number of new discoveries.[27] Some years later, Swammerdam was experimenting with acid injections to congeal the blood, and he attempted also to contrive a method by which the vessels of the body might be filled with molten pewter.[28]

The most consequential fruit of Swammerdam's efforts, however, was his discovery of a way to inject warm, tinted wax that then solidified in the vessels,[29] a technique then taken up and further developed by Ruysch. Although acknowledging Swammerdam's priority, Ruysch insisted on his own independent development of the technique: "What kind of substance this blessed man [Swammerdam] used to fill the vessels he never told me, and I never asked." The extent of his indebtedness to Swammerdam remains uncertain – Swammerdam himself noted that Ruysch had long been experimenting with methods to preserve the body – but Ruysch jealously guarded the details of his own procedures for years to come.[30]

Injection was a technique rich with aesthetic dividends, merging the pleasure of subtle technical mastery with vivid and unexpected spectacles. Bernard Albinus, while reticent about the microscope when he assumed his charge at Leiden, wondered at the labor and "scarcely credible dexterity" that lay behind injected preparations. His son and successor Bernhard Siegfried Albinus, one of the preeminent anatomists of the eighteenth century, focused emphatically, in his own oration in 1721 on the "true way" to explore the structure of the body,

26 Ruysch (1665), see plate facing p. 7; Vugs (1975), 162, 168n. 14; Bastholm (1950), 158n. 2; *DSB*, s.v. "Stensen, Niels."

27 JS (1672), 44; Graaf (1677), 473, 451, 462; Ruysch (1665), preface "Candido lectori"; Bils (1659b), 6; Deusing (1662), 129; idem (1664), preface.

28 Graaf (1677), 472, 714; *LJS*, 71, 73; Nordström (1954–5), 30–1, 34–5.

29 Having developed the method while working with Horne at Leiden in the late 1660s, Swammerdam made the technique public in JS (1672), esp. pp. 42–3.

30 Ruysch (1725), 2–6, 13–14 (quotation from p. 5); JS (1672), 36. Subsequent writers have been emphatic about Ruysch's debt to Swammerdam but have also stressed how much further Ruysch developed the technique; see Haller (1774–7), 1:529, 540; Cole (1921), 303–4. Regarding Ruysch's own secrecy, see also Haller (1774–7), 1:529, and Cole (1921), 307–9; cf. Scheltema (1886), 49–50.

on that "most beautiful skill [*artificium*]" that injected the vessels of the body with vividly colored fluids.[31] Deeply conscious of his own personal finesse – whereas Swammerdam had been able to inject the hairlike vasa capillaria, he noted, he himself could fill vessels as fine as down and spider webs – Ruysch likewise acclaimed wax injection as the "most beautiful" of skills.[32]

Like other new injection techniques, it offered a vivid visual experience. De Graaf delighted in such displays as the spread of green fluid through the branching arteries on the surface of the brain,[33] while a group of anatomists working together in Amsterdam took particular pleasure in the "extraordinary spectacle" of injected mercury that turned the bronchial artery of a calf to silver and spread throughout the network of small arteries in the lungs. Nuck was similarly moved by the spread of mercury through the branching vessels of the breast, and for Blanckaert, even the injection of ink into the vessels of the kidney produced an exhilarating sight.[34] The injection of colored wax into the furthest extremities of the blood vessels likewise offered an "elegant spectacle," and Ruysch's injected preparations would be remembered as well for their "beauty," to which Ruysch himself clearly responded, as for what then seemed their decisive importance to the medical sciences.[35]

The special attribute of wax as an injection medium was its ability also to fix and preserve the structures of the body it filled. Other fluids flowed back out after injection, but, once cooled and solidified, the wax remained.[36] Swammerdam spoke of a liver he had prepared that was greatly admired when shown at the Leiden anatomical amphitheater in 1667, but his prized example of his new technique was an injected human uterus he ultimately gave to the Royal Society.[37] By the turn of the century, Ruysch claimed to have developed wax injection to the point that it now seemed to call the whole body back to life – "as all bear witness," he wrote, noting as well that no one ascribed such a feat to Swammerdam. Others remarked indeed on the startling ability of the "Ruyschian art" to restore and preserve the lifelike appearance of the human body and its parts, and Fontenelle cited Ruysch's museum of elaborate and often bizarre preparations as one of the great wonders of the Netherlands.[38]

31 B. Albinus (1702), 25; B. S. Albinus (1721), 25–6. The younger Albinus had indeed mentioned the microscope in passing in this address (see n. 17 above), but the focus of his enthusiasm is clearly reserved for injection. Regarding his own impressive achievement in the use of injection techniques, see Elshout (1952), chap. 4.

32 Boerhaave & Ruysch (1751), 76; Ruysch (1725), 4–6.

33 Graaf (1677), 707–8, see also 42, 542.

34 Collegium Privatum, Amsterdam (1673), 15–16; Nuck (1691a), 12; Blanckaert (1686), 415.

35 Muys (1738), preface [136]; Haller (1757–66), 2:394; Ruysch (1701), 33; idem in Wedelius (1737), 14; see also Rau (1713), 30.

36 Ruysch (1725), 3–4.

37 JS (1672), 2, 33, 36–7, 41, 46; Oldenburg (1965–), 8:617–18, 9:367–9.

38 Ruysch (1744b), 4–5; idem in Vater (1727), 14; Ruysch (1725), 13; Rieger (1763), 2:54; Fontenelle (1818), 1:459.

Ruysch's museum exemplified a tendency to exploit the new technique as much for exotic display, to be sure, as for the benefit of science,[39] but injection was prized nonetheless as the key to the continuing advance of subtle anatomy. In the hands of skilled practitioners, it not only restored those fragile structures that disappeared with death but also exposed much that could not be seen even in vivisections.[40] Swammerdam declared that it made visible the most concealed and subtle things in anatomy and, if the difficulty was not too great, could bring to light whatever was significant and wondrous in the body. Although inclining more to mercury than wax, Nuck agreed that the most probing research in anatomy was to be achieved only by injection.[41]

To Ruysch, his wax injection method was the means by which the innermost recesses of nature could be seen and the structure of the smallest parts revealed. It alone, he exulted, recovered the thousands of tiny parts that, having been suppressed by death, now reappeared as if alive and cheerful – "as if they came to talk with us and say: 'you have talked and disputed mistakenly about us long enough; now you can see us from behind, in front, above and below; tell now who and what we are and how we are busy day and night to sustain you.'"[42] It was the only source of his continuing discoveries, Ruysch repeatedly insisted, discoveries that eventually prompted Fontenelle to declare in his eulogy of Ruysch that he had raised anatomy to a new level of perfection.[43]

To the skilled researcher, consequently, wax injection offered a richly gratifying experience that wove together not only the refined practice of an intricate technique and the spectacle of its effects – including its own magical illusion of life[44] – but also the exhilaration of discovery. His method was a source of new wonders every day, wrote Ruysch, and continuing discovery punctuated his long life with moments of heady joy and what he on at least one occasion described as "almost ecstatic wonder."[45] In his own preparations, Swammerdam felt that he was confronting the divine wisdom and workmanship that God had concealed in his creatures, and Ruysch's transports of joy were heightened as well by his conviction that, in revealing his discoveries to the world, he was serving as a chosen instrument of God.[46]

In contrast, hence, to the general reticence about the microscope in leading medical circles in the Netherlands, the rhetoric inspired by injection could reach

39 Cole (1921), 287. See the examples of Ruysch's collection depicted in his several *Thesauri anatomici.*

40 B. S. Albinus (1721), 25–6; Boerhaave & Ruysch (1751), 29; Ruysch (1744b), 18.

41 JS (1672), dedication, 2, 36, 38, 40; Nuck (1685), preface; see also Nuck (1691a), 12.

42 Ruysch (1729), preface; idem in Vater (1727), 14; Ruysch in Wedelius (1737), 9; Ruysch (1744b), 4, 6; Boerhaave & Ruysch (1751), 49–51. The quotation is from Ruysch (1725), 8–9.

43 Boerhaave & Ruysch (1751), 66, 78; Ruysch (1725), 15–16; idem (1726), 7; idem (1744b), 18; Fontenelle (1818), 1:455.

44 Ruysch claimed indeed that, when he first displayed his technique for making a whole cadaver appear as if it had returned to life, it was spoken of as if it were magic, *toverij* (Ruysch 1725, 14).

45 Ibid.; Boerhaave & Ruysch (1751), 65, 78; idem (1733), 12.

46 JS (1737–8), 2:886; idem (1672), 1; Ruysch (1733), 12; Boerhaave & Ruysch (1751), 80; Ruysch (1725), 10–11, 13, 15, 36; idem in Vater (1727), 14.

an extraordinary pitch. De Graaf protested that Swammerdam extolled the wax injection method ad nauseam, and Ruysch could get downright lyrical about it.[47] If other Dutch anatomists were not so exuberant in its praise as its inventors, they too, nonetheless, came to look upon injection as the new technology that was most dramatically providing access to the level of minute structure where the secrets of the living body were still concealed.[48] The high hopes it inspired by the early eighteenth century reflected not only the intrinsic attributes of the technique itself, however; they reflected as well the increasing grip of a theory of the body that united the century's most celebrated anatomical and physiological discoveries with the yearning for both system and mechanistic comprehensibility.

Pervasive in its influence, Harvey's revelation of the circulation of the blood emphasized a unified movement of fluid as the integrating foundation of the processes of the animal body. The continuously surging blood and its network of vessels united all the parts of the body in a single system. The discovery of the lacteals and lymphatics only accentuated the heightened consciousness of fluids and vascular networks, and the simplifying and systematizing compulsion of Dutch mechanists, Cartesian or otherwise, insisted on the integration of these networks (including the nerves) and increasingly portrayed this integrated system of vessels and fluids as the fundamental fact in physiology. For Blanckaert and others the human body had become a hydraulic machine whose life (but not its soul) consisted in the circulation of fluids.[49]

Research anatomists also reflected this view of the body in their interpretation of new discoveries. It was within this framework, for instance, that "glands" now acquired a specific and critical physiological function. A great variety of organs, including almost all the viscera, were at one time or another considered glands, but it was a classification that, reflecting only the appearance of the organs, signified no distinctive function; for Galen, the function of a given gland may have been only to support the vessels passing through it, or even simply to fill up space. Shortly after the middle of the seventeenth century, however, the glands, or at least those with ducts, were identified with the function of *secretio,* separation, by which the various fluids of the body were selectively extracted from the blood.[50] As for the ductless glands (in large part the lymphatic nodes), most were recognized as an integral part of the lymphatic and lacteal networks. The newly discovered glands – and in 1691 Nuck cited fifty-three different kinds[51] – were consequently also incorporated into the emerging fluid system of the body. With a few possible exceptions, notably milk, the

47 Graaf (1677), 466; Ruysch (1733), 23–4.
48 See also B. S. Albinus (1721), 25–6, 30; Cole (1921), 285.
49 Blanckaert (1685), 3; cf. idem (1701), 1:282. To be sure, not eveyone even in the Netherlands succumbed to this vascular physiology; see Bidloo (1715a), 3, 121, and Evertze (1706), 6–7.
50 M. Frank (1916), 302; Foster (1901), 102–3; Ruestow (1980), 266. With respect to the Netherlands in the seventeenth century, see, e.g., Voorde & Heide (1680), 389n. 3.
51 Nuck (1691a), 5–7.

blood became the common reservoir from which all the fluids of the body were derived. Indeed, according to Bontekoe, the sole purpose of the blood's circulation was the separating out of fluids by the glands.[52]

Together with ongoing discoveries, the central importance now of *secretio* in physiological speculation incited a virtual passion for glands in anatomy, and nowhere was this new "glandulous anatomy"[53] elaborated more enthusiastically than in the Netherlands. Malpighi made a substantial contribution as well, for he thought he saw minuscule glands packed together in almost every animal organ he examined, a conviction that many Dutch anatomists and physicians eagerly embraced. Thus Bontekoe proclaimed that the skin, liver, spleen, lungs, brain, pancreas, and almost everything else was a heap or tissue of glands, a fact he prized as one of the modern discoveries unknown to antiquity. Ruysch would emerge in time as a vigorous opponent of these ubiquitous glands, but he recalled that as a beginning student he too believed he saw them, for no one else thought otherwise.[54]

With *secretio* and glands so much in the foreground, the mechanist bent in seventeenth-century medical thought inevitably strove to grasp the ultimate nature of the process. Having revived the ancient imagery of shaped particles and pores, Descartes offered the analogy of a sieve that sifted out properly shaped particles in the blood, an idea that was echoed by such leading anatomists as Steno, de Graaf, and Nuck as well as by more doctrinaire mechanists.[55] Malpighi was more immediately identified with his model for the most elementary gland – a small, membranous sack surrounded by branching blood vessels and nerves and emptying through an excretory duct – but the ultimate apparatus of secretion he also conceived in terms of a sieve. Malpighi left no doubt, though, that the analogy was purely hypothetical, for he emphasized that the true structure of the tiny glands (the *acinos*) he perceived in the organs remained beyond the microscope.[56]

After two decades, however, the image in the Netherlands of the apparatus of secretion began to assume another shape: The sieve was transformed if not repudiated outright, and the context that wrought that change was the continuing reduction of physiology to no more than a system of fluids and vessels. In speculative anatomy, the final consequence was the conviction that the solid

52 Voorde & Heide (1680), 389 n. 3; Diemerbroeck (1685), *Anatomes,* 46; E. Nieuwentyt (1686), "Prooemium"; Blanckaert (1701), 1:202–3, 213; [Craanen] (1685), pt. 1:132; Bontekoe (1684), 59.

53 Ruysch in Ettmüller (1728), 20.

54 Clarke & Bearn (1968), 324; Bontekoe (1689), 1:65; Boerhaave & Ruysch (1751), 69, 75. See also Diemerbroeck (1685), *Anatomes,* 111 (the passage in question is not found in the earlier 1672 *Anatome corporis humani*), and see Baumann (1949), 180. For a negative reaction reflecting the rage for minute glands, see Kerckring (1670), 177–8; Nuck (1685), 104–5.

55 Descartes (1897–1910), 11:127–8; Boerhaave & Haller (1740–4), 2:457–8; Bontekoe (1689), 1:24; Blanckaert (1701), 1:203; Vugs (1975), 163; Graaf (1677), 549; Nuck (1685), 29–30. See also Diemerbroeck (1685), *Anatomes,* 46; and E. Nieuwentyt (1686), "Prooemium."

56 Malpighi (1697a), 1; idem (1686–7), 2:75; Duchesneau (1975), 112; Adelmann (1966), 2:866, 867n. 5; Ruestow (1980), 276–8.

parts of the body comprised only vessels and nothing more. This conception was not unique to the Dutch Republic,[57] but it flourished there as nowhere else.

The idea was inspired in some measure by empirical experience, including Swammerdam's observation of fluid oozing from newly unfolded insect wings and elytra when they were cut, from which he himself concluded that at least the skin and membranes of all animals were composed of vessels.[58] Behind its fuller elaboration, however, lay the drive for simplicity, comprehensibility, and system. Although influenced by Swammerdam, the Amsterdam physician Pieter Guenellon in 1680 felt the complete vascularity of the body confirmed by its capacity to resolve nagging problems in anatomy and physiology (such as the nature of parenchyma and the failure of the ends of the veins and arteries to join) and its compatibility with the presumption that God had made the body in the most simple and commodious way. Having acknowledged that the details of this vascular structure were beyond the eye and the microscope, he added, with a familiar ring, that they were nonetheless accessible to the mind and offered a clear and mechanical explanation of processes that lay beyond the senses.[59]

The vascular physiology was most taken up in the following decade by physicians with emphatic Cartesian links. In good mechanist fashion, Bontekoe referred to the body as a "miraculous machine" and a "divine clock," but subsequently he reflected that, the body being almost nothing but pipes, the pipe organ was a better analogy. No one familiar with anatomy would deny that the body consisted only of vessels and fluids, Blanckaert also contended, or that "whatever in the body is solid" was constructed of vessels alone.[60] The appeal of the concept extended beyond the ranks of the out-and-out mechanists, however, and by the end of the century Nuck too was offering a "tubulous" account of the body in his classes at Leiden.[61]

The sieves of glandular secretion were hence also transformed into complexes of vessels. The blood produced all kinds of fluids when it was strained through different kinds of pipes as if through sieves, wrote Blanckaert, who ascribed the diversity of these secreted fluids, as well as the variety of substances in the living body, solely to the size of particles and the capacity of pipes.[62] Nuck may have been also teaching that the differences in bodily fluids followed from the different ways in which tubules were arranged and twisted, and by the early eighteenth century Ruysch was insisting that whatever served to sustain the

57 See King (1666); Birch (1756–7), 3:179–80; Cole (1921), 310–11; Malebranche (1958–70), 12:240; Ridley (1695), 20, 90–8.

58 JS (1669), pt. 1:106; idem (1737–8), 1:275, 331–2, 2:717. As an example of the influence of Swammerdam's early remarks, see Guenellon (1680), 30–3.

59 Guenellon (1680), 31, 37.

60 Bontekoe (1689), 1:7, 58–9; idem (1684), 52; Blanckaert (1685), 2; idem (1686), 180; idem (1683a), 27; idem (1701), 1:198; see also Overkamp (1686), 170, 172.

61 Hoeve (1690), §9. It might be noted, however, that although Nuck himself does not betray a Cartesian commitment, his student Hoeve does. As further evidence of the infiltration at Leiden of the conception of a totally vascular bodily structure in the late seventeenth century, see Cocquis (1688), "Theoretica" §§3, 4.

62 Blanckaert (1686), 146; idem (1701), 1:39–40, 199, 202–3, 219, 276–7.

body – by which he meant its many fluids – was produced only by the different windings and intertwinings of the clustered ends of the arteries.[63]

Both the elaboration and the ascendancy of this vascular physiology reached their apogee in the teachings of Herman Boerhaave, the brightest star of the Leiden medical school at the height of its international celebrity in the early eighteenth century. His greatest pupil, Albrecht von Haller, testified that Boerhaave's *Institutiones medicae,* propounding the vascular system he taught at Leiden, was used in nearly every university in Europe. Appearing in five authorized editions during Boerhaave's lifetime and in a host of pirated and translated editions as well, it was indeed a seminal work in establishing the very discipline of physiology.[64] Looking back in 1801, Marie-François-Xavier Bichat, who was launching a new era in physiology himself, claimed that Boerhaave's system had dazzled all the *esprits* of the preceding century and had brought about a revolution akin to that produced by Descartes's vortices.[65]

The *Institutiones* described for its readers a system of interconnected levels of successively smaller vessels that, deriving initially from the arteries, removed successively more subtle fluids from the blood, from serum to lymph and on ultimately to the very "spirit" of the nerves. This succession of ever-smaller vessels producing their respective fluids continued, believed Boerhaave, until it reached vessels similar indeed in size and contents to the nerves.[66] To preserve the motion of the diverse fluids, moreover, the vessels and other solid parts of the body had to remain flexible, and that was only possible, Boerhaave maintained, if those vessels and other parts were themselves composed of still finer vessels also containing moving fluids. Insofar as it was pliant, consequently, the whole body was composed of vessels, and ultimately all the solid parts of the body consisted entirely of nerves.[67] Boerhaave conceived an even more schematic hierarchy of increasingly complex structural forms in the body that rose first from absolutely simple fibers to the simplest, most minute membranes woven of such fibers, to the smallest nerve vessels composed of such membranes, to the next level of membranes woven of these vessels, and so on through an unknown number of repetitions. As if it might still be unclear, his student Johannes de Gorter reiterated that it was from these increasingly compounded vessels and membranes that all the organs of the body were composed.[68]

63 Hoeve (1690), §9; Boerhaave & Ruysch (1751), 57, 53; Ruysch in Vater (1727), 12; Ruysch (1733), 12; idem (1726), 7. It is not clear whether Hoeve's reference to different fluids deriving from the different arrangement of vessels pertains to Nuck's instruction, to which Hoeve had alluded, or is Hoeve's own elaboration of that instruction; cf. Nuck (1691a), 54. Heidentryk Overkamp had also noted in passing that our bodies must contain as many different small pipes and vessels as there are different parts that have to be nourished (Overkamp 1686, 174).

64 Haller (1774–7), 1:756–7; Lindeboom (1968), 70–4; idem (1959), 37; Fulton (1938), 4861–2, 4864; Rothschuh (1973), 120.

65 Bichat (1801), 1:xxxviii.

66 Boerhaave (1734), 138–9, 158, 160, 164. See also: idem (1751), 1:445; Boerhaave & Haller (1740–4), 2:555–6.

67 Boerhaave (1734), 164, 227–9; idem (1751), 1:446; Boerhaave & Haller (1740–4), 2:154, 3:638.

68 Boerhaave (1734), 364–5; idem (1751), 1:446; Boerhaave & Haller (1740–4), 6:23–7; Gorter (1735–7), 1:10; see also Gaubius (1725), 20–2.

The successive levels of diminishing vessels were so interconnected, moreover, that all their varied fluids were driven in common by the heart.[69] Apart from the body's excretions and exhalations, these fluids also ultimately returned to the circulating blood, for, according to Boerhaave, every level of vessels deriving from the arteries had a corresponding set of vessels returning to the veins. In effect, nearly the entire physiology of the body (including growth, aging, disease, and death) had been gathered under the dominion of the beating heart; thus life itself, as Blanckaert had once remarked, was indeed reduced to little more than the circulation of fluids.[70]

All of this was depicted with the simple mechanistic imagery that had so appealed to preceding generations. The epistemological obsession of the purer Cartesians had been abandoned, and, under the influence of Newton, Boerhaave repudiated the delusion that science could rest on reason alone; rather, he now expanded at length on the insuperable limitations of human understanding.[71] Nonetheless, the passion for mechanism had long shaped the quest for physiological system,[72] and Boerhaave not only placed his own system under the rule of "hygrostatic, hygraulic, and mechanical" laws but also cast it in terms of firm structures and colliding shapes. To be sure, ill-defined "forces" made an occasional appearance, and he left the explanation of some effects to chemistry; but nutrition was still essentially a replacement of particles that had been borne or worn away, and the critical process of *secretio* was achieved by the motion, pressure, rubbing, and banging together of particles that drove them into the mouths of smaller vessels.[73]

The physiological system that eventually found its fullest expression in the teachings of Boerhaave was thus a seductive synthesis indeed.[74] It joined the prestigious aura of recent anatomical research to a sweeping explanatory power. It appeared as the full and final fruition of the great physiological discovery of the century, the circulation of the blood, presiding now over a unified, hierarchic system of processes and structures. The synthesis was a compelling one to which many of the leading lights of Dutch academic medicine, Boerhaave's students and compatriots, were particularly susceptible.

The advocates of the vascular physiology also claimed the support of both injection and the microscope. Blanckaert, Bontekoe, and Boerhaave agreed that injection – anatomy's most noble aid, said the last – confirmed the vascular structure of the body's solid parts. The younger Albinus, even after having himself

69 Boerhaave & Haller (1740–4), 3:627; Boerhaave (1734), 158, 205.

70 Boerhaave (1734), 138–9, 143, 160, 166, 227, 234–5, 364–7, 371–5, 392, 396, 465, 473; idem (1751), 1:445. Boerhaave also insisted that for all venous vessels returning to the heart there had to be corresponding arterial vessels (idem 1734, 139; idem 1751, 1:443–4). See n. 49 above.

71 Boerhaave (1738), 27.

72 See Folter (1978), 186.

73 Boerhaave (1734), 14, 138, 228, 230, 233, 364, 396.

74 Like Bichat (1801, 1:xxxviii), Boerhaave's notorious former student Julien Offray de La Mettrie also spoke of Boerhaave's system (or, more precisely, "systems") as "seductive," *séduisants* (La Mettrie 1751, 1:342). Cf. Fontenelle (1818), 1:498.

abandoned the belief in the total vascularity of the body, still likewise attributed its most compelling evidence to injection, particularly the work of Ruysch.[75] Boerhaave also appealed above all to the injections made by Ruysch,[76] his close if much older friend and, to Boerhaave, Holland's greatest anatomist.[77]

If Bontekoe emphasized the role of the injection syringe, however, he cited as well the microscope as having demonstrated the vascularity of the body, and he credited microscopic blood vessels with having provided the first evidence that, beyond the visible vessels of the body, almost beyond reckoning themselves, lay many more vessels that even the microscope could not reach. In 1675 Leeuwenhoek had alluded to animal membranes that appeared as if they were woven indeed of tiny vessels, and five years later de Heide had also described how the lungs and urinary bladder of a frog looked through his microscope like "an ingenious tissue" of minute vessels.[78] In the following years, Leeuwenhoek offered a series of vivid accounts of masses of microscopic blood vessels so thickly tangled at times that they seemed to leave room for nothing else. He reported such densely packed vessels in a variety of animal parts – the villi of the intestinal lining (apparently from cattle), the fins of eels, the wattle of a cock, the foot of a crab[79] – and found a great number of other tubules in bone, tusks, and teeth,[80] in animal brains, and, in later years, surrounding and pervading the muscle fiber.[81] He was struck as well by the abundance, in the bodies of insects, of ramifying vessels (the tracheae), vessels whose branchings became so fine they ultimately escaped his microscope.[82] Nor was Leeuwenhoek reluctant to relate such observations to human anatomy; how inconceivably great by comparison, he wrote after describing the looping capillaries that completed the circulation of the blood in the tails of small fish, must be the number of such loops in the human body.[83] It was not without good reason, then, that

75 Blanckaert (1685), 2; Bontekoe (1684), 52; Boerhaave & Haller (1740–4), 2:154; B. S. Albinus (1754–68), 3:3–6; see also idem (1721), 27; Elshout (1952), 60, 64.

76 Haller (1774–7), 1:529–30, 757; Boerhaave & Haller (1740–4), vol. 1, dedication; Schreiber (1732), 6. Ruysch is indeed repeatedly cited by Boerhaave throughout the *Institutiones*. See, in particular, Boerhaave (1734), 136, and Boerhaave & Haller (1740–4), 2:243, 252.

77 Boerhaave (1962), pt. 2:86. Regarding the friendship between Boerhaave & Ruysch, see ibid., p. 290; Ruysch (1733), 7; Scheltema (1886), 34–5.

78 Bontekoe (1684), 52; idem (1689), 1:60; *AB* 1:288; Voorde & Heide (1680), 496n. Malpighi had first reported capillaries in the lungs of a frog in 1661 (Belloni 1967, 288–90).

79 *AB* 4:180–9, 8:68–73, 194–5; AvL (1696), 15–16.

80 *AB* 6:20–9, 192–207.

81 *AB* 2:216–21, 4:258–79; AvL (1718), 447–8; AvL to RS, 9 Jan. 1720, AvL Letters, fols. 225r–v, 227r; AvL to RS, 24 Jan. 1721, ibid., fols. 246v–247r; AvL to RS, 11 April 1721, ibid., fols. 254r–v.

Leeuwenhoek also eventually considered human and animal skin to be composed of an inconceivable abundance of interwoven vessels, a conclusion he reached primarily as a result of observations of fish scales: *AB* 4:250–1, 5:336–9, 6:32–7; AvL (1718), 418, 426–7, 429. Cf. AvL to RS, 1 March 1712, AvL Letters, fol. 170v; Boerhaave & Haller (1740–4), 2:154; Boerhaave (1734), 222–3; Elshout (1952), 60.

82 *AB* 6:322–5; AvL (1693), 463–6, 531.

83 *AB* 8:48–51.

Boerhaave called upon Leeuwenhoek and the microscope to testify, along with Ruysch and injection, to the pervasive vascularity of the body.[84]

But the microscope was not a reliable ally: Leeuwenhoek, for instance, had found no vessels but only globules in early tadpoles. Boerhaave himself – even with a lens he reported to be as small as a sand grain – could not discover the multitude of smaller vessels he posited between the capillaries;[85] indeed, he could maintain his system only by disregarding Leeuwenhoek's repeated insistence (with one anomalous exception) that the capillaries remained a closed system of vessels from which fluids escaped only by seeping through their walls.[86] Like Boerhaave, Albrecht von Haller also continued to assume the presence of more minute vessels issuing from the arteries, but he acknowledged at midcentury that the most powerful microscope lenses – namely, those made and used by Leeuwenhoek – had failed to find them.[87]

The effect of injection, on the other hand, was conveniently selective; it revealed only vessels or related hollow structures. Variants of injection technique also provided the essential evidence for the interconnection of vascular networks, the critical assumption, after all, of the vascular physiology. Inflation, the injection of air rather than fluid, convinced Nuck (mistakenly) that the lymphatics and the arteries were connected in both the kidney and the spleen; but it was his mercury and a rare dye that finally succeeded, or so he claimed, in passing from the arteries of the breast to the lactiferous ducts of the nipple and thence back into the arteries. (On the other hand, whereas Nuck had forced air, he believed, from the splenic artery to the lymphatic vessels of the spleen, he had not been able to do so with mercury – a fact that, to him, reflected the selectivity of the mechanisms of secretion, since elsewhere, he asserted, mercury would pass where air would not.) Ruysch likewise boasted of having discovered and publicly demonstrated by injection a direct connection between the artery and the "urinary ducts" of the kidney.[88]

The reinforcement provided by the bond between the theory of vascular physiology and the technique of injection was in fact reciprocal; each with its own intrinsic appeal strengthened the grip of the other. If the revelations of injection necessarily conformed to the vascular preoccupation of the theory, the

84 Boerhaave (1734), 123; Boerhaave & Haller (1740–4), 2:253 (see also 3:639, where there are also echoes of Leeuwenhoek's letters); Boerhaave (1751), 1:250–1.

85 *AB* 8:12–15; Boerhaave (1751), 1:445–6. The passage in question in this last citation is an interesting example of how what would seem negative evidence for a theory can be made to serve as positive support; since in the foot of a frog a small lens reveals places where no blood is seen – which would seem to mean the spaces between the capillaries – there must then be smaller "lymphatic" arteries, concludes Boerhaave, that is, vessels removing a subtler fluid from the blood but too small for the red blood itself. According to Boerhaave, the presence of some fluid-filled vessels there was required by the fact of nutrition and growth. See also Boerhaave & Haller (1740–4), 2:427; Boerhaave (1751), 1:250–1, 444–5.

86 Ruestow (1980), 279–82.

87 Haller (1757–66), 1:98–9, 2:378–9.

88 Nuck (1691a), 11–13, 54, 61; Ruysch (1744b), 10–11, 22–3; idem (1701), 32–3; idem (1702), 64–5; idem (1703), 39–41. Ruysch's injection of the kidney had in fact been anticipated by Bartolomeo Eustachi more than a century before (Cole 1921, 289).

theory dictated a research program for which the technique was uniquely suited. Accentuating what was already an established focus of subtle anatomy in the Netherlands, the vascular physiology called for an intensified search for further vessels and the exploration of the origin, distribution, and function of those vessels that were already known.[89] The knowledge of the circulation of the blood kindles a bright light in our minds, wrote Bontekoe, and having happily sailed our ship of exploration on the great ocean of the blood, let us steer it now into the rivers of other fluids. His metaphor served to guide Bontekoe through a survey of contemporary physiological thinking, but it reflected as well the vista that was now unfolding before researchers.[90]

The vascular physiology was the purveyor indeed of an excited optimism, for it assured the ambitious anatomist that, with persistence, he too could play his part in the final attainment of the true and conclusive understanding of the living body. It was the ability of Ruysch's technique to demonstrate the vascular endings he took to be the mechanisms of secretion that transported him to the brink of ecstasy (and thrilled Boerhaave). That exultant joy surely stemmed in part from the sheer spectacle of the diverse structures of tiny vessels – now interwoven (in Boerhaave's vivid description) like a net, or gathered in a ball, bunched like a painter's brush, wound in a coil, or teased into a shaggy mass – that Ruysch believed he had brought to light. But Ruysch's joy clearly derived as well from the conviction that what he had now revealed was the basic foundation of physiology.[91] For Ruysch too this ostensible discovery bespoke a broader "system," a system he also claimed was new; in fact, it rehearsed a fantasy that had stirred Dutch medical speculation for at least two decades before it vested his injections with such exhilarating significance.[92]

Its promise and importance thus enhanced by theoretical anticipation, injection remained the reigning technique in anatomical and physiological research through the eighteenth century, while the contribution of the microscope to these fields, by contrast, declined into relative insignificance.[93] An openly and

89 See, for instance, Boerhaave's insistence that, since the lymphatics that were known were "venous," there must be an as-yet unseen network of corresponding "arterial" lymphatics (Boerhaave 1751, 444–5). Regarding the search for such arterial lymphatics, see also Académie Royale des Sciences (1744), 48. Nuck searched for similar vessels that he was convinced nourished the ovula in the ovaries (Nuck 1691a, 65–6).

90 Bontekoe (1684), 59. See also, for instance, Nuck (1691a), dedication and pp. 140–1.

91 Ruysch (1733), 12; Boerhaave & Ruysch (1751), 28–9. See also Ruysch (1701), 30–1, 42–3; idem (1702), 38; idem (1703), 11, 33–4, 39–42; idem (1704), 32, 41; idem (1705a), 33; idem in Campdomercus (1725), 6–7; Ruysch (1725), 16; idem (1726), 7. Regarding Ruysch's exhilaration over other such ostensible discoveries, see Boerhaave & Ruysch (1751), 65, 78.

92 Ruysch in Vater (1727), 12; see nn. 62, 63 above. In 1705, the French physician Raymond Vieussens published his own "new system" in which he also argued that all secretions (and excretions) derived from tiny vessels originating in the walls of the arteries, but he cited Bontekoe (who in turn also cited de Graaf's work on the testicle) as a source of inspiration (Vieussens 1705, preface and pp. 72–3, 78, 80–1, 88–90, 93, 95, 107, and esp. 138–40).

93 Cole (1917–21), 285, 287, 316–17, 334, 339–40 (and see Conclusion); Bradbury (1967a), 66; Fournier (1991), 18. Anatomists and physiologists in the eighteenth century did not by any means drop the microscope from their arsenal of instruments (see, e.g., I. Hahn 1728, 29–30; Haller 1757–66, 1:iv; Terrada Ferrandis 1969, 26–7, 49, 55), but as a reflection of the greater attention academic medical circles in the Netherlands still paid to injection as opposed to the

sometimes vigorously expressed distrust of the early microscope offers what, at first sight, might seem a reasonable explanation of the instrument's diminishing role in medical research following the deaths of Malpighi and Leeuwenhoek, for such doubts had quickly surfaced after Malpighi's first sorties into microscopic anatomy. A respected physician of German extraction but long resident in Amsterdam, Theodorus Kerckring had gone public with his own misgivings as early as 1670; provoked by the rage for minute glands and his own sense of the shortcomings and unreliability of microscopic images, he had urged that anything observed only by means of a microscope be held suspect. Although he was answered by no one less than Malpighi himself – we do have to be careful and use good sense, he said in effect – Kerckring's uneasiness over the fad for glands was not ill-founded, as Malpighi's own perception of tightly packed glands in the cerebral cortex amply illustrates.[94]

Injection, however, misled its devotees no less. Insisting, characteristically, that only injection could grasp the truth, Ruysch derived immense satisfaction from exposing Malpighi's error regarding the cerebral cortex; yet, in doing so, he replaced one celebrated mistake with another.[95] Take a small piece of a cortex whose blood vessels had been injected with wax, he instructed his fellow anatomists, and suspend it by a hair in a suitable fluid; by manipulating the thread, gently agitate the bit of cortex until it unravels and spreads out in infinite little branches (Fig. 11); and thus it would be revealed that the cortex was composed not of glands but solely of vessels.[96]

Ruysch's skill with injection added significantly to the knowledge of the vascular system throughout the tissues of the body,[97] but his demonstrations of his physiological system rested on errors and self-deception. Most injection mistakes followed from the use of mercury, and some of Nuck's erroneous demonstrations resulted simply from the leakage of mercury (and sometimes air as well, to be sure) from the vessels into which it had initially been injected. But Ruysch's wax also escaped into other vessels, obscured other structures in the surrounding flesh,[98] and left tiny amorphous clumps upon which his expectant imagination worked its will. Needless to say, such errors were hardly unique to Nuck and Ruysch, and using diverse injected fluids, anatomists across Europe continued to discover or confirm a variety of nonexistent vascular connections throughout the body.[99]

Footnote 93 *(cont.)*
microscope, see Munniks (1771), 17. Also compare the emphasis on injection to the brief references to the microscope and even to Leeuwenhoek in Rudolph Forsten's academic address on the contributions Netherlanders had made to physiology (1776, 12, 35–50, 59).

94 Kerckring (1670), 177–8; Malpighi (1697b), 31 in the first series in the pagination; idem (1686–7), 2:78. See also Adelmann (1966), 2:831–32, and Clarke & Bearn (1968).

95 Regarding the influence of Malpighi's misperception in the Netherlands, see Bidloo (1685), pl. 10 (fig. 2), and Blanckaert (1686), pl. 11 (fig. 3); see also Bidloo (1715d), 37–9.

96 Ruysch in Ettmüller (1728), 9–10; Boerhaave & Ruysch (1751), 66; Ruysch (1744b), 8, 39. See also idem (1703), 33–4; idem (1725), 11–12; idem (1726), 2.

97 Cole (1921), 304–5; Rothschuh (1973), 103; L. Wilson (1960), 179.

98 Cole (1921), 286, 305, 313, 337.

99 Ibid., 313, 315, 317–20, 323, 330–1, 336.

Nor was injection's susceptibility to error unrecognized. Ruysch was inclined in fact to dramatize the criticism his discoveries evoked, including not only denial, he said, but jeers and laughter. He ascribed to jealousy the charges that his technique distorted the vessels he injected and destroyed the glands he then said he could not find, but friendly critics too questioned whether his wax did not obliterate some vessels while distorting others. Even Boerhaave ventured that Ruysch's wax hid a variety of other marvelous structures in the tissues of the body.[100]

Nonetheless, the prestige and appeal of injection as the foremost method for researching the fine structure of the body persisted throughout much of the eighteenth century, arguing that a recognized capacity for error and deceptive artifacts does not in itself discount a new technique.[101] In this instance, a compelling theoretical predisposition enhanced the perceived potential of injection and hence promoted a tolerance for – and indeed susceptibility to – its characteristic errors.[102] Consequently, it seems to have been less the recognition of the microscope's potential for deception that curtailed its use in anatomical research after Malpighi's death than a theoretical commitment that vested greater promise and significance in an alternative route to discovery.

Simply to oppose injection and the microscope is misleading, however, for not only did both provide support for the vascular physiology, even if the one did so less reliably, but they were often used together. Nuck made no mention of the microscope even when he spoke of injected vessels that, in their smallness, disappeared from sight, but Swammerdam had at least in passing used a lens on the finer vessels he injected.[103] Indeed, as wax injection was increasingly refined, recourse to the lens became routine. By Ruysch's own account, a number of the subtler results of his preparations were visible only through a microscope,[104] and both he and Boerhaave (whom Ruysch lauded as expert with the instrument) had brought it to bear on those fictitious vascular apparatus through which the secretions of the body ostensibly oozed.[105] According to Ruysch, Boerhaave had remarked that to observe the finest vessels in Ruysch's preparations required some of the stronger lenses available, and Boerhaave himself recounted how, with the microscope, he had followed the shaggy endings of such

100 Boerhaave & Ruysch (1751), 50, 66–7, 35, 37; Bohl (1744), 11; Boerhaave & Haller (1740–4), 2:427; Boerhaave (1751), 1:251–2; Rau (1713), 30. See also Muys (1738), preface [137], 176–8.

101 See Conclusion, nn. 71, 73.

102 B. S. Albinus (1754–68), 3:3–6; Haller (1757–66), 1:98–9, 2:378–9; Cole (1921), 323, 330–1, 336.

103 Nuck (1691b), 11. An account of his investigation of muscle fibers does imply that Nuck did indeed use the microscope in at least some of his researches, though again the instrument remains unmentioned (see Sénac 1749, 1:56–7). JS (1672), 33, 56; see also Blanckaert (1686), 415.

104 Ruysch in Ettmüller (1728), legend to pl. 14 (fig. 1); Ruysch (1726), 7; idem (1733), 26; idem (1739a), 123, 125, 134, 139, 156.

105 Ruysch (1704), 42–3; Boerhaave & Ruysch (1751), 65, 71–2; see also the preceding note. Boerhaave himself had recourse to the microscope in medical diagnosis as well as research and teaching (Boerhaave & Haller 1740–4, 2:72–4, 542–3, 582, 3:12, 4:187, 6:176, 326; Haller 1774–7, 1:757).

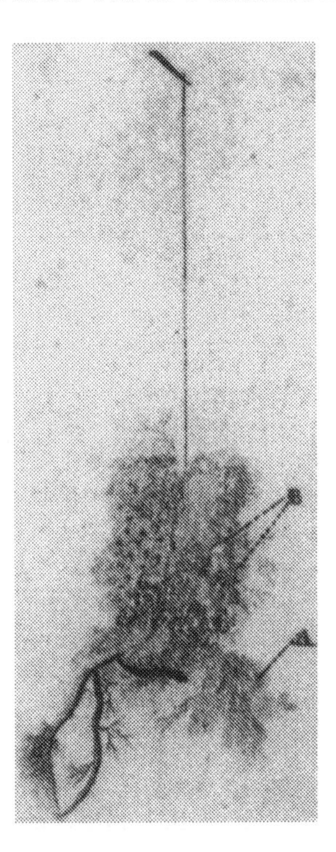

Figure 11. A preparation of a piece of the cortex of the brain by Frederik Ruysch. According to Ruysch, the extremities of the vessels composing the piece are so fine that they cannot be seen without a lens (Ettmüller 1728, pl. 14, fig. 1). Photo courtesy of the *Journal of the History of Medicine and Allied Sciences.*

preparations until they dissolved into teasing atoms that fled his sight (*ludentes, fugientesque atomos*).[106]

But even if the program of research shaped by injection and the vascular physiology made room for the lens, the ascendancy of that program profoundly limited the nature and scope of microscopic research in subtle anatomy. Al-

106 Boerhaave & Ruysch (1751), 29, 72. On the latter page (where indeed, in the passage in question, *assequi* has been misprinted as *asserui* in the edition that is otherwise cited throughout these notes, so see the same page in the 1722 edition), Ruysch cites Boerhaave as having said that microscopes of the "fifth kind," *quinti generis,* were needed to see the final endings of the arteries that Ruysch had prepared in livers. Although the numbering was subsequently reversed in the characteristic commercial microscopes of the eighteenth century, the popular Marshall microscope in the late seventeenth and eighteenth century offered six numbered objective lenses of different power, the strongest of which was number six (Bradbury 1967b, 160, 162–3). That this was the numbering scheme with which Ruysch and Boerhaave were familiar is indicated by Boerhaave's identification of the microscopes with which Leeuwenhoek observed the connections of the capillaries as lenses of the "sixth order," *vitra sexti ordinis* (Boerhaave & Haller 1740–4, 2:609–10).

though Ruysch wrote at one point of some Leeuwenhoek instruments to which he unaccountably had access, he otherwise said little about the microscopes he used. He confessed in 1705 that, however often he had referred to the microscope, he in fact had made use of the instrument only because of his failing vision; others with sharper eyes, he acknowledged in this instance, needed no lens to see the subtler disclosures of his injected wax, and the plates in his publications indicate "microscopes" of only ×2 to ×4![107] In 1725, six years before he died (at almost ninety-three years of age), the subtlety he boasted of achieving with his technique was the injection of vessels as fine as the threads of a spider web – and, fine though that may be, it is not truly microscopic.[108]

Although he was admired in his day as the master of the wax injection technique, moreover, Ruysch's preparations do not appear to have been able to sustain truly microscopic research. At midcentury, the Berlin anatomist Johannes Nathanael Lieberkühn did succeed (as did the younger Albinus) in injecting microscopic vessels; but when he had occasion to examine such vessels in examples of Ruysch's preparations, he found the pigment settled in splotches – Boerhaave's fleeting "atoms," perhaps – and the wax itself fragmented, uneven, or solidified in clumps outside the vessels from which it had escaped.[109] Even more successful injections could exact their price, however, as the tangle of wax-gorged vessels obscured other fine structures within the body.[110] Consequently, although the microscope was not excluded from injection research, its potential in late seventeenth- and eighteenth-century anatomy was stifled, given the infatuation with injection, to the extent that it was reduced to serving as an accessory to that research. (Even Lieberkühn, another of Boerhaave's students, devoted his professional life to injection and the exploration and preparation of vascular tissues.)

What at first sight appears a very different aspect of the fine structure of the body did inspire, however, one remarkably protracted effort in microscopic research by a Dutch academic. Having studied at Leiden, Wijer Willem Muys occupied a chair in medicine at the University of Franeker through the opening decades of the eighteenth century. Also an admirer of injection, he stood out among his academic colleagues in anatomy for his outspoken enthusiasm for the

107 Ruysch (1705b), 31, 62; idem (1701), pl. 4 (fig. 6B and legend); idem (1702), pl. 1 (fig. 5 and legend); idem (1744a), pl. 4 (figs. 2, 3 and legend); idem (1726), pl. 2 (fig. 3 and legend); Wedelius (1737), pl. 16 (figs. 13, 14, 17, 18); Ettmüller (1728), pl. 14 (fig. 1 and legend). The illustration of fig. 1 and the legend to fig. 2 in this last citation indicate that the object was suspended in a vial of fluid by a thread, and the frequent depiction of a similar suspending thread in many other illustrations in the previously cited works suggests that those objects were likewise observed in vials. Hence, lenses used on these objects in preparation would indeed have generally required a relatively long focal distance. Ruysch also speaks on one occasion of having used a lens the size of the crystalline humor of a small ass (Ruysch 1729, 38).

108 Ruysch (1725), 6, 16.

109 Cole (1921), 309, 329–30; Lieberkühn (1745), 8–9; Elshout (1952), 68. Few of Ruysch's preparations have survived.

110 Cole (1921), 305.

microscope as well – "never praised enough nor exalted with words too splendid."[111] By his own account, he was still a boy when he had first turned the microscope to the observation of muscle fibers, the focus of a research commitment that persisted throughout his career and culminated in 1738 (though with intimations of more to come) in the 430 pages of his *Investigatio fabricae, quae in partibus musculos componentibus extat.*[112] Ultimately, however, Muys's concluding tome also bore witness to the pervasive grip of the vascular physiology and hinted at how it could limit the prospects for microscopic discovery in still more subtle ways.

The nature of muscular action had emerged as another of the salient physiological problems of the late seventeenth century. In the Netherlands, the inclination to conceive that action as an inflation or expansion of muscular tissue corresponded to the mechanist drift of physiological thinking and readily accommodated itself to the vascular physiology. Indeed, in the 1680s Blanckaert proclaimed that he had not only succeeded in injecting from arteries through muscle fibers into the veins but that he could even demonstrate by injection that those fibers had valves. Early in his professorship at Franeker, Muys too had convinced himself that he had succeeded in injecting fluid into the muscle fibers from the artery.[113]

The microscope had contributed to another facet of the conceptualization of the muscle fiber, however, and it was this facet that was the central concern of Muys's *Investigatio fabricae.* Older notions had posited fibers in the muscle as a framework on which the flesh of the muscle collected as parenchyma; but in the general assault on the concept of structureless parenchyma, Steno had insisted that the muscle consisted of fiber alone.[114] Steno's observations, apparently with the naked eye,[115] were soon followed by microscopic studies that now also reported successive levels of muscle fibers, one level enclosed within the other.[116] Foremost, then, among the ostensible discoveries Muys now offered after twenty years of observation was the conclusion that the succession of fibers consisted of eight successive levels or "orders" encompassed one within the other. They stretched from three larger orders of fibra that could be seen without a lens to the ultimate and smallest fila that appeared through his most pow-

111 Muys (1738), preface [136–7]; idem (1714), 52–3, 55–6. See Napjus (1939–40).

112 Muys (1738), preface [142–3], 206–7. This volume is called a *Dissertatio prima* on the title page (and in running heads throughout), and future "dissertations" are indeed anticipated in the text (pp. 48, 286).

113 Blanckaert (1686), 500, 502; idem (1688a), 306–8; idem (1701), 1:243; *Journal literaire* (1732 ed.), 3(pt. 1) (Jan.–Feb. 1714): 238–41.

114 Berg (1942), 346–7, 359–60; Bastholm (1950), 81–2, 120–2, 125–6, 144–8.

115 Muys was of the opinion that Steno's observations were made with the naked eye (Muys 1738, 153), though it is not known to what extent Steno may in fact have made use of the microscope in his researches (*DSB,* s.v. "Stensen, Niels"). Girolamo Fabrici, for one, had already been stimulated by naked-eye observations to speak of the muscle fibers as resolving themselves into ever smaller fibers until they passed beyond the limits of vision (Berg 1942, 356, including n. 2).

116 Heide (1686), 32–4, 42; *AB* 3:392–6, 400–2; AvL (1718), 7, 20, 21.

erful microscopes – droplet lenses of "incredible" magnification, he wrote[117] – as linen threads appeared to the naked eye.[118]

Apart from whatever predispositions helped fabricate those eight orders, however, Muys's perceptions of the ultimate fila also betray the insidious influence of the vascular physiology. In arguing that the smallest fila (whose width he estimated to be less than a quarter the diameter of a red blood corpuscle)[119] ended the succession of diminishing muscle fibers, Muys referred to other ultimate functional parts of the body the microscope had reached: the extremities of the blood vessels and the supposed tubules that composed the nerves, the brain, and the varied membranes of the body.[120] Clearly himself an adherent of the vascular physiology, Muys had no doubt that those ultimate fibers in the muscle were also hollow tubules for conveying fluid.[121] What Muys perceived in the doubtful light and among the unfamiliar shapes confronted in high-powered lenses appears then to have been skewed by a readiness to find expected forms, for along the length of many of the smallest fila he frequently detected a succession of nodules that, as he illustrated them, recalled the successive swellings on the lymphatics denoting valves (Fig. 12).

Potentially as significant as what Muys's imagination added, however, was what he may have failed to see, or at least remark, in microscopic images, for the spell of both fiber theory and the vascular physiology limited what he had a mind to find. Fiber theory in itself anticipated little of interest beyond a diminishing repetition of form, and Muys, knowing as well that the muscle fibers ultimately reduced to vessels and a hydraulic physiology, was prepared to attend only to what he could recognize as pertaining to a minute hydraulic system. Apart from the likes of valves, consequently, there was not much more to look for. We can only speculate as to how much Muys failed to see because he was not prepared to see it, but it may well have constituted a significant further curtailment of the prospects for microscopic discovery.[122]

Muys's intensive quest for microscopic discovery hence also serves to underscore the pervasive and prejudicial influence of a physiological outlook that tended to limit severely the potential of the instrument. It was an outlook that, together with the preoccupation with injection, narrowed both what was sought and what was perhaps perceived through the lens; thus theory and an alternative

117 Muys (1738), 2–3.

118 Ibid., preface [143], 5, 21, 39–40, 206, 227–30, 237; *Journal literaire* (1732 ed.), 3 (pt. 1) (Jan.–Feb. 1714): 238.

119 Muys (1738), 279. The width of these final fila would thus have corresponded, indeed, to that of the real myofibrils within the cells that constitute the muscle "fibers" of today.

120 Muys (1738), 280–92, esp. 282–7.

121 Ibid., 287; see also p. 283 and preface [1], [137].

122 For an example in twentieth-century microscopy of how preconception and selective attention can "blind" an observer, see Barber & Fox (1958–9), 131–3, 135; see also Neisser (1976), 20, 43–4, 55, 80, 87. With respect to "instructional set," Ralph Norman Haber and Maurice Hershenson conclude that, in those situations in which such a set is efficacious, it manifests itself not as a greater sensitivity to attributes toward which the set is biased but as a diminished response to other attributes (Haber & Hershenson 1973, 236).

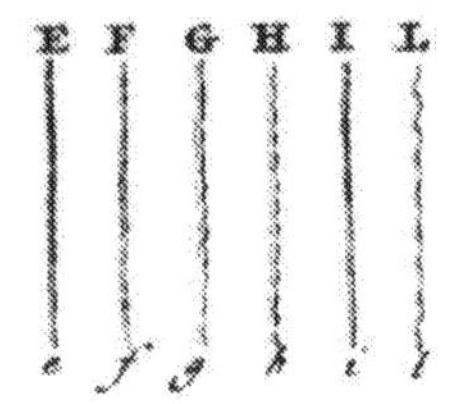

Figure 12. Ultimate muscle fila illustrated by Wijer Willem Muys (Muys 1738, pl. 1, fig. 17). Photo courtesy of the National Library of Medicine, Bethesda, Md.

research technique worked hand in hand to confine the use and the fruitfulness of the microscope in subtle anatomy.[123] The prejudicial aspects of the research program defined by injection and the vascular physiology were more clearly revealed, to be sure, in the course of its fuller development in the early eighteenth century, but the assumptions, the techniques, and the anticipations that underlay that program had been taking shape in the Dutch medical community during the preceding decades. The emerging outlines of the program presumably worked then as well to forestall Dutch anatomists of the late seventeenth century from assuming the lead in initiating a new era of microscopic discovery.

As was also the case, however, with Cartesian rationalism, naturalist miniature painting, and the theology of nature as revelation, the attitudes associated with the program of vascular research by no means posed a rigid and unyielding barrier to microscopic research. The Cartesians had regularly voiced their respect for the lens, and the mechanistic bent, when divorced from their epistemological preoccupation, could stimulate the search for the microscopic mechanisms it still posited throughout nature. Despite the perplexity of Leeuwenhoek's visitor over the purpose of microscopic structure that could not be seen, the anticipation of confronting God in his works could also become a meaningful incentive for the continuation of microscopic observation once truly microscopic realms had been breached. Nor, after all, did a commitment to injection or the vascular physiology preclude the use of the microscope in anatomy; to the contrary, improving injection techniques increasingly encouraged a recourse to the lens, although still tending to limit its horizons to injection effects.[124] The obstacles latent in these patterns of thought and activity were subtle and tenuous, susceptible to the vagaries of mood, personality, and circumstances.

Nevertheless, the time, effort, and even discomfort required for any consequential microscopic research made these subtle obstructions meaningful. Muys stressed that such research in anatomy required not only good instruments but

123 For an analogous twentieth-century example of how one research approach involving theoretical and instrumental commitments can work to impede or delay the development of another, see Edge & Mulkay (1976), 363, 391.

124 See Elshout (1952), 60, 66–9.

great care, diligence, and perseverance. Though his own example may not be a reassuring one, he would have also agreed that even the ability to observe effectively through a microscope was acquired only through experience[125] – and such prolonged experience with the best research instruments of his day remained in itself something of an ordeal. In addition, there was the difficulty of the perpetual and necessary search for new preparation techniques and, frequently, the frustrating uncertainty of what was seen. Thus attitudes too insubstantial to impinge upon the commonplace observations in a flea glass became a serious challenge to the conviction and sense of purpose needed to sustain an extended and often wearying effort.

To be sure, those who finally launched the age of early microscopic discovery in the Netherlands interacted critically with one or more of the groups of scholars or craftsmen who, while focusing on nature's minutiae, nurtured outlooks prejudicial to the microscope. But in a significant sense, those pioneering microscopists were also socially or emotionally outsiders to these groups, and it was rather on the periphery that, stimulated as well by foreign influences, they applied themselves to the microscope with a persisting intensity. That very marginality, indeed, did more than fend off the biases of more traditional pursuits; it focused their commitment to the microscope and helped engender the intensity and perseverance that opened the way to discovery.

125 Muys (1738), 182–3, 222–3, 287; Hacking (1983), 189, 191; Polanyi (1969), 106–7.

CHAPTER FIVE

Swammerdam

Jan Swammerdam was a distressed soul. In later years, he repeatedly lamented the misery of human life, which he likened to a dark and fleeting night.[1] Of either the joys or trials of his youth we know very little, but he later remarked in passing that childhood and its miseries derived from the Fall and the curse of God.[2] Though also a source of absorbing pleasure, his scientfic research testifies to a driven man continually striving against the brevity of time and the inescapable incompleteness of all he undertook.[3] His research thrilled him with the awareness of his own achievement, but at the same time it perpetually reminded him of the hopeless ignorance and impotence of humanity. Although he believed, indeed, that it brought him face to face with things divine, he convinced himself that it also entailed deep spiritual peril.

His anguish was not without its positive consequences, however. The compounding of his restless drive and emotional difficulties influenced the direction of his scientific efforts and led to his pathbreaking use of the lens. Had Swammerdam never taken up the microscope, he would have remained nonetheless a master anatomist, a brilliant experimenter in physiology, and an extraordinary naturalist. He had a compulsion to probe and query nature's varied forms and delighted in the invention and mastery of techniques. He was an intense and creative scientist with or without the microscope. Together with the example and challenge of Malpighi, however, the stresses of his emotional life focused his efforts on an unconventional field of research, the anatomy of insects and other small animals, that not only called forth a fuller development of his remarkable skills but also drew him into an unprecedented program of persistent and innovative microscopic discovery.

Although the term "insect" was still used very loosely in the seventeenth and even eighteenth century, it nevertheless suggested a very distinctive form of

1 For instance, see JS (1675a), "Ernstige aanspraak," [1]–[2], 3, 15; also JS (1737–8), 2:592–3, 598, 705; *LJS* 114, 116, 120 (see also p. 148).

2 JS (1737–8), 2:593.

3 See, for instance: JS (1669), pt. 1:106, 129–30, pt. 2:1; idem (1675a), 70, 86–7; idem (1737–8), 1:57, 175, 193, 334, 2:465–6, 601, 634, 659, 836; *LJS* 53, 131–2.

animal life.[4] It was commonly held to be "imperfect" in some problematic way and marked by both a radical simplicity of organic structure and a susceptibility to spontaneous generation. Aristotle and Pliny were the classical authorities most often cited, and the former granted the insects something analogous to a heart (and sometimes perhaps even more than one), a gut running from mouth to anus, and a "sack" of sorts that served for a stomach, but otherwise no other viscera. Their flesh was also not quite flesh but something between flesh and shell, so that neither bone nor shell was needed; hence, according to Aristotle, the insect body was hard throughout and, excepting a very thin skin, homogeneous. Aristotle also granted them eyes and something like a tongue, and Pliny, in some well-known lines, was inclined to grant them a full range of senses. Notwithstanding, Pliny otherwise agreed that insects had no other internal organs beyond a gut, and even that was sometimes missing.[5]

Such diverse influences as the tide of mechanism and the rush to glorify God's handiwork in all his creatures contributed in the seventeenth century to a very contrary drift as well, however. The mechanistic penchant allowed no function or process that was not produced by moving parts; hence, when conceived now as "narrow Engines" and "puny automata," insects acquired their own "curious Mechanism and organical Contrivance." Indeed, Robert Boyle and Pierre Gassendi cited the microscopic images of insects as a demonstration of atomism, since the tiny organs thus observed could only function, they reasoned, if composed of still smaller parts. In Holland at midcentury, Jan Fokkens Holwarda, professor of medicine at Franeker, similarly passed in his imagination from the beak, legs, and hairy body of a mite to the parts of its digestive tract and the membranes, muscles, heart, brain, blood vessels, and nerves it presumably contained, all in order to grasp the dimensions of atoms.[6]

By revealing the continuing differentiation and articulation of even the external parts of such minute bodies, the lens also prodded speculation about a

4 Aristotle used the term translated from the Greek as "insects" (from the Latin *insecta*, derived from *inseco*, "to cut into, cut up") explicitly to mean "such creatures as have notches on their bodies, either on their bellies or on both backs and bellies" (*Hist. An.* 1.1.487^{a}35; see also 4.1.523^{b}14–15). Réaumur would note, however, that the name *insectes* was no longer limited in his day to those animals whose bodies were composed of rings separated by a kind of cut or incision that first occasioned the name (Réaumur 1734–1929, 1:57). Unwilling even to limit it to animals of a small size, he himself was ready to extend the name to all animals whose forms did not fit the classifications of "ordinary" quadrupeds, birds, or fish; hence, he included not only slugs, earthworms, starfish, frogs, and toads among the insects, but was willing to gather in the crocodile as well, though a furious insect it would be, he conceded (ibid., 57–8). Stopping well short of crocodiles, Swammerdam nonetheless also counted slugs and frogs as insects (JS 1669, pt. 1:84; JS 1737–8, 2:828), although in Dutch he preferred "bloodless animals" as an equivalent designation, as in the alternative title of the *Historia insectorum generalis: Algemeene verhandeling van de bloedeloose dierkens*. "Bloodless," to be sure, alluded specifically to the absence of red blood, for an analogous or equivalent fluid, although perhaps of a pale or whitish color, was assumed in all animals (see *Parts An.* 2.1.648^{a}2–3, 6, 20–1; also JS (1737–8, 1:119–20).

5 Aldrovandi (1602), 3–4; Jonston (1657), preface [1]; Burgersdijck (1642), 260–1; *Parts An.* 2.1.647^{a}32, 3.650^{a}24, 5.651^{b}5, 8.653^{b}20–1, 8.654^{a}27–30, 17.661^{a}15–25, 3.4.665^{a}30–5, 4.5.678^{a}2–3 29, 682^{a}13–19; *Hist. An.* 4.7.531^{b}27 (cf. 532^{b}8–9), 4.7.532^{a}5–14, 4.7.532^{a}29–532^{b}3, 4.7.532^{b}5–8; Morge (1973), 41; Pliny *HN* 11.1.2, 2.7–8, 3.9–10.

6 Power (1664), preface [7], [10], and 58; R. Frank (1980), 91, 94–5; Holwarda (1651), 9–11.

similarly diminishing intricacy in internal anatomies.[7] But it jarred the imagination even more with images of the actual moving intestines and pulsing heart within the transparent flea or louse, a sight that continued to fascinate microscopic observers throughout the century.[8] Swammerdam pointed out that many other insects were similarly transparent, however, and Harvey, with a lens, had also observed the pulsating heart through the skin of wasps and hornets.[9] Together with the silkworm's remarkable role in a valued industry, it was very likely its transparency before metamorphosis – as clear as a crystal, noted Swammerdam – that, at the end of the preceding century, had inspired the German scholar Andreas Libavius to dissect the silkworm, its chrysalis, and the male and female moth.[10]

Nonetheless, despite the fuller recognition now of the powers of the lens – of which Libavius in 1599 had in fact made no mention – neither the picturesque speculations about minute anatomies nor the images of working innards in transparent bodies inspired any further efforts comparable to those of Libavius for over half a century. On rare occasions, anatomists had been urged to apply their skills to insects, but the attempts to do so remained few and superficial and made little real use of the microscope.[11] Nor were the devotees of the instrument itself quick to try their hand at such anatomies. Hooke, who in addition followed Odierna's example in undertaking the dissection of insect eyes, sliced open the abdomens of flies in search of the vessels found in larger animals and, discovering them in abundance, concluded that there was no less "curious contrivance" here than in the larger animals; but his forays were abrupt and crude, and further thoughts of dissecting a spider and a mite remained just thoughts, apparently deemed unrealistic.[12] Even the lure of "excellent contrivances" and "internal curiosities" to be discovered could not overcome doubts, it would seem, about how much more could be achieved in insect anatomies even with the lens.[13]

Nor did more compelling reasons for undertaking such a difficult and unusual enterprise come readily to hand, although animal dissections were fundamental to the anatomical and physiological advances of the seventeenth century. Lead-

7 Kircher (1658), 45; Borel (1656), 16–17; Goedaert (1700), 3:89–90.

8 Moffett et al. (1634) 276, "Epistola," [6]; Fontana (1646), 148–9; Power (1934), 71–2; idem (1664), 9, 59; Hooke (1665), 212–13; AvL (1694), 589; JS (1669), pt. 1:72–3; idem (1737–8), 1:57–8, 75, 77, 79–80.

9 JS (1675a), 78; Harvey (1628), 28, 64.

10 JS (1675a), 78; Libavius (1632), 377–405. To be sure, Libavius had also seen the pulsing heart through the transparent skin of the silkworm before dissecting it, although he failed to recognize it as indeed the heart (ibid. 383, 387). Thomas Moffett and Ulisse Aldrovandi had also crudely depicted the silk-producing organs of the silkworm (Moffett et al. 1634, 181; Aldrovandi 1602, 282; Freeman 1962, 175).

11 Moffett et al. (1634), "Epistola," [4]–[5]; Severino (1645), 343–7; Cole (1944), 140, 148.

12 Hooke [1665], 177, 184, 199, 205–6, 208. Henry Power also cut open diverse insects, in some cases by simply cutting off the head and observing the beating heart even with the naked eye (Power 1664, 4, 7, 24, 30).

13 Hooke (1665), 208.

ing researchers stressed the importance of comparative anatomy,[14] and Blanckaert's insistence that the science of anatomy required the dissection of all kinds of animals was hence widely accepted as a commonplace truism; but so too was his implicit assumption that the ultimate concern of that science was still knowledge pertinent to the study of humankind.[15] Thus for comparative anatomy's usual rationale to provide a stimulating incentive to insect dissection, some initial assumptions would have had to be stretched to the breaking point.

The recourse to animal dissection generally presupposed basic anatomical and physiological similarities in man and animals, and the attitude was often reflected in early insect dissections as well. Whereas analogies between humans and the higher animals might be difficult not to see, however, insects demanded decidedly more imagination. To be sure, early microscopic observers were quick to evoke the larger animals in their descriptions of the new magnified images of insects (and Odierna's comparison of the proboscis of the fly to the elephant's trunk found graphic expression in his illustrations). More specific to anatomy, Leeuwenhoek later remarked that, with the microscope, animal structure was more clearly and conveniently explored in the smaller creatures than in the larger, whereas Malpighi explained his own dissection of insects as a search for an anatomical simplicity that could shed light on the complexity of higher animals.[16]

That the insides of insects might have much to reveal about the insides of humans was far from obvious to all, however. Hartsoeker would protest against Leeuwenhoek's researches that the comparative anatomy of insects, unlike that of the larger animals, had little to offer man and medicine, for even those few insect parts that could be discovered had no relation to the parts of the human body. It was an outlook Ruysch seems to have taken to be the prevailing one. During the debate over the nature of the glands, he once responded that if Malpighi in his comparative anatomy began with the smallest animals, he himself would begin with the largest, and who was proceeding more soundly, he wrote with an apparent touch of smugness, the reader could decide. The lesson to be learned from what Malpighi had found in insects, added Ruysch, was rather that God had constructed the viscera of animals in infinitely diverse ways.[17] From such a perspective, insect anatomies themselves testified to their irrelevance to a science ultimately concerned with humanity.

14 Without it, wrote Swammerdam, the true knowledge of the function of the entrails would be beyond us, for what is obscure and difficult to observe in one animal is often seen clearly in another (JS 1737–8, 2:499, 830). R. Frank (1980), 16; Cole (1944), 12.

15 Blanckaert (1686), 2; Roger (1980), 258–9; Cole (1944), 23.

16 Kircher (1646), 834; Borel (1656), 16. Regarding Odierna, see Singer (1915), 333 and fig. 45; AvL (1718), 140; Malpighi (1686–7), *Anat. plant. idea,* 1. See also JS (1737–8), 2:855; idem (1669), pt. 1:121.

17 Hartsoeker (1730), "Ext. crit." 32, 42, 61; idem (1710), 83; Boerhaave & Ruysch (1751), 68–9. Lyonet, the greatest insect anatomist of the eighteenth century (Freeman 1962, 175, 178), would also pointedly warn against the dangers of analogies drawn between insects and higher animals (Lyonet 1762, xiii; see also xi, 77–9; Lesser 1745, 2:89n).

The radical difference between the eyes of insects and of higher animals prompted Swammerdam as well to attest that God could achieve his goals in thousands of different ways,[18] but Swammerdam, like Leeuwenhoek and Malpighi, was also susceptible to a very contrary perception. On other occasions, he saw the anatomy of insects as testifying rather to an analogy among all creatures,[19] and so much so that at one point he proposed that God had formed but a single animal, hidden though it was in an infinite variety of distortions.[20] The insect eye underscored just how precarious that perception was, but Swammerdam's simultaneous interest in both comparative anatomy and entomology did lead, after all, to some early and important initiatives in insect dissection.

The context was the research on animal reproduction that was of such acute interest to the members or former members of his Leiden circle.[21] It was while working at Leiden with Horne that Swammerdam, having turned to dissecting bees, made the quite startling discovery – since their sex life had been a mystery for centuries – of egg masses within what had almost universally been assumed to be a "king" not a queen.[22] Although he searched in vain for the sex organs of the worker, by 1668 he had also discovered the testicles within the drone, and he displayed them together with the penis in a collection of insects and anatomical parts that now boasted egg masses from grasshoppers, mayflies, and dragonflies.[23]

A preoccupation throughout his life, that collection itself testified to the depth of Swammerdam's passion for insects and his fascination with working on their bodies. By the end of the 1660s, the collection already contained some twelve hundred items that, according to Christiaan Huygens, included some astonish-

18 JS (1737–8), 2:397, 405, 501–2; see also idem (1675a), 119–20.

19 JS (1737–8), 1:232; 2:368; see also 2:855 and idem (1669), pt. 1:121.

20 JS (1737–8), 2:713.

21 Although eggs had long been observed within insects (see Moffett et al. 1634, 276; Libavius 1632, 401; Severino 1645, 345–6), Swammerdam would later remark that it was the discovery of eggs in human ovaries that first led him to think that eggs were to be found in all animals (JS 1737–8, 1:305).

22 JS (1669), pt. 1:[105]. Dirck Outgerszoon Cluyt had already affirmed, to be sure, that the "king" did lay the "seed," *saet,* of the bees in the cells of the comb (1597, 10); yet Thévenot, for one, still found Swammerdam's discovery of the eggs within the queen a most surprising revelation (Harcourt Brown, 280–1).

23 Regarding the sex organs of the drone bee, see JS (1669), pt. 1:[105]; idem (1737–8), 1:275, 2:456, 506; cf. Harcourt Brown (1934), 280–1. With respect to other organs and other insects, see JS (1669), pt. 1:93, 97, [105]–6, 129, pt. 2:18.

In the text above, "egg masses" is a rendering of the word *eyerstok,* or *eierstok* in modern orthography. The standard translation is now "ovary," but Swammerdam likens the *eyerstok* of the mayfly and dragonfly to the "roe," *kuit,* of fish and relates how both insects "shoot" their *eyerstok* into the water (JS 1669, pt. 1:97, and pt. 2:18; idem 1675a, 5–6; see as well the longish but ill-defined "double" *eyerstok* of the dragonfly illustrated in JS 1669, pl. 8 [fig. 1]). Whereas Swammerdam also wrote of having found the *eyerstok* in the queen bee, Thévenot, to whom Swammerdam had shown this discovery, described it as the discovery of "un grand nombre de ces oeufs" (Harcourt Brown 1934, 280, 281).

For a listing of the principal anatomical parts from man and other higher animals in Swammerdam's collection, consult Sinia (1878), 70–5.

ing things previously unseen.[24] By Swammerdam's own account, the creation and continuing expansion of this collection exacted a heavy toll in time, trouble, and expense,[25] but it also indulged his passion for experimenting with ingenious techniques and exploring his considerable manipulative skills. Indeed, his method of wax injection very likely derived from a technique developed for preparing caterpillars for display.[26]

Apart from the pressing question of animal reproduction, however, Swammerdam does not appear to have entered as yet in the 1660s upon any systematic program of insect dissection, and it is even more improbable that any such effort seriously involved the microscope. Besides sex organs and egg masses, further bits and pieces of insect anatomy in his collection – such as the intestines and stomach of the butterfly – may have been but added dividends of the search for reproductive organs, whereas other items were perhaps no more than the by-products of Swammerdam's preparation techniques. Those techniques often involved the removal of the insides of insects, even if only in a crude and summary way, and the organs, for instance, of the alimentary canal of the grasshopper in Swammerdam's collection may reflect little more than his discovery that, to preserve that insect in particular, he had to remove its entrails.[27] Consequently, such anatomical parts are not clear evidence of the purposeful practice of insect dissection.

Nor in 1669 did Swammerdam refer to the microscope when speaking of his early forays into insect anatomies in his *Historia insectorum generalis;* in this, his first major work, he emphasized rather the easy visibility of such anatomical parts as the sex organs of the bee.[28] He did remark in passing on the prettiness of the other entrails in the bee, but declined to elaborate further since he intended to deal with the bee as the subject of a separate treatise. His brief description of that future treatise, however, indicated only that it would deal with the eggs, the larvae, and their metamorphosis. Although he promised "infinite further details [*naukeurigheeden*]," nothing else suggested that a fuller account of the bee's internal anatomy was yet in hand or anticipated.[29]

24 JS (1669), pt. 1:168; *OCCH* 7:45; see also *LJS* 57.

25 *LJS* 70. Swammerdam also asserted that by 1673 his collection included some three thousand insect specimens (JS 1737–8, 2:534).

26 JS (1672), 43; idem (1737–8), 1:318. Swammerdam also apparently tried injecting insects with poison (F. Schrader 1679, §X). An admiring Boerhaave later briefly described some of Swammerdam's techniques for preserving as well as dissecting insects (JS 1737–8, "Het leven van den schryver," [8]).

27 JS (1669), pt. 1:93, 129; Nordström (1954–5), 30–1. Swammerdam also speaks of making a small hole in the larva of the rhinoceros beetle and simply squeezing out the intestines in order to prepare it for preservation (JS 1737–8, 1:318).

28 JS (1669), pt. 1:[105]. Regarding the testicles of the drone, see also JS (1737–8), 2:456, 506; and Harcourt Brown (1934), 280–1. Swammerdam also stresses the visibility of the "lungs" – the abdominal air sacs (Cole 1944, 291) – of the bee (JS 1669, pt. 1:[105]) and the easy recognizability of parts within the digestive system of the grasshopper (ibid., pt. 1:93).

29 JS (1669), pt. 1:106. Swammerdam remarks here as well that he had been unable to give as much time as he had anticipated to preparing the *Historia insectorum generalis* for publication, but he speaks in this context only of "reasonings," *redenen,* he was consequently compelled to abbreviate and not of observations omitted.

Swammerdam noted as well that he had observed the muscles in grasshoppers and beetles and had been struck by the similarity of their structure (*maaksel*) to what Steno had found in larger animals; but what aspects of the muscles he had in mind – there is no explicit mention even of fibers – is by no means clear.[30] The one microscopic observation of hidden muscles (as well as veins and entrails) offered in the *Historia* is that of the transparent louse, which Swammerdam cites (as had Hooke) as testifying to the utility of the microscope in revealing the working organs without dissection. Swammerdam also provided magnified illustrations of other insects and their eggs, but apart from the egg masses from the body of the dragonfly – and these are shown as apparently seen with the naked eye – there are no figures of internal anatomies or their parts.[31]

Swammerdam was among the many who now marveled over imaginings of minute anatomies, but when he wrote the *Historia* he still emphasized not the prospect but the apparent hopelessness of ever actually observing one. The disposition of the limbs, muscles, veins, and nerves in the anatomy of the larger animals was astonishing enough, he wrote, but to find the same in animals whose whole bodies were smaller than the point of one's knife was stupefying. In fact, Swammerdam had yet to face so small an anatomy and really did not expect to. Since we lack the eye and the hand for even the slightest dissection of

30 Ibid., pt. 1:121; cf. his later comments on muscle tissue from the louse and the frog (JS 1737–8, 1:70, 2:834).

31 JS (1669), pt. 1:72–3, pl. 7 (fig. 6), pl. 8 (fig. 1); Hooke (1665), 186. To be sure, Boerhaave would later write that, in Thévenot's gatherings in France in what must have been 1664 or 1665, Swammerdam had already amazed his onlookers by dissecting "little animals" and thus displaying their entrails (JS 1737–8, "Het leven van den schryver," [3]). It is strange, then, that the journal of Olaus Borrichius, who was in Paris at the same time and commented in that journal both on Swammerdam's doings and on the meetings of Thévenot's group, fails to allude to such dissections (Nordström 1954–5, 36). Johan Nordström, who has edited excerpts from Borrichius's journal, nonetheless declares that Boerhaave's statement must be true because Swammerdam himself speaks of his "dissection" (*dissection* in French) of insects in a letter to Thévenot in 1665 (ibid., 36, 40). But Swammerdam also associates "dissection" (now the Dutch *ontleeding*) with not only the simple exposure of eggs within the female insect (JS 1669, pt. 2:25, 32) but, more important, with the removal of the outer integument of pupae and chrysalides to reveal the future imagoes within (ibid., pt. 1:20, 153) – an operation to which Swammerdam attached the greatest significance during these years (as is subsequently noted in the text) and one that, he says in the *Historia insectorum generalis,* he had indeed already demonstrated for Thévenot (ibid., pt. 1:26–7, pt. 2:41–2). Thévenot has himself been persuasively identified as the author of an anonymous letter dated 1669 in which he is clearly writing that, in the *Historia* (although both book and author also remain unnamed), Swammerdam had now published many experiments he had indeed earlier shown before Thévenot's gatherings; but the letter also indicates that the most advanced dissection in that book, the investigations of the sex organs of the bee, was in fact a much more recent accomplishment (Harcourt Brown 1934, 280–1). Swammerdam himself said he had first discovered the eggs in the queen while working with Horne, which would have been after his return from France, and 1668 is the earliest year he mentions in which he showed the testicles in the drone to anyone (see nn. 22, 23, above). Consequently, not only is there no compelling evidence that Swammerdam had in fact been performing true insect dissections in the mid-1660s in France (i.e., beyond the disclosure of the imago within the pupa), but also there is good reason to believe that, in his work on the bee, the *Historia* in fact presented the most advanced point he had reached in the practice of such dissections.

these parts, he continued more realistically, that inner form will remain beyond our reach.[32]

Although the interest in comparative anatomy had prompted Swammerdam to turn to insect dissections, those efforts had as yet produced no significant initiatives with the microscope and showed little prospect of doing so. Moreover, the incentive provided by comparative anatomy focused only on specific questions, notably regarding the reproductive organs, that had been raised concerning the anatomy of humans and higher animals, and there is little in the *Historia* to suggest that Swammerdam as yet even contemplated a general, systematic program of insect dissection with or without the aid of the lens. Hence, in 1669, despite the promising mix of Swammerdam's interests in entomology, anatomy, and the microscope, he had yet to venture into that field of potentially endless microscopic exploration in small anatomies. Discovering that prospect would require both a headier vision of what was possible and an intensity of commitment capable of sustaining a persisting assault on formidable technical obstacles. Malpighi inspired the vision, and the ordeals of Swammerdam's troubled personality produced the focused passion.

The *Historia insectorum generalis* was in press when Swammerdam in 1669 received a copy of Malpighi's *Dissertatio epistolica de bombyce,* published that same year by the Royal Society. In that remarkable work, Malpighi joined a study of the life history of the silkworm to an account and illustrations of the anatomy of both the moth and the caterpillar. Swammerdam was deeply impressed.[33] He was also driven to repeat and, if possible, surpass Malpighi's dissections.

The attempt to duplicate those dissections proved a profoundly educational experience for Swammerdam. Malpighi himself later stressed the great difficulty and wearisomeness of the undertaking, for it was completely new, he wrote, and the anatomical parts so small, fragile, and intertwined as to demand their own special method of dissection; it was so exhausting that after several months his eyes became inflamed and he succumbed to fevers. Swammerdam likewise described it as the most trying kind of dissection. Indeed, at first it seemed hopeless, he confessed, for he was ignorant of Malpighi's method; but in time and through chance he discovered a method of his own.[34]

The silkworm was only the beginning, however, and as Swammerdam tried his newly acquired techniques on other insects his skills developed rapidly. He dissected the nymph of the mayfly in 1670 with great finesse; but when five years later he published his *Ephemeri vita* on the mayfly, he regretted that, with his increased knowledge and skill, he could not do it all over again.[35] He

32 JS (1669), pt. 1:2. The Latin version in JS (1737–8), 1:1–2, was also used for this abbreviated rendering of Swammerdam's passage.

33 JS (1737–8), 2:554, see also pp. 520, 704; Cole (1944), 183–97; Freeman (1962), 176.

34 Malpighi (1697b), 56 in the first series in the pagination; JS (1672), 16, 18.

35 JS (1675a), 86–7; see Cole (1944), 278, 280. Regarding his increased skills, see also: JS (1737–8), 1:194; *LJS* 138–9.

would later stress the "ingenious inventions" (*kunstige uytvindingen*) – presumably new techniques – he had to devise and the variety of aids (*hulpmiddelen*) to which he was forced to turn, among them being necessarily now the microscope (of which Malpighi, however, had made little mention in his own *De bombyce*).[36] Swammerdam now dissected directly below a lens, and even the lancets and styluses he used – though a fine pair of scissors was his key instrument – were so small that he sharpened them under the lens as well. The more delicate and difficult parts of an insect's anatomy he removed from the body and placed on a small, sometimes colored piece of glass "as thin as can be blown at the lamp." The bit of insect anatomy having dried on the fragment of glass, it was pasted to a bit of cork and the cork stuck on a needle's point, there to be observed more closely with the microscope.[37]

Such fine dissections allowed Swammerdam's talents their fullest expression. It was above all in the anatomy of insects that he had excelled and surpassed all others, declared his successor Pierre Lyonet, adding that Swammerdam's dexterity in such dissections "exceeds the imagination and partakes of the prodigious." In the seventeenth century, only Malpighi approached his achievement, and in the eighteenth, only Lyonet himself. Hence, it was with a new and uncommon authority that Swammerdam reaffirmed that as much was locked within the narrow confines of the least of animals as in the colossal viscera of the greatest. Experience of an unprecedented kind now infused Swammerdam's insistence that nature had to be explored in all her littleness in order to reveal her greatness; "and though her subtle smallness may threaten to turn us back, we must spare no labor, for the smaller she is, the greater and more gloriously she resides in her invisible parts. . . ."[38]

Despite his desire to redo it all, his description of the mayfly nymph in the *Ephemeri vita* already testified to exquisite dissection; yet the study he himself prized above all others was the treatise on the bee he had promised in the *Historia*. Over the course of the following years, it now came indeed to include the most accurate and detailed of his insect anatomies. Whereas in the *Historia* he had reported, for instance, only a mass of eggs in the queen, he now offered a precise description of the ovaries with their strings of developing eggs – the oocytes in the ovarioles – in both the mature and immature queen (Fig. 13). His anatomical studies of the mayfly and the bee are but the masterpieces among a host of diverse dissections of other insects, however, and he found precise though diminishing anatomical structures everywhere.[39]

36 JS (1737–8), 1:300. Swammerdam now explicitly declared that all the parts of the mayfly nymph he described were observed and drawn through the microscope (JS 1675a, 413–14).

37 JS (1675a), "Ernstige aanspraak," [11], 72–3, 415; idem (1737–8), 1:70, 2:373–4, 408, 410–11, 455, 656.

38 Lesser (1745), 1:41n; JS (1737–8), 1:300. F. J. Cole also includes Leeuwenhoek with Malpighi, Swammerdam, and Lyonet to make an exceptional foursome (Cole 1951), but cf. Freeman (1962), 175, 178.

39 Cole (1944), 280, 288, 291; Sinia (1878), 141, 144, 149. On his deathbed, Swammerdam singled out his treatise on the bee to urge that it be published in Dutch (Swammerdam's "Testament," 25 Jan. 1680, JS Papers, fol. 1v).

Epitomizing such newfound structures, and in the true insects encountered literally everywhere, were the fine and intricate networks of air vessels, the tracheae. Hooke as well as Swammerdam had earlier remarked on the many "milk-white" or "silver" vessels within the transparent louse, and while Hooke also found them abundant in the abdomens of flies, Swammerdam in his *Historia* remarked on the silver-white threads woven through the egg mass of a grasshopper. Like Hooke, he was initially inclined to take them for veins and arteries. Malpighi, however, describing the ramifying system in the silkworm, noted how they were strengthened by solid rings to prevent their collapse and deduced their true respiratory function.[40]

Swammerdam was astonished by the pervasiveness and abundance of these vessels. Through the microscope, he recounted, the whole abdomen of the mayfly nymph appeared shot with silver threads that, spreading throughout the thorax, head, legs, and wings of the adult fly as well, supplied the intestines, muscles, nerves, and indeed all the furthest parts of the insect's body with air. Not only were they thick about the eggs and ovary of the queen bee and the testicles of the drone, they were intertwined about the components (the ommatidia) of the bee's compound eye and perhaps, thought Swammerdam, composed the very framework of its grid of hexagonal shapes.[41] The outer membrane of the stomach of the louse was likewise endowed with an "inexpressibly great" number of these "lung pipes," the smallest of which could be seen only with a good microscope. Whole systems of these vessels seemed in fact to recede into inaccessible smallness, for through the microscope he saw them shining within the bodies of "worms" so small that they were nearly invisible themselves, he wrote – worms he had found as parasites within the body of a larva of the cheese skipper fly (*Piophila càsei* [Linnaeus]), itself less than a centimeter long.[42]

The structure of the tracheal vessel proved a marvel in itself. As Swammerdam described it, an individual vessel was composed of delicate membranes wrapped about a stiff, tightly wound spiraling thread (the taenidium) that kept the vessels open even when wrenched about in the body of a writhing larva. To explore that structure further, Swammerdam threaded a hair through a segment of vessel and, fastening down the ends of the hair with wax, pulled apart the spiral rings with pincers and fine needles. He stretched out a piece of such a spiral to some two feet in length, so that it looked, he said, like silver thread that had been wound about a needle.[43] All in all, the glory of the finest human weavers paled beside the structure of the tracheal vessel, declared Swammerdam, and he was only further astounded to discover that, when the silkworm

40 Hooke (1665), 184, 212–13; JS (1669), pt. 1:93, pt. 2:17; Malpighi (1686–7), *Diss. epist. de bombyce*, 12–13.
41 JS (1675a), 75–6, 84; idem (1737–8), 1:71, 249–50, 2:473, 494–5, 507, 662.
42 JS (1737–8), 1:76, 2:708.
43 JS (1675a), 80; idem (1737–8), 1:339, 2:410, 412, 663, 704–5.

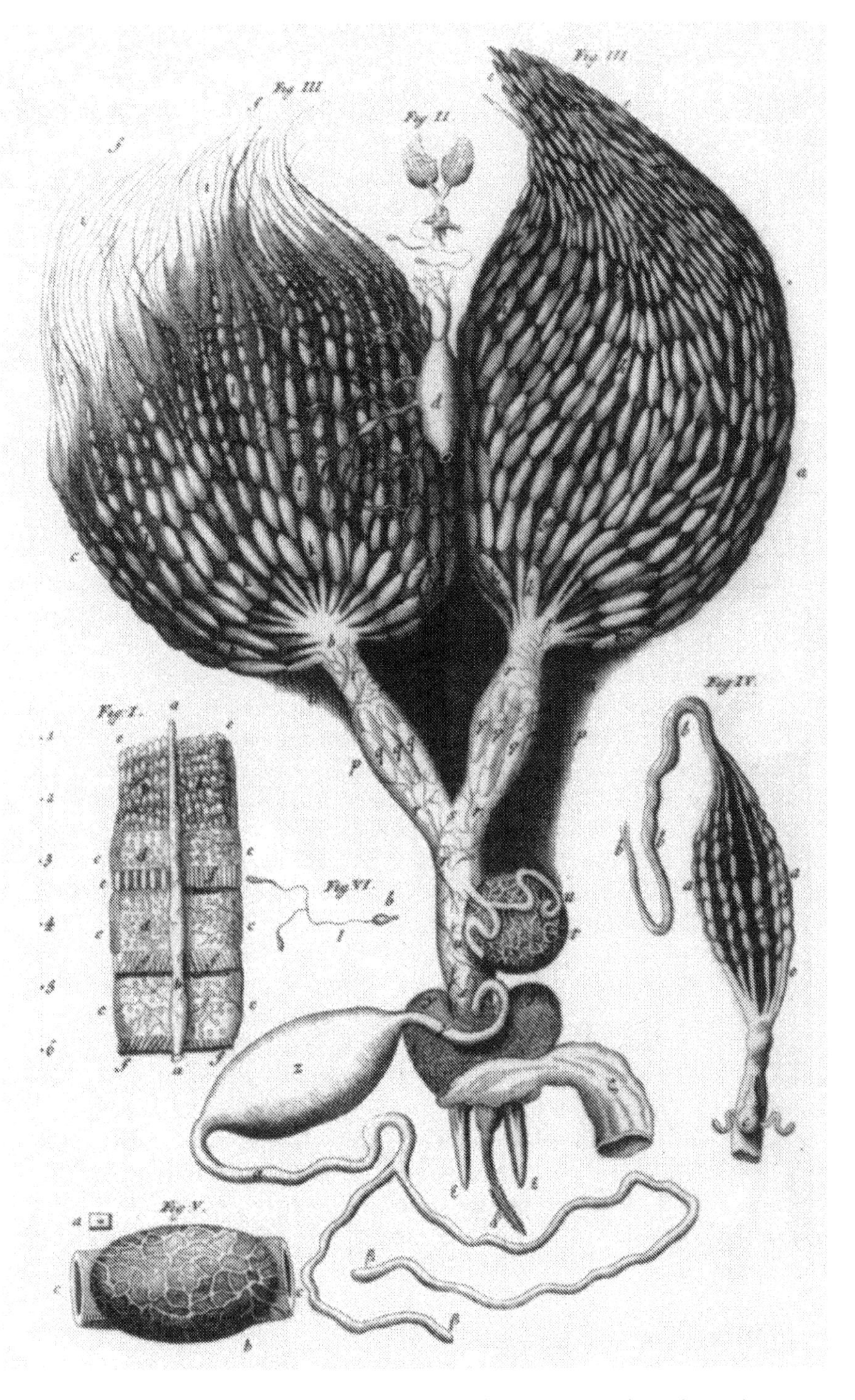

Figure 13. Fig. III represents the ovaries of a mature (on the right) and immature queen bee by Swammerdam (JS 1737–8, 2:pl. 19, fig. 3). Fig. II represents the ovaries as they appear to the naked eye. Photo courtesy of the Department of Special Collections, University of Chicago Library.

and other larvae molted, these vessels shed a delicate inner membrane that still bore the impress of the spirals in their walls.[44]

But such extraordinary observations entailed extraordinary labors, and, wondrous as he found them, the tracheae themselves burdened Swammerdam's dissections with exceptional difficulties. Densely interwoven with the testicles of the drone, they posed a major obstacle to his investigation of these organs in the bee, and so similarly entangled were they about the intestinal system that Swammerdam anticipated months of work to overcome them.[45] Indeed, he wrote, the study of the bee in particular involved a period of unrelenting research, from 5:30 in the morning to noon without a break for months, and then, after the dissections of the day, he took up his pen at night. His skills continued to grow, however, and few of his studies were so extended: He seems to have completed his later study of the louse in just six days, although he dissected no fewer than forty specimens.[46] Nevertheless, his aspirations kept pace with his skills so that his periods of concentrated research appear to have remained no less intense. In later years, he still wrote of sweating over his dissections in the blazing sun until the rain and darkness of autumn drove him from the fields and denied him the light he needed for dissections he continued in his room.[47]

Having shown what could be done, Malpighi himself had soon turned away from insect dissection;[48] why, then, did Swammerdam persist with such ardent commitment in a field of such difficult and unorthodox research? Swammerdam would surely have answered that it was further to reveal the glory of God, for he was virtually obsessed with the religious significance of scientific research and rehearsed the familiar themes with greater persistence than perhaps any other major scientist of his day. The only goal of scientific labors, he repeatedly insisted, was to demonstrate the attributes of God in his works so that mankind would revere and glorify him all the more. For this reason alone these works were given to us to contemplate, and we were bound, hence, to use them to climb toward their author, no matter how small or trifling they might be.[49]

Swammerdam especially embraced the theme that God was as great and almighty in his smallest creations as in the large, as astonishing in the louse as in Behemoth and Leviathan.[50] It was thus the fulfillment of what Swammerdam insisted was the only purpose of his science when, shown the molted membrane shed by the tracheae of the silkworm, his good friend Johan Oort cried out (or so Swammerdam assures us): "Oh wondrous God! Who would not know you from this and, knowing you, not love you!" Swammerdam was indeed confiden that those who denied high providence would be thrown into confusion by

44 JS (1737–8), 1:301, 309; idem (1672), 17; idem (1675a), 80–1.
45 JS (1737–8), 2:454, 507, see also p. 588.
46 JS (1675a), 248–9; idem (1737–8), 1:82–3, 84, 2:518.
47 *LJS* 110, 111, 138; JS (1737–8), 2:595, 597, 601.
48 Malpighi (1686–7), *Anat. plant. idea,* 1.
49 See, for instance, JS (1737–8), 1:57, 65, 193, 2:502–3, 505, 836; idem (1675a), "Ernstige aanspraak," [3], 70–1.
50 JS (1737–8), 1:301, 332, 2:394; idem (1672), 21; idem (1675a), 109–10.

the tracheae in the mayfly nymph and abandon their arrogant reasonings before the structure – as he put it – of a worm.[51] Having followed Malpighi more deeply into the innards of insects, Swammerdam had discovered how forceful a witness was to be found in that elaborate intricacy.

Other more orthodox avenues of research were also revealing reflections of divinity in nature's finer constructions, however. The anatomy of large bodies offered a subtlety of construction that Swammerdam likewise extolled as evidence of their Maker, and, even when well advanced in his program of microscopic dissection, he paused in the midst of acclaiming God's handiwork in small animals to acknowledge that their entire bodies in fact contained fewer wonders than did a single particle of their larger counterparts. The human uterus he prized as an example of wax injection presented the small parts of the body in such a manner, said Swammerdam, that it was as if one could touch God's incomprehensible wisdom with their hands.[52]

Nor was their anatomy the only aspect of insects that stirred Swammerdam's religious sensibilities. To the contrary, insect life suggested divinity in many different ways, from the religious symbolism he also ascribed to metamorphosis,[53] to the social behavior of bees, to the patterned regularity of insect life cycles that manifested God's providence and omnipotence.[54] The unfolding of a butterfly's wings upon emergence from the chrysalis he considered one of the rarest sights in nature; but why shouldn't it be astonishing, he asked, since it came from God, who thus makes himself visible and palpable in his works. "And one may and must truthfully say," he continued, "that our reason and understanding feel and grasp God better if we can feel and handle tangible things with our natural limbs" – a facet of experience that was lost, to be sure, in the realm of microscopic anatomy.[55]

Swammerdam's identification of science with piety vested his insect anatomies with heightened meaning (about which more will be said) but did not distinguish those anatomies from other fields of research that had engaged him for years. The yearning to discover God in his creatures was too all-encompassing and too indifferently responsive to diverse fields of research to refocus Swammerdam's energies as was needed on a new field of insect anatomy. Moreover, his thoughts about the religious implications of his science would in time yield troubling contrary currents, and he himself would call into question the sincerity of his insistent avowals of religious motivation. In baring his religious anguish, however, he pointed to other influences shaping his dedication to microscopic dissection that were potentially more decisively selective.

During the very years, the early 1670s, in which he was first honing his new skills in insect dissection, Swammerdam was gripped by emotional difficulties

51 JS (1675a), 78–9, 81.
52 JS (1672), 1, 12, conclusion to appendix; idem (1675a), "Ernstige aanspraak," [12].
53 JS (1737–8), 1:207–8, 333, 2:593–4, 571–2, 666–7, 684, 829.
54 Ibid. 2:394, 625; idem (1669), pt. 1:57, 147.
55 JS (1737–8), 2:425; idem (1675a), 115; see also idem (1669), pt. 2:48.

that assumed the form of a religious crisis. The despondency and spiritual panic that overcame him led to the renunciation and abandonment of his science. He not only now acknowledged that he had pursued his researches for purposes other than piety, but perceived those other purposes as so powerful and ascendant as to pervert his scientific labors and render depraved what should have been sanctifying. His self-accusations starkly exposed the turbulent interplay of diverse and often conflicting emotions that underlay the intensity of his program of microscopic research.

Swammerdam never qualified his insistence on what science ought to be; if anything, he was increasingly unwilling to accept it on any other than religious grounds. In the *Ephemeri vita,* in which he announced his abandonment of his researches, nature remained a "natural Bible" wherein were to be read "the great and amazing deeds of the Almighty." All created things were still for Swammerdam reflected images of God given to men so that they might see that reflection,[56] and it was God himself who also moved men to investigate and describe those creatures. Even his own talent for research Swammerdam continued to acknowledge as a gift from God.[57]

Nonetheless, Swammerdam now saw in his researches the danger of personal spiritual jeopardy, "So that you make me a dangerous proposal, Sir," he responded to an unwelcome entreaty, "when you press me to continue my inquisitive practices [*curieuse oeffeningen*]." He included "curiosity of the eyes" along with ambition, avarice, pride, and lust among the devil's snares, and a "hankering for learning [*weetenschappen*]" belonged as well (or so he implied in a rhetorical question) among the moral failings to which one who had Christ in his heart "and is nailed with Him to the cross" was not likely to succumb.[58] Man, to be sure, was entirely free to choose his calling, asserted Swammerdam, on the condition that he maintained his love of God; but because that love must be undivided,[59] Swammerdam's researches now placed him in mortal danger.

As Calvin had warned, Swammerdam had reached the point at which, by his own account, he pursued those researches for his own pleasure rather than for the love and glorification of God. Because of the unrelenting tempo of his investigations, Swammerdam explained with remorse, he had repeatedly neglected his prayers and other religious exercises, an omission that so distressed him that he often gave way to tears. "For it was as if a warring host were there within my spirit, the one party compelling me to cling to God, the other, with infinite arguments, to go on in my [pursuit of] curiosities [*in mijne curieusheeden voort te gaan*]."[60] He offered indeed a remarkable public confession:

56 JS (1675a), 209, 324–5.
57 Ibid., "Ernstige aanspraak," [15]–[16], 241.
58 JS (1675a), 180, 244–5, 284.
59 Ibid., 218, 241.
60 Ibid., 249. Regarding Swammerdam's use of the word *curieusheeden* at the time, see also ibid., 218, 219, where he lists such "curiosities" along with "sciences" (*weetenschappen*), "amusements," honors, riches, and even friends and fatherland among the distractions that competed for the love one owed to God. Years later, he would speak perhaps more narrowly of the *curieusheeden* of which his collection was composed (*LJS* 70).

For I am so fixed and attached to [these researches] that I talk about them, write about them, practice them, and long for them almost before I know what I am doing. My affections consequently fasten on them and rush to the things they love without giving a thought to God, my perfection [i.e., the effort to perfect himself?], or those who are close to me. Add to this that it was self-love and self-gratification that brought me to such a high level of knowledge in these things, for I have never aimed at money, the greatest idol of Christians today, but indeed at the idols of my own pleasure and delight and the honor and praise of men, for which I have striven night and day to surpass others and to raise myself above them with ingenious inventions and subtle techniques. I was driven to this still more by the wondrousness in these things that made me see, touch, and feel God on all sides, but I have not loved him purely and exclusively therein, but only incidentally and in so far as it gave me pleasure and a rationale among my friends and acquaintances for continuing and persevering in these foolish pursuits.[61]

Hence, though he still maintained that his scientific studies had brought him closer to God, Swammerdam decided to end those studies and to turn away from "the fruit of this forbidden tree of science [*weetenschappen*]" so that he might love God more completely.[62] His soul's ultimate choice was God alone, he wrote:

And I therefore no longer want to look for you in the woods, the groves, or in the mountains; in the fields, valleys, or among the hills; or in the rivers, waters, or the seas. I wish no more to seek you by turning nature upside down, nor [by searching] in the incomprehensible wonders of the embroidered entrails of animals, both great and small. For their inscrutable origin, oh my God, is in you alone.[63]

In his own mind, his science had been perverted by both the pleasure it provided and the ambition it aroused and served. These were the true motives that had shouldered God aside, he testified against himself, and now drove him so

61 "Want ick ben soo vast ende gekleeft aan de selve; dat ick daar van spreeck, daar van schrijf, die oeffen ende die begeer, eer dat ick het haast merck of weet: soo dat mijne geneegentheeden haar hechten ende loopen tot de dingen die sy beminnen, sonder dat ick aan Godt, mijn volmaacktheyt, of mijn naasten kom te dencken. Doet hier nu by, dat de eyge liefde, ende mijn voldoeningh, my tot sulck een hoogen trap van kennis, in de selve gevoert heeft; want het gelt, den grootsten Afgodt der huydige Christenen, heb ick daar nooyt in beooght, maar wel de Afgooden [*sic*] van mijn eyge plaisier ende vermaack; te gelijck met de eer en lof der menschen; waar door ick by dagh ende nacht getracht hebbe, booven andere te willen uytmunten, ende om my met kunstige uytvindingen, ende subtiele handt-greepen, daar over te verheffen. Waar toe my noch meer aandreef de wonderlijckheyt deeser saacken, die my aan alle kanten Godt deeden sien, tasten ende voelen; maar die ick daar niet suyverlijck ende eenighlijck in bemint heb, maar alleen toevalligh, voor soo verre my dat vermaack gaf, ende dat ick daar door by mijne vrienden ende bekenden reeden hadt, om in die dwaase beesigheeden te blijven ende te volherden" (JS 1675a, 245–6).

62 Ibid., "Ernstige aanspraak," [9]–[10], 245. As a manifestation of this renunciation, Swammerdam destroyed the text of a treatise he himself had composed on the silkworm and sent the illustrations to Malpighi (Malpighi 1975, 2:715).

63 "Hierom soo en wil ick u dan niet meer soecken, in Bosschen, Bosschagien, of op Bergen; in Velden, Valleyen, of op Heuvelen; in Rivieren, Waateren of Zeen; ick en wil u oock niet meer soecken, in het onderste booven te keeren van de Natuur; ofte in de onbegrijpelijcke Wonderen der geboorduurde Ingewanden van de groote ofte de kleenste dierkens; want haaren ondersoeckelijcken Oorspronck is in u alleen, ô mijn Godt" (JS 1675a, 322–3).

compulsively to his research. So powerful was their grip that he saw no other recourse than to abandon his science altogether to be rid of them.

Of that driving ambition, however, Swammerdam's earlier years have left us little evidence. At Leiden, to be sure, his regard for niceties in matters of priority and his apparent circumspection in what he did and did not share with others bespoke a sensitivity, at the least, to the challenge of competition.[64] It was a sensitivity, however, that hardly set him apart, and certainly not in the circle in which he moved at Leiden. Nor does it yet suggest either the compulsive drive he would later lament or the jealous thirst for fame that others would also soon ascribe to him. He seemed amiable enough even when he let slip the suspicion that Ruysch had sneaked an unacknowledged look at Swammerdam's own drawings of the lymphatic valves.[65]

Contrary evidence in fact bespeaks a more unusual disinterest in acquiring a reputation or keeping up with rivals. Some years behind Steno, Ruysch, and de Graaf in publishing, Swammerdam brought out his first work, a dissertation on the mechanics of respiration, only when the procedures of graduation compelled him to do so in 1667. Following a more developed version of that dissertation and some anatomical studies published anonymously, the *Historia insectorum generalis* seems to have been elicited only by the importunings of Swammerdam's friend and patron, the French savant Melchisédech Thévenot. Although even Christiaan Huygens added his voice to those urging him to publish more,[66] Swammerdam continued to display a remarkable casualness about the future of his work. Since he had not thought seriously, he wrote, about publishing his original dissections of the mayfly nymph, by the time he did publish the *Ephemeri vita* some notes and drawings had been lost and his recollection of some aspects of his observations was consequently hazy. Having labored so hard on his study of the bee, he gave what he thought was the finished manuscript to an acquaintance in 1673 or 1674 and left it "to the care of God alone whether it would see the light of day."[67] At that time and in that form, it would not.

In the light of his approaching confession in the *Ephemeri vita*, this apparent indifference to publication is perplexing. Did Swammerdam merely succumb in that confession to the morbidly excessive self-accusation that often accompanies

64 Having earlier seen the "seminal tubules" of the testicles of a mouse in Steno's collection at Leiden in 1662, Swammerdam had later written Steno to inquire if he would take it ill if Horne dealt with related material in print before Steno himself (JS 1672, 51). Swammerdam also asserted that he was accustomed to communicating everything with his friends, including specifically his wax injection method (ibid., 43, 45, see also p. 38); however, Ruysch maintained otherwise (Ruysch 1725, 5), and de Graaf accused Swammerdam of even lying to him about techniques with which he was experimenting (Graaf 1677, 472–3). It must be remembered, however, that Ruysch was defending the originality of his own achievement, and that de Graaf was writing in the heat of a bitter exchange with Swammerdam.

65 Nordström (1954–5), 45.

66 JS (1667a, 1667b); Schierbeek (1947), 24, 103, see also p. 111; Harcourt Brown (1934), 280–1; *LJS* 57.

67 JS (1675a), 85–6, 120.

depression? Or did his attitudes toward his research suddenly and radically change during the course of the early 1670s? Alternatively, had the real force of his ambition simply been effectively concealed, after all, for over a decade? An argument can be made, indeed, that what seemed indifference in one significant instance at least – the abandonment of a completed treatise left in the hands of a second party – in fact entailed a personal strategy by which Swammerdam sought to deal with the ambition he found repellent in himself. His slowness in publishing may reflect as well less disinterest than an unwillingness to settle for what he found incomplete and imperfect, an attitude perhaps nurtured itself by the very scope of his ambition as well as by scientific and aesthetic sensibilities.

These are speculations of a most inconclusive kind, to be sure, and it is doubtless misguided to attempt too fine an understanding of behavior known to us so fragmentarily. Nonetheless, only a year before he left his manuscript on the bee in someone else's hands, Swammerdam had already betrayed his disappointment that a previous second party, Horne, had failed to make public some of his former student's discoveries in anatomy. In the last years of his life, moreover, Swammerdam would repeatedly speak of the accumulated insect studies he was finally readying for publication as his "great work," and there are indications that he had long been cultivating the grand aspirations the phrase suggests. Indeed, while parrying earlier urgings that he publish more, he had already vaguely written in 1670 of "my book," as yet unpublished but already adumbrated in his thoughts. Even in the *Historia* the year before, he had promised a future volume and further studies "if we can but find the time and opportunity for so great a work."[68] Thus, the very reach of his aspiration may have resisted the further publication of what he viewed as lesser, incomplete works in natural history.

The true scope and history of Swammerdam's ambition remains beyond our grasp, but, even apart from his own confession in the *Ephemeri vita,* some of his contemporaries also found his behavior in the early 1670s indicative of a vigorous and indeed larcenous lust for glory. Malpighi, for one, perceived at least an aggressively competitive spirit in Swammerdam's comparison of their dissections of the silkworm, even though the Dutchman had written admiringly of both Malpighi and his treatise. Having recounted how he himself had ultimately found his way through the difficult dissection, Swammerdam pointedly reaffirmed that the laurels still belonged to the "first and glorious discoverer," to whom Swammerdam declared himself still indebted for having revealed such wonders. In explaining why he chose to repeat Malpighi's dissection, however, Swammerdam declared that he accepted no one's word in matters of anatomy, nor would he blindly acquiesce in matters he thought he himself had some ability to explore. He made it clear that he had undertaken his own dissection in no small part to check the accuracy of Malpighi's plates, and in addition to offer-

68 See n. 71 below; *LJS* 70–1, 95, 132, 141, 57; JS (1669), pt. 2:1.

ing praises, hence, Swammerdam emphasized that he had found significant mistakes. Malpighi bristled at the tone and felt mistreated. Not only had Swammerdam alone criticized his book, he groused, but he had done so with uncalled-for harshness. Reminding his readers that he himself had indeed been the first to enter this new and unknown territory, Malpighi likened Swammerdam to a runner-up who nonetheless tried to snatch away the prize.[69]

Malpighi's offended response reflected in part his own touchy sensitivity,[70] but Swammerdam's comments on his treatise had appeared in the broader framework of a truly embittered outburst of resentment and jealousy in Swammerdam's *Miraculum naturae,* occasioned in 1672 not by Malpighi but by de Graaf's treatise on the female sexual organs. Animal reproduction had emerged as a prominent area of research, and Swammerdam had relied on Horne, with whom he had collaborated in the late 1660s, to make known Swammerdam's own achievements in the field. Horne, however, had died in 1670 having published no more on the subject than the letter he had hastily written in anticipation of de Graaf's earlier book on the male organs in 1668 – a letter wherein Swammerdam went unmentioned. In the *Miraculum naturae,* Swammerdam noted that earlier silence about his own part in Horne's research with agreeable acceptance; nor would he himself have now gone public, he wrote, if what he had to say could still have been expected from Horne's own hand. What he did have to say, though, included not only the claim that he had shared in that earlier research but that he, not Horne, had indeed been responsible for most of it. Turning then to de Graaf, he questioned his integrity and accused him of claiming the work of others for his own. De Graaf's second book on the reproductive organs had thus driven Swammerdam to disparage his late mentor, whose memory he nonetheless insisted he still revered, and to grapple publicly with a former friend and colleague in a contest for personal recognition.[71]

In the midst of his attack on de Graaf, to be sure, Swammerdam urged that they set aside the concern for personal renown and amicably explore the secrets of nature together; but de Graaf answered the *Miraculum naturae* in part by berating Swammerdam for *his* avid appetite for glory and hunger for honors that rightly belonged to others. If de Graaf had also allowed his own resentment to get the best of him, Swammerdam had nonetheless exposed just how intensely he was driven by longings he himself feared and detested.[72] There is something plaintive in the closing lines of the *Miraculum naturae,* in which Swammerdam implicitly acknowledged not only how much he had counted on Horne

69 JS (1737–8), 1:77, 123, 2:408, 410, 520, 554, 704; idem (1672), 16–17, 18; Malpighi (1697b), 59 in the first series in the pagination. Through Steno, Swammerdam had informed Malpighi as early as 1671 both that he held him in high regard and that he had found some mistakes in Malpighi's treatise (Malpighi 1975, 2:598).

70 Adelmann (1966), 1:399.

71 JS (1672), 2, 41, 51, 55.

72 JS (1672), 17–18; Graaf (1677), 449, 477, 488. Swammerdam explicitly opposed "our own conceited glory" to the honor of God as the ultimate purpose of scientific research (JS 1737–8, 2:860).

to tell the story of their joint researches but how much, as well, he was bedeviled by ambition and tormented by its disappointment:

> Meanwhile, this untimely death of such a man [Horne] taught me that nothing in human affairs can be relied upon but virtue, and that anything, no matter what its claims, that does not lead to that end is vain. Hence, I now consider to be happy only those who, far from anxiety and ambition, live content with their lot and determine their fate themselves.[73]

As he acknowledged three years later in the *Ephemeri vita,* from competitiveness and a yearning for recognition came indeed much of the intensity with which Swammerdam gave himself to his research.

Powerful as these compulsions were, however, they were entangled in an intricate interplay of subtler emotions that appear in the 1670s to have focused his ardent motivation more sharply on natural history and on his new microscopic dissections in particular. We are told by his first biographer, Boerhaave, that Swammerdam's fascination with insects began in childhood, and that as a new student at Leiden's medical school he already possessed the beginnings of his insect collection. During the following years, however, he was caught up in the excitement of the advances being made in anatomy; an assiduous student, he soon revealed his abilities in this thriving and prestigious field.[74] Nonetheless, his researches on insects and kindred creatures persisted and, once his studies at Leiden were completed, began pushing his other scientific interests aside.

This withdrawal from the traditional practice of anatomy surely reflected a variety of circumstances, from the unexpected death of Horne to perhaps growing doubts about the value and propriety of anatomy, whose relevance to medicine he now denied in the *Ephemeri vita,* where he cited Scripture against it as well. There had been no intimation of such doubts in the *Miraculum naturae,* however, and they appear, consequently, more an embellishment of his renunciation of his scientific past in toto than a contributing cause of his earlier, limited retreat from conventional anatomy. Nor did Horne's death end Swammerdam's opportunities to work even on the human body: Only weeks after Horne's passing, permission was granted Swammerdam to dissect corpses "from time to time" in the municipal hospital in Amsterdam. That he had requested such permission bespeaks a lingering interest in more orthodox anatomical research, to be sure; but, already in 1668 and 1669 by his own account, just fol-

73 "Interim haec tanti Viri praematura mors documento mihi fuit, nihil in rebus humanis esse, cui tuto inniti possis, praeter virtutem, & vana esse omnia, quae huc nos non perducunt, quocunque etiam titulo sese commendent; ut jam illos tantummodo beatos aestimen, qui procul a metu & ambitione sua sorte contenti vivunt, & ipsi fortunae suae proram & puppim faciunt" (JS 1672, 53).

74 JS (1737–8), "Het leven van den schryver," [1]–[2]. Olaus Borrichius recorded having seen Swammerdam's insect collection at Leiden in 1662 (Nordström 1954–5, 25–6), the year after Swammerdam first matriculated there; see also *LJS* 70. Sylvius (1679), 34; Faller (1986), 242.

lowing his graduation, he had turned away even from his anatomical work with Horne to give himself more completely to his studies of insects.[75]

This narrowing focus of research followed in no small part from the growing excitement and significance of these studies during the immediately preceding years. Not only had Swammerdam recently exposed the sex organs of the bee, he had discovered as well that beneath the skin of a mature caterpillar could be found the parts of the future butterfly – a discovery he prized above all others, according to Thévenot in 1669.[76] These and other discoveries not only overturned the accepted understanding of both the generation and metamorphosis of insects; as interpreted by Swammerdam, they also cast a dramatic new light on the most fundamental questions regarding the generation of all animals and touched, as they did so, on grave religious issues to boot. Thus even before his encounter with Malpighi's treatise on the silkworm, Swammerdam already faced the prospect of a field of endless discoveries of potentially profound significance. It was a prospect that spoke to that yearning for recognition that, by the 1670s at least, plagued him so insistently.

It was more indirectly, however, that Swammerdam's competitiveness most likely worked to incline him away from conventional anatomy and toward a more absorbing involvement with insects. His competitive sensibilities dismayed him, but a field was opening before him that now promised renown without competitors. In addition to the prospect of both exciting and distinguishing research, his work with insects offered an escape from the tense give-and-take among ambitious anatomists and from the frustrations and bitterness of both shared and contested discoveries – that is, from the pressures, anxieties, and disappointments of competing in the established field of anatomy. How unnerving it probably was, then, when Malpighi's *De bombyce* suddenly appeared, displaying skills in Swammerdam's chosen field of which Swammerdam had scarcely dreamed! It was presumably a threatened as well as acutely competitive Swammerdam who set out with such determination to equal and surpass Malpighi's remarkable achievement. But Malpighi was a distant rival, and his challenge, if impressive, proved a passing one.

An acute sensitivity not only to the social strains engendered by competition but also to his own competitive arousal conforms to a more general trait in Swammerdam's character suggested by the sources that touch upon his life – a trait that even in its broader implications appears to have urged him toward a more complete and exclusive commitment to his insect researches. Swammerdam was remembered as a difficult personality who was ill at ease in society. Boerhaave would cite him in his medical lectures as an example par excellence

75 G. A. Lindeboom underscores the opportunity to practice anatomy that Horne indeed had provided Swammerdam (Lindeboom 1975, 15); JS (1675a), 243–4; Engel (1950), 145. Swammerdam's correspondence with Henry Oldenburg bespeaks not only a continuing interest in comparative anatomy in the following years (Oldenburg 1965–, 9:584) but in postmortem dissections as well (Birch 1756–7, 3:153; JS 1675b, 273–4; see also *LJS* 68–9); JS (1672), 52.

76 JS (1669), pt. 1:27, 53, pt. 2:41–2, 47; idem (1737–8), 1:272–3, 2:603, 612; Harcourt Brown (1934), 281.

of the aloof and angry melancholic temperament, and stories that were sometimes clearly romanticized exaggerations memorialized a tongue-tied or withdrawn taciturnity. Although Swammerdam often graciously expressed his admiration for other scientists, his clash with de Graaf was not the only example of his capacity for animosity.[77]

His personal life in the years following his graduation from Leiden had been scarred by a difficult and persisting struggle with his father over Jan's continuing and impractical passion for research and his inability or refusal to begin a career. His father's death in 1678 (his mother had died seventeen years before) then initiated a trying conflict between Swammerdam and his sister over their father's legacy. Continuing family tensions thus added to what had emerged as an increasing aversion to society. "It is my misfortune," he wrote at this time, "that I am compelled to live with these people, for I love quiet and solitude."[78] Indeed, he longed in these later years for isolation and anonymity and, in a moment of acute disillusionment, declared Thévenot in distant Paris to be his only friend, to whom he also wrote that men were now so corrupt that to speak to them was to defile one's soul.[79] Wasted by fever and unresponsive to the solicitude of those who had gathered about him, he died in 1680 at the age of forty-three in the grip, according to Boerhaave, of a "melancholic madness [*furor*]."[80]

Such a litany of unhappy social interactions during the last dozen years of Swammerdam's life suggests that the narrowing of his scientific focus reflected in part the appeal of the distinctively private character of much of his work with insects. When traveling in France in 1664–5 as a medical student, Swam-

77 Boerhaave (1734), 432; see also Boerhaave's account of Swammerdam's life in JS (1737–8), "Het leven van den schryver," [3], [12]. The picture of Swammerdam in Olaus Borrichius's journal during the former's days in Paris does not conform to Boerhaave's sketch of a young man who could be made to speak only through persistent entreaty (see Nordström 1954–5, passim); nor, for that matter, does the impression left by Boerhaave that Swammerdam never returned to his science after his religious crisis (JS 1737–8, "Het leven van den schryver," [11]–[12]) conform to what we know from Swammerdam's correspondence. Boerhaave also related that Swammerdam was unable or unwilling to respond in the routine debate that followed his doctoral disputation (Boerhaave & Haller 1740–4, 6:275), conjuring up an intriguing scene, to be sure, but one to which, given the examples of Boerhaave's inclination to romanticized distortion, we cannot give unreserved credence. That Boerhaave was apparently confident that such characterizations of Swammerdam would generally be found acceptable, however, suggests indeed the collective memory of Swammerdam that was being passed on. For other inaccuracies in Boerhaave's biographical sketch of Swammerdam, see n. 31 above.

Having befriended both the biographer and his subject, Ruysch was doubtless an important source for Boerhaave, but he also very likely contributed to Boerhaave's distortions. For instance, Ruysch himself wrote without qualification that, after winning celebrity through his treatises on respiration, the bee, and the mayfly, Swammerdam stopped dissecting (*het anatomiseeren*) and "abandoned the world," *verliet de Wereld* (Ruysch 1725, 4–5).

Swammerdam also angrily attacked Caspar Bartholin (JS 1737–8, 2:675–6, 734, 800); however, apart from his expressed admiration for Malpighi, he spoke appreciatively of such naturalists as Francesco Redi and Hooke, although he did not refrain, of course, from correcting them as well (see, e.g., ibid. 1:349, 352, 2:456, 734, 876).

78 See below regarding nn. 87, 88, 91; *LJS* 120, 139.

79 *LJS* 110–11. Regarding his longing for isolation, see *LJS* 75, 78, 86, 120.

80 *LJS* 164; Boerhaave & Haller (1740–4), 6:275.

merdam had presented some of his early efforts before the gatherings of savants regularly hosted in Paris by Thévenot,[81] and, as noted, he subsequently dissected bees while working with Horne. Swammerdam also had a handful of very special friends – Thévenot, Steno (who wrote Malpighi that, after God, nothing was holier than friendship), the anatomist Mattheus Sladus (his best friend in Amsterdam, Swammerdam volunteered in 1672), and Johan Oort, Lord of Nieuwenrode – who at one time or another also took part in his insect researches.[82] For Swammerdam, nonetheless, those researches became intensely and emphatically private. They brooked no company, he wrote Thévenot, sometimes not even a second person.[83] Perhaps they were once the retreat of a very private youth as well, and now in later, presumably tenser years, they offered relief and escape to a personality that too often found social relations difficult and disturbing.[84]

Swammerdam's difficulties with those about him were rooted in a deeper social estrangement, however, one that was likely to have enhanced the appeal of the demanding, manipulative activity of his insect dissections in particular.[85] Indeed, the emotional implications of Swammerdam's unstable life suggest a further critical source of the intense commitment that ultimately underlay his microscopic discoveries.

81 Such at least seems to be the implication of some remarks by Thévenot (Harcourt Brown 1934, 280–1).

82 Regarding Thévenot as a friend: *LJS* 100–1, 110–11. With respect to his sharing in Swammerdam's insect studies: JS (1667b), dedication; idem (1669), pt. 1:27, 53, pt. 2:41–2; idem (1737–8), 1:149, 195, 272–3, 2:750; Harcourt Brown (1934), 280–1.

Steno as a friend: JS (1737–8), 1:145, 195; Malpighi (1975), 2:478. Re sharing in Swammerdam's insect studies: JS (1667b), dedication; idem (1737–8), 1:149, 195, 2:750.

Sladus as a friend: JS (1667b), 84; idem (1672), 45. Re sharing in Swammerdam's insect dissections: JS (1737–8), 1:301.

Johan Oort and his wife as friends: JS (1737–8), 1:139, 194–5, 448, 2:763. Re sharing in his studies in natural history: ibid. 2:770.

83 *LJS* 86; JS (1737–8), 1:79.

84 Although concerned with a very different time and a very different field of research than Swammerdam's, David C. McClelland concludes from a survey of studies of the psychological makeup of creative scientists in twentieth-century experimental physics that outstanding scientists avoid interpersonal contact and enjoy being alone, probably because people and human relations seem both difficult and uninteresting to them. They are disturbed by complex human emotions, and perhaps particularly interpersonal aggression: "Scientists react emotionally to human emotions and try to avoid them." McClelland also suggests that the scientists' withdrawal from people can perhaps be viewed as a mode of defense against conflict over aggression, and he argues that they turn to nature because interpersonal relations have proved so laden with frustration and anxiety (McClelland 1962, 146–9, 165, 169, and passim).

McClelland also notes that the unsociability of scientists appears as early as the age of ten, by which time their scientific interests have typically emerged as well (ibid., 145, 150; see also A. Roe 1953, 91–2; Eiduson 1962, 46–7, 50). On the other hand, although Swammerdam indeed seems to offer an extreme example of some of the major character traits McClelland ascribes to creative scientists, Chapter 6 stresses the importance to Leeuwenhoek and his science of extensive and continuing social interaction.

85 Mihaly Csikszentmihalyi's comments regarding what he calls "microflow" activities do suggest, at least, that socially alienated personalities depend more than others on the "flow" experience provided by manipulative activities (1975, 152, 159, 174–5, 178).

In addition to the social difficulties and a penchant for self-reproach, what we know of Swammerdam's life bespeaks an inability to come to terms with the expected patterns of life in seventeenth-century Dutch society. Having completed his medical studies in 1667, Swammerdam balked before the prospect of actual medical practice,[86] and, as the years passed, he declined to undertake any self-supporting vocation. He remained dependent on his chagrined and doubtless disappointed father, who continued, by Swammerdam's account, to provide him room and board but little more. In 1670, Swammerdam lamented to Thévenot that his father, continually pressing him to begin medical practice, now refused him further clothes and money. Three years later he addressed his first letter to Antoinette Bourignon, the peripatetic French mystic who had acquired no small notoriety during her recent residency in Amsterdam and whose subsequent correspondence with Swammerdam did much to shape his emotional turmoil into a religious crisis. His letter has not survived, but her response, which has, attests that among the several personal problems he laid before her was still that of beginning a career, be it now in the ministry (as, according to Boerhaave, his father once had wished), commerce, or medicine. But ultimately he chose none of these, and, in the late summer of 1675, having published the *Ephemeri vita,* he elected rather to abandon Amsterdam and Holland and to join Bourignon's small band of followers on an island in the North Sea off Denmark.[87]

Within nine months, however, he was back in Amsterdam and, soon resuming his researches on insects and other small animals, once again dependent on

86 Engel (1950), 145; cf. *DSB,* s.v. "Swammerdam, Jan."

87 *LJS* 57, 53; Lindeboom (1974), 187–94; JS (1737–8), "Het leven van den schryver," [1]; Lindeboom (1975), 16–17. Regarding Bourignon's stay in Amsterdam and the following she acquired there, see Lievense-Pelser (1977), 212ff.

Among the circumstances perhaps also contributing to Swammerdam's intensifying emotional difficulties during the first half of the decade was the shock of the French invasion of the Dutch Republic in 1672 and the political and social upheaval that followed. The French army remained camped on Dutch soil not far from both Leiden and Amsterdam through much of 1673, and, though all French troops were withdrawn by the spring of 1674, the new regime of the restored Stadholder William III continued an increasingly unpopular war for several more years (see Geyl 1964, 121–61). Swammerdam makes no explicit reference to these events, but they may well be echoed in a note of antipathy for governmental authority in general and in his consciousness of power politics in a world he depicted as grim and threatening. In the *Ephemeri vita,* just before alluding to "the darkness of this century," he included among the sorrows and afflictions humanity suffered in the world "the continual struggle we have against the governments [*Overheeden*], against the powers, against those wielding force in the world . . ." (JS 1675a, 180). He had earlier voiced his admiration for the social life of bees and ants, founded as it was on love and the inclination to reproduce, he wrote, with only a minimum of ruling power or authority (JS 1669, pt. 1:106, 136; pt. 2:24–5).

This is not to say, however, that Swammerdam did not have useful contacts with important political authorities. With the usual lavish praise, he had indeed dedicated the *Historia insectorum generalis* to the burgomasters of Amsterdam and in particular Koenraad van Beuningen, who also played an important role in the diplomacy of international power politics. It was through van Beuningen's influence that Swammerdam succeeded in acquiring permission in 1670 to dissect corpses in the Amsterdam hospital (Lindeboom 1975, 15). It was also to Johan Hudde, a magistrate and future burgomaster of the city as well, that Swammerdam was indebted for techniques regarding the microscope (JS 1669, pt. 1:81).

grudging paternal support. So grudging was it that during the winter of 1677–8 his father (and presumably his sister, with whom either his father was going to live or both father and son already lived) finally compelled his son to find new lodgings of his own.[88] Swammerdam turned to his friend Johan Oort, who had once invited him to come and live on his estate, but that offer was now withdrawn, stoking Swammerdam's pessimism about the reliability of human relations.[89] His father's death in the spring offered the prospect of a final resolution of Swammerdam's predicament: a modest but continuous income from his inheritance to sustain him henceforward.[90] His sister contested the terms of the legacy, however, and, still deeply unhappy, Swammerdam was tempted by Thévenot's suggestion that he come to France – although if he did so, he made it clear, he would wish to live in isolation.[91] Before leaving the Netherlands again, he died, never having found a way to function successfully as a member of the society that surrounded him.

That failure argues the likelihood of a self-perception afflicted by more than he managed to express even in his excesses of self-reproach, and it testifies as well to a social estrangement of broader scope than simply a difficulty in dealing with others. It is not unlikely that his inability to adapt to the society in which he found himself was accompanied by feelings of ineffectiveness. Is it unreasonable to propose, for instance, that his failure to escape from his dependence on the resentful support of his father fostered a sense of his own powerlessness in controlling the circumstances of his life? The resulting damage to his self-perception in the years following his graduation might in part explain a tenser edge in his competitiveness and a greater need for the reassurance of recognition. It would also have rendered him more susceptible to the emotional returns of microscopic dissections.

Swammerdam's social failure may intimate a further dimension of estrangement as well. His avoidance of what might narrowly be called a socially productive vocation was not a flight from the prospect of work, which he seems rather to have actively sought within the framework of his scientific pursuits. An alternative explanation suggests itself: that he was gripped by a sense of the meaninglessness of the established and expected routines of life, routines that would have appeared particularly bleak in the light of the excitement and self-fulfillment he found in his research.[92] (An additional concern he had raised in

88 Lindeboom (1974), 197; *LJS* 18–19, 88n. 1, 87. Regarding Swammerdam's resumption of his dissections and natural history studies after his sojourn with Bourignon, see, for instance, *LJS* 75, 80, 81; see also n. 139 below.

89 *LJS* 110.

90 Swammerdam hoped for an income of 400 guilders a year – twice what his father had given him, on which, he wrote, he could not live (*LJS* 107).

91 With respect to his struggle with his sister over their father's inheritance, see *LJS* 120, 139. Regarding the prospect of going to France: *LJS* 71, 82, 100–1, 139, 157. Swammerdam repeatedly voiced his desire in these later years to live alone and unrecognized: *LJS* 75, 78, 86, 120, 110–11. On his unhappiness, see n. 1 above.

92 See Deci & Ryan (1985), 23, 27, 29, 34–5; Csikszentmihalyi (1975), 33, 99, 129, 198.

his first letter to Bourignon, his desire to free himself from apparent commitments to an unnamed young woman, raises the possibility that he was faltering as well before the prospect of familial domesticity.)[93] If such were indeed the roots of his feelings of estrangement in these years, it casts further light on the intensity of his commitment to insect dissection.

In many ways – in the progressive development of intricate skills, the gradual elaboration of challenges, the successive discoveries in both technique and observations, and even in the very narrowness and oddity of focus[94] – Swammerdam's insect dissections displayed the hallmarks of those absorbing activities that have the capacity to engulf the practitioner in an experience profoundly contrary to the sensations of helplessness and of the emptiness of the routines of everyday life. Such activities offer not only escape, after all – and escape from self-consciousness as well as from the vexations of social entanglements[95] – but positive feelings of interest and excitement, of continuing personal development, of control and effectiveness, and of clarity and meaningfulness of purpose.[96] The very fact that insect dissection remained outside the normal practice of anatomy – and not only removed hence from the immediacy of competition but from established disciplinary guidelines – would have at the same time diminished pressures and heightened Swammerdam's sense of his freedom of action.[97] How such feelings would have contrasted with the frustration and despair that otherwise enveloped his later years!

Indeed, while so often bemoaning the pervasive misery of human life, he also wrote of the joy he found in his researches. Those had become spiritually perilous to him not only because they were rooted in ambition, after all, but because they were a source of such pleasure as well, and he left little doubt as to how great that pleasure had been. He wrote of his study of the eye of the bee, undertaken in 1673, as having been accomplished with "the greatest joy in the world" and "with more pleasure than if I had been granted a few hundred years." In the *Ephemeri vita* he described his researches as his total pleasure and

93 Lindeboom (1974), 195.

94 See Csikszentmihalyi (1975), 16, 18, 26, 30, 33, 35–48, 128–9, 181–2, 192. The very fact that insect dissection was so unusual an activity may have enhanced Swammerdam's sense of self-determination and autonomy, essential to the emotional rewards alluded to here (see Deci & Ryan 1985, 29–32, 35–6, 38, 55–62, 85).

95 Csikszentmihalyi (1975), 36, 38, 194–5, 42–4, 182, 194–5. Compare these passages to McClelland (1962), 168.

96 Csikszentmihalyi (1975), 5, 26, 30, 33, 44–6, 80, 99, 134, 158, 177, 182, 184–5, 191–2, 195, 198; Deci & Ryan (1985), 28–9, 34–5.

97 Deci & Ryan (1985), 43–56, 80–1, 321–6; Csikszentmihalyi (1975), 22, 25; and see n. 94 above. Swammerdam was not indifferent to the sensation of free will: When reflecting on his scientific labors in the *Ephemeri vita,* he declared that, as long as one maintained one's love of God, an individual's choice of calling was absolutely free, and he emphasized that it was now also of his own free will that he gave those labors up (JS 1675a, 241, 245). Though the context is indeed radically different, it is not without interest that, in her study of forty eminent twentieth-century scientists, all academically affiliated, Bernice T. Eiduson was struck by the importance in their reflections on happiness of "their expansive feelings of freedom, and of the potentialities for choice" (1962, 163–4).

delight, his "only and beloved son" whom he now drove out like another Ishmael.[98]

According to Boerhaave, the melancholic temperament was not only uncongenial, it was disposed as well to that perseverance and concentration that were most likely to advance the frontiers of science.[99] The melancholic temperament may no longer serve as an acceptable explanation, but personal estrangement and scientific achievement appear nonetheless to have been so reciprocally intertwined in Swammerdam's life that they cannot be disentangled. To attempt to explore the world of Swammerdam's emotions is to depart from more reliable paths of historical argument, but to decline to do so is to accept his intense and critical absorption in microscopic research, and hence a decisive element in the history of discovery, as incomprehensible by default.

His was an emotional history shaped fundamentally by his response to society and marked by his retreat to its periphery, where he sought isolation while still clinging to the prospect of eventual recognition. The interplay of his yearning for recognition and a growing aversion to the stress of social interactions helped focus his research drives more exclusively on his ongoing work with insects,[100] in which, following the example and challenge of Malpighi, microscopic dissection acquired a central place. That Swammerdam's achievement in microdissection rested upon the uncommon intensity of his commitment to this uncommon direction of research seems obvious enough, and hardly less obvious is the presumption that such intensity derived from some unusual emotional source. That his life in general was warped out of any expected pattern by recurrent emotional turmoil is clear from the record that has survived; it is only somewhat less clear that this emotional turmoil also reflected the difficulty of his interactions with society and those about him. It was a kind of difficulty and a kind of turmoil, moreover, that would have greatly enhanced the emotional advantages of immersion in the skilled activity of microdissection.

This scenario argues, then, that the magnifying lens became a meaningful scientific instrument in part, at least, because of a potent interplay of very personal and indeed antisocial emotions. With the microscope now at hand, persisting interests and traditions in Dutch culture – subtle anatomy, natural history, and the expectation of revelation in nature's minutiae – had placed Swammerdam on the brink of a new field of scientific research; but, at that brink, those inter-

98 JS (1737–8), 2:502; idem (1675a), "Ernstige aanspraak," [5], 240, 248.
99 Boerhaave & Haller (1740–4), 6:275; see also Boerhaave (1734), 432.
100 Max Delbrück proclaimed at the time of his Nobel prize: "The scientist has in common with the artist only this: that he can find no better retreat from the world than his work and also no stronger link with the world than his work" (Delbrück 1970, 1314). Regardless of what Delbrück had in mind when he spoke of a link with the world, Swammerdam himself still recognized that his insect studies were tied to a desire for fame when he wrote (c. 1678) that he was not at all attached now to natural history, for he had learned that the pursuit of glory – "la recherce [*sic*] de la gloire" – was the greatest foolishness in the world (*LJS* 110). With respect to the extent to which he *was* still attached, however, to his work in natural history, see, for instance, his letter to Thévenot, apparently the preceding month; there he included his "anatomies des insectes" among the "attachemens [*sic*]" that were holding him fast where he was and preventing his going to Paris (*LJS* 100–1).

ests and traditions lacked the capacity to elicit, and very likely inhibited, a decisive commitment. Hence, it was rather Swammerdam's emotional response to social circumstances and the reassuring balm of skilled activity that ultimately drew him into a new and unfolding realm of discovery.

No sooner had Swammerdam entered upon this new field of research, however, than his cultural inheritance began to enrich the motivations that helped sustain it. Images and ideas from older cultural traditions colored both his understanding of what he saw and his experience of microscopic discovery itself. They shaped the description, depiction, interpretation, and perhaps even, at times, the perception of what he now confronted through the lens. They also impinged upon his conception of what he was about. Beyond the easement of his "melancholic" and ambitious temperament, consequently, Swammerdam's cultural legacy vested his microscopic research with further levels of meaning.

Swammerdam's reactions to microscopic marvels reveal a mind richly diversified in its cultural affinities. He was, of course, deeply engaged in both the established discipline of anatomy and the still ill-defined field of natural history. He had emerged, too, as an eminent representative of the aggressive new experimental science and was a committed participant in the epistemological debates it inspired. While so troublingly susceptible as well to the current of overwrought religious anxiety that coursed through the century, he also responded to the more purely aesthetic expressions of seventeenth-century Dutch culture – art, belles lettres, and objets d'art – flourishing above all in Amsterdam. Such varied interests and sensibilities – he was himself a finicky draftsman and a poet – shaped a complex and multifaceted response to strange new microscopic images.[101]

In a more distinctively "scientific" frame of mind, concerned with providing a systematic and potentially secular accounting of nature, Swammerdam was expecting to discover in his insect dissections the reflection of some unifying order below the apparent confusion of the observable world. The "laws" characteristic of the century's new physics may or may not have been appropriate to biology,[102] but in the *Historia insectorum generalis* Swammerdam emphasized an underlying uniformity that he posited in the generation and development of living things. He insisted that the development of both plants and animals was no more than the emergence and growth of previously hidden parts, so that, governed by identical rules, all the "natural changes" of plants and animals were basically alike. The most compelling evidence he offered was the revelation of the adult insect – most dramatically, the butterfly – concealed below the larval

101 Swammerdam referred in several places to his having done his own drawings (JS 1737–8, 2: 890; *LJS* 71, 82; see also Lindeboom 1982, 121), and on at least one occasion expressed his reluctance to disfigure one of his illustrations with letters to help identify the different insect parts (JS 1737–8, 1: 353). A number of his poems are to be found in the *Ephemeri vita*.

102 For an introduction to the discussion of the question of "laws" in biology, see Hull (1974), 70–100.

skin; but, exploiting his enhanced dissecting skills, he now also searched for continuities persisting through metamorphic change within the deeper anatomy of insects.[103] Not all that he observed was reassuring,[104] but images that did satisfy his expectations he looked upon as manifestations of nature's most fundamental processes.

His observations in insect anatomies generally assumed another, very different kind of underlying uniformity as well, a basic if often clouded similarity in the way all animals were put together. Presumably testing and enjoying the reach of his skills, Swammerdam seems to have ranged about freely in the innards of insects to see what he could see, and, convinced now that those innards were indeed as remarkable as those of any other animal,[105] he expected a further similarity in anatomical structure and physiology. Moved more by the appeal of analogy than the distinctiveness of such parts as the insect eye, he imposed on these new anatomies a rough but familiar schema of organs.[106] It does not appear to have been his purpose to argue this analogy, however; nor did he suggest (as did Malpighi) that he had undertaken his insect dissections in order to throw further light on the anatomy of mankind and the higher animals.[107] Rather, the assumption of analogy served Swammerdam's needs; it helped make sense of unfamiliar structures and, at the same time, emphasized the association between these unusual studies and an established and recognized natural science, traditional anatomy.

Swammerdam presented his dissections to the public in the pictorial style, indeed, of the anatomical tradition. He was intimately familiar with the work of the naturalist miniature painters and in the *Historia* urged the continuing value of the genre to natural history.[108] But in illustrating his observations in insect anatomies – and the drawings were his own[109] – he turned to the stark and diagrammatic figures of the anatomists. For the sake of an emphatic and unnatural clarity, scientific illustration characteristically suppresses the devices of illusionist mimicry while resorting to explanatory texts and accessory reference marks (numbers, letters, arrows, and the like). Such unfamiliar forms as insect viscera demand such aids, especially when magnification expunges any familiar framework for judging the context and dimensions (Fig. 14). Swammerdam conse-

103 JS (1669), pt. 1:30, pt. 2:40; idem (1737–8), 1:319, 340, 345, 2:584ff. For other passages in which Swammerdam asserts that all the processes of change in living things are founded on the same rules, see JS (1669), pt. 1, dedication [3], 28; and idem (1737–8), 1:20, 2:792, 861–3, 867.

104 See JS (1737–8), 2:666ff.

105 See, for instance, JS (1737–8), 2:713.

106 Needless to say, his identification of specific insect organs with those in higher animals sometimes missed the mark; see Cole (1944), 290–1. Nor, by any means, did he presume to try to identify the function of all the parts he discovered in insect anatomies (JS 1737–8, 2:694).

107 Malpighi (1686–7), *Anat. plant. idea,* 1.

108 JS (1669), pt. 1:70–1. On the other hand, he also remarked, with respect to the study of metamorphosis, that the description of a single case as an example of all the others would be more useful than painting the changes that many varied caterpillars, with their colors, underwent (ibid., 131).

109 See n. 101 above.

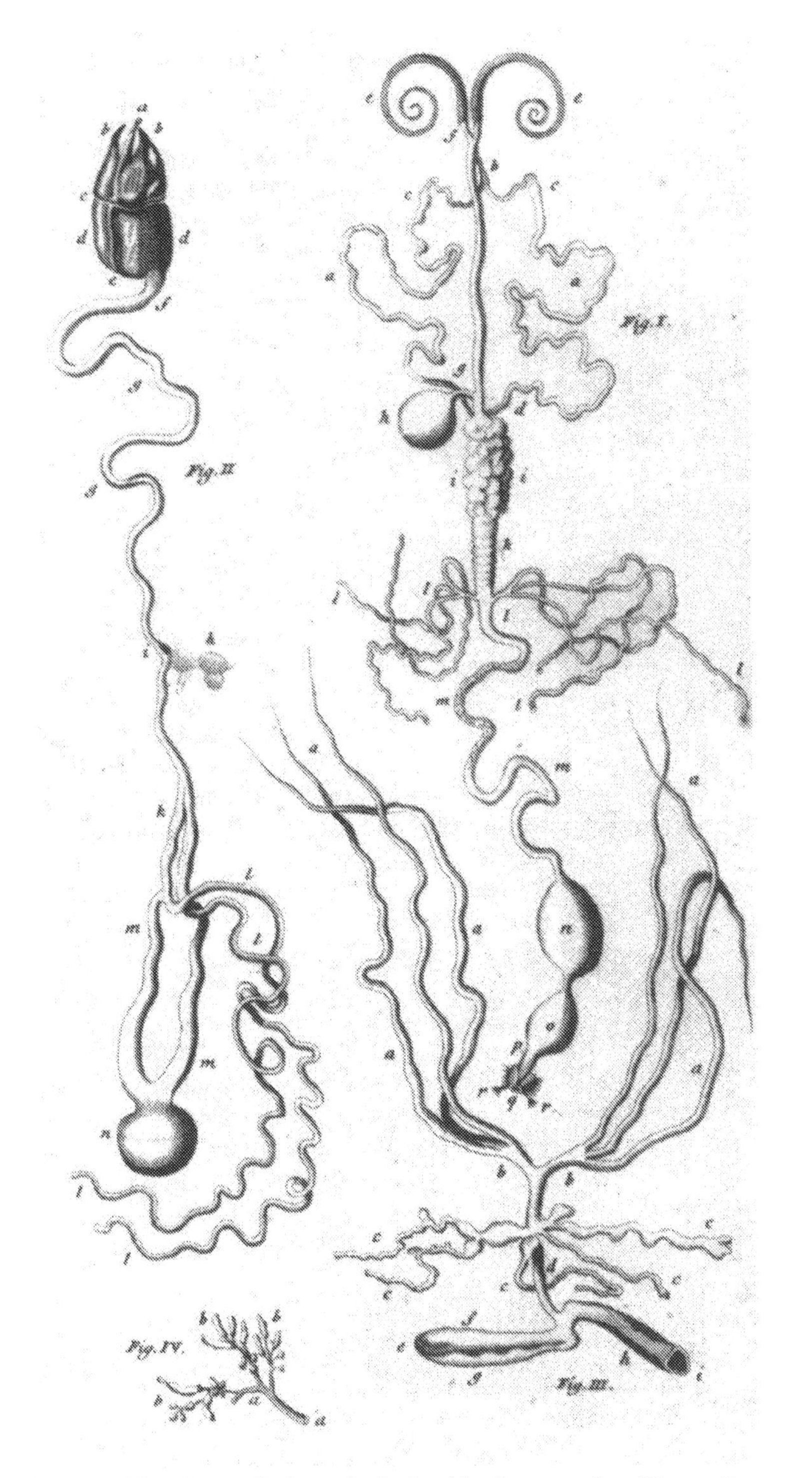

Figure 14. The viscera of a butterfly depicted by Swammerdam (JS 1737–8 2:pl. 36). Photo courtesy of the Department of Special Collections, Uni versity of Chicago Library.

quently chose an illustrative style that, while associating his new observations with the discipline of anatomy, denuded them of the sensual richness of reality and the evocative appeal of seventeenth-century Dutch artistic tradition.[110]

It was Swammerdam's text, however, that now ranged freely beyond the boundaries of strict scientific relevance and revealed how much richer, aesthetically and emotionally, the experience of these observations had truly been. He reacted, after all, to a wide range of visual properties, and his aesthetic and religious sensibilities detected other, more exalted qualities as well. His efforts to articulate such sensations betrayed how broadly his cultural inheritance was reflected in his perceptions of this strange new realm, but its very strangeness, in turn, left its own imprint on that reflected legacy.

Most obviously cast aside in Swammerdam's illustrations were the color, texture, and changing play of light in which Netherlandish painters had so delighted. In part, to be sure, the technology of printing dictated the disregard for these qualities, but at times, despite the work of Hoefnagel and the continuing line of his successors, the developing science of entomology displayed not only an indifference but an outright distaste for even the use of color. The miniaturist Goedaert had indeed personally tinted the engravings in some copies of his *Metamorphosis naturalis;* but Swammerdam, to illustrate the *Historia,* deliberately chose insects that could serve his purposes without such embellishment, and opined that the illustrations he offered were so well done that adding colors would have spoiled them. In the following century, Réaumur as well explicitly discouraged the use of color in illustrating even the external aspect of insects; only in the nineteenth century would it be used in the depiction of their internal anatomies, and then as a means to emphasize organ systems rather than to simulate their appearances.[111]

Nonetheless, Swammerdam in fact responded enthusiastically to the colors and light effects he encountered in his researches in natural history. His insect studies are studded with vivid, even lyric passages that, evoking images of precious stones, polished surfaces, and on occasion works of fine craftsmanship, recall both his father's collection of rarities and the elegant *pronk* still lifes painted and marketed in late seventeenth-century Amsterdam.[112] He spoke of the interior of an oak gall as itself a little still life[113] and rhapsodized over the "richly glittering beauty" of butterflies (found in fields he elsewhere described as "natural and richly ornamented tapestries"), whose wings looked like mother-of-pearl and

110 Swammerdam did resort to simple shading to help convey form even in his anatomical illustrations, it is true, but, for contrast, see the echo of the miniaturist tradition in the depictions of whole creatures in pl. 3 (fig. 3), pl. 30 (figs. 2, 3), and pl. 46 (fig. 6). For an echo of that tradition in illustrations done through the lens as well, see the drawing of the leg of a louse done for Leeuwenhoek (*AB* 1:pl. 32 [fig. 2]; see also p. 293n. 18).

111 Bol (1959), 4; JS (1669), pt. 2:14; Réaumur (1734–1929), 1:53–4; Freeman (1962), 183. To be sure, some illustrations done by Swammerdam of the silkworm and its anatomy that he passed on to Malpighi are partly colored (Belloni 1968, 178–9), although I have not had the opportunity to see the original illustrations and to determine how Swammerdam used the color.

112 See Bergström (1956), 260–1; Alpers (1983), 114–15.

113 JS (1737–8), 2:764.

silver bestrewn with diamonds, pearls, rubies, turquoise, and countless sapphires. With more restraint, he likened a fly's wing to glistening mother-of-pearl.[114]

Although such descriptions were far removed from the austerity of Swammerdam's plates, similar sensibilities echoed, after all, in his textual accounts of microscopic anatomies. The fat that filled the body of the larva of the soldier fly (*Stratiomyia furcata* [Fabricius]) was to Swammerdam as white as new-fallen snow and inexpressibly pretty.[115] He took great pleasure too in the changing colors in the fly's body during metamorphosis[116] and responded likewise to the changes within the chrysalis of the butterfly, in which the succession of colors, he said, was too wondrous to describe.[117] The microscopic globules of fat in the chrysalis mimicked yellow grapes, but those within the cheese skipper larva took on a dull white so lovely that it also defied description. Inside the eye of a dung fly was to be found the prettiest purplish red one could ever want to see, and on the inside lining of the stomach of the hornet larva, the prettiest purple net.[118]

Whether he observed them through the skin or in the dissected body, Swammerdam repeatedly remarked on the silver or pearly tracheae that took on an ever-brighter white with every molt.[119] Wonderfully pretty were the silver and mother-of-pearl tracheae seen beneath the blue translucence of the skin of the larva of the rhinoceros beetle, and so also were those that, together with the whitish particles of fat, shone within the dark-green body of the newly molted pupa of the soldier fly.[120] Even the spirals of the tracheae he likened to little rings of silver thread.[121] Seeming for all its clouded translucency like the foot of a crystalline glass, the penis of the bee offered a pretty sight as well, and, when semen settled in the hollows in its base, it looked like glassy marble or agate flecked with white. Within the penis of the butterfly Swammerdam detected a granular "fluid" that, when spilled from a damaged organ, glistened like a silver sand.[122] Indeed, the glint of an elegant prettiness might be encountered anywhere, even in the excrement of the hornet larva, where fragments of flies it had eaten now shone like gold.[123]

If Swammerdam persistently found "prettiness" and "elegance" in insect anatomies,[124] however, he also responded aesthetically to other qualities whose precise nature, if they had one, is more difficult to grasp. "Order" was a term

114 JS (1669), pt. 1:134; idem (1737–8), 1:273, 2:761, 784.

115 JS (1737–8), 2:659, 663; Schierbeek (1947), 267 (re pl. 39). The fly in question has also been identified as *Stratiomyia chamaeleon* Linnaeus (ibid., pl. 19 caption; see also Sinia 1878, 154).

116 JS (1737–8), 2:682; also 2:674–5, 678–9, 691, 693.

117 Ibid., 2:586.

118 Ibid., 2:407–8, 500, 588, 703–4.

119 JS (1675a), 81; idem (1737–8), 1:71, 2:549.

120 JS (1737–8), 1:308, 2:672–3.

121 Ibid., 1:339, 2:663.

122 Ibid., 2:509, 513, 599.

123 Ibid., 1:320.

124 The adjectives Swammerdam most frequently used in expressing these qualities were *fraay* (*fraai*), *aardig*, and *cierlyk* (*sierlijk*).

with different meanings, or at least shades of meaning, for Swammerdam. The "rules" that guided all of nature embodied a kind of order that in natural history was most strikingly manifest to Swammerdam in the unchanging patterns of insect life cycles.[125] But he also saw order embedded in the tiny structures and anatomies he observed through the lens. Already in the *Historia* he had assumed a "wondrous order" in the fine anatomical structure of all animals both large and small, and, predictably, he later felt compelled to emphasize again that it was to be found no less in the small than in the large.[126] He observed this order in the unfolding parts of the butterfly emerging from the chrysalis, in the way the eggs were set within the ovariole of the "flying water scorpion" (*Nepa cinerea* Linnaeus), and, turning to the plant kingdom, in the opening of the sporangia and the ejection of the "seeds" of the fern.[127]

What constituted this order in these small structures and operations is far from clear, however. In the ovarioles of the flying water scorpion, it seems to have resided in the intricate construction that held the eggs together and in the neat manner in which they were fit in a line within the narrow organ. In the emerging butterfly, it apparently resided in the way its diverse visible parts were arranged and fit together, but also in the progression of movements by which those parts unfolded as the butterfly worked its way out of the chrysalis. As an observable quality, such order seems to have had an affinity in Swammerdam's mind with *regulierheyd,* "regularity," and *maat,* meaning presumably the quality of being "measured" in the sense of being restrained, duly proportioned, or, again, regular. *Regulierheyd* was observed in the opening of the fern sporangia and *maat* in the emergence of the butterfly, which Swammerdam also described as a *reguliere,* "regular," pattern of movement.[128] But these terms also appear to be the bearers of little more than a vague suggestiveness reflecting a basic, perhaps, but poorly articulated aesthetic.

More often Swammerdam also associated the order he observed through the lens with *inventie* and *kunst.*[129] *Kunst* was "art," "skill," and "craft," and *inventie,* although defined in Hendrick Hexham's seventeenth-century Dutch–English dictionary as "Invention, or a Finding Out," would seem better rendered in Swammerdam's passages as "inventiveness."[130] To Swammerdam, the unfolding of the parts of the butterfly was also *kunstig* – translated by Hexham as "Artificial, Ingenious, Skillfull" – while the placement of the eggs within the ovarioles of the flying water scorpion was so *kunstig* as well – and the

125 Ruestow (1985), 233–5.
126 JS (1669), pt. 1:3; idem (1737–8), 1:300, 2:713.
127 JS (1737–8), 1:232, 2:583, 909.
128 Ibid.; see also 1:300.
129 Ibid. 1:232, 2:708, 713, 886.
130 Hexham (1648). Hieronymus David Gaubius's early eighteenth-century Latin translation of Swammerdam's Dutch text of the *Bybel der natuure* characteristically renders *kunst* as *artificium* and *inventie* as *inventio,* though at one point, at least, *meerder inventie* is translated as *subtilius excogitatum* (JS 1737–8, 1:232).

construction of the eggs themselves so unusual – that Swammerdam confessed to having seen nothing "wherein more *inventie* and order lay hidden."[131]

Although also enveloped in their own haziness of meaning, *inventie* and *kunst* share at least one clear attribute: They both imply a prior process of creating and crafting, a shaping hand wherein resided the skill, artistry, or ingenuity and the searching purpose that alone "invents" or innovates. Hence, to express his response to a cluster of such qualities as might in some other time, place, or context be described as intricacy, precision, articulation, and unexpectedness, Swammerdam turned to a familiar way of speaking about art and craftsmanship. What he perceived through the lens was interpreted and represented in terms of qualities prized in objets d'art. This way of viewing or, at a minimum, describing what he saw not only reflected the taste for human artistry and craftsmanship so richly cultivated in the Netherlands, however; it also either derived from or was immediately assimilated to the biblical understanding of nature as created, and among the other qualities he often cited in microscopic anatomies alongside his references to order was now "wisdom" too.[132]

Indeed, it was an "all-seeing" and "omniscient Architect" who, with incomprehensible wisdom and understanding, ordained the operation of the fern sporangia and thoughtfully and skillfully (*kunstryk*) placed the eggs of the water scorpion within the ovariole. It was as well a supremely inventive (*alles uytvindende*) Architect who had "carpentered" (*getimmert*) such work as the muscles of the retractable eyestalk of the snail.[133] Cited repeatedly throughout Swammerdam's accounts, the great Architect also bestrode those pages as the great "Artist" (*Kunstenaar*) whose works included those branching tracheae that, thirty times thinner than a hair, never collapsed even in the thrashing body of a cheese skipper larva.[134] In such hidden places as the anatomy of the mayfly and the louse (and even, Swammerdam could still only presume, within the invisible smallness of the spores of the fern), the artistry of that "Artist of all artists" put even the celebrated subtlety of the legendary lines of Apelles to shame.[135]

Swammerdam also invoked the Deity in attempts to grasp and convey other, even more ill-defined qualities that he detected in what he saw through the lens, but now, instead of citing divine attributes that echoed human skills, he pointed rather to the profoundly inhuman essence of divinity. Like the aesthetic qualities, these further qualities were Swammerdam's own responsive feelings, which he perceived as attributes of the microscopic anatomies themselves; but

131 JS (1737–8), 1:232; 2:583; Hexham (1648). Gaubius (see preceding note) translates *kunstig* as *artificiosus* and *artificiose* on some occasions (ibid. 1:231, 2:762), *kunstig wys* as *artificiose* or *sapienter* (1:232), and *kunstryk* as *stupenda quadam arte* (1:233). In at least one case, however, Gaubius just disregarded *kunstig* (2:583), and in another *verwonderlyk kunstig* is rendered simply *admirabilis* (2:707).

132 JS (1669), pt. 1:98; idem (1737–8), 1:300, 2:708, 713.

133 JS (1737–8), 1:104, 233, 2:909; see also 1:74, 104, 137–8, 150, 166, 233, 2:385, 411, 564, 654, 768, 770, 796, 886, 897, 909.

134 Ibid. 2:705; see also 2:385, 505, 650, 703, 705, 718.

135 JS (1675a), 95; idem (1737–8), 1:67, 2:908 (see also 1:301, 2:650). The allusion is to the story told of Apelles of Cos in Pliny *HN* 35.36.81–3.

these were feelings engendered now by the discovery of so much that, though seemingly crafted in ways reminiscent of human skills, was at the same time profoundly unexpected and alien to the familiar world. The elder Huygens had earlier remarked how the microscope aroused that sense of wonder rich with religious suggestiveness that too much familiarity with the everyday marvels of nature had deadened, and the initial pious response to early microscopic images was surely in part charged with the shock of familiarity suddenly stripped away and the fascination of strange and often grotesque forms rendered all the eerier by the optical effects of the lens.[136] Nor did the seriousness and intensity with which Swammerdam now worked the lens diminish his own capacity for such wonder. Absorbed as he was in the even stranger images of insect dissection, and always searching for signs of the transcendental, he exalted the sense of wonder as the very goal of his researches.

To articulate the sense of the uncanny in what he saw, Swammerdam seized upon the more awesome attributes of the God of scriptural tradition. Such themes were already evident in Swammerdam's science before Malpighi plunged him more deeply into microscopic dissection,[137] but in elaborating that idiom to express yet another level of his experience with the microscope, Swammerdam vested what he saw not only with artistry and ingenuity but with more unnerving intimations of divinity as well – intimations of both the immediacy of its presence and, at the same time, its infinite remoteness from all that was human.

How full of wonder his researches truly were is suggested nowhere more forcefully than in his account of what he dubbed the "miraculous, viviparous, crystalline" snail (*Viviparus viviparus* [Linnaeus]).[138] Begun apparently in 1677,[139]

136 See Otto (1931), 26, 27–9, 40, 65–6; ConHs (1897), 120.

137 In the *Historia insectorum generalis,* Swammerdam had indeed already played upon several of the motifs, to be emphasized in the continuing text above, through which he later expressed the sense of religious awe occasioned by his microdissections. That earlier book opens with the pronouncement that God is inscrutable and incomprehensible in his creatures (JS 1669, pt. 1:2; see also pt. 1:28) and closes by asserting how the invisible things of God, and notably his eternal power and divinity, are to be seen and understood in those creatures (pt. 2:48). Already voicing an awestruck consciousness of humanity's technical inadequacy before the challenge of researching nature (pt. 1:2–3), Swammerdam also remarked that, though the wonders of God were so clearly revealed that nothing could obscure them, human depictions captured only the outline of the shadow of these wonders (pt. 1:147), which were difficult to express as well (pt. 1:83, 97).

138 JS (1737–8), 1:169–80; Sinia (1878, 136) and Schierbeek (1947, 143) identify this snail as *Paludina vivipara,* whereas C. A. Regteren Altena is not so sure (Schierbeek 1947, 265 [re pl. 9, fig. 5]). Vera Fretter and Alastair Graham identify *Paludina vivipara* with *Viviparus viviparus* (Linnaeus) (1962, 694). Regarding the diversity of names with which the species is or has been associated, see also Kessel (1933), 130.

139 Swammerdam sent a letter dated 10 Sept. 1677 to the Royal Society relating his discovery "of a sort of snails, that are viviparous" (Birch 1756–7, 3:372). In a letter that G. A. Lindeboom dates January 1678, Swammerdam wrote enthusiastically to Thévenot of his observations in this snail and told him of his intention to write a detailed description of the same (*LJS* 80, 82). Later in that same year, Swammerdam also wrote of having shown the tiny embryonic snails he had found within his newly discovered viviparous snail (see a few paragraphs below in the text) to Christiaan Huygens (*LJS* 123).

Swammerdam's dissections of this common snail, flourishing in the ditches and rivers of Holland, provided one of his most intoxicating encounters with the unexpectedness of God's hidden works. He had begun dissecting snails years before and thought he knew them well, only to be astounded now, however, by the "many wonders and unheard-of things that perhaps have never even been imagined." Having seen these wonders, he remained amazed – "I think about it almost every moment" – by "the unfathomable wisdom, inventions [*uitvindingen*], and omnipotence" in works that he increasingly found to be as inscrutable and incomprehensible as God himself.[140]

When he began his dissections of the snail, Swammerdam was first astonished to find parts of its body so packed with clusters of hard, transparent, "crystalline" globules that his knife and scissors cracked and grated as he cut. The quantity of these globules in the tentacles alone, he exclaimed, revealed "what ingenious constructions [*inventies*] and unheard-of wonders the all-seeing Architect had hidden in the Bible of his creatures." That the muscles, nerves, and vessels ran then between these globules was in itself "a wondrous business," he wrote Thévenot. Who, he asked, could describe how the arteries, the veins, and the nerves had been so constructed? "No one, in truth," he answered himself, "save he who made it all."[141]

More wonders were to follow. Proceeding further in his dissection – into the snail's uterus, according to his account – he further discovered an abundance of tiny, oblong "worms," and, dissecting these, he found within each of them two or three and sometimes four still smaller worms.[142] Shaped like tadpoles, these smaller worms swam vigorously when placed in water and, when viewed with a lens against the sky, appeared as if composed themselves of grains of sand. He was again astounded, having had no idea, he related, that he would discover so many marvels in a small animal, and marvels, moreover, that so forcefully compelled him to recognize "my ignorance and blindness as to the reason [*oorsaaken*] for all of this."[143]

Continuing with the dissection of what he identified as the oviduct, he was stunned once more to find there a little snail just like the larger in which it was contained. Observed with a lens, he noted, its shell offered the prettiest sight one could imagine, and, placed in water, it also began to swim and crawl about. Swammerdam (and Blanckaert as well) had earlier discovered the eggs laid by oviparous snails, but now he concluded that the miraculous, crystalline snail was also viviparous. It bore living young as large as a common pea, he wrote, and perfect in all their parts.[144]

140 JS (1669), pt. 1:84; idem (1737–8), 1:169, 171.

141 JS (1737–8), 1:172, 174; *LJS* 82. These are apparently the "calcareous spherules" that are found, ranging generally from 0.5 to 200 microns in diameter, in the soft tissues of many invertebrates (Watabe et al. 1976, 283). In his letter to Thévenot, Swammerdam also reported that these globules effervesced in acid.

142 The cercariae in the sporocysts of flukes, whose larval stages characteristically parasitize snails (Schierbeek 1947, 265 [re pl. 9, figs. 7, 8]).

143 JS (1737–8), 1:173–4.

144 JS (1737–8), 1:174, see also p. 168; Blanckaert (1688b), 155–6, see also pp. 151–2.

Within another of the miraculous, viviparous snails, however, he also discovered a dozen transparent eggs containing tiny embryonic snails that diminished in size until, in the smaller six eggs, they could not be seen at all. In the course of the following year, further dissections revealed increasing numbers of such eggs and embryos succeeding each other in similar series. Visited by Christiaan Huygens – apparently in the spring or early summer of 1678 – Swammerdam showed him sixty eggs or embryos in a single snail and related that he himself had encountered as many as seventy-four. Sixty, however, sufficed to amaze his guest, recounted Swammerdam, "because of nature's unheard-of works of art [*kunstwerken*], so astonishing and so inscrutable."[145] Within the crystal-clear eggs of oviparous snails, he had already watched the embryo with its shell turning slowly in the fluid in which it was suspended; now, against the candlelight in a darkened room, he again watched the living embryos of the miraculous, viviparous snail turning "elegantly" within their eggs. The wonder of it all swept him away, and he declared that whoever read his account must grant that God had nowhere shown himself more clearly or forcibly than in this little animal.[146]

Although such religious intimations struck Swammerdam as extraordinary in this remarkable snail, they were woven through all his descriptions of microscopic anatomies and were rooted, despite the recurrent allusions to artistry, in a sense of the remoteness of these anatomies from human capacities. The more he searched the wonderworks of God, Swammerdam protested, the more he found they surpassed the comprehension of man. Though the miraculous, viviparous snail inspired these particular lines as well, the theme had already sounded in the *Historia* and echoed throughout the accounts of his microscopic dissections.[147] Our understanding is too feeble to fully comprehend even one of the numberless works of God, he wrote with reference to his extended study of the mayfly; and even were he to devote his entire life to perfecting a single anatomical discovery, he remarked elsewhere, in the end he would still discover only his own ignorance.[148]

Not only were the works of God in insect anatomies incomprehensible; they could also be neither described nor portrayed, despite Swammerdam's having devoted much of his life to doing just that. Apart from repeated allusions to the inadequacy of even Apelles's lines, Swammerdam asked what artist could picture the entrails of the mayfly or the louse, or even the structure of a single tracheal vessel – the implicit answer was none – and, for that matter, what intellect or pen could describe them?[149] He challenged anyone to describe completely even the least particle of the very least animal and declared that, no matter how great

145 JS (1737–8), 1:175, 179, 180; cf. *LJS* 123. Huygens had left Holland for Paris in late June 1678 (*OCCH* 22:703).
146 JS (1737–8), 1:168, 176, 179.
147 JS (1737–8), 1:169; idem (1669), pt. 2:30. In addition to what follows in the text, see JS (1675a), 70–71, and JS (1737–8), 1:250.
148 JS (1675a), 87; idem(1737–8), 2:836.
149 JS (1675a), 79; idem (1737–8), 1:67, 301.

their understanding or learning, all the finest minds together could not describe the wonders in the two-and-a-half inches of a mayfly.[150] He likened his own description of the living embryo within the miraculous snail to a charcoal drawing of the sun.[151]

His consciousness of the presence of far more in what he saw than he could grasp or convey was for Swammerdam deeply suggestive in itself of divinity. He repeatedly linked the incomprehensibility and inscrutability of insect anatomies to the incomprehensibility and inscrutability of God, and at times explicitly identified his sense of something ungraspable looming before him in the lens with the divine aura still lingering about God's works. What could be seen and touched and handled in the body of a mayfly enclosed so high and hidden a wisdom, Swammerdam declared, that it was as incomprehensible and unsearchable as its creator;[152] in turn, the inability to describe, depict, dissect, or even truly see those entrails revealed God as infinite and incomprehensible in that body. The internal changes observed in the metamorphosis of the solder fly were also as unsearchable as their author, but, surpassing all human understanding as it did, that metamorphosis was at the same time a visible display of divinity.[153]

If we may judge from his accounts, Swammerdam's sensation of the presence of divinity was often quite a vivid one, and he maintained that he could not only see God in his handiwork[154] but could touch him there as well.[155] Swammerdam was speaking loosely, but the straining to find things transcendental hidden in the experiences of the senses was a hallmark of the age. True to his time and place as well as to his own overwrought temperament, Swammerdam seized upon the sensation of the profoundly unfamiliar as the perception of something holy. Hence, he identified what he believed he perceived with the essence of holiness as exalted in biblical tradition – with such attributes of God himself, that is, as divinity, majesty, goodness, eternity, power, and wisdom.[156]

Even in the diminutive realm of the microscope, however, confronting Jehovah was not a wholly reassuring experience, and Swammerdam continued to insist that the revelations encountered in God's works should incite not only a greater love but a greater fear of their creator.[157] Wonder, after all, arouses apprehensiveness as well, and the majesty Swammerdam discovered was sometimes the "terrible" (*verschrikkelyke*) majesty of a "dreadful" (*vreeschelijcken*) God so "formidable" (*ontsaggelijck*) in his works – in this instance, the anatomy of the louse – that "strong, brave men must cleanse themselves of their sins before the aspect of his creatures." "Indeed," Swammerdam had God query Job,

150 JS (1737–8), 2:705; idem (1675a), "Ernstige aanspraak," [7]–[8].
151 JS (1737–8), 1:177.
152 JS (1675a), 92, 116.
153 Ibid., 79; JS (1737–8), 2:666, 680.
154 JS (1672), 1; idem (1675a), 115; idem (1737–8), 2:367, 394, 451, 495, 591, 762, 873.
155 JS (1672), 1; idem (1675a), 115; idem (1737–8), 2:762.
156 JS (1672), 21; idem (1675a), 79, 92, 420; idem (1737–8), 1:169, 300, 327, 332, 2:367, 394, 451, 495, 505, 591, 666, 762.
157 JS (1737–8), 1:57, 65, 2:598, 694.

"who then will stand before my countenance [*aansicht*]?"[158] When reducing the divinity he saw in microscopic anatomies to more specific qualities, Swammerdam cited most often the omnipotence as well as the wisdom of God – both, in one instance, shining visibly before him like a bright and glittering diamond – and even that wisdom was not without its touch of fearsomeness. Whoever in this vale of tears and ignorance did presume to describe completely even the least particle of the least of animals, Swammerdam cautioned, the divine sun of the truths in God's creatures would blind his understanding in punishment for his rashness.[159]

Thus the suggestiveness of Swammerdam's cultural background forged from his reaction to microscopic images a diversity of perceived qualities, ranging from a rule-governed orderliness to prettiness and, ultimately, to an intimidating holiness. The correspondence of rhetoric to true sentiment always remains uncertain, but Swammerdam's aesthetic and religious susceptibility cannot be doubted. His microscopic observations of insect anatomies were thus laden with many levels of meaning with arresting associations.

The significance with which Swammerdam endowed his microscopic researches therefore made them fit to serve what he continued to insist was the only purpose of science: to glorify God and to move humanity to love and fear him.[160] Swammerdam's inability to anatomize the proboscis of the butterfly exposed all human science, he felt, as but a deep abyss of ignorance from which the only knowledge gained was that of a higher being; but nothing more was needed, he wrote, and all else was but vanity of vanities.[161] The only message that mattered was the beauty, wisdom, and incomprehensibility of insect anatomies that so clearly and endlessly manifested God's majesty. Laden with such religious import, Swammerdam's studies gloriously exemplified the oft-quoted passage from Paul cited in the Dutch Reformed confession, a variant of which served Swammerdam as a fitting conclusion to his accumulated insect researches: If we use our senses well, we can learn of invisible things from the visible.[162]

Nor should Swammerdam's confession of hypocrisy in the *Ephemeri vita* persuade us to dismiss these protestations as merely cant. The anxious godly folk of the century were often plagued by a consciousness of their own hypocrisy, accentuated in this instance by the despairing state of mind that produced the *Ephemeri vita*.[163] Swammerdam surely exploited the conventions of piety to disguise ambition, but the very agonizing that yielded his confession bespeaks the deep importance no less than the precariousness of the religious framework of his life and science. To be sure, Swammerdam often resorted to a conventionalized rhetoric to characterize his experience with the microscope, but the ex-

158 Ibid. 2:394; JS (1675a), "Ernstige aanspraak," [13]. See Otto (1931), 17–19.
159 JS (1737–8), 2:705, 762; see also idem (1675a), 115–16. Regarding Swammerdam's references to the wisdom and omnipotence of God, see n. 156 above.
160 See, for instance, JS (1737–8), 1:57, 65, 193, 2:502–3, 505, 694, 836.
161 JS (1737–8), 2:598.
162 Ibid., 2:873; see also 1:106, 2:394.
163 Seaver (1985), 18, 42.

citement of discovery and the evocative force of strange new images could restore affective meaning even to well-worn clichés. An edifying justification could easily pass, with the moment, into motivating inspiration.

Infusing other commonplace expressions in danger of becoming platitudes, the thrill of discovery added a further religious dimension to the very pursuit of microscopic research, and even provided Swammerdam with brief but exhilarating moments of reassurance of personal grace – no small boon in a life of such troubled self-scrutiny. He maintained (as had his master Sylvius) that, having hidden so many of his works, God chose at his pleasure when to make them known.[164] It was even God, according to Swammerdam, who first moved one to investigate insect anatomies, and only through his grace could such researches be carried out.[165]

God revealed his hidden works to whom he wished as well,[166] and Swammerdam spoke of his own observations as having been indeed granted him by God. All true sciences and discoveries were a gift from God and the work of his grace,[167] after all, and, astonished perhaps by his own ingenuity, Swammerdam ascribed even his novel techniques to a divine generosity (though his early methods for using the lens he attributed more prosaically to Hudde).[168] It was God's grace, he wrote, that had enabled him to contrive the "experiments" (*experimenten*) that revealed the dispersal of the tracheae throughout the louse: "For we have nothing from ourselves, since we ourselves are an instrument prepared by the supreme Creator whereby – oh wonder! – his creation [*maaksel*] knows its creator."[169] Swammerdam's discoveries and his talents testified that he was a chosen instrument of God, and although that conviction perhaps truly gripped him only rarely, the memory of those moments could only enrich the emotional import of his microscopic research.

His adaptation of the conventional religious rhetoric of science to his microscopic discoveries also helped fashion a novel vision of nature itself that now positively encouraged the continued and aggressive use of the microscope. For nature now loomed before Swammerdam as a boundless repository of unpredictable and endless marvels that, hidden within the receding dimensions of smallness, waited to astonish their discoverer with the infinitely creative genius of God. Swammerdam now spoke of the remarkable examples of God's handiwork in minute anatomies as "hidden wonders" or "secrets" (*verborgentheeden*) that their Creator had "locked up" in these tiny bodies.[170] The metaphor of nature's secrets was a common and flexible one that had proved adaptable to a variety of contexts,[171] and when in the *Historia insectorum generalis* Swammer-

164 JS (1737–8), 2:467, 860, 902. See Sylvius's letter at the beginning of Graaf (1677).
165 JS (1737–8), 1:67; idem (1675a), "Ernstige aanspraak," [15]–[16], 95.
166 JS (1737–8), 2:860.
167 JS (1675a), 95; idem (1737–8), 2:571, 860, 902, 910; *LJS* 80.
168 JS (1672), 1–2; *LJS* 106; JS (1669), pt. 1:81.
169 JS (1737–8), 1:73.
170 JS (1737–8), 1:101, 176–7, 300, 2:517, 583; idem (1675a), "Ernstige aanspraak," [13].
171 See, for instance, Eamon (1985), 26, 45–6, and passim.

dam himself had earlier written of such secrets – and of nature's sealed treasure house of wonders – he had had in mind the social life of bees and the life cycles and metamorphoses of caterpillars;[172] but his subsequent microscopic dissections gave these rhetorical flourishes a new and very concrete embodiment. In such small anatomies as the miraculous snail, where "in the space of a tiny point, the great Master Builder [*Boumeester* now rather than Architect] has hidden and locked up so many astonishing and wonderfully ingenious contrivances [*overkonstige uytvindingen*],"[173] the concealment was now literal and physical as well as metaphorical. Another cliché had acquired a vivid new meaning.

To be sure, larger anatomies offered no less concrete a version of buried secrets, and Swammerdam himself affirmed that the bodies of all animals including humans were incomprehensible and inscrutable because of the multitude of wonders they enclosed.[174] But the secrets discovered in the microscopic anatomies of lower animals had a special aura. The strangeness and endless diversity of structures so alien to the familiar anatomy of higher animals – the very extravagance of the *inventie,* as it were – seemed to bespeak an inexhaustible display of creativity that far exceeded and at times appeared to defy the needs of organic function. Contemplating the miraculous, viviparous snail, Swammerdam was also struck by God's having made an animal that, though filled with "stones" – the crystalline globules – could nonetheless move and contract its parts;[175] since he found such stone-encumbered movement so remarkable, however, it is unlikely that Swammerdam could conceive of those crystalline globules as having been embedded in the snail for the sake of the snail itself. For what purpose then, or for whom, had God created and hidden them? Swammerdam's frequent recourse to the idiom of the arts in characterizing what he saw through the lens may reveal a readiness to account for the sights that confronted him as a display of creativity for the sake of the display – for the sake, that is, of an eventual observer. Had these "works of arts" as "wonders" not been hidden there precisely, then, to astonish their discoverer?[176] Swammerdam did not commit himself, and, in the instance of the marvels within the miraculous, viviparous snail, he stressed rather that he had no inkling why they were there;[177] but the conception of nature as intended revelation would have suggested that he himself was among the reasons.

As secrets intended for eventual discovery, God's wonders hidden in small anatomies were concealed by their diminishing dimensions as well as by the

172 JS (1669), pt. 1:106–7, 131.

173 JS (1737–8), 1:176–7.

174 Ibid. 1:300, 2:669, 902; Ruysch (1733), 12.

175 JS (1737–8), 1:172.

176 In addition to his reference to the "works of art" within the miraculous snail (JS 1737–8, 1:180), Swammerdam also on occasion spoke not only of some minute structure but also of the insect itself or even an aspect of its behavior as an "artwork," "masterpiece" (*konst-stuk* or *proefstuk*), or "showpiece" (*pronkstuk*) of God (ibid. 1:309, 2:456, 773, 908). To be sure, he likewise wrote of the human body (though with inconsistent spelling) as a *kunst-stuk* (JS 1675a, 227) in addition to likening it to a machine (JS 1737–8, 2:859).

177 JS (1737–8), 1:173.

body parts around them. Together with the seemingly endless diversity of these anatomies, the sheer abundance of the wonders that could hence be packed "in the space of a tiny point" implied a bottomless reservoir of hidden treasures, a hoard one easily stumbled upon. When recalling what he had encountered within the pupa of the soldier fly, Swammerdam remarked how the wonders of God in such small creatures were virtually thrown before us in a heap.[178] Yet the experience of research also bred an anticipation of further marvels that perpetually lay just beyond the advancing frontier of improving techniques. Two years (apparently) before his death, Swammerdam attested that he could always discover something new in what he had researched before. The prodigious works of God were inexhaustible, he noted as well after remarking on the mayfly's spreading tracheae, and creatures like the miraculous, viviparous snail encompassed so many marvels that dissecting the smallest animals could consume a lifetime. Given the incalculable number of such animals, it was a truly limitless prospect that now stretched before him; "I believe that were one to live eternally," he wrote Thévenot, "he would eternally discover new wonders."[179] A cluster of traditional, essentially religious ideas had been transformed by microscopic experience itself into a promise of endless microscopic discovery.

Refashioned in the light of his extraordinary endeavors, conventional religious themes provided the idiom through which Swammerdam articulated the experience of discovery, and his perception of research as piety contributed to the persistence and intensity with which he explored minute anatomies. It was a wavering perception, however, for Swammerdam also prized his research as an achievement that would establish his stature as a scientist.[180] As he briefly but vividly acknowledged, piety and the yearning for renown thus contended as incentives, endowing his research with both added tension and added intensity. But Swammerdam's decisive commitment to microscopic dissection, after Malpighi had revealed what was possible, is best understood as having sprung as well from more elusive emotions that remain unmentioned even in his unusually personal scientific writings. Diverse and compounding sources of social discomfort fostered a desire for withdrawal and a shunning of orthodox fields of research, so that the need to accommodate that withdrawal with his yearning for recognition decisively shaped his efforts. He had been drawn to natural history in childhood, and that early infatuation was subsequently fueled by the promptings of both religion and ambition; but his pathbreaking microscopic research was the product as well of an afflictive difficulty with personal and social interactions.

178 Ibid. 2:517, 681.

179 *LJS* 80; JS (1737–8), 1:176–7, 194, 250.

180 He declared that he had in fact succeeded in his studies in revealing matters that the finest minds had been blindly seeking for some two thousand years, or from the time, that is, of Aristotle (JS 1737–8, 2:873), and his consciousness of the pioneering character of his insect dissections was explicitly voiced in the case at least of the mayfly (JS 1675a, 69–70).

CHAPTER SIX

Leeuwenhoek I: A clever burgher

The lives and temperaments of Jan Swammerdam and Antoni van Leeuwenhoek were strikingly different. Swammerdam was increasingly ill at ease in the Dutch middling burgher society from which they both derived, and in his last years he was looking to escape it. By contrast, Leeuwenhoek remained secure and comfortable in its midst, mixing in its business, profiting from its practices of municipal government, and cultivating in its small homes and bustling society the blend of privacy and company he enjoyed. Unlike Swammerdam, Leeuwenhoek also married – twice – and fathered five children. Only his daughter Maria survived infancy, but she remained with him to the end of his life.

Although the younger by five years, Swammerdam had developed his scientific interests much earlier, had soon made his mark in the world of international learning, and carried through his design for a great work in the feverish and exhausting years before his death just days after reaching the age of forty-three. Not until Leeuwenhoek was forty do we first hear of his microscopes and observations, which he then continued steadily and persistently, although without Swammerdam's pressing sense of insufficient time, through the remaining half-century of his life. Like Swammerdam, he would also marvel at the handiwork of God discovered through the lens, but never so effusively, and without the accompanying anxiety over the state of his soul. Nor did he share Swammerdam's responsiveness to the arts.

Despite striking contrasts in their scientific biographies, however, an essential similarity remains: For Leeuwenhoek as for Swammerdam, a sensitivity to social relations and an eventual positioning on a social margin appear to have been critically related to the sustained commitment to microscopic research that underlay his ultimate achievement. The idiosyncrasies of personality and social situation again played a decisive role. Idiosyncrasies being what they are, however, even this similarity in their scientific lives reflected the differences in their personalities and situations. Swammerdam's emotional makeup compelled him to retreat to the fringes of a scientific social group of which he had once been more solidly a member, but Leeuwenhoek was an original outsider who, unexpectedly drawn to the fringes, found that his background prevented any closer approach. Whereas Swammerdam's intensive recourse to the microscope was an aspect of the very process of withdrawal, Leeuwenhoek's persistent use of the instrument appears a response to finding himself on a social boundary. That response also betrayed a sensitivity less to the discomforts of social interaction than to the linkage of social identity and self-esteem.

Leeuwenhoek was born in 1632 in Delft, which, sharing in the rapid urban growth in the province of Holland that continued into the 1670s, numbered perhaps thirty thousand inhabitants by midcentury, roughly half the size of Leiden and a fourth that of Amsterdam.[1] His mother came from a well-connected family of Delft brewers, whereas his father (and *his* father before him) was, as a basket maker, an apparently substantial artisan. Antoni was the fifth of seven children. When still a boy, his father having died, he went to live for an uncertain time with an uncle just outside the city before departing, at sixteen, to learn the draper's trade in Amsterdam. Marrying his first wife, a draper's daughter, in 1654, he returned to Delft, set himself up in business, and bought the house in which he would live for the rest of his life. Six years later, he acquired the first of several minor municipal offices, largely sinecures, that were to provide him with a further and growing income; the salary of this first office was soon raised to 400 guilders (the annual income Swammerdam had hoped to live on from his father's legacy), and by his death his municipal salaries and related incomes had risen perhaps to well over twice that sum. At what point he abandoned his drapery business is unknown – although the last transaction of which we know was in 1660, he still wrote in 1676 that circumstances, whatever they may have been, allowed him only limited spare time to pursue his microscopic observations – but that he was comfortably off is suggested by the number of gold and silver microscopes he left at his death in 1723.[2]

The earliest surviving reference to Leeuwenhoek's interest in microscopes and microscopic observation came in the spring of 1673 from the hand of de Graaf, for some years now also a resident of Delft. The occasion was a brief letter written to Henry Oldenburg, secretary of the Royal Society, in the context of de Graaf's clash with Swammerdam; in the closing lines de Graaf informed Oldenburg of "a certain very ingenious man named Leeuwenhoek" who had made some excellent microscopes. When and why this tradesman and minor official first became involved with the instrument remains a puzzle. The earliest microscopic observations to which he himself unambiguously refers had taken place only two years before, and those de Graaf forwarded with his letter as examples of Leeuwenhoek's efforts – mold sporangia, the sting, mouthparts, and eye of the bee, and varied details of the louse – do not yet bespeak a prolonged,

1 Faber et al. (1965), 56–8; Mols (1954–6), 2:253, 522–3; Montias (1982), 220.

2 Dobell (1932), 19–37; Schierbeek (1950–1), 1:12–30, and the genealogical table facing p. 278. (Although Schierbeek's genealogical table also indicates that Leeuwenhoek's mother died in 1644, Dobell [p. 22] is very definite in placing her death in 1664.) *AB* 2:158–9; Haaxman (1875), 17. We know of three gold and over 170 silver microscopes (*Catalogus;* Folkes 1723).

intensive, or very original program of microscopic observation.[3] Nonetheless, years later Leeuwenhoek was to write that he had in fact been making tiny lenses, apparently beads, as early perhaps as 1659![4]

Indeed, it would not have been out of character had Leeuwenhoek originally been more intrigued by the technical challenge of making the instruments than by any thought yet of their serious application. He always delighted, it appears, in exercising his ingenuity and self-sufficiency in technological matters, and having, for instance, taught himself such technical skills as glass blowing and metal working, he would even in later years extract the silver for his microscopes directly from the ore.[5] Consequently, whatever the initial source of inspiration – perhaps Hudde, as Swammerdam suggests[6] – the prospect of making so extraordinary an instrument as a powerful microscope by simply fusing beads of glass would surely have appealed to such a penchant for technological dabbling.

In any event, Leeuwenhoek himself was conscious of his technical innovations as the background to the beginnings of his celebrity – he alluded later in 1673 to the interest that had been aroused by "my recently invented microscope" (*mijn nieuw gevonden microscopix*) – and the first of the visitors who soon began to descend upon him were struck indeed by his "various and very interesting [*curiosissimus*]" microscopes and the skill with which they were made. Although de Graaf, to be sure, had touted the accuracy of Leeuwenhoek's observations as well, he too placed more stress on the excellence of the microscopes and included the observations with his letter to Oldenburg only as an example, a specimen, of the quality of those instruments. Oldenburg echoed this emphasis when he concluded the first account of Leeuwenhoek's observations in the *Philosophical Transactions* with a note of expectation that more was yet to come, "the better to evince the goodness of these his Glasses."[7]

Leeuwenhoek's introduction to the Royal Society marked a turning point in his life and the beginning of that purposeful exploitation of his instruments that turned into a program of microscopic research unprecedented in its scope and persistence. It was a development that directly reflected the society's early and continuing interest. In 1679, after six years of having had observations published in the *Philosophical Transactions* (and six months, however, of not having heard from the Royal Society), he wrote that he would not know to whom else to turn. The following year, when the society elected him a Fellow, he declared

3 Oldenburg (1965–), 9:602–3; *AB* 1:44–5, 28–39. Dobell has argued, however, that Leeuwenhoek had a microscope with him when he visited England in 1668 (Dobell 1932, 51), whereas Brian J. Ford has more recently proposed that it was Leeuwenhoek's exposure to Hooke's *Micrographia* during that trip that first aroused Leeuwenhoek's interest in the microscope (Ford 1985, 39–40). Cf. Berkel (1982), 189.

4 AvL (1702), 91.

5 AvL (1696), 127; *AB* 8:114–17; AvL (1702), 97, cf. 5–7, 12, 382, 386; Dobell (1932), 96.

6 JS (1737), 2:377.

7 *AB* 1:42–3; Bartholin (1673–80), 3:7; Jacobaeus (1910), 84; Oldenburg (1965–), 9:602; *Philosophical Transactions*, no. 94 (19 May 1673), p. 6038.

himself bound for life by the honor and pledged to dedicate himself henceforward to making himself more worthy, a commitment he reiterated four years later in gratitude for other expressions of appreciation.[8] The influence of the Royal Society hence underscores the critical importance of the social context of Leeuwenhoek's microscopic efforts from the first moment they become known to us in the surviving historical records.

Leeuwenhoek was not without support and encouragement closer to home, however. It was, after all, de Graaf who had initiated his association with the Royal Society and, according to Leeuwenhoek, had first persuaded him to commit his observations to paper. De Graaf was to die soon after in August of 1673, but by then the elder Constantijn Huygens, now well into his seventies, had also developed an interest in Leeuwenhoek. When Huygens and Leeuwenhoek first communicated remains unknown, but, even before de Graaf's death, Leeuwenhoek was already consulting the eminent poet and cosmopolitan courtier about observations for the Royal Society.[9] They visited and corresponded often during the next few years, Constantijn exhorting him to persevere and assuring him of the pioneering nature of his efforts.[10] Indeed, several of the Republic's most prominent personages would in time number among Leeuwenhoek's correspondents, who also came to include such admiring foreign luminaries as Leibniz.[11]

Yet it was the Royal Society that provided the most significant and enduring of Leeuwenhoek's scientific relationships, continuing, despite significant breaks, literally to his deathbed. It was through his letters that he communicated all his observations, and the great majority of these letters (over three hundred, if one reckons mere polite exchanges as well) were written to the society. It served as more than a passive recipient, however: The interest shown by such members as Oldenburg, Hooke, and Nehemiah Grew helped to fashion the nature and direction of his research as well as to sustain his commitment through the early years. In the long run, the *Philosophical Transactions* would also prove a major source of his belated scientific education.[12]

De Graaf had proposed in his original letter to Oldenburg that, if the Royal Society chose to encourage Leeuwenhoek and to test his skill, it should put to him some more difficult questions regarding the subjects he had observed, and, as the years passed, Leeuwenhoek continued to follow the many suggestions from the society as to where he should next turn his lenses. The course of his observations after 1673 was thus in no small measure a response not only to the

8 *AB* 3:100–3, 230–1, 220–3, 4:254–5.
9 *AB* 1:42–3; ConHs (1911–17), 6:330–1.
10 *AB* 1:66–7, 122–3, 206–7, 2:228–9, 3:82–3, 7:360–3; AvL (1693), 476.
11 Christiaan Huygens soon came to number among Leeuwenhoek's correspondents, and so likewise did Boerhaave, Antonie Heinsius, who was the Grand Pensionary of Holland, Nicolaas Witsen, burgomaster of Amsterdam, and, still closer to home, Hendrik van Bleyswijk and Jan Meerman, both burgomasters of Delft.
12 Schierbeek (1950–1), 1:42; Palm (1989a), 152; idem (1989b), 200–3. On the vicissitudes of Leeuwenhoek's correspondence with the Royal Society, see Pas (1975).

encouragement but to the specific interests of the Royal Society.[13] In time, however, he proved similarly receptive to the suggestions and requests of additional correspondents, not to mention the many who simply showed up at his door with all kinds of oddities for him to examine through his microscopes.[14]

That he offered written accounts of his observations only in his correspondence in itself underscores the decisive importance of continuing social interaction to his microscopic investigations. Translations or translated extracts of 116 of these letters ultimately appeared in the *Philosophical Transactions,* and another half-dozen were printed in the Dutch journal *De Boekzaal van Europe*. In 1684, Leeuwenhoek was also persuaded to publish the first of a series of volumes, initially in Dutch and subsequently in Latin translation, that ultimately reproduced 165 of his letters to the Royal Society and others.[15] Together with the illustrations that often accompanied them, these letters remain the only surviving record of his observations.

Another remarkable expression of his attentiveness to an interested audience is lost to us: a unique collection of microscopic preparations that he, like Swammerdam, was forever building. Unlike Swammerdam's, however, Leeuwenhoek's collection was emphatically and almost exclusively microscopic and constantly in use for the sake of a steady stream of visitors, one that in time became oppressive. Were he to receive everyone who came to his house, he remarked in 1710, "I would have no freedom and would be like a slave." Although by then he required letters of introduction, he still wrote the following year that in the space of four days he had received twenty-six persons – all, except for a duke, a count, and their tutor, provided with such letters. Oppressive as the numbers became, however, his desire to share the images he encountered in his microscopes never waned.[16]

He had, indeed, very early begun fixing the more striking of his observations before their separate microscopes in anticipation of visitors, and by 1676 could offer some fifty different images of wood alone. He continued to augment and

13 Oldenburg (1965–), 9:602–3. Regarding suggestions from the Royal Society and Leeuwenhoek's response: *AB* 1:92ff., 2:206–7, 210ff., 290–1, 326–7, 3:146–7, 422–3, 4:152–3, 6:18–19, 312–13, 7:136–9; AvL to J. Chamberlayne, 10 Sept. 1709, AvL Letters, fol. 114r; AvL to J. Jurin, 19 March 1723, ibid., fol. 316v–317r. Leeuwenhoek at times solicited such suggestions: *AB* 6:306–7; AvL (1702), 222.

Concerning Leeuwenhoek's consciousness of the society's appreciation of his work, see in addition to nn. 8, 60: *AB* 3:142–3, 348–51, 384–5, 418–19, 4:254–5, 5:216–17, 6:4–5, 16–19. But there were less happy instances when Leeuwenhoek felt he was being neglected as well: *AB* 3:106–7, 7:132–3, 8:177, 186–7; AvL to J. Chamberlayne, 17 May 1707, AvL Letters, fol. 27v; AvL (1718), 69; AvL to RS, 27 June 1721, AvL Letters, fol. 265v; Pas (1975), 5–9.

14 As examples of other correspondents whose questions or concerns he took up: *AB* 7:192ff., 11:26–9, 80–3; AvL to RS, 21 March 1704, AvL Letters, fol. 282r, 288v; Belloni (1963), 328; AvL (1718), 290, 308–13.

The odds and ends people brought to Leeuwenhoek for him to observe varied from ditch-water to vomit; see, for instance: *AB* 7:98ff.; AvL (1694), 68, 621–31; idem (1697), 298, 320.

15 Dobell (1932), 389–97; Palm (1989a), 148.

16 AvL to RS, 14 Jan. 1710, AvL Letters, fol. 125r; AvL to J. Petiver, 18 Aug. 1711, ibid., fol. 140r. As examples of his desire to share his observations even in these later years: AvL to RS, 5 Feb. 1703, ibid., fol. 221v–222r; AvL (1718), 118–19, 316.

replace these preparations through his very last years, constructing new microscopes specifically for that purpose.[17] In 1700 he spoke of having hundreds of the instruments, and when Zacharias von Uffenbach visited ten years later, he was shown not only a number of microscopes with a variety of objects set before them, but some hundred and fifty small cases wherein such microscopes were kept, usually two by two. In 1747, following the death of his daughter and a quarter-century after the death of Leeuwenhoek himself, a legacy of 173 such cases and small boxes were put up for sale by auction. Two hundred and forty-eight of his typical microscopes, constructed to hold objects before the lens or lenses, were among the contents of these cases, 167 of the instruments still having their object intact before them. With these should be reckoned the twenty-six microscopes willed to the Royal Society, each of which had also once been provided with its specific object.[18]

Preparing and fixing the objects themselves was in itself no small undertaking. In 1688, Leeuwenhoek wrote of having killed more than a hundred mosquitoes over the course of several days in an attempt to display the mouth parts before a microscope for others to observe.[19] Some solid objects he simply glued to the pins that held them before his lenses,[20] whereas others – from precipitated gold to insect muscle fibers and even spermatozoa – he allowed to dry on pieces of mica or of extremely thin glass and then fixed the pieces to the pins.[21] Fluids, such as blood or the alcohol containing embryonic oysters, he often set before the lens in fine glass tubes that he drew out on occasion, he claimed, to the thickness of a hair.[22] To show the actual circulation of the blood as well, however, he constructed specially designed microscopes to be used with finger-sized eels, a supply of which he tried for a while to keep alive at home for the sake of visitors.[23]

His collection of preparations was the means by which Leeuwenhoek hoped to share the experience of microscopic observation. Like Swammerdam, he was conscious of the inadequacy of words to convey what he saw through the lens;[24] but he was also troubled, more acutely than Swammerdam, that it might therefore remain only a private experience. It distressed him when a technical diffi-

17 *AB* 1:142–3, 156–7; 2:34–5; 8:177, 186–7. See also, for instance: AvL (1694), 536; idem (1718), 16–17, 142; AvL to RS, 9 Jan. 1720, AvL Letters, fol. 225r; AvL to RS, 21 April 1722, ibid., fol. 269v.

18 AvL (1702), 305; Uffenbach (1753–4), 3:358; *Catalogus*; Folkes (1723), 447, 449–50.

19 *AB* 7:348–9, 358–9.

20 AvL (1694), 543, 602–6; idem (1696), 93; idem (1702), 289; AvL to RS, 22 Sept. 1711, AvL Letters, fol. 148v, 149v–150r.

21 Uffenbach (1753–4), 3:356; *Catalogus*, item 53; AvL (1718), 118, 295; idem (1702), 99; Folkes (1723), 450.

22 *AB* 1:96–7, 108–9, 118–19, 120–5; Uffenbach (1753–4), 3:354–5; see also *AB* 2:258–9; *OCCH* 8:27.

23 *AB* 8:80–95; AvL (1696), 20; idem (1702), 376–7; Uffenbach (1753–4), 3:354.

24 *AB* 2:12–13, 34–5; AvL (1718), 103, 399. Leeuwenhoek repeatedly remarked that what he had seen could only be grasped by those who saw it for themselves: *AB* 7:12–13, 8:72–3, 112–13, 256–9; AvL (1696), 10; idem (1702), 297–8; idem (1718), 68, 118–19.

culty (the rapid desiccation of a bit of nerve) prevented his showing an observation to others, and he went out of his way on occasion to entice them to come and share an extraordinary sight.[25] To be sure, he also derived great pleasure from long stretches of private observing,[26] yet he struggled to share the extraordinary images that his instruments and his skills had brought to light.

Indeed, doing so had quickly acquired a very pressing significance. Although it was specifically the "wonders" that Leeuwenhoek had first displayed for his visitors,[27] his preparations and particularly his demonstrations for special visitors served as more than a showcase of marvels: They pertained as well to the question of credibility. The very reach of Leeuwenhoek's instruments and researches now engendered an atmosphere not only of astonishment but also, too often, of disbelief. Indeed, few others could duplicate his more difficult observations, and it was surely not without a touch of frustration, if not bitterness, that he alluded to those who took issue with him because they thought their microscopes as good as his. Nor was it only a question of his instruments, however; his experience, ingenuity, skill, and patience – even the hostile Hartsoeker granted him the patience of an angel – forged this dilemma.[28]

His secrecy regarding both his methods and his microscopes compounded the problem,[29] but it was a very selective secrecy, and what he did reveal suffices to confirm that he was no less inventive in his techniques of observation than in his instruments. Swammerdam recognized Leeuwenhoek as the first to use small glass tubes with the microscope, and Hartsoeker, in a similar moment of grudging admiration, acknowledged that no one before Leeuwenhoek had thought of using a bead lens to observe transparent objects against the light.[30] Both practices provided Leeuwenhoek rich rewards, and for the latter he also pioneered the technique of sectioning very thin slices, cut with a razor-sharp blade and laid wet upon a piece of glass.[31]

He also brought an uncommon skill to bear in such techniques. (Uffenbach was struck by the continuing steadiness of his hand at the age of seventy-

25 AvL (1718), 316, 118–19, 287; *AB* 8:80; AvL (1702), 299.

26 See nn. 49, 51.

27 *AB* 2:34–5.

28 AvL to RS, 24 Jan. 1721, AvL Letters, fol. 247v; NH (1730), "Ext. crit.," 42. Boerhaave also described Leeuwenhoek as a man of infinite patience (Boerhaave & Haller 1740–4, 3:717).

On the basis of the instruments with prepared objects left to the Royal Society, Martin Folkes, vice-president of the Royal Society at the time Leeuwenhoek died, remarked that the latter's skill in preparing objects for observation was itself difficult to duplicate and had contributed to his ability to discover what others could not; Folkes stressed Leeuwenhoek's unflagging effort and experience as well (1723, 452). What Leeuwenhoek was able to achieve through his methods of observing has also astonished modern microscopists (Ford 1973, 56, 58; Dobell 1923, 311).

29 *AB* 1:210–11, 292–3, 2:104–7, 114–15, 200–1, 204–5; Birch (1756–7), 4:365, 386; Molyneux (1692), 281; Uffenbach (1753–4), 3:356, 358.

30 *LJS*, 106; NH (1730), "Ext. crit.," 44–5.

31 Dobell (1932), 333. For examples of Leeuwenhoek's practice in sectioning slices, see, for instance: *AB* 1:270–1; AvL to RS, 6 Dec. 1707, AvL Letters, fol. 67v–68r; AvL (1718), 128, 361, 367–8. Leeuwenhoek may have also been the first to have tried staining a subject – muscle fibers, in this instance – for microscopic examination (Seters 1933, 4584; J. Baker 1945, 4–5).

eight.)[32] He claimed that he cut slices of animal tissues – nerves, muscle, fat, sinew, and skin from his own hand and arm – as thin as a hair; indeed, recently discovered samples of his slices from various plant materials compare favorably with the best hand-cut sections that would be prepared today. The skills he developed in insect dissection have also, in instances, challenged twentieth-century observers and their modern instruments.[33]

Such extraordinary manipulative dexterity employed in such novel ways inevitably added to the incredulity that often greeted his observational accounts. Specifically questioning Leeuwenhoek's insect dissections, Hartsoeker provoked what must have been a picturesque moment. Half in the street, as Hartsoeker himself tells it, and half in the entryway to Leeuwenhoek's home (from which he was being unceremoniously ushered out), he challenged Leeuwenhoek as to how he could possibly dissect a flea, much less a mite. (Swammerdam's work on the flea remained as yet unpublished, and Leeuwenhoek did indeed have trouble with the mite.) What kind of lens could one use, queried Hartsoeker, for with a small lens, the observer himself would block the light, whereas a large lens would not be strong enough. And what kind of knife would do? Wouldn't the knife itself block the observer's view and at the same time push the exposed anatomy out of focus? To such questions, Hartsoeker's account concludes, Leeuwenhoek had no answer.[34]

The broader problem was of greater moment than the aggressiveness of a would-be rival, however. Observations that depended on such exceptional abilities, techniques, and instruments – particularly when partially shrouded in secrecy – were all the more difficult for others to reproduce. Given the novelty of Leeuwenhoek's researches, the unexpectedness of his claims, and the general waywardness of the imagination, even those more kindly disposed toward Leeuwenhoek often received his reports with skepticism. In the circumstances of the day, his best recourse was indeed the supporting confirmation of witnesses with whom he had shared his observations, and witnesses of social standing and reputation – those "who I trust will surely merit the fullest confidence" – a social nuance that was not misplaced.[35]

He was early aware of the difficulty. He knew in the spring of 1674 that others, including even Christiaan Huygens, were still unable to see the microscopic globules he himself had by now observed in the blood, and early the next

32 Uffenbach (1753–4), 3:350; cf. AvL (1718), 106, 429, 459; AvL to RS, 9 Jan. 1720, AvL Letters, fol. 225r.

33 AvL (1718), 316–17. See also, for instance: AvL to RS, 24 Jan. 1721, AvL Letters, fol. 246r; AvL to RS, 1 May 1722, ibid., fol. 277r–v; AvL (1718), 329, 418, 421–2, 425–6; Ford (1991), 51ff.; Cole (1937), 29–30.

34 NH (1730), "Ext. crit.," 7–8; see also idem (1710), 84. Regarding Leeuwenhoek's dissection of the flea and the mite, see, for instance: *AB* 3:324–9, 4:20–3; AvL to RS, 22 Sept. 1711, AvL Letters, fol. 150r; AvL (1718), 108.

35 *AB* 8:54–7. In addition to citing their knowledgeability and good judgment, Leeuwenhoek spoke of these witnesses as "gentlemen," *Heeren*, a bit of information not without significance to the leading members of the Royal Society, who were fully conscious of the importance of proper witnesses in early modern science (Shapin & Schaffer 1985, 55–9).

year Oldenburg informed him of those "of great judgment" in Paris and elsewhere who did not believe in the globules he claimed to have seen in a variety of other substances as well. Having provided Christiaan's father, Constantijn, Sr., with some of his small glass tubes and an explanation of how he himself had observed the globules in the blood, Leeuwenhoek also replied to Oldenburg that, if they lived in the Netherlands, he would invite those men of great judgment to come see the globules in things for themselves. In the following year, he offered to do a study of wood in the presence of some learned gentlemen and to send the Royal Society an attested report of what they saw.[36]

Leeuwenhoek's celebrated "little animals" (*diertgens*) in spice infusions occasioned more notorious doubts about his credibility. In October 1676, he sent Oldenburg a famous letter recounting his systematic, often daily observations of microscopic creatures that displayed the most bizarre forms and behavior and often appeared in incredible numbers. Those who had not seen these things would indeed find them difficult to conceive, he later granted; but in his October letter he either declined or neglected to describe his techniques for observing them – almost a study in itself, he noted at one point – and explicitly remarked that at least some of these techniques he had no intention of sharing.[37]

The translated letter was read during the course of three meetings of the Royal Society in early 1677, and Oldenburg was instructed to inquire further into Leeuwenhoek's "method of observing" so that the society might confirm his observations. Leeuwenhoek replied in March that, at least when showing the animals to others, he drew a small amount of the water to be observed into a capillary tube to set before his microscope, but he still kept to himself other methods – and other lenses – by means of which, he said, he saw the smallest of these creatures. (Those methods remain to this day a subject of speculation.) He added in conclusion, however, that, in order to reassure the society as to the multitude of living things he had observed in such a small quantity of water, he would procure an attestation (*attestatie*) when he encountered them in great numbers again.[38] This he did in October, forwarding the signed testimonials of eight respectable visitors to satisfy the "Gentlemen Philosophers" as to the existence of thousands of animals in a sample of water the size of a millet grain. In the interim, both Hooke and Grew had attempted without success to duplicate Leeuwenhoek's observations, and only in the following month did Hooke finally succeed in providing the society with its own observations of Leeuwenhoek's little animals.[39]

36 *AB* 1:122–5, 278–9, 2:38–9.

37 *AB* 1:64–161 (esp. pp. 114–15, 118–19, and also 80–1, where he comments on the difficulty of the techniques he employed), 2:198–9.

38 Birch (1756–7), 3:332–4; *AB* 2:198–201, 204–7; re Leeuwenhoek's use of capillary tubes in his observations of microorganisms, see also *AB* 2:258–9, 266–71. Regarding speculation as to the method he would not reveal for seeing the smallest of the microorganisms he observed, see: *OCCH* 13:842; Dobell (1932), 331–2; Seters (1933), 4585–9; B. Cohen (1937), 344–6; Ford (1973), 52–62; cf. Cittert (1934a), [13]; Kingma Boltjes (1941), 72–4.

39 *AB* 2:252–71, esp. pp. 262–3, 270–1; Birch (1756–7), 3:338, 346–7, 349, 352.

In the summer of 1688, when he made his first observations of the blood circulating in the capillaries of tadpoles and other lower animals, Leeuwenhoek began inviting various gentlemen to come see this new spectacle as well. His initial purpose seems to have been simply to share the "beautiful sight," but near the end of the summer he was reminded of the doubts he often faced and was advised to procure again an attestation by eyewitnesses – advice he took "that I might suffer less contradiction" about these new discoveries. In 1711, he still wrote of having shown spermatozoa to others because of accusations that he did not tell the truth about such things. His continuing awareness of such skepticism sharpened his incentive to enable others to share his experience with the microscope more directly, and the doubters and witnesses together thus constituted one more elaboration of the social context of Leeuwenhoek's microscopy.[40]

Leeuwenhoek's implicit allusion to the social warrant for the credibility of his witnesses, moreover, reminds us of yet another complexity in that context: the diversity of the social strata that composed it. His world of correspondents, favored visitors, and witnesses was largely one indeed of "gentlemen" and men of letters,[41] but the multifaceted society that enveloped him more closely in Delft and its environs was also very much a part of his new adventure with the microscope. Not only did the attitudes and beliefs of that society continue to encompass him, but he drew upon the experience of its diverse members as a resource and involved them, witting or not, as collaborators in his researches.

He lived but a stone's throw from the municipal fish market and butchers' hall, where he watched the butchers and mongers at work, pressed them with questions, and counted on them for the anatomical parts and specimens he needed.[42] It was from the butchers and the pieces of animals they provided that he had learned what he knew in anatomy, he stressed.[43] But he also debated different kinds of wood with carpenters, discussed eels and oysters with fishermen and grubs with farmers, and queried seamen about such things as whale skin and coconuts.[44] No one, moreover, shared more closely in Leeuwenhoek's researches than did the members of Delft's artistic community who served as his

40 *AB* 8:80–1, 7:276–7, 8:23–5, 34–7, 48–9, and esp. pp. 54–7. AvL to J. Petiver, 18 Aug. 1711, AvL Letters, fol. 140v; see also AvL to RS, 14 Jan. 1710, ibid., fol. 125r.

For examples of Leeuwenhoek's continuing awareness of persisting doubts about the truthfulness of his claims, see, for instance: *AB* 3:332–3, 5:304–5, 6:16–17, 7:388–9; AvL (1702), 54; idem (1718), 166–8; AvL to RS, 24 Jan. 1721, AvL Letters, fol. 247v; AvL to RS, 21 April 1722, ibid., fol. 269r. In 1710, Uffenbach was told by Leeuwenhoek's daughter that her father no longer wished to publish any more of his observations; prominent among his reasons, she said, was his having been accused of seeing more through his imagination than his lenses (Uffenbach 1753–4, 3:349).

41 See n. 35 and text later in this chapter.

42 *AB* 2:418–19; AvL (1718), 210, 408–10, 412; idem (1697), 321, 332; idem (1702), 423–4, 426, 297; idem (1718), 320, 322; idem (1694), 614.

43 AvL (1702), 137–8.

44 *AB* 6:136–39 (and see 156ff.), 9:148–9; AvL (1694), 694–5, 574; AvL to RS, 1 March 1712, AvL Letters, fol. 170r–v; AvL (1718), 29, 263. See also: AvL to J. Chamberlayne, 3 March 1705, AvL Letters, fol. 340r–v; AvL to RS, 29 Dec. 1705, ibid., fol. 395v.

draftsmen. Often immediate participants in his exploratory observations, they were a voice at his side even in the analysis of microscopic images.[45]

Local merchants, meanwhile, collected infested grain for him and allowed him to look for pests in their stores of cheese, and workers as well as governors of the East India Company (whose wharf loomed large on the city waterfront) were on the lookout for exotic vermin to pass on to him.[46] He recruited a whaler to bring back the eye and penis of a whale, asked neighbors to save the shavings from their visits to the barber, and even persuaded laborers to part company with their calluses.[47] His maidservant not only hunted fleas for him but literally gave her blood for his researches. He was disappointed and perhaps a bit resentful when his fellow townsmen declined to volunteer their blood while ill; but in giving what they did, sharing what they knew, and even in resisting him,[48] they too played their part in shaping Leeuwenhoek's microscopic explorations. Together with his attentiveness to the interests of his correspondents and his need to share his more striking observations, his pursuit of his inquiries among the townsfolk of Delft and the sailors, fishermen, and farmers of greater Holland enmeshed his researches in a dense and complex social network. To this extent, those researches entailed not withdrawal but expansive social interaction, and his own consciousness of the social diversity of those with whom he interacted played a critical role in the unfolding of his discoveries.

There was also a deeply private side to Leeuwenhoek's microscopy. At the core of such research remained inescapably a very insular experience, and, though he strove so persistently to overcome that insularity, the isolated absorption it afforded was a constant source of joy. He spent long hours simply looking – just staring, he wrote – sometimes with his daughter or an accompanying draftsman but still merely for the pleasure of the spectacle.[49] His letters describe a continuing succession of sights that delighted him, from the more dramatic of his discoveries, the masses of spermatozoa or the blood pulsing in the capillaries, to the less momentous images of the transparent "globules" in a grain of wheat, the crisp impression of an aphid's parts upon its molted skin, or a mite that, stuck on its back to the pin of his microscope, kept passing an egg back and forth from foot to foot.[50] He could watch such sights for hours until his eyes could simply

45 Leeuwenhoek's letters indeed reflect years of close interaction with the artists who served as his draftsmen; see, for instance: *AB* 1:42–3, 7:378–9; AvL (1694), 543; idem (1696), 151; idem (1702), 54, 56, 445, 448; AvL to RS, 21 March 1704, AvL Letters, fols. 285r–v, 288v; AvL to RS, 6 Dec. 1707, ibid., fols. 68v–69r, 70r–v; AvL to RS, 24 Jan. 1721, ibid., fol. 246v.

46 *AB* 8:278–9, 310–11, 328–9, 7:128–9; AvL (1702), 168–80, 182; Bleyswijck (1667), plate following p. 66.

47 AvL (1718), 38; AvL to RS, 1 March 1712, AvL Letters, fol. 168r; AvL to J. Chamberlayne, 22 Nov. 1709, ibid., fol. 121v–122r; AvL to RS, 7 July 1722, ibid., fol. 300v.

48 *AB* 2:412–13; AvL (1694), 542, 565; Haaxman (1875), 69. See Chapter 8, nn. 96–8, 101.

49 AvL (1696), 151–2; idem (1718), 64–5. See also, for instance: *AB* 4:266–7, 5:106–7, 7:318–19; AvL (1694), 619, 702; idem (1718), 102, 107, 243.

50 See Chapter 7 concerning the spermatozoa and the capillary circulation. *AB* 2:132–3; AvL (1696), 85; AvL to RS, 22 Sept. 1711, AvL Letters, fol. 149 v.

endure no more,[51] and his personal enjoyment of these endlessly varied images continued to echo through fifty years of letters. Nonetheless, throughout this same half-century, Leeuwenhoek's efforts and observations continued to reflect as well the busy and varied society in the streets of Delft and beyond. Despite the intensity of his private absorption in microscopic images, his microscopic research remained the product of the social contexts and circumstances that encompassed it.

Not surprisingly, the drives at work in that research appear in fact to have been numerous and diverse. Late in his life, Leeuwenhoek insisted that his microscopic efforts derived only from his inquisitiveness and an inclination to explore the principles (*beginselen*) of things. Though apparently something more, after all, than a simple delight in extraordinary images, that inquisitive drive was nonetheless also emphatically a very personal one that Leeuwenhoek characterized as an innate passion (*in geschapene drift*).[52] On other occasions, however, he also acknowledged more socially oriented purposes. In addition to instructing himself, he declared in 1691, his intent was to separate the world from its errors, and, three years later, he portrayed his efforts as having no other goal than that of leading the world to truth and freeing it from ancient superstition.[53] Often, an outright utilitarian commitment surfaced as well,[54] while, on a more personal level, he continued to respond to the individual interests of visitors and correspondents virtually to the end of his life.[55] Hence, beyond the sheer pleasure of new microscopic images, a fluctuating weave of diverse motivations, both personal and social, supplied Leeuwenhoek's own understanding – or, at a minimum, his rationalization – of what he was about.

Other incentives he was less likely to acknowledge, and perhaps to recognize, were also at work, however. His occasional retreat into secrecy betrays an incentive of a less edifying or congenial stamp; dedicated as it was to self-aggrandizement, it conflated social self-consciousness with very personal priorities. A case can be made that such unacknowledged additional incentives provided a critical, indeed decisive, stimulus to a continued commitment to the microscope itself, whatever the subject matter that came to hand. They can explain fifty years of dedication to an instrument that, as Hooke himself observed, could no longer claim any other major devotee by the time the seventeenth century drew to a close.[56]

Leeuwenhoek's commitment to the microscope was indeed unique. Unlike Malpighi and Swammerdam (or, for that matter, Nehemiah Grew), he gave

51 *AB* 3:408–9; AvL (1696), 107; idem (1702), 323; AvL to RS, 5 Feb. 1703, AvL Letters, fol. 218r; AvL (1718), 399.

52 *AB* 1:42–3; AvL (1718), 235, 222; AvL to RS, 21 April 1722, AvL Letters, fol. 269r; see also AvL to RS, 27 June 1721, ibid., fol. 265v.

53 *AB* 8:176, 184–5; AvL (1694), 672.

54 See, for instance, *AB* 6:218–19, 8:292–7, 312–13; AvL (1696), 78.

55 See nn. 13, 14 above. See also: AvL to RS, 16 Sept. 1704, AvL Letters, fol. 303r ff.; Leeuwenhoek to the Royal Society, 27 March 1705, ibid., fol. 354r ff.

56 Hooke (1726), 261, 268.

himself first of all to exploring the potential of the instrument itself rather than to pursuing a specific field – natural history or anatomy (or botany) – to which it brought an important, often groundbreaking new capacity, but to which it nevertheless remained only an adjunct. Like Hooke, Leeuwenhoek cast about for whatever could be set before his lenses; unlike Hooke, for whom the microscope was only a sporadic and occasional interest after the *Micrographia,* Leeuwenhoek held fast to microscopic research as the major preoccupation of the second half of his long and very active life.

Significantly, this was despite the fact that Leeuwenhoek too had varied scientific interests. In addition to his bent for acquiring diverse technological skills, he was early inclined to mathematics and, four years before de Graaf's letter to Oldenburg, had taken and passed the surveyors' exam, though he seems never to have practiced the profession. He had testified at the time that he had long "exercised himself in the art of Geometry," and his letters are liberally sprinkled with mathematical calculations carried out in extenso. These are often intended to convey the smallness and abundance of microscopic structures, such as the 3,181 still smaller fibers he posited within a muscle fiber or the 3.2 billion vertical vessels he ascribed to an oak tree with a four-foot diameter. In the latter instance, however, he went on to calculate the weight of the volume of water that would rise in these vessels during the course of a day, and other mathematical sallies dealt with such questions as atmospheric pressure, the resistance of the air to falling bodies, and the water pressure on the eye of a whale.[57]

Leeuwenhoek's own first letter to Oldenburg was a response to the Royal Society's request for drawings pertaining to the initial account of his observations forwarded by de Graaf, and Leeuwenhoek now commented as well on further microscopic observations of wood and of blood passing through a louse. But in addition to calculating (in the context of his observations on wood) the force required for a piston to raise the water in a vertical, hair-fine tube to a height of one hundred feet, he wrote that he could not refrain from also including his thoughts on the compression of air, and he proceeded to describe experiments with apparatus he had made composed of sealed glass pipes and plungers.[58] On the very first occasion on which he himself directly addressed the members of the Royal Society, consequently, he let them know that his scientific interests and capacities stretched beyond the skill in making microscopes and microscopic observations that had brought him to their attention.

Three letters later, while relating some observations of blood in fine glass tubes, he also volunteered a theory of capillary action. He had gathered from the figures in books by Hooke and Robert Boyle he had been shown that both authors had dealt with the issue. Unable to read English, however, he had not understood what they said, "and thus having never heard or read any demon-

57 Dobell (1932), 33–5; Schierbeek (1950–1), 1:22–3; *AB* 3:400–3, 152–7, and see pp. 160–3 as well; Dijksterhuis (1948).

58 *AB* 1:32–3, 42–3, 48–51, 54–61.

stration thereof, I shall here, with your indulgence, include my thoughts." Toward the end of 1675, he also mentioned a small glass apparatus he had invented to determine the specific gravity of different "waters," and expressed his pleasure at having inferred from the figures in the *Philosophical Transactions* that Boyle had contrived a similar device. The following spring, having now managed with the aid of an English–Dutch dictionary to gather the gist of some chemical experiments reported in the *Philosophical Transactions,* he related a series of his own subsequent experiments in which the microscope, though involved, played a very minor role.[59]

This was on the eve, however, of his October letter on the little animals in infusions, which, together with his letter a year later on spermatozoa, initiated the period of the most intense general interest in his microscopy. In the meantime, Oldenburg's response to Leeuwenhoek's nonmicroscopic offerings had been decidedly ambivalent. *All* that Oldenburg had published in the *Philosophical Transactions* from the text of Leeuwenhoek's first letter had been his account of his experiments with compressed air and his briefer description of the feeding louse, whereas Leeuwenhoek's reflections on capillary action were then deleted from later extracts. Those letters recounting both Leeuwenhoek's chemical experiments and his instruments for determining specific gravities never appeared in the *Philosophical Transactions* at all.[60]

Nonetheless, in time Leeuwenhoek's concern with nonmicroscopic questions would resurface. In 1686 he informed the Royal Society of experiments he had performed with the aid of various glass vessels to determine the amount of "air" and "expansion" produced by the combustion of gunpowder, and in 1692 he recounted another series of experiments with small glass air pumps to establish how much air was absorbed in blood and urine. He wrote five years later of experiments with magnetic fragments set afloat in variously shaped glass vessels, and, at the turn of the century, of barometric experiments and the question of capillarity again. In 1690, an apparatus he had constructed to demonstrate phenomena occasioned by a spinning Earth had roused the interest of Christiaan Huygens, and in 1703 he described another device he had contrived to illustrate barometric effects. Other devices he designed included a liquid level to be used in surveying and an inhalator for medicinal vapors.[61]

59 *AB* 1:98–101, 336–7, 2:44–59. The manuscript and contents of one of Leeuwenhoek's first three letters to the Royal Society have been lost.

60 *Philosophical Transactions,* no. 102 (27 April 1674), pp. 21–5; no. 106 (21 Sept. 1674), pp. 121–8. Conscious of what did and did not appear in the *Philosophical Transactions,* Leeuwenhoek did at time express his disappointment and irritation in later years: *AB* 2:206–7; AvL (1718), 95; AvL to J. Jurin, 19 March 1723, AvL Letters, fol. 316v.

61 *AB* 5:368–95, 9:4–27, 38–47; AvL (1718), 2–20; idem (1702), 252–60; *OCCH* 9:390 (see also AvL 1697, 263–8); AvL to RS, 5 Feb. 1703, AvL Letters, fol. 224v; AvL (1702), 380–6, 428–9. Again, the response to these efforts was not overly encouraging. From the four letters Leeuwenhoek sent to the Royal Society regarding the several series of experiments, only excerpts from the two dealing with gunpowder (after several years delay, at that) and magnetic fragments ultimately appeared in the *Philosophical Transactions* (no. 200 [May 1693]: 758–60; no. 227 [April 1697]: 512–18).

Leeuwenhoek's scientific tastes thus ran in many directions, and judging from a history of Delft written shortly after his death, his fellow townspeople considered him an expert in navigation, astronomy, mathematics, and philosophy as well as in "natural science" in some narrower sense.[62] It may well be, therefore, that if a fuller range of his scientific interests had found more encouragement, Leeuwenhoek's microscopic observations would have proved a more occasional and sporadic pursuit, as they did in the case of Hooke, Hartsoeker, and Christiaan Huygens. Given his enduring taste for mathematical calculations and his penchant for experimental and demonstrative apparatus, it is not at all clear, indeed, that his microscopic researches accurately reflected his most instinctive scientific inclinations.[63]

Hence, his extraordinary half-century focus on the microscope implies some further incentives more specific to the instrument than were either the personal enjoyment of scientific research or the ideals of general enlightenment or utility. Nor, in itself, does an ingratiating desire to accommodate visitors and correspondents measure up to the emotional investment Leeuwenhoek's difficult and persisting labors would seem to entail. Rather, Leeuwenhoek's letters suggest that his microscopic researches were tied additionally to an acute concern with a socially gauged self-perception.

Delft was a conservative city deeply conscious of class divisions, and its consciousness was one that Leeuwenhoek shared. Among the reasons he had given Oldenburg for his early reluctance to write down his microscopic observations was that he had been educated only for trade, and he initially asked even sympathetic supporters among the social elite to "always remember who I am" when passing judgment on what he offered.[64] When elected a member of the Royal Society in 1680, however, he asked the elder Constantijn Huygens if, having been invested with this quality, he would still be obliged to give way to a doctor in medicine. Having himself succeeded his father as secretary to the Prince of Orange, the younger Constantijn observed with patronizing amusement that Leeuwenhoek's election had indeed puffed up his vanity a bit. (It was in the mid-1680s that Leeuwenhoek also introduced the "van" into his name, although the name itself referred only to a house and street corner in Delft.)[55]

The social consequences of Leeuwenhoek's unexpected celebrity transcended mere questions of formal deference, however. All the world sought him out as the great man of the century, noted the younger Constantijn Huygens in the same slightly derisive remarks in 1680, and in the following years an exalted line

62 [Boitet] (1729), 765.

63 Anne Roe has noted suggestively that, among the eminent modern scientists she studied, many of those who became physical scientists and almost none of the others were taken up in their youth, at least, with one kind of "gadgeteering" or another (1953, 78, 232).

64 Montias (1982), 179; *AB* 1:42–3, 202–3, 206–7, 2:142–3, 3:136–9.

65 *OCCH* 8:296. In his surviving letters, the "van" first appears in Leeuwenhoek's signature in 1686 (*AB* 6:42–3); Schierbeek (1950–1), 1:12–15.

of visitors indeed descended upon Delft to peer through Leeuwenhoek's microscopes. A bright moment for Swammerdam had been the visit in 1668 of the future Grand Duke of Tuscany to see his own and his father's collections; but among the steady stream of Leeuwenhoek's visitors were aristocrats, princes, and royalty from all over Europe.[66] It was heady fare for a small-time former draper!

Although Leeuwenhoek appears to have shown no further interest in questions of social precedence, in later years he did betray a consciousness of the distance between himself and the ordinary run of humankind that peopled the streets and fields about him. If he had begun by urging those he perceived as his social betters to remember "who I am," he was also soon shaking his head over the misconceptions of the "common man" (*gemene man*).[67] The span of years over which such references are scattered testifies to a social distancing that, in contrast to the question of decorum regarding who deferred to whom, was of enduring significance to Leeuwenhoek. The emotional import of this distancing is suggested by Leeuwenhoek's later identification of the common man, whom he otherwise typically cited with tolerant understanding, with a distressing moment in the earlier years of his own growing celebrity.

In a letter to Oldenburg in 1674, Leeuwenhoek recounted how his ignorance of the difference between nerves and tendons – to both of which the same Dutch word, *zenuwen,* applied – had been exposed by one of his early visitors. Clearly embarrassed by the incident, Leeuwenhoek asked Oldenburg to keep it to himself, and, volunteering now that he had not applied himself to his studies as a youth, he closed his letter with an apology for his boldness in sending the Royal Society what he did. When, years later, he remarked that the common man called the tendons *senuwen* (an orthographical variant of the time) since he had no knowledge of the nerves (*senuwen*),[68] Leeuwenhoek – whether or not he consciously recalled his earlier humiliation – revealed that, for him, being a common man was painful. Acutely conscious of the shortcomings of his own background, Leeuwenhoek was much in need of reassurance as to where he stood on this social divide, and his new social affiliations offered the most compelling affirmation.

66 Lindeboom (1975), 13. Leeuwenhoek's visitors included the Electors of the Palatine and Bavaria, the Landgrave of Hessen-Kassel with his heir, the future James II of England and, subsequently, his daughter and usurper, Queen Mary; Czar Peter I invited Leeuwenhoek to join him on his boat as it passed through Delft, and the Prince of Lichtenstein arranged to have him show his discoveries to the Holy Roman Emperor in The Hague, although changing winds bore the latter off too soon to Spain (*AB* 11:86–7, 8:180–1, 3:106–9, see also 6:184–7; AvL 1702, 431–2; Folkes 1723, 450–1; cf. AvL 1702, "Opdragt"; Dobell 1932, 55–6). Haaxman also includes Charles II and George I of England, Augustus of Poland, and Frederick I of Prussia in the list of Leeuwenhoek's visitors (Haaxman, 1875, 111). See also Schierbeek (1950–1), 1: 52–4.

67 *AB* 2:390–3, 6:128–9; AvL (1697), 186; AvL to RS, 25 Dec. 1702, AvL Letters, fol. 205r; AvL to RS, [late winter or spring 1708], ibid., fol. 82v; AvL to J. Jurin, 13 June 1722, ibid., fol. 285r; AvL (1718), 372, 453.

68 *AB* 1:144–5, 166–7; AvL (1718), 453, see also p. 372.

It was not the visits of princes and crowned heads, of which he seldom spoke, that meant most to him, however, but rather his apparent acceptance as one of their own by the "gentlemen philosophers"[69] of the Royal Society and of the "Republic of Letters" at large. Fellows of the Royal Society appear to have visited Leeuwenhoek as early as 1675, and he declared his own election as a member to have been the greatest honor in the world. He was touched when the society subsequently hung his portrait in its meeting hall, and expressions of esteem from the society in his closing years was enough to move him to tears.[70] Closer to home, not only the attentions but the friendship of the elder Constantijn Huygens – who signed himself Leeuwenhoek's close (*bysondere goede*) friend – must also have touched him deeply, and it was doubtless with no small pride that Leeuwenhoek later spoke of the deceased Christiaan too as, in his time, "my great friend."[71]

Leeuwenhoek surely treasured the mutual esteem and even affection of such relationships, but they also signified a new and elevated social identification for the erstwhile draper, for all concerned seem to have been aware of the breaching of class lines that was entailed. On Leeuwenhoek's side, the elder Huygens had been among those he had urged to remember his origins, while Constantijn (who had also emphasized the humble social origins of another, earlier protégé, Rembrandt van Rijn) spoke of Leeuwenhoek to Christiaan as "our bourgeois philosopher at Delft."[72] Unlike the gracious but inevitably patronizing visits of princes and sovereigns, however, friendship implied enduring personal closeness and, hence, some degree of social assimilation. An exchange of correspondence did not in itself betoken such personal ties; but, like membership, it did represent, if continued, a measure of social inclusion. Hence, Leeuwenhoek's new associations offered him an enhanced and deeply gratifying new self-perception. The importance to Leeuwenhoek of this image is perhaps directly reflected in the style and presentation of his research: By always reporting his observations in the form of personal letters to the Royal Society or other "eminent," "learned," or "highly placed" persons (as he advertised on the title page of the volumes of letters he published himself), was he not implicitly underscoring a social context and its implications?[73]

Important as these new social ties became to Leeuwenhoek, however, they also remained necessarily tenuous, for the prejudicial stamp of his social background could not be erased. There was, in the first place, what seems to have

69 *AB* 2:206, 252, 254, 3:198. *Heeren liefhebberen* was a phrase Leeuwenhoek also commonly applied to the members of the Royal Society; see, for instance, *AB* 1:286–7, 294–5.

70 *AB* 1:336–7, 3:196–9, 6:86–9; AvL to RS, 21 April 1722, AvL Letters, fol. 269r.

71 AvL (1702), 68, 372; *AB* 7:360–3.

72 *AB* 1:206–7; *OCCH* 8:159; Slive (1953), 14.

73 In response to the inclusion of translations of his recent letters in Hooke's (1678) *Lectures and Collections*, Leeuwenhoek wrote Nehemiah Grew that he preferred to have his observations published in the *Philosophical Transactions*, in part, he said, because he was not himself inclined to publish a book, although he was repeatedly advised to do so (*AB* 2:360–1). In 1691, Leeuwenhoek spoke of his delaying the publication of certain observations until he had decided to whom to address or dedicate them (*AB* 8:177, 186–7).

been an uncertain grasp of social graces and a lack of savoir-faire. Swammerdam groused that Leeuwenhoek was so opinionated (*partiaal*) and his reasoning so barbaric that it was impossible to talk to him, and Hartsoeker commented as well on Leeuwenhoek's discourtesy if one presumed to criticize. Swammerdam's own manner could be difficult, however, and Hartsoeker doubtless in part asked for what he got; others, after all, found Leeuwenhoek a "very civil complaisant man."[74] But there were moments when Leeuwenhoek succumbed to what may indeed have been a streak of boorishness. When the visiting Landgrave of Hessen-Kassel apparently suggested that Leeuwenhoek give him some of his microscopes, the latter reportedly responded rather haughtily that he had never given any to anyone and did not intend to do so. Allowing the landgrave to see only three or four of his instruments with their preparations at a time, Leeuwenhoek further explained that he did so for fear of losing one, for he did not trust his visitors and especially not Germans, a point he repeated two or three times! "O che bestia!" wrote the younger Constantijn Huygens to his brother Christiaan, both of whom had hopes of the landgrave's becoming a Maecenas of the sciences.[75]

Hartsoeker was also appalled by what he took to be the bad taste displayed in Leeuwenhoek's letters. Although he declared at one point that unclean sights and strong stenches made him sick, Leeuwenhoek in time became the embodiment of Francis Bacon's injunction that lowly and even nasty things not be neglected. A potential source of awkwardness, however, lay in Leeuwenhoek's chosen form of communicating his observations: those letters to the social elite. Thus Antonie Heinsius, Grand Pensionary of Holland (but also a friend from Delft), was the recipient of a letter about what Leeuwenhoek had found between his toes after two weeks without having removed his stockings, and the Imperial Historiographer, Jan Gerard Kerkherdere, was honored with an account of Leeuwenhoek's bowel movements and the microscopic study of his excrement. Hartsoeker found the first letter, with the "most grotesque observations in the world," merely ludicrous, but the second he condemned as filthy.[76] How Heinsius and Kerkherdere reacted to Leeuwenhoek's contribution to their celebrity remains unknown.

Lapses in delicacy and decorum were not of great moment, however – at least not to Leeuwenhoek, who seldom betrayed any doubts or embarrassment about his social comportment. Gaps in his education were quite another matter; they were a more decisive barrier to the social identification to which he aspired, and he was very much aware of it. The ultimate distinction, after all, that separated the common man from the likes of Leeuwenhoek's idealized mem-

74 *LJS*, 106; NH (1710), 83–4. (Regarding Hartsoeker's inviting rough handling from Leeuwenhoek, see also Chapter 1, n. 91.) Birch (1756–7), 4:366.

75 *OCCH* 9:38, 31–2. Christiaan, to be sure, also found the landgrave's requests for instruments somewhat excessive.

76 *AB* 7:230–1; Bacon (1877–89), 1:214–15; AvL (1696), 103, 105, 109–10; idem (1718), 384–6; NH (1730), "Ext. crit.," 37, 65.

bers of the Royal Society – not just "gentlemen" but "learned gentleman philosophers"[77] – was one indeed of intellectual competence and commitment. Although he cited the "eminent" and "highly-placed" along with the "learned" among the recipients of his letters, it was specifically with other "inquisitive philosophical eyes" that, in 1676, he wished to share the wonders he had seen, and almost four decades later he still avowed that it was for the "philosophical" alone that he wrote.[78]

There is little doubt, to be sure, that he savored the attention of Europe's social luminaries regardless of their intellectual stature. In 1691, responding to a letter from the Florentine Antonio Magliabechi, he not only warmly thanked his correspondent for his expressions of esteem but proceeded, having brought up Leibniz's interest in his work as well, to mention the visits paid him by the King of England (apparently James II when still Duke of York) and other prominent personages such as the Bishop of London and the Elector of Bavaria. It is difficult not to detect a note of self-congratulation in this account of such illustrious recognition; and yet Leeuwenhoek, having then gone on to remark that the frequency of such visits had often in fact become an intolerable burden, concluded by lamenting how few of such visitors proved capable of the concentration required in observing what he showed them.[79] He could not but take pleasure in the attention paid him by the eminent, but it was a narrower intellectual elite among them with whom he wished to identify. That intellectual elite was distinguished by the range of its shared learning, however, and in his interactions with its widely dispersed members, the deficiencies of Leeuwenhoek's own education became only too apparent.

Though genuinely admiring *ce bon Leeuwenhoeck* for his curiosity and industriousness, the elder Constantijn Huygens had early described him to Hooke as a person "unlearned both in sciences and languages," a quality that inevitably struck others from the social elite who visited him. In a letter to the Royal Society from Holland in 1685, Sir Thomas Molyneux still expressed surprise to find Leeuwenhoek

> a stranger to letters, master neither of Latin, French or English, or any other of the modern tongues besides his own, which is a great hinderance to him in his reasonings upon his observations; for being ignorant of all other mens thoughts, he is wholly trusting to his own, which, I observe, now and then lead him into extravagancies, and suggest very odd accounts of things, nay, sometimes such, as are wholly irreconcileable with all truth.

Swammerdam also ascribed Leeuwenhoek's "barbaric" reasoning to his lack of schooling, and Hartsoeker, mixing reluctant admiration with hostility, voiced his astonishment at the technical originality of a "man unschooled and without genius like Mr. Leeuwenhoek."[80]

77 *AB* 2:254, 3:198–9. 78 *AB* 2:34–5; AvL (1718), 22. 79 *AB* 8:174–5, 180–1.

80 *OCCH* 8:159; ConHs (1911–17), 6:330; Birch (1756–7), 4:366; *LJS* 106; NH (1730), "Ext. crit.," 45; cf. Damsteegt (1982), 16–17.

The most obvious evidence of his educational inadequacy was indeed his ignorance of other languages. Latin still persisted as a language of international learning and scholarship, and an ability to handle it was a hallmark of the cultivated elite in the Netherlands as elsewhere. Such was also increasingly the case with French: Already the language of much Dutch international intercourse, its use was now a rising fashion in polished Dutch society.[81] Leeuwenhoek's linguistic provincialism was thus a social as well as intellectual stigma.

But even Leeuwenhoek's Dutch betrayed the deficiencies of his background. The grammar of his letters was frequently confused, and his spelling, at least initially, was old-fashioned and inconsistent. Colloquialisms and the usages of the tradesmen and craftsmen with whom he interacted – the common man, after all – abounded.[82] Hartsoeker abusively derided Leeuwenhoek's writing style as servile, tedious, and base, and even Boerhaave observed that Leeuwenhoek's writings would have been bought up with gold had he but written gracefully.[83]

Beyond his stylistic and linguistic limitations – although a necessary corollary of the latter, as Molyneux pointed out – Leeuwenhoek's unfamiliarity with the literature and state of learned science must often have jarred his new acquaintances as well. His early confusion as to nerves and tendons warned of further embarrassments to come, and although years of exchanges with visitors and correspondents provided a continuing education,[84] a willful disregard for much that he was told was still often perceived as ignorance. As late as 1710 he insisted that the pulse at the wrist beat downward rather than upward, and he continued to maintain that extension rather than contraction was the true action of the muscle fiber. In part, such notions represented Leeuwenhoek's assertion of his independence of mind, but to such visitors as Uffenbach they testified rather to the poverty of his understanding of anatomy.[85] They were perceived as examples of those "extravagancies" – of those "peculiar ideas that are typically more ingenious than well-founded," in Uffenbach's phrase[86] – that followed from an unfamiliarity with learned discourse and literature.

81 Zumthor (1962), 108–9, 112, 236; J. Price (1974), 224.
82 Schierbeek (1950–1), 1:65–6; Mendels (1952), 314–15, 318. See also: Damsteegt (1976); Jongejan (1940); Dobell (1932), 305–13.
83 NH (1710), 82; Boerhaave & Haller (1740–4), 3:717. In addition to their frequent stylistic raggedness, Leeuwenhoek's letters also characteristically jump abruptly from one subject or observation to another, displaying little sense of any guiding organization. That lack of organization, and to some extent doubtless that stylistic raggedness as well, reflects his practice of directly transcribing his letters from notes taken down sometimes years before (see *AB* 6:18–19, 8:236, 262–3; AvL 1702, 435; AvL to RS, 14 Jan. 1710, AvL Letters, 125v; AvL 1718, 69). That practice – as well as his general approach to his notes, which he declined to integrate when he took up a previous subject of study again (AvL 1718, 236, 321–2; AvL to RS, 20 Nov. 1720, AvL Letters, fol. 232r) – itself reflects a marked indifference to the systematization of observations and research efforts. Although this disregard for system did not elicit much explicit comment, it was one more aspect of his letters that clashed with the approach to scientific matters inculcated by the universities and nurtured by the international circle of the learned (see also Berkel 1982, 203–5).
84 See n. 105 below.
85 Uffenbach (1753–4), 3:350; AvL to RS, 1 March 1712, AvL Letters, fol. 170r; AvL (1718), 102.
86 Uffenbach (1753–4), 3:358. The occasion for Uffenbach's comment was ill chosen, since he was rejecting Leeuwenhoek's quite correct affirmation of the compound structure of a fly's eye.

Leeuwenhoek was fully conscious of the difficulties posed by his limited education, which indeed he was hardly allowed to forget. In his explanation in 1673 of his prior reluctance to record his observations, he had briefly elaborated on the consequences of an upbringing focused only on trade; he had himself pointed out the neglect of any preparation in languages and cited as well his lack of the style "or pen" to express his thoughts. Three years later he found it necessary to emphasize his telltale inadequacy with languages once again when, because Oldenburg still assumed he could at least handle French, he had to reaffirm that the only language he knew was Dutch.[87] The reminders of this deficiency were perpetual, whether he was struggling to read the *Philosophical Transactions* with the help of a dictionary or searching for others who could translate these and other works.[88] At times, he also worried about how well his Dutch communicated his thoughts to foreign correspondents, and the year before he died he wrote of the difficulty of finding someone to translate his letters into Latin.[89]

Consciously or unconsciously, Leeuwenhoek appears to have attempted to narrow the obvious cultural gap in various ways, some more constructive than others. A recourse at times to uncommon words and phrasings in his letters may represent a misguided effort to impress, while the early 1680s also saw the beginning of a continuing effort to update and regularize his spelling. Together with changes in his handwriting, this orthographic effort provided his letters with a more cultivated veneer, but a lack of organization and the slipshod grammar persisted throughout.[90]

Beginning also in the 1680s, a more metaphysical and religious rhetoric began to embellish Leeuwenhoek's accounts. His letters are now sprinkled with references to the perfection, order, and wisdom found in created things[91] and to a "provident nature" that was consistent in its methods and invariably purposeful.[92] Ultimately, that provident nature assumed the visage of God the omniscient creator of the universe,[93] and, although the identification of science with worship was notably absent from Leeuwenhoek's explanations of why he pursued his researches, he did proclaim in 1696 that there was no better way to glorify God than to observe in amazement his omniscience and perfection in all living things no matter how small.[94]

The perfection, order, and wisdom Leeuwenhoek extolled he also declared to be incomprehensible and inscrutable,[95] and he spoke now as well of the se-

87 *AB* 1:42–3, 342–3.
88 *AB* 2:44–5, 1:342–3; AvL to RS, 21 March 1704, AvL Letters, fol. 282v.
89 AvL to RS, 22 June 1714, AvL Letters, fol. 208v; AvL to J. Jurin, 13 June 1722, ibid., fol. 284r.
90 Mendels (1952), 315–16, 318–20; Dobell (1932), 309.
91 See, for instance: *AB* 6:306–9, 338–9, 7:358–9; AvL (1693), 474; idem (1696), 7, 132; idem (1702), 180, 340; idem (1718), 67.
92 *AB* 6:106–9; AvL (1693), 480; idem (1696), 12; idem (1702), 137.
93 See, e.g.: *AB* 3:396–7, 6:306–7, 7:116–17, 378–9; AvL (1718), 294, 303, 337, 448.
94 AvL (1697), 235.
95 See, for instance: *AB* 6:306–9, 338–9; AvL (1693), 474, 480; idem (1702), 340, 401, 410; idem (1718), 67.

crets (*verborgentheden*) in things,[96] secrets the human mind by itself could often never conceive.[97] Of the silkworm and its thread, he exclaimed in 1702 – not altogether intelligibly himself – that we might well cry out: How inscrutable and incomprehensible are the works of the mysteries (*verborgentheden*)![98] As early as 1679, to be sure, but particularly in the 1690s and thereafter, Leeuwenhoek's letters repeatedly emphasized the incomprehensible smallness and enduring concealment of nature's parts.[99] During these later years, the theme of "how little we know" also became a commonplace in his letters.[100]

Such language in part reflected the cumulative impact of his extraordinary observations. When he wrote that what the microscope had revealed in living things was but a shadow compared to what still lay hidden, or asserted that the smallness and fabric of the parts of living things defied comprehension no less than the broad expanse of the universe, his words were informed by an immediacy and intensity of experience that only Swammerdam and Malpighi had approached.[101] Yet, with their decidedly familiar ring, such expressions reflected as well the rhetoric of the learned circles he had come to know – a rhetoric that, after all, predated Leeuwenhoek's (and, for that matter, Swammerdam's) microscopes. The Royal Society had scarcely heard of Leeuwenhoek (and Swammerdam had barely begun his microscopic dissections) when Nicolas Malebranche, in the manuscript of his *Recherche de la vérité* (1674–8), elaborated his speculations on an infinite series of diminishing mites on mites and, with further echoes of Pascal's even earlier reverie, conjured up whole universes of similarly dramatic differences in scale. Malebranche's purpose was to evoke an awestruck awareness of the limits not only of the senses but of the human imagination, and to convey a more astonishing sense of the power and grandeur of God.[102]

Although seldom, presumably, with such expansive and vivid fantasy, Leeuwenhoek's educated visitors must have often resorted to their own, more summary flights of metaphysical and religious allusion when responding to the sights Leeuwenhoek set before them. It was an "illustrious personage," wrote Leeuwenhoek in 1695, who had exclaimed: "Oh, depth of wisdom, how inscrutable are your works. Shall anyone still be found who says there is no God?" – or "some such words," Leeuwenhoek hedged. The exclamation seemed then to echo in variant phrases (in "some such words," that is) of Leeuwenhoek's own in subsequent letters: "Oh, depth of wisdom, how inscrutable is the construction of your works"; "How inscrutable is the depth of wisdom"; "Oh, depth of mysteries [*verborgentheden*], how unsearchable is the structure of all creatures, and how little it is that we know, even though we search with finely ground and

96 See, for instance: *AB* 3:66–7, 396–7; AvL (1693), 473; idem (1718), 21, 67.
97 AvL to RS, 11 April 1721, AvL Letters, fol. 257r.
98 AvL (1702), 435.
99 See, for instance: *AB* 3:66–7; AvL (1693), 473, 492; idem (1694), 590–1, 686; idem (1696), 7, 45; idem (1702), 306, 359–60; idem (1718), 62, 316.
100 AvL (1696), 172; idem (1702), 51, 201, 359–60; idem (1718), 21, 105, 182.
101 AvL (1697), 235; idem (1693), 483–4.
102 Malebranche (1958–70), 1:xxv, 80–9.

sharp-seeing microscopes."[103] Also, it was apparently Francis Ashton, a member and, for a time, secretary of the Royal Society, who had referred in a letter in 1683 to the "great secret" of generation – a phrase that would reappear in Leeuwenhoek's later letters with a more specific meaning of great importance to the Dutch microscopist.[104]

To be sure, it was not merely a rhetoric of conventional if exalted allusions that Leeuwenhoek learned from his exchanges with the better educated. He also learned more about both old and new currents of scientific thought[105] and, inevitably, of new ways not only of talking but of thinking about his experiences with the microscope. It was surely more than rhetoric that he gained from such conversations as that with Christiaan Huygens that turned to how much was still found that could not be explained.[106] The loftier – or deeper – expressions that began to find their way into Leeuwenhoek's letters bespeak not only the accumulating years of observation, therefore, but an enriched response to them as well. Nonetheless, another contributing factor continues to suggest itself: that his loftier new expressiveness reflects as well an aspiration to a fuller cultural and social assimilation.

Despite Leeuwenhoek's efforts and his intellectual growth, however, a more complete assimilation in the society of Europe's learned elite was still closed to him. Baron von Tschirnhaus, another of Leeuwenhoek's correspondents and a recognized scientist himself, ranked Leeuwenhoek with those "very learned gentlemen" – Leeuwenhoek's phrase – Malpighi and Hooke, and Leeuwenhoek's panegyrist Arnold Hoogvliet proclaimed in his verses that Leeuwenhoek had surpassed even Huygens and Descartes.[107] Yet Leeuwenhoek's scientific achievement – quite apart from Hoogvliet's extravagant claims – was one thing, and joining the social community of learned gentlemen was another. Wittingly or not, Huygens himself occasioned one instance of many that underscored the point.

Subsequent to their talk about Leeuwenhoek's apparatus for displaying the effects of a spinning Earth, Christiaan presented Leeuwenhoek in 1690 with a copy of his newly published *Traité de la lumière* and, bound with it and more pertinent to that recent conversation, his *Discours de la cause de la pesanteur.*[108] Christiaan's gift was a gesture of esteem and a seeming affirmation, after all, of Leeuwenhoek's place in the community of scientists, but both knew that Leeuwenhoek could not read what he had just received. As in 1674, when he had

103 AvL (1696), 62; AvL to RS, 21 March 1704, AvL Letters, fol. 283v; AvL (1702), 106; AvL to RS, 26 Feb. 1703, AvL Letters, fol. 235v; AvL (1718), 21. See also AvL to RS, 13 Dec. 1704, AvL Letters, fol. 335v.

104 *AB* 4:122–3; see Chapter 9, n. 143.

105 Leeuwenhoek's references to what he was learning from others about the state and progress of learned science began very early in his letters: *AB* 1:144–5, 148–9, 154–5, 192–5, 206–7, 2: 4–5; see also *AB* 8:175, 180–3.

106 *AB* 7:280–1.

107 AvL (1702), 129; Hoogvliet (1738), 2:512.

108 *OCCH* 9:390, and see pp. 379–80n. 1.

been unable to read Hooke's and Boyle's thoughts on capillarity, Leeuwenhoek would not be deterred from his own speculations; but the bestowal of Christiaan's treatise implicitly underscored how isolated from learned discussions Leeuwenhoek's efforts remained and how uninformed he inevitably was when he turned to matters other than his microscopes. Thus while seeming to confirm Leeuwenhoek's inclusion, Christiaan's gift in fact symbolized the extremely narrow limits of Leeuwenhoek's ability to participate in the shared elaboration of ideas around which the community of the learned revolved. To that extent, indeed, he remained irremediably an outsider.

Whatever his other inclinations and abilities, consequently, Leeuwenhoek's association with the learned elite ultimately rested on their interest in his microscopes and what he managed to do with them. His ingenuity and cleverness in this endeavor had first opened the perhaps unexpected and unlooked-for prospect of such gratifying social contacts, and his ability to produce continuing new observations sustained them. It was an endeavor in which his lack of learning loomed less prejudicially, for his observations, in their novelty, constituted a unique field of research about which – leaving his interpretations aside – the received corpus of learned literature had little as yet to say. It was a field, moreover, in which the learned world itself had produced few adepts and none who plied the new instrument with Leeuwenhoek's intensity and relentless application. Granted his skills, therefore, it was only his uncommon persistence in an uncommon line of research that created what Europe's learned elite acknowledged as interesting though unschooled science, and hence that also kept alive the prospect of the social and personal identification to which Leeuwenhoek aspired. Would he have failed to recognize this?

Although Leeuwenhoek's attitude toward the learned elite was not without ambivalence, that ambivalence, as reflected in his letters, also bore witness to his sensitivity in matters of self-perception and self-esteem. Beyond his desire for closer association, Leeuwenhoek's interactions with the learned elite betrayed an ego at the same time easily wounded and assertive. His distress over his ignorance regarding nerves and tendons revealed his vulnerability, but that was only the most salient of several difficult moments that almost persuaded him, or so he later wrote, to cease his observations. Already in 1674, he complained not only of the upsetting (*aenstotelijcke*) way in which his errors were pointed out but also of the reticence of those to whom he showed his observations. Feeling unappreciated, unaided, and perhaps exploited by visitors (among whom Swammerdam may have been a major offender), Leeuwenhoek contemplated bringing his observations to a halt, even if only for a while.[109] That he was still susceptible to bouts of hurt petulance in later years was suggested in 1707 when he blamed the Royal Society's failure to answer his letters for his own failure to record certain recent discoveries – "so they are as much forgotten, as if I never had seen them" – and three years later, his daughter remarked to Uffenbach that her father was reluctant to publish any more because of the abuse he had

109 *AB* 1:142–5, 192–3, 298–9, 3:80–1.

received. Very near the end of his life, indeed, he still alluded to his distress over unspecified affronts (*smaatheeden*).[110]

But Leeuwenhoek's first letter to Oldenburg had also already suggested the aggressiveness mingled with his vulnerability. While explaining his initial reluctance to write about his observations, Leeuwenhoek had cited not only his background, after all, but his impatience with contradiction. The picture that emerges through time is a complex one. In his own thumbnail sketch of Leeuwenhoek, the elder Huygens described him as modestly submitting his observations and his thoughts about them for the learned to correct, and, true to that portrayal, Leeuwenhoek solicited the Royal Society's objections and repeatedly assured its members that he would openly acknowledge and correct his mistakes.[111] Not long after complaining of his treatment at the hands of some early visitors, however, Leeuwenhoek also wrote the society of his resolve to stand by his own speculations until better informed or provided with better evidence,[112] and he was soon to reveal a capacity for determined insistence that sheds more light on Swammerdam's portrayal of him as opinionated and unreasonable. Indeed, in his response to some elements of the learned elite, Leeuwenhoek betrayed not only a need to affirm his abilities but an aggressiveness colored by resentment – a resentment that also helped sustain his singular commitment to the microscope.[113]

It was the encouragement of "learned gentlemen" that had buoyed his spirits when, dismayed by the manner of those early visitors, he had considered bringing his observations to an end; but, in resolving to persist, he also resolved to have nothing to do henceforward with the physicians and surgeons of his land.[114] Indeed, the ambivalence in his feelings about the learned elite surfaced most dramatically in his relations with those who bore the mark of the medical schools. To be sure, de Graaf had first called him to the attention of the Royal Society, and others among the Republic's most celebrated physicians joined the growing ranks of his visitors, including (besides both de Graaf and Swammerdam), Craanen, Bontekoe, Ruysch, and Boerhaave.[115] Cornelis 's Gravesande,

110 AvL to J. Chamberlayne, 17 May 1707, AvL Letters, fol. 27v (see also *AB* 8:177, 186–7); AvL to RS, 24 Jan. 1721, AvL Letters, fol. 246v. Regarding his declared intention of publishing no more of his work, see also AvL to J. Chamberlayne, 10 Sept. 1709, AvL Letters, fol. 116r; AvL (1718), 168. In addition to further letters that appeared excerpted in the *Philosophical Transactions,* however (although there was another period [1713–20] during which, with very few exceptions, his letters were not being published in that periodical [see Schierbeek 1950–1, 493–4; Pas 1975, 8]), Leeuwenhoek published the *Send-brieven* in 1718 and at his death left the plates and Latin translations prepared for another publication (*Catalogus,* 43).

111 *AB* 1:42–3; ConHs (1911–17), 6:331; *AB* 2:164–5, 356–7, 3:384–5; AvL to RS, 21 March 1704, AvL Letters, fol. 289r; AvL to RS, 4 Nov. 1704, ibid., fol. 320r; AvL (1693), 507; idem (1702), 274; see also *AB* 1:330–1.

112 *AB* 1:202–3.

113 David Hull notes the vigorous spur to scientific research often provided by what Donald Schon also calls "productive hostility" (Hull 1988, 353; Schon 1963, 99–101).

114 *AB* 3:80–1, see also 1:192–3.

115 AvL (1694), 642, 670–1; *AB* 2:280–1; AvL to RS, 20 April 1706, AvL Letters, fol. 7r; AvL (1718), 254, 261. Even Bernhard Siegfried Albinus visited Leeuwenhoek as a child (B. S. Albinus 1754–68, 3:6), presumably accompanied by his father, Bernard Albinus.

Delft's municipal instructor in anatomy (and, apparently, once Leeuwenhoek's neighbor), also provided him assistance in anatomical matters and participated in some of his observations;[116] 's Gravesande's successor, Abraham van Bleyswijk, of whom Leeuwenhoek spoke as both a friend and kinsman, drew even closer.[117] Although Leeuwenhoek expressed a high regard for other physicians as well, nothing was more likely to win that regard, even in his later years, than a physician's admission that he really knew very little about medicine.[118]

Despite the friendships, the recognition, and the praise, after all, the medical profession at large brought out the combativeness in Leeuwenhoek. He challenged the medical practitioners on a host of issues and caustically refuted leading medical figures of the day. He denounced physicians and surgeons who talked about the body and medicines "as the blind do about colors," and groused about those who, when it came to such matters, were "bent only on contradicting."[119]

In the notorious dispute that arose over whether the male semen or de Graaf's mammalian "egg" was primarily responsible for mammalian generation, Leeuwenhoek championed the semen with an unyielding conviction that bordered at times on defiance of the whole tradition of academic learning. In 1678, he wrote Nehemiah Grew that he remained undaunted by the contrary authority of both "your Harvey and our de Graaf," and he expected the Royal Society to trust what he himself had written – mistakenly, in this instance – about the constituents of the semen. Add Harvey and de Graaf to seventy other authors with similar views, or to seventy times seventy authors, he blustered seven years later, and he would still declare them all to be wrong.[120] He spoke of his own views as irrefutable and decried the opposing opinions as "silly" or "stupid."[121] Nor was that assertive disregard for the accumulated weight of learned authority limited to this dispute alone. "But I know, most noble gentlemen," he wrote to the fellows of the Royal Society in 1685, "that I should not trouble myself about those who contradict me, for if someone gathered all the old and new authors who have written about the flesh, muscular movement, blood, milk, fat, brains, skin, hair, bone, phlegm, etc., seldom would any be found who agreed with what I have written."[122] He left little doubt, however, as to who, to his mind, was right.

116 *AB* 1:398–401, 144–5, 148–9, 154–5, 192–5, 8:56–7; AvL (1718), 209. In the light of this last citation, see also *AB* 4:6–7.

117 AvL (1718), 131, 254, 285, 305, 315, 322; see also Schierbeek (1950–1), 1:14.

118 AvL to RS, 25 July 1707, AvL Letters, fol. 44r.

119 *AB* 8:228–9, 262–3. For further examples of Leeuwenhoek's critical and, on occasion, downright disdainful response to medical men and practices, see also: *AB* 4:38–41, 94–5, 9:46–9; AvL (1697), 328–9, 332; idem (1702), 128–9, 136–7; idem (1718), 382; AvL to RS, 18 Oct. 1707, AvL Letters, fols. 55v, 59r–v; AvL to RS, 31 May 1723, ibid., fol. 324v. Steven Blanckaert seems to be clearly at issue in *AB* 3:390–1, 8:202–3, 206–7. Leeuwenhoek could be particularly blunt about out-and-out quacks as well: AvL (1702), 16–20.

120 *AB* 2:332–9, 5:150–3. With its seventy sections of diverse recorded references to eggs in animals, it was very likely Charles Drelincourt's recent *De foeminarum ovis* that inpired Leeuwenhoek's allusion to seventy (and seventy times seventy) authors (Drelincourt 1684).

121 *AB* 5:236–7, 262–3, 170–1, 6:122–3.

122 *AB* 5:172–3.

At the end of the century, he would still affirm that he lacked any knowledge of the arts of medicine and surgery and had spent no more than two or three hours observing the dissection of the human body. But this avowal of innocence of any formal medical instruction only prefaced an assertion of how much he had learned on his own.[123] Leeuwenhoek clearly believed that his self-directed studies had provided him with an understanding of diseases, medicines, and the body itself that surpassed the learning of the schools, and the foundation of that confidence, ultimately, was the extensiveness and uniqueness of his microscopic observations.

Leeuwenhoek's insistence on the preeminent role of the semen – and subsequently of the spermatozoa in particular – rested on wide-ranging microscopic studies of animal semen, and observations such as those of the microscopic crystals derived from vesical stones and the tophus of gout lay behind his indignant challenges to medical practitioners.[124] In 1685, in justifying his speculations on salt particles in the body, he informed the Royal Society that he was better fit to draw conclusions from his observations than those who, even if they had heard of such things, had never seen them.[125] While preserving his ties with the learned elite, consequently, Leeuwenhoek's microscopic studies also provided a defiant riposte to the tradition of learning that obstructed his closer identification with that elite. Like his self-taught skills in glass- and metalworking, but with far higher emotional stakes, his insistence on the value of his microscopical researches over the learning of the schools was an assertion of the worth of his own native wit and abilities, called perpetually into question by the anomalous and unresolvable social situation in which he found himself.

By linking his self-esteem so closely to his microscopic efforts, Leeuwenhoek's social circumstances played a decisive role in producing his career of discovery. His achievement demanded dexterity, cleverness, and exceptional instruments, but no less essential was the unusual tenacity with which he employed them. He remarked two years before his death that no one he had met

123 AvL (1702), 137–8. Leeuwenhoek had observed some of de Graaf's experimental demonstrations with animals (*AB* 2:310–11), but, with de Graaf's particular interest in the reproductive organs in mind, cf. *AB* 5:158–61. B. W. Th. Nuyens maintains that Leeuwenhoek was in fact no stranger to the anatomical demonstrations at Delft (*AB* 3:295n. 15; see also Schierbeek 1950–1, 1:43–4), but the evidence hardly sustains the assertion that Leeuwenhoek "generally attended" those demonstrations.

124 See Chapter 8, nn. 79, 81; *AB* 8:228–9. Regarding Leeuwenhoek's studies of vesical stones and tophus, see *AB* 3:90–101, 114–23, 4:284–7, 7:192–237, 258–69, 8:214–25.

Leeuwenhoek's conception of the role of the spermatozoa challenged not only traditional conceptions of generation, but also an influential new doctrine that was enjoying great favor in up-to-date medical circles (Ruestow 1983, 193–6). Also based on recent observations, that doctrine – which assumed that nothing more substantial than a spirit, vapor, odor, or "irradiation" from the semen made contact with the egg – had been partly forged in the Dutch Republic and was widely embraced by prominent Dutch physicians ([Grew] 1678; Overkamp 1686, 170; Blanckaert 1701, 1:219; Nuck 1691a, 69). Hence, while also garnering only limited support abroad (Roger 1971, 309–19), Leeuwenhoek's claims for the spermatozoa encountered widespread resistance in the Dutch medical community (see also Chapter 9, n. 2, and Sterre 1687).

125 *AB* 5:64–5.

seemed to have given the time and effort that he had given to researching natural things,[126] and we may be sure that he had met no one who had done so with the microscope. A conviction as to the real promise of microscopic research was still wanting even among those who had experienced the power of the tiny new lenses and had shared in the excitement of their first discoveries. Having returned to Holland after the flurry in Paris over his and Huygens's new bead-lens instruments, Hartsoeker wrote Christiaan in 1679 that it would be a long time, he believed, before the microscope made anyone much wiser.[127] When Hartsoeker's own ambitions revived, he turned largely to other scientific interests, as did almost everyone who had been involved in those events. Leeuwenhoek's anomalous social situation and its relevance to a problem of self-perception appear to have made the crucial difference.

Late in his life, Leeuwenhoek insisted that his years of labor sprang primarily from curiosity rather than the pursuit of praise,[128] yet his sensitivity to the response of others, and particularly the learned elite, clearly and significantly affected his microscopic research. Most obviously, the interest and encouragement of those he saw as his social and intellectual betters first enticed him to make his efforts public and then helped sustain him through the difficult moments that followed. On an admittedly less palpable level, socially mediated influences also presumably enriched Leeuwenhoek's very experience of microscopic observation, for the new rhetoric and expressiveness he acquired doubtless embellished not only his letters, after all, but his conception of what he was achieving as well as what he observed. In further apprising Leeuwenhoek of both the traditions and the progress of learned science, his new acquaintances made him more keenly aware that his observations included discoveries that, as he forewarned a correspondent in 1683, would astonish the world. Two years later the elder Huygens assured him that generations to come would be inspired by his investigation of mysteries seldom even conceived before.[129]

Five decades of difficult and unconventional but nonetheless persistent microscopic research argue an involvement charged with more emotion than sympathetic encouragement alone could arouse, and the enhancement of Leeuwenhoek's own understanding of what he was about, although less accessible to the historian, broached a potential source of such charged commitment. That enhanced understanding aggrandized Leeuwenhoek's self-perception and offered a sense of extraordinary self-development and fulfillment. To the end of his life, his letters to the Royal Society continued to speak of his "trifling" notes and observations; although largely an affectation in later years, this recalled genuine self-doubts with which he had wrestled.[130] Leeuwenhoek became con-

126 AvL to RS, 21 April 1722, AvL Letters, fol. 269r.

127 *OCCH* 8:224–5.

128 AvL (1718), 222.

129 *AB* 4:110–11, 7:360–1.

130 See, for instance: *AB* 7:340–1; AvL to RS, 5 Feb. 1703, AvL Letters, fol. 224v; AvL to RS, 6 Dec. 1707, ibid., fol. 65r, 70v; AvL to RS, 11 April 1721, ibid., fol. 253r.

scious in time, however, of having had an exceptional intellectual life as well as of having made momentous discoveries, of having not only observed but meditated about things that not even the thousandth person, as he put it to Leibniz in 1715, had ever conceived.[131] Molded by social interactions as well as by years of unprecedented observations, his enriched perception of what he was achieving and experiencing confirmed that he was no common man.

Nonetheless, continuing tension is a likelier source of enduring effort than is self-satisfaction, and the tensions engendered by Leeuwenhoek's equivocal relations with the learned elite also tapped the emotional potential of a sensitive but assertive ego. His ties to that elite provided reassurance that he stood apart from the social milieu in which he had been raised and amid which he lived out his life; but contact with the learned also constantly reminded him how tenuous the ties really were and, in doing so, posed its own perpetual challenge to his self-esteem. Having early decided against withdrawal, Leeuwenhoek must soon have grasped that only his continued use of his microscopes could sustain those reassuring ties and, at the same time, provide him the means to strike back at the implicit denigration they also entailed. As with Swammerdam, Leeuwenhoek's sensitivity to social contexts focused his energies, talents, and ingenuity on microscopic research, heightened its emotional import, and thus engendered the long-term intensity of commitment that produced his extraordinary achievement.

131 AvL (1718), 177.

CHAPTER SEVEN

Leeuwenhoek II: Images and ideas

To have argued that what sustained Leeuwenhoek's commitment to microscopic research derived from his sensitivity to social circumstances and a solicitude for self-esteem is not to deny that the experience of microscopy offered its own intrinsic and deeply gratifying rewards. The most celebrated of Leeuwenhoek's observations in particular – those pertaining to the red corpuscles of the blood, the circulation of the blood in the capillaries, the abundance and variety of microscopic life, and the spermatozoa – leave a trail of personal absorption and excitement throughout his letters. The images he confronted in these researches clearly made a deep and enduring impact. But Leeuwenhoek was not simply a passive observer, and, bringing curiously eccentric expectations to bear along with inclinations he shared more broadly, he endowed the most stirring of those images with significant attributes. The resulting interplay of images and ideas forged a consciousness of startling confrontations both with nature's unexpectedness and, conversely, the vivid embodiment of expectant intuitions; from such beginnings also sprang at times an exhilarating sense of having garnered a profound new understanding. Thus constructed, the experience of "discovery" vested microscopic research with heady dividends, enhancing the appeal of both the instrument itself and those chosen lines of research.

Very early on, Leeuwenhoek had been struck by the *wonderen* in the microscopic structure of wood,[1] but spectacular images of a very different order were yet to come. In 1688, Leeuwenhoek first encountered what was to remain perhaps the most exhilarating microscopic spectacle of his life: the circulation of the blood in the capillaries. He was engaged in observing the embryonic development of the frog when he noticed a rapidly pulsing stream of blood corpuscles circulating in the exposed and transparent gills of the larva. Having repeatedly watched this flow over the course of several days, Leeuwenhoek began searching the larva's body for another view of the moving blood after a developing flap of skin obscured it in the gills, and in the tail he now discovered a sight that he described as surpassing anything he had ever seen before. In a multitude of vessels so thin that, at points, only one corpuscle at a time could pass through,

1 *AB* 2:34–5.

he saw the blood flowing from the interior of the tail toward its outer surface or periphery and then, looping back, returning to the interior again.[2] Turning then to diverse parts, from wings to antennae, of a wide variety of other animals, from bats to millipedes, he found that several small, cold-blooded animals offered similar spectacles, to which he responded with no less enthusiasm (Fig. 15).[3]

Something of the gripping vividness of these images was captured in his letters. In 1688 he described the turning corpuscles as they passed through capillaries in a fish's tail that were as transparent as glass. At the end of the century it was again in the extreme capillaries of a tadpole's tail that he saw the circulation of the blood as clearly "as when, with the naked eye, we see the water leaping high out of a fountain and then falling down again . . ."; but no fountain provided as much delight, he later added.[4]

If the clarity of such images struck him so forcefully, however, so did the sheer abundance of the capillaries and the torrent of movement that engulfed the spectacle. In his initial observations in the tail of a tadpole, it was in no small part because the circulation could be seen in so many looping vessels that the sight exceeded everything Leeuwenhoek had previously seen; he was hardly less astonished in the following year by the immense number of capillaries running through and over each other in the fins of an eel, fins that in places appeared to consist of nothing else. He was amazed as well by the fine arterial branchings that, spreading out and then returning and regathering as veins in the tail fin of a fish, created an "inconceivable" movement of blood to the end of the fin and back again.[5]

Like other early observers of the capillary circulation,[6] Leeuwenhoek was continually struck as well by the racing velocity of the blood (although he also recognized it as an illusion of the microscope).[7] Whereas in 1689 it was the great multitude of vessels in which the blood could be seen coursing in the tail fin of an eel that rendered that observation unsurpassed fare for a "philosophic eye," seven years later it was the rapidly pulsing motion in the gathering veins that outdid all the amazing and delightful sights that had preceded it.[8] This even though among those sights had also been the colorless corpuscles of a crab moving with such speed and in such inconceivable numbers that they seemed to Leeuwenhoek like wind-driven snow.[9]

2 *AB* 7:276–7, 8:8–35. Ironically, Swammerdam had also earlier observed the gills in the tadpole and had wondered indeed whether the blood circulated there, but he was distracted from pursuing these thoughts by illness, he noted, and other unhappy events that befell him at the time (JS 1737–8, 2:815–19, 823).

3 *AB* 8:36–57, 68–113, 156–69, 194–5, 200–3; AvL (1694), 721, 555; idem (1702), 182, 332, 349, 362.

4 *AB* 8:52–7, see also pp. 74–5; AvL (1702), 54, 123.

5 *AB* 8:26–7, 42–5, 68–73.

6 See, for instance: Voorde & Heide (1680), 149n. 6; Molyneux (1692), 281; idem (1685), 1237; B. Nieuwentyt (1717), 101; Porras (1716), 351.

7 *AB* 8:132–3. On the perceived velocity of the blood as an illusion, see Lumsden (1980) 1:132–3. See also Verheyen (1710a), 2:68.

8 *AB* 8:70–3; AvL (1697), 274. Regarding Leeuwenhoek's response to the speed of the blood in microscopic images of the capillaries, see also: *AB* 8:132–5, 36–7, 74–5, 112–13, 158–9, 194–5.

9 AvL (1696), 3.

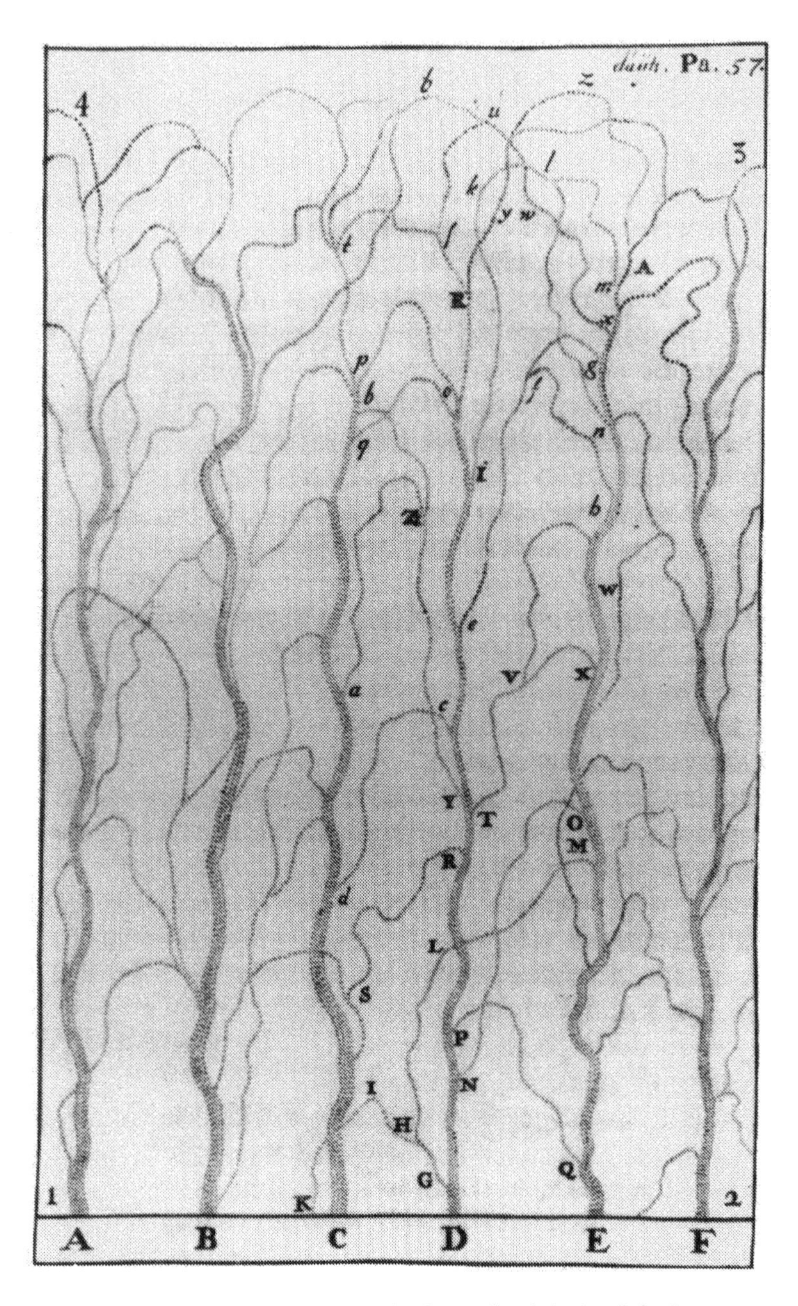

Figure 15. The capillaries in the tail of a small eel depicted for Leeuwenhoek by an artist whose observational powers Leeuwenhoek particularly commended (AvL 1702, plate facing p. 57; pp. 54–8). Denoted by dots (actually tiny circles) representing blood corpuscles, the looping vessels correspond to an area that, according to Leeuwenhoek, was smaller than a large grain of sand, a millionth part of which, he writes, would still have been too large to pass through the narrower capillary passages. Photo courtesy of the National Library of Medicine, Bethesda, Md.

There can be no doubt about Leeuwenhoek's enthrallment with such images. From the very beginning in 1688, he made it clear how often he had continued, repeated, or resumed his observations for no other reason than the sheer pleasure of the spectacle,[10] and a decade later that pleasure remained undiminished. Never satisfied with the wondrous sight, he remarked in 1699, he lingered, still looking at the circulation of the blood in the tadpole after the visitors to whom he had been showing it had departed. He wrote soon after that watching the blood circulating in the web between the toes of a frog had also brought him so much pleasure that he had continued these observations for days.[11] Indeed, Leeuwenhoek spent many private hours absorbed in such images, and the delight they provided helped sustain such observations through at least the last dozen years of the century.[12]

The spectacle of moving masses of spermatozoa exercised a similar grip. In 1679, the motion and multitude of the spermatozoa he discovered in the vasa deferentia of a rooster – over fifty thousand, he thought, in the space of a grain of sand – provided one of the earlier of those sights that he found more astonishing than anything he had encountered before.[13] After a pause during the 1690s, the observations of the great swarms of spermatozoa that reappeared in his letters after 1700 still evoked the same amazement. In a specimen from a ram's testicle, he described them all swimming one way together and then another, whereas in a sample from the epididymis they moved in a thousand different directions in a manner he thought no words could convey. The descriptive image that came most readily to his pen was that of cloudlike masses of spermatozoa passing through each other, but he still contended that only those who had seen it could believe it.[14]

Inseparable from the delight of these spectacles was wonder, and no images cast so pure a spell of wonder as did his observations of microorganisms. Unabating, that spell remained the hallmark of Leeuwenhoek's persisting studies of microscopic life throughout a half-century of observing, and its enduring fascination for him also derived in no small part from its pervasive activity and swarming abundance in so many diverse and often startling places. He first encountered microscopic life in astonishing multitudes in the spring and summer of 1676 (which gave rise to his long letter to Oldenburg in October). He had been amazed to discover an incredible number of different microscopic creatures in water in which he had been soaking some pepper (intending to explore the cause of its "heat"). Along with rain water, well water, and various other kinds of water, Leeuwenhoek now incorporated infusions of pepper and other spices

10 *AB* 8:22–4, 32–3, 44–5, 72–3, 196–7; AvL (1702), 124–5.

11 AvL (1702), 115, 162; see also pp. 170, 224–5, 235.

12 *AB* 8:32–3, 40–3, 72–3, 80–1, 158–61, 196–7; AvL (1697), 274; idem (1702), 115, 117–18, 124–5, 162, 170, 224, 235. See Miller (1968).

13 *AB* 3:14–17, 24–35; see also 3:202–9, 5:154–7, 7:10–13, 392–3.

14 AvL (1718), 286–7, 399; idem (1702), 297–8 (and see *AB* 7:10–13), 392; AvL to J. Petiver, 18 Aug. 1711, AvL Letters, fol. 141v; see also AvL (1718), [296], 308, 399.

into what became a systematic study of microscopic life that continued for at least another two years.[15] The results were startling, and sporadic later observations continued to discover new multitudes of microscopic creatures over the course of the next several decades.

Leeuwenhoek was struck in 1676 (as in fact he had been earlier) by the variety of strange motions he observed in such creatures, motions that he again struggled to convey: the creatures trembled, tumbled and turned, wriggled like eels, spun like tops, raced forward in spurts, mingled placidly like gnats in the air, hopped in the water like magpies or fleas, or simply wallowed on their backs and bellies.[16] Years later, he would similarly wrestle with descriptions of the rotating corona of the rotifer, a source, indeed, of even greater amazement.[17]

But the mass movement of microscopic multitudes again provided a persistently astonishing spectacle as well. Leeuwenhoek was amazed when he first realized that the tiny particles he had seen floating about in countless numbers, and which he had likened to bits of shaven hair, were in fact also alive. Other microscopic entities, so small that he had previously been unable even to determine their shape, he came to see as small, thickly teeming worms or eels. He likened the sight to a tub of water so full of tiny eels that the water everywhere seemed alive with animals. No sight had given him more pleasure than so many thousands of living things milling about in a tiny drop of water, he wrote, and of all the wonders he had seen (though both the spermatozoa and the circulation in the capillaries were still ahead of him), he prized this as the most astounding.[18]

This was still in his 1676 pepper infusions, however, and in the following years he would find such microscopic creatures abounding no less in a variety of unexpected and perhaps unnerving places. In 1683 he reported the inconceivable multitudes he had encountered in the accumulation about people's teeth (including his own), where they teemed so thickly in one instance that the whole fluid particle seemed to be alive.[19] They turned up as well in his rotted molar, the saps of trees, the intestines of flies, and often in animal excrement;[20] indeed, they swarmed so thickly in a particle of his own diarrhetic feces that it seemed to consist of nothing else, a sight that also hence delighted him.[21] Such images of pervasive motion and rushing multitudes – among them, the spermatozoa and the circulating corpuscles of the blood – highlighted Leeuwenhoek's fifty years of observing with a succession of microscopic displays that continued to astonish him.

15 *AB* 2:64–161 (esp. p. 90), 194–5, 318–19, 402–3.
16 *AB* 2:72–3, 76–7, 88–9, 92–3, 96–7, 108–9, 112–17, 122–3, 134–5, 1:164–5.
17 *AB* 7:94–5; AvL (1702), 405–6; AvL to RS, 25 Dec. 1702, AvL Letters, fol. 206v; AvL to RS, 4 Nov. 1704, ibid., fol. 320r–321r, 326r; AvL (1718), 64, 67, 69.
18 *AB* 2:100–1, 108–9, 114–15.
19 *AB* 4:126–33 (esp. pp. 130–3); see also AvL (1693), 509–13.
20 AvL (1702), 41; *AB* 3:168–9, 196–7, 322–3, 372–5, 4:76–7.
21 *AB* 3:366–9.

The wonder evoked was also the work, however, of what Leeuwenhoek himself brought to these spectacular visions. It was the clash with expectations rooted in a more familiar world that rendered the microorganisms so astounding in the first place, of course, but more precisely developed ideas also contributed to the aura of wonderment. The most salient instance was the very concept of an "animal" and the cluster of preconceptions it entailed; for Leeuwenhoek was persuaded by the nature of the motion of microscopic organisms – its irregularity and apparent innateness – that they were indeed animals.[22] It was a conclusion evidently too obvious to warrant much further explanation, but Leeuwenhoek elaborated its implications as a source of deeper astonishment.

He had related the organisms he observed in 1676 to a scale of littleness that, having descended to creatures fifty times smaller than a louse's eye, moved on to others one millionth the size of a grain of sand and, finally, to those just on the edge of visibility through the microscope.[23] In subsequent years he wrote of creatures one hundred million and ultimately more than a billion times smaller than a grain of sand – or more precisely in the latter case, a "coarse" grain of sand – approaching the dimensions, that is, of the smallest microscopic particles he had ever managed to detect.[24] But Leeuwenhoek also shared Swammerdam's belief in a basic similarity in the anatomical structure of all animals, and although able to see little more than a watery fluid and globules within the microorganisms he observed, he insisted nonetheless that even the smallest of these creatures necessarily had a mouth, intestines, vessels, muscles, nerves, and indeed all the "perfection" of a larger animal.[25]

Already in 1679 that imaginative prospect had induced Leeuwenhoek to indulge his penchant for calculation in an effort to convey this wondrously diminishing microcosm. In order to suggest the scale of a particle of water, he undertook to calculate the minuteness of the vessels in an animal twenty-seven million times smaller than a grain of sand. By assuming that the proportion between the whole body and its smallest vessels was the same as that in the human body, he reckoned (and not altogether flawlessly) that more than 5.88 trillion of the vessels in such an animal could be gathered together within the cross section of that

22 *AB* 3:332–3, see also 1:164–5; AvL to RS, 5 Feb. 1703, AvL Letters, fol. 218r; AvL to RS, 3 June 1712, ibid., fol. 186v.

23 *AB* 2:72–5, 88–91, 94–5, 112–13, 150–1.

24 *AB* 3:192–3, 332–3, 5:20–1; AvL (1702), 41. He spoke of other entities – such as crystalline particles, described as less than a billionth the size of a coarse grain of sand – as being visible only to his finest microscopes and floating at the very limits of their power: *AB* 5:10–11, 88–9, 7: 196–9, 294–5, 8:254–7; AvL to RS, 5 Feb. 1703, AvL Letters, fol. 223v; AvL (1718), 395–6.

Intimately familiar to Leeuwenhoek in the form of the sand used in the endless scouring that went on in Dutch households (AvL 1694, 724; AvL 1718, 238; Zumthor 1962, 137, 139), sand grains were one of Leeuwenhoek's favorite standards for attempting to convey microscopic dimensions. The editors of Leeuwenhoek's letters have concluded that one billionth of a coarse grain of sand roughly corresponds in fact to a diameter of 1 μm (see, e.g., *AB* 7:199n. 31), which is approximately the resolving power of Leeuwenhoek's best surviving microscope (see Chapter 1, n. 43).

25 *AB* 2:68–71, 88–91, 98–9, 144–5, 5:20–1, 3:396–7.

same grain of sand.[26] Leeuwenhoek spoke often indeed of the incomprehensible smallness he discovered through the microscope;[27] but those diminutive dimensions that were the greatest source of wonder were fashioned, in this instance at least, as much by speculation and calculation as by observation.

Thoughts of such fantastic anatomies could not have been too far from Leeuwenhoek's thoughts when he had his eye pressed against his microscopes. He had very early noted the "legs" of certain (ciliated) microorganisms and was convinced by what he saw of the existence of still other legs he could not see. Could he then conceive such limbs without the vessels and moving parts he not long after explicitly ascribed to them?[28] His unhesitating assumption of other, still invisible legs suggests that his imagination was indeed readily engaged by the sight of such creatures, and it is therefore not unlikely that an awareness of their necessary inner intricacy also flickered through such observations, accentuating the wonder of the moment and perhaps entailing other evocations.

The assumption of such anatomical intricacy within the newly discovered microorganisms was typical of the age, after all, and contemporaries were not slow to tie it to more exalted themes. Malebranche, arguing the existence of parts that neither sense nor imagination but only reason could reach, embraced those imagined anatomies as a means to cultivate a greater awareness of the greatness of God.[29] Leeuwenhoek, in what is in effect his earliest reference to God in his letters, likewise cited the unseen but necessary anatomy in the vinegar eel to underscore the wondrousness of the Creator. In 1699 he reminded the members of the Royal Society of creatures a billionth the size of a grain of sand; unable to grasp the smallness of their parts, he wrote, we must stand back in amazement and take heed of the impenetrability of the depth of wisdom.[30] For Leeuwenhoek, such fantasies of smallness were often indeed linked not only with wonder but with mystery, wisdom, and God's inscrutability.[31]

But the surreptitious impingement of ideas could also turn images into the seeming vehicles of extraordinarily enhanced understanding, thus imbuing visual spectacles with the capacity to arouse an even more gratifying exhilaration. Such a capacity may indeed have permeated Leeuwenhoek's observations of capillary circulation. Leeuwenhoek was fully aware of their significance for the continuing debate over the circulation of the blood, and he referred to his ob-

26 *AB* 3:54–67.

27 See, for example: AvL (1693), 483–4, 492; idem (1702), 306, 359–60; AvL to RS, 24 April 1705, AvL Letters, fol. 369r.

28 *AB* 2:68–9, 98–9, 106–7, 390–1, 3:54–5.

29 Malebranche (1958–70), 1:80–1, 83.

30 *AB* 3:396–7; AvL (1702), 106. Leeuwenhoek himself had cited God in his correspondence before his 1682 letter only in what is in effect no more than a closing – and uncharacteristic – valediction two years before (*AB* 3:222–3). The very first mention of God in association with his letters comes rather from the pen of two of the witnesses to his observations of infusoria; both pastors, they referred as well to the passage from Romans (1:20) echoed in the Dutch Confession (*AB* 2:256–7; also see Chapter 2, n. 125).

31 *AB* 3:66–7; AvL (1693), 473–4; idem (1694), 591; idem (1696), 53–4, 132; idem (1702), 106; AvL to RS, 13 Dec. 1704, AvL Letters, fol. 335v; AvL (1718), 21.

servations in 1688 as a demonstration he had long been seeking.[32] He was not the first to observe capillary anastomoses,[33] but from his own studies he now drew the key conclusions more decisively and more insistently than any before him.[34] Added to the pleasure of the spectacle, therefore, was his consciousness of these observations as the capstone to the century's foremost physiological innovation. Nonetheless, those images of pale blood corpuscles jostling their way hurriedly through transparent capillaries appear to have embodied for Leeuwenhoek an even deeper level of discovery.

Among the differences between Swammerdam and Leeuwenhoek as observers was the artistic sensibilities of the one and the mechanistic inclinations of the other. Leeuwenhoek rarely if ever compared what he observed to human artistry, and his letters lack Swammerdam's constant allusions to the ornamental, the elegant, or the pretty.[35] Leeuwenhoek, unlike Swammer-

32 *AB* 8:54–5, see also pp. 30–3 and 1:288–9. The doctrine of the circulation of the blood was indeed still warmly contested in various parts of Europe on the eve of Leeuwenhoek's initial efforts with the microscope (see Roger 1971, 42–3; Adelmann 1966, 1:269, 273).

In 1680, Leeuwenhoek had himself remarked that his initial belief in the continuity of the ends of the veins and arteries had been shaken by observations that now inclined him to suspect that the ends of the arteries and veins were not joined, and that the blood poured out of the ends of the arteries to be either gathered up again by the veins or excreted from the body (*AB* 3:306–9). Indeed, even while subsequently insisting that the circulatory system was a closed one, he still for some time doubted whether the capillaries actually had their own walls (see n. 34 below).

33 Malpighi (1686–7), 2:140–3; Voorde & Heide (1680), 557n. 1, 149n. 6, see also 496n; JS (1737–8), 2:831–2; Palm (1978), 170–1.

34 *AB* 8:32–3, 192–3; AvL (1718), 184, 290, 344; cf. Malpighi (1686–7), 2:143; and Voorde & Heide (1680), 149n. 6. Leeuwenhoek's notion of a closed circulatory system was long qualified, however, by his uncertainty as to whether the capillaries in fact had their own vascular walls (*AB* 8:50–3, 96–9). His observations of crossing capillaries finally convinced him that they did indeed have such walls (AvL 1702, 125–6).

35 There is also only one instance I can cite at this moment in which Leeuwenhoek, in contrast to Swammerdam, likened the microscopic structures he observed to anything that was *kunstig:* AvL (1696), 54–5.

Delft was not Amsterdam, to be sure, but its own midcentury efflorescence of local painting appears to have touched the great bulk of its citizenry (Montias 1982, 111, 177, 179–82, 220, 270, 332), and Leeuwenhoek certainly had more than a passing acquaintance with at least one of the enduring masterpieces of Dutch painting, Jan Vermeer's *The Art of Painting* (Montias 1989, 229–30, 350; regarding Leeuwenhoek's acquaintance with other paintings by Vermeer, see also pp. 219, 221, 228). Leeuwenhoek also had extensive, indeed intimate, contacts with the artistic community, for his microscopic researches involved years of close interaction with the artists who served as his draftsmen (see Chapter 6, n. 45). Artists also appeared among the visitors who came to see his observations (AvL to RS, 4 Dec. 1703, AvL Letters, fol. 261r), and his stepfather had been a painter as well (Dobell 1932, 22). It is noteworthy, hence, and perhaps a reflection indeed of a lack of responsiveness to the arts that he never mentioned the name of even the draftsman who worked with him for years at the end of his life and whose eye and judgment he respected (AvL to RS, 20 Nov. 1720, AvL Letters, fol. 234v; AvL to RS, 24 Jan. 1721, ibid., fol. 246v; see Dobell 1932, 343–4).

The intriguing fact that in 1676 Leeuwenhoek was appointed trustee of the bankrupt estate of Vermeer's widow has inclined some historians to believe that Leeuwenhoek and Delft's greatest artist in fact were friends (Dobell 1932, 35–6, 345; Wheelock 1981, 13–15, 136–8). In his detailed studies of the vicissitudes of Vermeer's family, however, John Michael Montias declares that "there is not a trace of documentary connection between Van Leeuwenhoek

dam,[36] yearned rather for the kind of understanding of nature's operations represented by seventeenth-century mechanism, a yearning that on occasion clearly shaped his accounts of the images he saw.

There is no mistaking Leeuwenhoek's mechanistic bent, even though the precise outlines of that bias are far from clear. It finds expression in a variety of assumptions and attitudes that range from a disdain, voiced in 1691, for the "ignorant and simple" person's preoccupation with magic to the assertion of the abstract principle that body could not be annihilated but only divided into invisible parts. He was explicit as to the quantitative implications of that principle as well: No body could be made smaller unless some body was removed from it, he wrote, and no body could be made larger unless further particles were added.[37] Another visage of mechanism appeared in the explanations he offered of physiological processes: He repeatedly cited the unseen shapes of unseen particles as the cause of different tastes and of medicinal and pathological effects, and he proposed that even vision was transmitted to the brain as motion through the globules of the optic nerve.[38]

There are echoes as well of the specifically Cartesian strain of mechanism, but echoes that suggest how distorted or garbled the Cartesian influence might be. (They also bespeak just how widely such Cartesianism, garbled though it may have been, permeated Dutch society. Like other salient aspects of Leeuwenhoek's mechanist outlook, those echoes appeared in his letters very early, before his extended exchanges with the learned world.) In 1673 Leeuwenhoek, like Descartes, posited three kinds of matter that corresponded to three grades of differently sized particles, and he acquiesced in the "common opinion" that the particles of the first or finest matter passed through solid bodies. Leeuwenhoek, however, identified those three kinds of matter specifically with the air, and he conceived them in a way that would have dumbfounded any true Cartesian: The first matter of the air, he wrote, consisted of particles ranging from the very smallest to those the size of the largest grains of sand; the second, of particles from the size of sand grains to that of red currants; and the third, from red currants to cherries or pebbles![39] Even apart from the homeliness of the units of his scale, Leeuwenhoek's three degrees of the matter of the air were a far cry, after all, from the three elements of Descartes's universe.

Footnote 35 *(cont.)*
and Vermeer or between their families," and seems decidedly doubtful about any ties of friendship before the city aldermen appointed Leeuwenhoek the curator of the estate (Montias 1989, 226).

36 See chap. 9, n. 90.

37 *AB* 8:210–11, 3:300–3, 1:56–7.

38 Regarding taste, see, for instance: *AB* 1:306–17, 2:16–19, 3:134–5, 5:114–17; AvL (1718), 203. With respect to medicinal and pathological effects: *AB* 1:314–15, 2:138–43, 3:114–21, 136–7, 4:188–9, 5:16–17. Re vision: *AB* 1:198–9, 5:320–3.

39 *AB* 1:56–61; see also Snelders (1982), 61–5. For another echo of Descartes's subtle matter, see also AvL (1718), 192. Leeuwenhoek related it to the suspension of microorganisms in water as well: AvL to RS, 4 Nov. 1704, AvL Letters, fols. 322v–323r.

Nor was Leeuwenhoek attuned to the epistemological preoccupation that lay at the heart of Cartesian mechanism. In 1687, Newton's *Principia* shocked Cartesian sensibilities by its recourse to an inexplicable universal gravitation, and Christiaan Huygens was among those who deplored the absurdity of Newton's "principle of Attraction." Nonetheless, Leeuwenhoek blithely continued to posit an attraction – or, more precisely, an inclination (*neijginge*) toward each other – between the particles of microscopic crystals, and wrote as well of the many "sympathies" – if he might call them such, he did add self-consciously – he had often observed in various substances.[40] Although not very rigorously thought out, Leeuwenhoek's mechanism entailed two essential characteristics of seventeenth-century mechanism in general: a conviction in the enduring consistency of quantity in matter and the tendency to assume that even organic processes could be reduced to a question of structure and of the motion of parts and particles.

Both characteristics were evident in Leeuwenhoek's conception of the circulation of the blood. To be sure, like Descartes, Leeuwenhoek believed that the expansion and beating of the heart was caused by the heating there of the venous blood; but he was nonetheless also emphatic that the major role of the heart was, in contracting, to drive the blood through the arteries and ultimately, with the help of the valves of the veins, back to the heart again. At the other end of the system were the capillary walls, whose fibrous, porous, or spongy structure allowed the thinner fluids in the blood to seep through to nourish the parts of the body.[41] It was a system to which Leeuwenhoek also applied the principles of hydraulics, and he attempted a rough calculation both of the time it took the blood to circulate in various regions of the human body and of the quantity of blood being pumped through the capillaries.[42]

These efforts were also tied to images of the capillary circulation. Leeuwenhoek's attempt to determine the time it took the blood to circulate in the human body rested on a rough initial measurement of the flow in the capillary of an eel, and his calculation of the quantity of blood being pumped through the capillaries was framed between enthusiastic encomia, including the analogy of a fountain, of the spectacle of the circulating blood in a tadpole. Such images also conveyed not only an illusion of velocity, moreover, but a sense of the surging thrust of the heart. In observations of the capillaries in the tail fin of a fish, Leeuwenhoek thought he saw the blood in the smallest capillaries (which he then conceived as lacking vessel walls) forging new paths "by force" (*met gewelt*) through the surrounding tissue.[43]

40 I. Cohen (1980), 69, 80–1; *AB* 1:234–5, 5:102–5; AvL to RS, 5 Feb. 1703, AvL Letters, fols. 220r, 222r–v; AvL to J. Chamberlayne, 3 Oct. 1704, ibid., fol. 316v; AvL to RS, 14 Jan. 1710, ibid., fol. 126v; AvL (1718), 442–3. Leeuwenhoek's reference to "sympathies" is to be found at *AB* 5:350.

41 *AB* 3:302–6, 8:96–9, 206–7.

42 *AB* 8:132–45; AvL (1702), 119–20.

43 *AB* 8:132–5; AvL (1702), 114–15, 123–5; *AB* 8:50–3, 96–7; see n. 34 above.

These images, to the extent that they seemed to answer to Leeuwenhoek's mechanistic expectations (and their linkage to his attempts at hydraulic quantification suggests that they did), doubtless stirred him all the more deeply. Two centuries later, when the Spanish neuroanatomist Santiago Ramon y Cajal first observed the capillary circulation in the mesentery of a frog and was abruptly converted to a mechanistic perception of life, the impact was no less than that of a revelation. That "sublime spectacle" struck him as if a veil had been lifted within his soul, he wrote, revealing a sweeping new understanding in which the living body was conceived as a hydraulic machine.[44] For Leeuwenhoek, to be sure, the effect would have been less one of conversion than of recognition and confirmation; but it was an understanding still very much at odds with the more familiar appearances of animals, and to discover it exemplified vividly below the cloak of those appearances would have been a gratifying discovery indeed. For Leeuwenhoek, his delight with the images of capillary circulation would have acquired a deeper and more thrilling resonance.

That such conceptualizations infused microscopic images and colored Leeuwenhoek's immediate experience of observing is still conjectural, however; it is far from obvious that ideas ever impinge on images at all. But other instances in Leeuwenhoek's accounts reveal in more manifest ways that the conjurings of his mind did indeed work upon the images he perceived. In his studies of the muscle fiber, in particular, Leeuwenhoek's mechanistic inclination more conspicuously molded – indeed, reshaped – what he at least reported he had seen, and it undeniably did so for the sake of an enhanced sense of understanding. At issue as well was yet another facet of his mechanistic bent, however: a tendency to conceive the operations of nature, when possible, in terms of contrivances that could be made by human hands.

Leeuwenhoek's observations of muscle fibers were also of many years' duration, and his perception of the striations in particular passed through three distinct phases. Like Swammerdam and Hooke, Leeuwenhoek initially saw the microscopic fiber of striated muscle as a string of globules, a misperception apparently due in large part to the quality of his early microscopes.[45] By 1682, however, he saw the striations as a series of rings around the fiber and proposed that they were in fact wrinkles (*rimpels*) directly related to the fiber's function. He believed that he now understood the stretching and contracting of muscle, he wrote, for he conceived the contracted fiber to be full of such wrinkles that disappeared when the fiber stretched itself out again (the true action of the fiber, he thought). The truer image of the striations was thus accompanied by a structural interpretation in terms of function, and Leeuwenhoek's perception of microscopic images conformed to that interpretation for over thirty years. This

44 Ramon y Cajal (1920), 104–6.

45 JS (1737–8), 2:834, see also 1:70; Birch (1756–7), 3:402, 406, 4:135, 140; *AB* 1:110–11, 182–3, 2:212–15. Even after having later recognized the error, indeed, Leeuwenhoek testified that he still saw the striated muscle fiber as globules through "ordinary" instruments (*AB* 3:385–7, 392–3, see also pp. 422–9).

perception was abandoned only when an alternative interpretation proved more compelling: In 1714, he now convinced himself that the striations were not a series of circular folds (*inkrimpingen* now), but a spiral that wound about the length of the fiber, the best construction one could imagine, he emphasized, for stretching out and contracting.[46]

Leeuwenhoek was also not the first to have perceived the striations as a coil (perhaps even suggested at times by the distortions caused by cutting or tearing muscle fibers); but the ascendancy of that perception in his observations appears to have been connected to his personal penchant for making things with his hands and a taste, hence, for analogies deriving from such handiwork. Indeed, Leeuwenhoek may have arrived at the coil in the muscle fiber precisely as a consequence of reflecting on just such an analogy. In 1712, two years before his change of heart, he had written of having often found the rings on the fibers so close together that it was as if he were looking with the naked eye at a very thin iron wire that had been wrapped around another wire. The analogy re-emerged prominently in 1714 when, in explaining his new conception of the fiber, he did so by likening it again to a construction in which a fine wire was wound about a needle.[47]

The following year Leeuwenhoek offered an account of the actual making of such a coil (of which he also included a drawing). Reiterating again, and not for the last time, that such a construction was the most perfect the mind could conceive for the action of stretching out and contracting, he proceeded to illustrate the point by having a length of copper wire wound around a quill and then slipped off as a coil. If the ends of that coil were then pulled and released, he wrote, one saw the stretching and contracting, from which we must conclude, he continued, that this is what also happens in the muscle fiber.[48] Leeuwenhoek's seemingly gratuitous recourse to describing the making of a coil merely accentuates an imaginative propensity most befitting a dogged student of technical skills and inveterate builder of contrivances.

Be that as it may, the fusion of image and mechanical imagery – and fused they were, for Leeuwenhoek continued henceforth to see the striations as a spiral, and to see them as such quite clearly, he wrote (Fig. 16) – amounted to a vivid and compelling new understanding of how a muscle operated. He had also written in 1712 that, having studied the fibers in cod for over twenty-five years in the hope of discovering how muscles were constructed and how they worked, he had in fact given up, having gotten nowhere. Two years later, however, asserting that his recognition of the spiral in the muscle fiber had shed more light on the fiber than all his previous observations, he boasted of having at last unraveled the knot about which so much had been written and argued.[49]

46 *AB* 3:394–5; AvL (1702), 50; AvL to RS, 1 March 1712, AvL Letters, fol. 170r; AvL (1718), 122–3.

47 Birch (1756–7), 3:396–7, 401; Sénac (1749), 1:57; AvL to RS, 1 March 1712, AvL Letters, fol. 169v–170r; AvL (1718), 122–3.

48 AvL (1718), 153–4 (and fig. 3 from the plate facing p. 150), 183, 330.

49 Ibid., 122–3, 13.

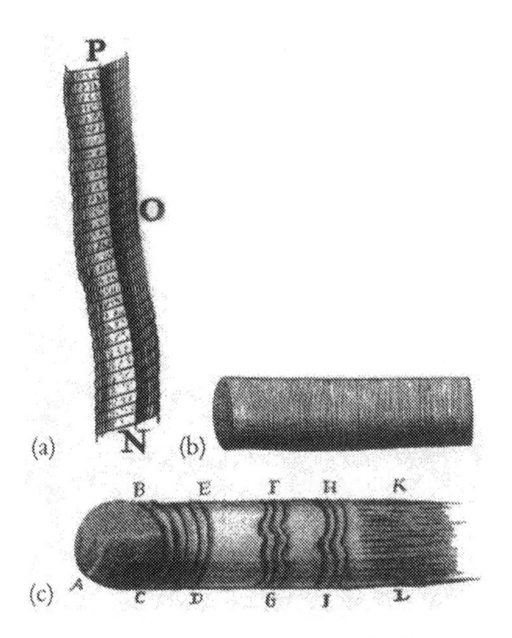

Figure 16. (a) Leeuwenhoek's depiction of muscle fiber striations in 1714. (AvL 1718, fig. 5 in the plate facing p. 113). Photo courtesy of the National Library of Medicine, Bethesda, Md. (b) His depiction (reduced) of such striations in 1683 (*Philosophical Transactions,* no. 152 [1683], fig. I). Photo courtesy of the Special Collections Department, University of Colorado at Boulder Libraries. (c) A still earlier, less schematic red chalk rendering (also reduced) of the striations by Leeuwenhoek in 1682. The separate groupings of striations marked by BCDE, FG, and HI in fact represent different patterns of striations observed at different times (*AB* 3:398–9). By permission of the President and Council of the Royal Society.

Leeuwenhoek's desire for a mechanistic understanding of what he saw through the lens was in itself a goad to more intensive observations as he searched for images – or perceptions – that better conformed to his expectations. His remarks on cod muscle in 1712 reveal that the supposed wrinkles in the fiber had never, after all, wholly satisfied him. He had searched for more than twenty-five years for a better understanding, and although he claimed to have given up, he went on nonetheless to find the spiral. Hence, to the extent that he ultimately did succeed in discovering or imposing a reflection of his mechanistic expectations on images, he eased a nagging uneasiness and forged a sense of clarified understanding. A give and take between image and idea had ultimately reached a deeply gratifying, if misguided, resolution.

The effort to render images comprehensible, however, may involve several levels of such give and take, and the potential diversity as well as the potential impact of interacting influences is nowhere better revealed than in Leeuwenhoek's studies of the red corpuscle of the blood, the erythrocyte. In addition again to seductive imaginings of diminishing structure, these accounts reveal the influence of two not wholly compatible prototypes: one, the generalized "globule" that, with the vessel and fiber, loomed large in Leeuwenhoek's mind as a basic element of organic structure; the other, another simple construction that Leeuwenhoek could and did fashion with his hands. Aesthetic prejudices appear to have also played a role, and very likely as well, something akin to Gestalt dynamics of perception.[50] Enveloping it all was also a disposition to assume that nature encompassed an ever-diminishing structure of "parts." For all the complexity of the resulting interplay among such varied influences, however, they shaped how Leeuwenhoek perceived the images of red corpuscles he confronted through his microscopes and, in doing so, provided them an entrancing if fictitious significance.

The erythrocyte was among the first of Leeuwenhoek's more dramatic discoveries. Again, there were intriguing precedents, for the blood had early come to mind as a fitting subject for the microscope. Kircher had already reported minute "worms" in the blood, and Pierre Borel described whale- or dolphin-shaped "insects" that swam in the blood as if in a red sea. What they had seen were perhaps stacks of adhering corpuscles (known as rouleaux), but, in a far less doubtful account in 1666, Malpighi wrote of the "red atoms" he saw driven through the transparent blood vessels of a frog.[51] Leeuwenhoek was surely unacquainted with Malpighi's Latin treatise, however, when he set out in 1673 to discover the parts that composed the blood.[52] Drawing blood from his thumb into capillary tubes, he observed numerous small globules floating in what appeared at first a cloudy but subsequently clear and crystalline fluid.[53] Despite the initial difficulties experienced by other observers, the observation soon became a commonplace, the continual motion of the turning corpuscles drifting in a clear serum providing what Swammerdam considered as pretty a sight as the microscope could provide.[54]

50 Gestalt psychologists argue that there are tendencies not only to form closed rather than mere linear patterns in visual perception but also to see groups as simple and symmetrical (Köhler 1969, 52, 57–8, 98). Although his focus is on rectilinear forms, D. N. Perkins also remarks that "the perceiver uses regularity assumptions plus a geometric capacity to achieve a spatial encoding of an essentially indeterminate stimulus" (1982, 73).

51 Bacon (1877–89), 1:307–8; Singer (1915), 335–8; Adelmann (1966), 1:266–7; Malpighi (1686–7), 2:130. Swammerdam has also been cited as possibly the first to have seen erythrocytes, though the precise date of the passage in question (JS 1737–8, 2:819) remains uncertain (J. Baker 1948–55, 89:111; Schierbeek 1947, 126; cf. Lindeboom 1981, 104–5).

52 *AB* 1:74–5, 300–1, see also 4:240–1, 241n. 40.

53 *AB* 1:96–7, 102–5, 66–7, 74–5, 120–3.

54 Provided even with Leeuwenhoek's own capillary tubes, Christiaan Huygens, for one, initially had a difficult time seeing the red corpuscles (*AB* 1:122–3; *OCCH* 7:400); by 1678, however, the corpuscles could also be observed at the Musschenbroek shop in Leiden (Bartholin 1673–80, 5:24), whereas others like Swammerdam (see n. 51 above, and *LJS* 80–1) and ultimately Huygens as well managed for themselves (*OCCH* 13:698). See also Birch (1756–7), 3:379–81.

Leeuwenhoek had also soon begun showing them to intrigued visitors: Bartholin (1673–80), 3:7; Jacobaeus (1910), 84; F. Schrader (1681), 32.

Nonetheless, no one beside Leeuwenhoek found them worthy of systematic investigation, and at the medical school in Leiden, the response to the new-found "globules" in the blood was decidedly ambivalent. Student disputations contested whether the corpuscles did or did not pertain to the nature of the blood and whether they facilitated or impeded its flow. Inclined rather to emphasize still the Cartesian shaped particles and diverse chemical constituents of the blood, the new generation of physicians being trained at Leiden tended to treat the new globules as peripheral.[55] In 1686, a graduating medical student giving his final disputation specifically on the blood made only a single allusion to the corpuscles – an allusion so glancing and uncertain, in fact, that it may be no such allusion at all.[56]

What the corpuscles actually were, after all, remained very much in question. They were widely held to have derived from the chyle, but Hartsoeker, for one, thought them only droplets of immiscible fluid, and Swammerdam, inclined to believe they were coagulated particles, doubted that they would be found in the living animal. Even Leeuwenhoek himself had early doubts as to whether they were typical of the blood in general, and he emphasized to Oldenburg that he was in fact writing only about blood he had drawn from his hand.[57] In time, he clearly came to see them as important to the physiology of the body,[58] but he never specified what precisely it was that they did. The blood itself circulated in order to nourish the body, but under normal circumstances, he was convinced, only the thinner fluids of the blood provided the body with sustenance.[59] In the face of starvation, to be sure, the corpuscles also served as nourishment,[60] but what other purposes Leeuwenhoek thought they might serve remain obscure, and he continued to speak of them as simply "the globules that make the blood red."[61] Nonetheless, he continued to observe them over the years with, if anything, increasing interest, for it was not the question of their function that most intrigued him. He conceived them as a kind of basic natural particle with a structure and origin pertaining hence to some fundamental natural processes.

Leeuwenhoek's explicit concern with the "parts" of the blood in his initial observations is perhaps an early expression of the interest he later avowed in

55 Hoeve (1690), §18; Beaumont (1698), cor. 4, 5; L. Jongh (1685), §1; Gerstmann (1687), §1; Burger (1694), §§2–4; E. Nieuwentyt (1686), §§11, 13–19.

56 E. Nieuwentyt (1686), §20.

57 See below regarding n. 82; cf. Blanckaert (1701), 1:197; Bidloo (1715b), §95. NH (1710), 83; idem (1730), "Ext. crit.," 5; JS (1737–8), 1:69–70; *LJS* 95–6 (see also F. Schrader 1681, 32); *AB* 1:74–5. Even in the early eighteenth century, there were still observers who denied that the red corpuscles existed at all (Adams 1710, 26–7).

58 The physiological importance Leeuwenhoek attached to the erythrocytes is evident from the fact that he continued to assess ostensible medicines and poisons and the like according to their effect on the erythrocytes; in addition to n. 81 below, see: *AB* 7:156–67; AvL to RS, 19 March 1706, AvL Letters, fol. 1r–3r; AvL (1718), 435.

59 *AB* 8:148–51, 206–9; AvL (1718), 344–5.

60 AvL (1702), 234–5, 377; idem (1718), 290.

61 See, for instance: AvL to RS, 29 June 1708, AvL Letters, fol. 89v; AvL (1718), 289, 435.

the "principles" – or indeed "beginnings" (*beginselen*) – of things,[62] and the erythrocytes he encountered appear to have quickly melded with other observations in an implicit theory of globular corpuscles as nature's universal building blocks.[63] Presumably in part the product of his inexperience as of yet with the microscope, however, the nascent theory seems to have proved short-lived; but, even after years of observing, a more elaborate figment of his imagination would again persuade him that, in his observations of the red corpuscle in particular, he was indeed on the trail of a basic new truth about nature's microscopic constructions.

Having early sought and found these "parts" of the blood – less than a millionth the bulk of a large grain of sand, he judged[64] – Leeuwenhoek had then begun to search for the parts that composed the red corpuscles themselves; indeed, he informed Hooke in 1678 that this had been the only reason for many of his repeated observations of the blood. In 1674 he had written in passing of the soft, fluid bodies of the corpuscles, but four years later he was pleased to be able to report that, "to my great satisfaction," he had now seen smaller globules enclosed within the membrane of the corpuscle itself, and he saw them as clearly, he said, as if he were observing the eggs in cod roe with his naked eye. Flexible like the corpuscle that contained them, these inner globules shifted about easily as the corpuscle changed its shape, and when some violent agitation burst the membrane that enclosed them, Leeuwenhoek claimed to have also seen those inner globules pouring out. Straining to determine their number, he at last concluded that there were six to each corpuscle, and throughout the following years he continued to see those six in distressed or disintegrating corpuscles and in their residue.[65]

Having detected these smaller globules, Leeuwenhoek now undertook to inquire into how the blood was formed by examining other substances that produced globules of a similar size and structure. In time, he turned to fermenting beer and, in the images of the yeast, found – or thought he found – what he was clearly looking for: globules floating in a crystalline fluid that were the same size as the corpuscles of the blood and composed of six smaller globules. He saw these yeast globules with their six constituent globules as distinctly, he wrote, as if with his naked eye he had seen a small, transparent bladder filled with six other bladders rolling down an inclined plank. There is little question that Leeuwenhoek considered these observations important ones, and three years later he concluded a letter to Christopher Wren, recently president of the Royal Society, by noting that the account of the globules in beer had yet to appear in the *Philosophical Transactions*. In fact, it never would, but even in the last weeks of

62 See Chapter 6 regarding n. 52.

63 At the time of his earliest observations of the erythrocytes, Leeuwenhoek was also discovering microscopic "globules" in a variety of other substances, from milk, hair, and bone to metals and pearls; see Schierbeek (1939).

64 *AB* 4:266–7, 8:32–3, 98–9, cf. 8:204–7.

65 *AB* 2:306–11, 1:92–3, 3:286–7, 4:32–5, 7:158–61, 164–5, 246–7, 256–7.

his life, with the origin of the red corpuscles again at issue, Leeuwenhoek would hark back to similar observations he had made on the dregs of wine.[66]

In his own mind, such observations pertained to one of the most contentious physiological questions that seventeenth-century anatomical discoveries had raised: Where did the blood acquire its final form, in the liver or the heart?[67] Leeuwenhoek's own proposal was a radically simple and unorthodox one. Believing that the globules of the yeast had come in fact from the grain, he argued by analogy that the red corpuscles came directly from the chyle. Reduced to a fluid consistency in the stomach and intestines, food was forced as chyle into narrow blood vessels, and there, under the influence of the motion in these vessels, it coagulated into globules one-sixth the size of the blood corpuscles, six of which globules then merged to form a corpuscle. Leeuwenhoek did not attempt to explain how the corpuscles acquired their redness (which he believed they often lost), but he was convinced nonetheless that his observations had revealed that neither the liver nor the heart was needed to make the blood (though he elsewhere wrote that the thinner fluids of the blood that nourished the body were in fact fashioned in the heart), for the blood vessels could produce blood anywhere.[68] To Leeuwenhoek, the images of the globules within the erythrocyte thus embodied a fundamental new truth about the nature and origin of the blood, and every time he managed to see them again – and since unreal, they were inevitably elusive – they reaffirmed the great step forward he had made in searching out the principles of things.

The pleasure of the spectacle added to the appeal of these observations. Leeuwenhoek had been delighted by his first sighting of the internal globules, but his experience in 1687 while observing the effect on the blood of an extract of cinchona bark was something exceptional. The corpuscles appeared now with an extraordinary brightness and transparency, and within them, he said, he could see the smaller globules and how they touched, as if they lay against each other. It was a fine sight, he wrote, one more of those indeed that gave him more pleasure than anything he had seen before, so he repeated it day after day "only for the great pleasure I had. . . ." The satisfaction of seeing the vivid embodiment of his expectations was thus dramatically enhanced by more purely aesthetic effects, the translucent shapes saturated with light and the motion Leeuwenhoek purposefully imparted to the scene.[69] What Leeuwenhoek was very likely in fact observing in such instances were erythrocytes distorted (and eventually destroyed) by the varying dilution of the fluids in which they floated, wrinkling them in dehydration or swelling them to the point of burst-

66 *AB* 2:388–91, 3:244–53, 4:42–3; AvL to J. Jurin, [Aug. 1723], AvL Letters (vol. 4, letter no. 86 [47]), fols. 325v–326r. (See also *Philosophical Transactions*, no. 380 [Nov.–Dec. 1723]: 435.) Leeuwenhoek made similar observations in the dregs or lees of wine and in fermenting syrups, and claimed to have even seen globules and globular constructions similar to the blood corpuscles in new-fallen rain (*AB* 3:280–93).

67 See Chapter 2 regarding nn. 35, 43, 44; cf. Horne (1652), [27]; R. Frank (1980), 168.

68 *AB* 3:250–5, 4:40–3, 278–81, see also 2:316–17; AvL (1702), 60; *AB* 3:302–3.

69 *AB* 2:308–9, 7:164–5.

ing,[70] and while the stunning clarity of the illusion in 1687 clearly accentuated the beauty of the spectacle, that half-conjured beauty could only in turn have tightened the grip of the preconception Leeuwenhoek thus imposed on the misshapen corpuscles and their remains.

So gripping was the preconception, indeed, that it spawned further elaborations. Apart from having also coerced the images of yeast cells to conform,[71] Leeuwenhoek now discovered multitudes of globules one-sixth the size of the red corpuscle in a variety of fluids and vapors, including not only excretions (and concretions) of the human and animal body, but fog, rainwater, and the airborne residue of candle smoke![72] One of the compelling features of Leeuwenhoek's conception of the globular composition of globules, moreover, was the prospect it offered, like vessels made of vessels and fibers of fibers, of repetitively diminishing structure,[73] and if he never saw a further level of still smaller globules within the six globules of the red corpuscle itself, he managed to find them elsewhere. Soon after his observations of yeast and the resultant conclusions in 1680 about the making of the blood from chyle, he turned to the chyle itself. Placing a sample from the lacteal vessels of a lamb before his microscope, he found clusters of globules of up to six globules each, each of which globules, not very surprisingly, was one-sixth the size of a red corpuscle. But he also now reported an "inconceivable" abundance of yet smaller globules and raised the question whether each of the six globules composing the corpuscle was not itself composed of six globules. Not long thereafter, he was now also discovering globules 1/36 the size of the blood corpuscle in concretions from the human body, in human and animal wastes, and in the residue of destroyed corpuscles themselves.[74]

Other conceptual predispositions besides the sexpartite globular cluster were likewise at work in Leeuwenhoek's perception of the erythrocyte itself, for its very shape was also at issue. Although elastic and tolerant of great distortion, the healthy mammalian erythrocyte is a swollen disk with a concave depression in the middle of each side (Fig. 17). Only in one profile is it round, but it lent itself too easily to Leeuwenhoek's early notion of the "globules" he thought he encountered everywhere and that he characterized in general as looking and behaving like essentially round but very soft and easily deformed fluid-filled

70 *AB* 7:195n. 29.

71 "There can be very little doubt that what Leeuwenhoek saw were aggregates of yeast cells due to rapid budding, and perhaps, in some cases, to small collections of cells which merely happened to be lying side by side. It is well known that the cells of some yeasts show a marked tendency to stick together and are not easily separated by the movement of the liquid in which they are suspended" (Chapman 1931, 435). See also n. 66 above.

72 *AB* 2:388–9, 3:98–9, 286–93, 298–303, 366–7, 370–3, 4:72–3, 76–7, 146–7, 256–7, 278–1, 286–7, 5:8–9, 316–17, 7:92–3, 260–1.

73 Leeuwenhoek was also fascinated by the thought of diminishing fibers in the muscle fiber: *AB* 3:394–7, 402–3; AvL to RS, 6 Dec. 1707, AvL Letters, fol. 69v; AvL (1718), 7, 21.

74 *AB* 3:296–7, 366–7, 370–1, 372–5, 4:256–7, 286–7, 7:158–61, 188–9, 204–5. Leeuwenhoek was also fully aware, however, that there was no obvious end to the series of repetitive structures he was positing in the corpuscle: *AB* 3:296–7; AvL (1702), 234.

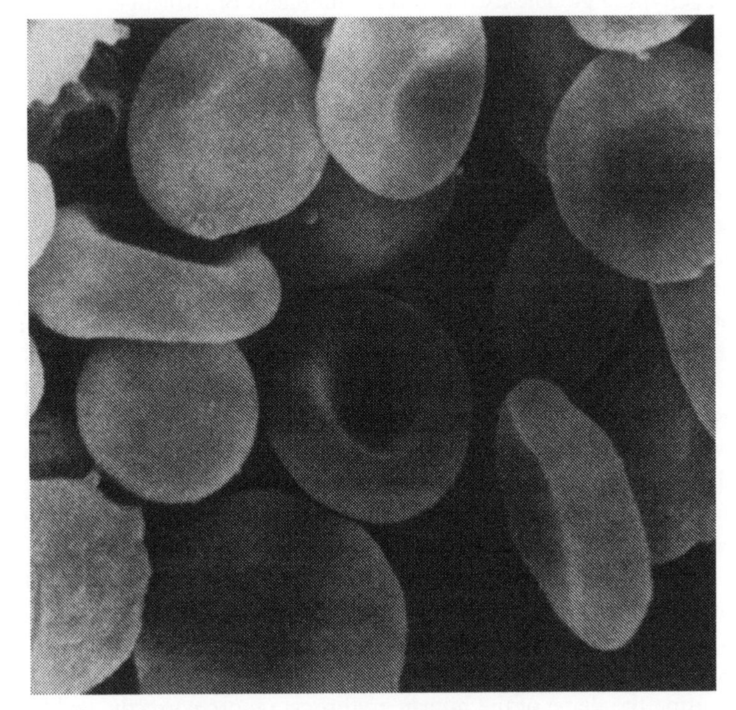

Figure 17. Rat erythrocytes imaged by scanning electron microscopy (Porter & Bonneville 1973, fig. 33a; © Lea & Febiger).

bladders. Hence, though he watched them spinning and turning before his gaze, he from the beginning perceived human and mammalian erythrocytes as precisely such globules and continued to consider them such throughout his life.[75]

Underscoring the peculiarly selective hold of this bias over his perception of human and mammalian corpuscles in particular, however, Leeuwenhoek early recognized the red corpuscles of fish, frogs, and birds as in fact both flat and oval. But it was a shape he now clearly had not expected, and it continued to strike him as something of an oddity, in contrast to the spherical. In 1700, while observing the corpuscles of a salmon, he made a point of checking that he had not deceived himself in seeing them as he did; "in order to satisfy myself about this," he had his draftsman draw the corpuscles as he himself saw them – without the guidance, that is, that Leeuwenhoek had previously provided (see Fig. 18).[76] Leeuwenhoek's wariness was, of course, well warranted but wholly misdirected, for at moments, after all, fish corpuscles too teetered on the brink of becoming globules.[77]

75 *AB* 1:66–7, 74–5, 96–7, 102–5, 120–3, 334–5, 2:304–11; see also n. 61 above.

76 *AB* 3:404 (the meaning of the word *vreemt* in the Dutch text is not adequately conveyed in the English translation on p. 405), 4:72–3, 240–3, 258–9, 8:52–5; AvL (1702), 229–30.

77 *AB* 8:52–3; AvL (1702), 226–7.

The prototypical globule itself also faced moments of challenge even in mammalian blood, it is true, and in later years what Leeuwenhoek had once described as an apparently unambiguous image (when not deformed by contact, that is) was increasingly supplanted by what was more a decision, rather, as to what was and was not to be considered normal. In 1678, Leeuwenhoek had already reported "stretched out" (*uijt gereckt*) corpuscles in his own blood – perhaps in fact a view of their flatter profile – and others that looked indeed like those he had seen in the blood of eels. Also remarking corpuscles distorted by their mutual contact, however, he commented only that they were more pliable than he thought, and it was subsequently with great pleasure, he continued, that he saw them regain their globular shapes in less confining spaces.[78]

Almost forty years later, Leeuwenhoek again described his own corpuscles as he had now observed them dispersed in a medicinal infusion. Spread out as he had never seen them before, the corpuscles now indeed looked flattish (and oval as well), and what now struck him most was a depression he noted in each corpuscle, as if a finger were pressed into a water-filled bladder – an apt description, indeed, of the concavity in a healthy human erythrocyte. Predictably, however, he ascribed the flatness, and hence presumably the depression, to the softness of the corpuscles and apparently to the effect of the medicinal infusion; four years later, in 1721, when he spoke again of flattened human corpuscles that had lost their rotundity, his thoughts now turned to problems he was having with his vision.[79] His belief in the blood globule had not wavered, but he was now reporting the shape of the human erythrocyte more correctly, and his decision that it was a distorted shape, mistaken though it was, was at least a conscious one. That he found that correct shape so worthy of comment after forty years of studious observation, however, strongly suggests that his perception as well as his judgment had long been misled.

The perception of the normal red corpuscle as spherical was hardly uncommon, after all; to the contrary, it had quickly become almost universal. Immediately after Leeuwenhoek's announcement of his discovery of the corpuscles, observers at the Royal Society as well as at the Musschenbroek shop in Leiden had also reported their globular form. Even Swammerdam, who, like Leeuwenhoek, watched spinning human corpuscles and recognized the flat, oval shape of the corpuscles of frogs, described and depicted the former as globules. Muys too observed the oval erythrocytes of a variety of animals with sufficient clarity to see not only their "elliptical" shape but apparently nuclei as well; yet the corpuscles of man and "four-footed animals" he also still emphatically described as globules. By his own account, Albrecht von Haller could no longer manage to see the corpuscles of warm-blooded animals at all, but with a solar microscope he now observed the corpuscles of cold-blooded animals as spherical as well. Indeed, Leeuwenhoek's own doubts about his perception of fish cor-

78 *AB* 2:304–7.
79 AvL (1718), 435; AvL to J. Jurin, 19 March 1723, AvL Letters, fol. 317r.

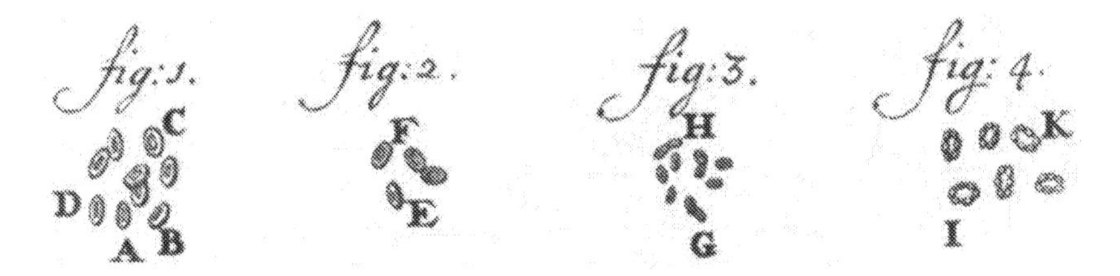

Figure 18. Fish corpuscles as seen and illustrated by Leeuwenhoek's draftsman in 1700. Within the oval corpuscles, both Leeuwenhoek and the draftsman usually observed a "little light," *ligje,* apparently indeed the nucleus (AvL 1702, 230; J. Baker 1948–55, 90:99). Fig. 4, however, now surely under Leeuwenhoek's guidance or direction, is the draftsman's representation of the fish corpuscle as composed of a ring of six small corpuscles (see further on in the text). Photo courtesy of the National Library of Medicine, Bethesda, Md.

puscles had been sharpened by visitors who, like Haller, had also seen them as round.[80]

To be sure, the potential sources of the error were many, from the optical shortcomings of early microscopes, apt to produce small aberrational disks of light in their images, to the tendency in visual perception for ill-defined stimuli to assume as simple a shape as possible. Preparation methods could easily mislead as well: The eighteenth-century English observer William Hewson vividly described how erythrocytes bloated into spheres when their fluid environment was further diluted with water, a sight Leeuwenhoek himself often confronted when testing the effect on the blood of diverse solutions of "salts."[81]

But ideas were also at work. For the many who believed the corpuscles to be fluid – and that was the common opinion, Hewson assures us – it was logical to assume the globular shape of an immiscible drop; whereas those of a more mechanistic bent, like Boerhaave, assumed that the original particles of chyle were inevitably rounded into spheres as they tumbled through the vascular sys-

80 Birch (1756–7), 3:380; Bartholin (1673–80), 5:24; *LJS* 80–1; JS (1737–8), 1:69–70, 2:835; Muys (1738), 299–303 (where the dark image of the nucleus of nonmammalian erythrocytes appears to have been misinterpreted as either a little bulge or depression – although the flattened sides of the erythrocytes of nonmammalian vertebrates are not concave – together with a ring around the corpuscle's edge [ibid., 300]); Haller (1757–66), 2:50–1, 53–4; AvL (1702), 229.

81 J. Baker (1948–55), 89:117, 120–1; Arnheim (1974), 53, 63, 67; Hewson (1774), 311–12; *AB* 7:195n. 29. Apparently viewing all white or transparent crystals as "salts," Leeuwenhoek seems to have also shared, to some extent, the belief that salts were the most decisively active of chemical substances, at least insofar as the animal or human body was concerned (*AB* 5:33n. 43; Diemerbroeck 1685, 161–2; Mort 1696, 109; H. Baker 1753, 17–18). Though he considered them a source of pain and disease, he assumed that salts were the basis of medicinal efficacy as well, and the crystalline particles he observed were often extracted, indeed, from ostensible medicines (see, e.g.: *AB* 1:236–41, 246–9, 2:138–9, 5:16–19, 90–1, 7:122–5, 130–1, 194ff., 220–3, 242ff., 274–5, 8:245ff., 264ff.). He explored the effects of such extracted salts on the red corpuscles of the blood in studies that the Royal Society emphatically encouraged (*AB* 5:342–3, 6:16–19, 7:182–3, 186–7, 190–1, 246–57).

tem. Often compelling as well was the special aptness of the spherical shape for motion and hence for the circulation of the blood. Graduating from Leiden in 1693, one Johannes Haberkorn thus argued that the red particles of the blood were found to be round globules through *rationalis argumentatio* as well as through the microscope.[82]

In supposing even blood serum and water itself to be ultimately composed of globules, Leeuwenhoek himself betrayed a similar train of thought. More tellingly, given his notion of the origin of the blood corpuscles, he also argued as early as 1674 that solid particles made from thin fluids in the body had to be round.[83] Ranged beside his more idiosyncratic and apparently more compelling attachment to the bladderlike globule (which he emphatically found apt even for the particles of water), such echoes of a broader inclination to link fluidity to spherical shapes only further disposed him to see the blood corpuscles as essentially round whenever he could. What they may also have prevented him from grasping may perhaps serve as testimony to the potential efficacy of such preconceptions as well: Despite fifty years of observing, Leeuwenhoek never recognized the irregularly shaped, amoeboid leukocytes as another family of particles to be identified with the blood.[84]

The web of ideas dictating the round corpuscle also implicated Leeuwenhoek's sexpartite clusters of globules, so redolent of sphericity, indeed, that they resisted being fit into those oval corpuscles he did acknowledge. Initially, he wrote, he could detect no parts at all in fish corpuscles no matter how hard he tried; when, in time, he did see smaller globules within their oval shapes – and now surely often including the nucleus, which is absent in mammalian erythrocytes – the prototypical six still eluded him. By the turn of the century, however, he was bringing such observations into line. A search for the parts in the corpuscles of a crab (whose blood cells do not in fact include erythrocytes) had left him fairly certain – *beelde ik my sekerlijk in* – that he had seen the six globules from which they were made. Not long after he also finally managed, "in so far as could be seen," to distinguish the ostensible six globules of fish corpuscles as well.[85]

In fact, however, this delayed success prodded his resurgent doubts in 1700 as to whether the blood corpuscles of fish were really oval after all and thus deprived of the "perfection" of sphericity. His draftsman depicted the six globules within the oval corpuscle as linked together in a ring (see Fig. 18), whereas Leeuwenhoek described them as lying in two rows, three on three.[86] It was in any event a structure that lacked the compactness and the multidimensional symmetry of the arrangement of globules he had envisioned for the spherical

82 Hewson (1774), 310, 315; NH (1710), 83; idem (1730), "Ext. crit.," 5; Hoeve (1690), §18; Schwencke (1743), 110; Boerhaave & Haller (1740–4), 2:280–91, 3:660; Bidloo (1715b), §104 (cf. §95); Haberkorn (1693), §8; see also Blanckaert (1701), 1:197.

83 *AB* 1:284–5, 298–301, 5:24–7, 1:110–13.

84 See, for instance, *AB* 3:93n. 13.

85 *AB* 3:404–7; AvL (1696), 17–18; idem (1702), 232–3; J. Baker (1948–55), 89:112, 90:99.

86 AvL (1702), 227–8, 232.

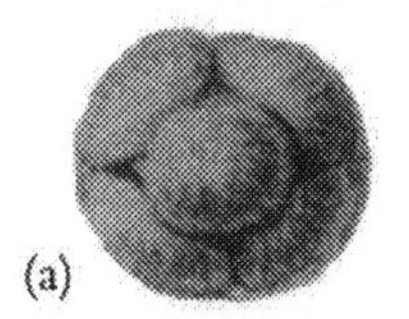

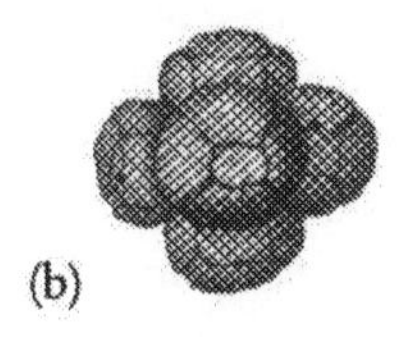

Figure 19. (a) Leeuwenhoek's red chalk drawings of a wax model of a yeast globule. By permission of the President and Council of the Royal Society. (b) Leeuwenhoek's illustration of a corpuscle composed of thirty-six smaller globules (AvL 1702, plate facing p. 230). Photo courtesy of the National Library of Medicine, Bethesda, Md.

corpuscle, an arrangement so appealing to Leeuwenhoek that he worked it up in models. In 1680, he had made and sketched a wax model of his analogous conception of the yeast globule with its six smaller constituent globules, and by 1700 he now had a model composed of not six but thirty-six globules for the sake of visitors inquiring about the corpuscles of the blood (Fig. 19).[87]

A complex and perhaps not wholly congruous tangle of ideas and predilections, some curiously idiosyncratic, others more widely shared, thus worked to shape the images of the spherical mammalian erythrocyte – a shaping, moreover, that vested those images with evocative significance. The geometry imposed on the corpuscle and its clustered parts was further ground won against arbitrariness in nature. Though he insisted on his deathbed that the why and wherefore would remain unknown,[88] the discovery of that structure, fictitious as it was, embodied for Leeuwenhoek a new understanding gained or, in another frame of mind, a previously hidden wisdom now revealed. In the light of other similarly molded observations – of yeast, in particular – the image in the corpuscle was the disclosure as well of a previously unrecognized universal natural process undergirding life itself.[89]

But may we presume such fabricated intimations in images only when manifest deceptions – the spherical erythrocytes, the sexpartite clusters, or indeed the coil in the muscle fiber – advertise the sway of preconceptions? Or may we look for laden images in the absence of such telltale signs as well? Though leaving no detectable distortions in Leeuwenhoek's descriptions, did a recognition of mechanism nonetheless color Leeuwenhoek's perception of the capillary circulation, and did imaginings of fantastic anatomical articulations still hover around

87 *AB* 3:250–1; AvL (1702), 233–4.

88 AvL to J. Jurin, [Aug. 1723], AvL Letters (vol. 4, letter no. 86), fol. 325v–326r. This letter has on occasion been interpreted as an encouragement to further research on the beginnings of the blood corpuscles (Cole 1937, 29; Schierbeek 1950–1, 2:373), but this would seem to be the result of a failure to recognize the reiteration of a passage from Jurin's previous letter with the first-person pronoun unchanged. For similar examples of such potentially misleading citations by Leeuwenhoek, see: *AB* 2:328–30, 3:420–1; AvL to RS, 25 July 1707, AvL Letters, fol. 49v; see also *AB* 8:17n. 26.

89 Leeuwenhoek believed indeed that he had even been able to produce the process in rainwater (*AB* 3:290–3); see also nn. 66, 72 above.

prolonged observations of busy microorganisms? If so, as in the case of the mammalian erythrocyte and the muscle fiber, they also enveloped these images of rushing shapes and pervasive illumination in a web of evocations that helped to fashion the microscopic wonders that Leeuwenhoek prized the most.

Leeuwenhoek himself was by no means unsuspecting of the encroachment of preconceptions upon observations. Acutely conscious indeed of the danger of self-deception, he often emphasized how many times he repeated observations before feeling secure about what he saw.[90] If he succumbed precipitously to the thought of the coil in the muscle fiber, it was only after years of observing that he convinced himself in 1717 that he had seen the much-heralded hollow in the supposed vessels in the nerve.[91] The language in his letters bespeaks as well an awareness of the need to recognize his more tenuous judgments about images and to distinguish them from the more certain. He had vigorously and very early insisted (in disagreement with Nehemiah Grew) that patterns of wood cells were indeed vessels he had clearly "seen," but he continued to speak more tentatively of what he "called," "judged," or "imagined" to be valves in the larger vessels of the wood, even though he had observed indications of such structures and believed the hydraulics of the tree demanded them.[92] Apparently on guard against the prejudicial tug even of earlier perceptions, in one instance at least he purposefully declined even to consult his earlier notes, since he anticipated not only seeing more in later observations but seeing things differently.[93] He was aware too, however, that expectation might also enable him to see what otherwise would go unnoticed.[94]

Nevertheless, he was unlikely to have acknowledged that the most exhilarating of his discoveries, of whose truth he had no doubt, were also constructed in part of preconceptions. He would not have recognized how much he himself had bestowed upon those evocative spectacles that so enriched his experience of microscopy and encouraged his pursuit of those lines of research that most ungrudgingly yielded such returns.

That experience in itself, however, was unlikely to give rise to a sustained program of inquiry, and the quickly waning interest of other early observers of the wonders, for instance, of the newly discovered microorganisms underscores the additional social and self-aggrandizing roots of Leeuwenhoek's more general commitment to the microscope. Leeuwenhoek's letters in late 1676 sparked several months of lively discussion in the Royal Society; but after the members (and the king) finally succeeded in seeing the "pepper mites" for themselves a

90 He had observed the smaller filaments composing the muscle fibers of a fish twenty-five times, he told Hooke, before he chose to write about them (*AB* 3:430–1). See also: *AB* 1:110–11, 3:208–9; AvL (1696), 153; AvL to RS, 22 Sept. 1711, AvL Letters, fol. 148r, 149v, 152r; AvL (1718), 118.

91 AvL (1718), 315–19, 323; E. Clarke (1968), 135–7.

92 *AB* 2:32–7, 1:48–53, 2:6–9, 12–13, 3:158–61, 170–1, cf. 180–1; Baas (1982), 84, 87.

93 AvL (1718), 321–2, see also p. 236.

94 *AB* 3:430–1, 8:38–9; AvL (1702), 390, 448.

year later, the society's own observations barely survived the following spring.[95] Christiaan Huygens also caused a stir in Paris in 1678 with the "infinity of little animals" discovered through his own new microscopes, but, after a brief period during which scientists and instrument makers around the court and the Académie Royale des Sciences busily exchanged different "waters" and infusions, such observations soon lost their appeal in Paris as well.[96] Piqued by Christiaan, Hartsoeker's interest was already flagging even before their trip together to Paris, and Christiaan's own observations, or at least those he chose to record, came to an end by the end of 1679. To be sure, he would resume them in the future in a remarkable series of observations that rival Leeuwenhoek's own, but this was only after a lapse of thirteen years and very near Christiaan's death.[97]

Social influences and interactions, including both the urgings and doubts of correspondents, visitors, and friends, often also critically influenced the specific direction of Leeuwenhoek's researches, decisively orienting as well as helping to sustain his efforts along even those lines that offered the most striking images. After Hooke had in fact nudged Leeuwenhoek toward the observations of the frog embryo that led to his discovery of the capillary circulation in the first place, it was Christiaan Huygens who early encouraged him to extend his efforts to warm-blooded animals.[98] Eager from the very first to show the capillary circulation to gentlemen visitors, Leeuwenhoek continued to remark their amazed reactions[99] and their assurances as to the exceptional importance of what they saw.[100] Negative reactions too had their stimulating effect, for Leeuwenhoek was also spurred to new observations of the capillary circulation by those who refused to believe him.[101]

The line of research most decisively shaped by such social and outside intrusions revolved around the spermatozoa. From the very beginning, the encouragement and urgings as well as the outspoken doubts of his correspondents pushed Leeuwenhoek in directions he was initially often reluctant to pursue. Members of the Royal Society in particular provoked or pressed him into continuing his observations of the spermatozoa at a time when he himself was not inclined to grant them much significance, and the society's promptings still helped to guide and sustain these observations even after his outlook on the

95 Birch (1756–7), 3:332–4, 338, 346–7, 349, 352, 364, 366–7, 383, 391, 393, 430; Hooke (1935), 335–6. There was, to be sure, something of a limited revival of interest at the Royal Society in the spring and summer of 1680, but it took the form in large part of a few reports of observations by Thomas Henshaw, after which that interest appears to have rapidly faded again (Birch 1756–7, 4:37, 44, 46, 47, 49).

96 *OCCH* 22:256, 13:704, 708, 711–12, 714–17, 8:92. In the long run, indeed, the academy showed little interest in either Leeuwenhoek or the microscope (Berkel 1982, 202; Roger 1971, 183). See Chapter 1 concerning the initial impact of Huygens's microscopes in Paris.

97 *OCCH* 8:70, 13:698–732.

98 *AB* 6:18–19; *OCCH* 9:310.

99 *AB* 8:34–7, 48–9, 80–1; AvL (1696), 16; idem (1702), 55, 115; AvL to RS, 28 Aug. 1708, AvL Letters, fol. 98r; AvL to RS, 14 Jan. 1710, ibid., fol. 125r–v.

100 *AB* 8:23–5, 60–1, 80–1. Regarding the enthusiasm for this sight expressed by Leeuwenhoek's friend and admirer Pieter Rabus, see also Rabus & Slaart (1692–1701), 8:455, 4:160.

101 AvL (1696), 2; idem (1702), 54.

spermatozoa had dramatically changed.[102] That change, however, soon plunged him into controversy, and it was now his own defiance that fueled what became an aggressive research campaign;[103] but when that combativeness waned, it was again his friends and sympathizers who were most likely to call him back to resumed observations of spermatozoa.[104]

The spectacle itself of teeming spermatozoa continued to delight Leeuwenhoek throughout, nonetheless,[105] and both the Royal Society's early interest and his later campaign hinged, after all, on an arresting set of ideas. These ideas framed the prospect of what would have been for Leeuwenhoek his greatest discovery, but they also entangled his observations of the spermatozoa with broader religious and philosophical concerns. Hence, while these ideas raised the tantalizing possibility of a great personal triumph, they also transformed Leeuwenhoek into a combatant in a struggle between contending speculative commitments. His studies of the spermatozoa thus underscored, as did no other line of research, how images, ideas, personal aspirations, and social interactions could all become inextricably interwoven; for those studies were inevitably caught up in the general question of the generation of living things – an issue whose wide-ranging implications involved the microscope in ongoing and sometimes fervent disputes that inspired their own social alignments and personal crusades.

102 It was in response to the proddings of Nehemiah Grew in particular that Leeuwenhoek had gone on to look for spermatozoa in other animals: [Grew] 1678; *AB* 2:326–7, 334–7, 346–7. See also: *AB* 6:316–19; Birch (1756–7), 4:217, 407; AvL (1694), 650–1.

103 *AB* 4:56–7, 68–71, 5:144–53, 172–3, 6:120–3; Ruestow (1983), 198ff.

104 AvL (1702), 250, see also p. 388; AvL (1718), 285, 290–1.

105 See n. 14 above; AvL (1702), 323, 397.

CHAPTER EIGHT

Generation I: Turning against a tradition

In the second half of the seventeenth century, a growing disenchantment with spontaneous generation invited the microscope to move center stage. What stirred this repudiation of a pervasive and age-old belief is not altogether clear, but philosophical and religious convictions were very much at stake. Prominent among the leaders of this movement of denial, Swammerdam and Leeuwenhoek reflect the array of concerns at issue, but they also reveal how such concerns could acquire a very personal stamp. Only the more engaged as a consequence, the two Dutchmen elaborated vivid new arguments for the assault on spontaneous generation, though, in doing so, they disregarded the inconclusiveness and ambivalence of what their microscopes had indeed to offer.

Deriving from both popular belief and a venerable classical legacy, the lore of spontaneous generation was extensive, elaborate, and richly specific. According to Kircher, one of the century's most enthusiastic students of spontaneous generation, all putrefaction and hence the decay of all plants and animals produced new life in the form of worms and insects, and the specificity ascribed to the process assured a comparable diversity. While nearly every plant produced its own particular insect, wrote Ulisse Aldrovandi, most animals similarly generated their own distinctive lice.[1]

Indeed, the variety of putrefaction and the offspring it might produce was limitless. Diverse kinds of mosquitoes and gnats derived from decaying matter in stagnant water as well as from certain plants and rotting moisture itself, including the dew under certain kinds of leaves.[2] Ancient tales about the origin of bees in the carcasses of oxen, bulls, cows, calves, and sometimes lions still echoed widely, rousing the interest of even the newly founded Royal Society in 1663.[3] Shellfish were ascribed to mud and slime, and snails to the putrefaction of fallen leaves, but the "common folk" had snails and mussels dropping from the sky as well, apparently engendered by ocean vapors,[4] while frogs and

1 Kircher (1658), 38–41, 48–9; idem (1665), 2:347; Aldrovandi (1602), 5, 545.
2 Aldrovandi (1602), 390; Jonston (1657), 59; Goedaert (1662?–9), 2:84.
3 Aldrovandi (1602), 58; D. Cluyt (1597), 6–8; Birch (1756–7), 1:270; Nylant & Hextor (1672), 176; Blanckaert (1688b), 202; JS (1737–8), 2:530.
4 AvL (1694), 728–9; idem (1696), 119, 154; idem (1702), 35; JS (1737–8), 1:98, 133.

tadpoles seem to have fallen with the rain everywhere.[5] Although their origins were only rarely so picturesque, to be sure, the broad range of animals that arose from spontaneous generation – from all kinds of worms and insects to species of fish, reptiles, occasional birds, and even rats – was emphasized repeatedly.[6]

Such an origin, however, usually implied creatures of a diminished, incomplete, or even sinister nature. In the preceding century, Paracelsus had described them as monstrous, poisonous, usually short-lived, and hated by those of their kind that had been properly born. More typical, however, was the Aristotelian outlook taught by Franco Burgersdijck at Leiden that associated spontaneous generation with "imperfect" animals marked not only by their small size and lack of elaborate structure but by their stunted ability to propagate as well. As in the case of many varieties of worms, he taught, animals generated spontaneously were often unable to reproduce at all, while others so engendered brought forth offspring that were then barren and different from those that spawned them, such as the worms produced by flies. Still others bore offspring similar to themselves and fertile, but for only a limited succession of generations – four, for instance, in the case of mice.[7]

If not all the bizarre tales, the fact of spontaneous generation, at least, was accepted by all the major schools of philosophy in the early seventeenth century. While Burgersdijck explained it in terms of an occult nature concealed in matter, Johannes Baptista van Helmont, preeminent advocate of a "chemical" philosophy, cited the "ferments" distributed by God at the Creation throughout the earth, air, and water, and the atomist Nathaniel Highmore wrote of the "seminal principles" unleashed by a "discerning corruption." Descartes ventured that it really did not take too much to make an animal, and his disciple Regius described how the heart, circulatory system, and other organs of a living animal emerged from an initial drop of lifeless fluid, heat, agitated particles, and the laws of motion alone.[8]

Conscientious observation confirmed the prevalence of spontaneous generation as well. The account of the observer who, though "very exact in noting such things," watched a rotting stick changing over the course of several days into a caterpillar that finally crept away may only amuse; but Goedaert also experimented with such substances as bran and urine to produce maggots and flies and, now over the space of several years, observed the emergence of spiders and harvestmen from decaying mushrooms. After his own sustained study of the generation of insects, William Harvey as well had no doubts about their susceptibility to spontaneous generation, and in 1678 Robert Hooke still reported

5 Bacon (1877–89), 1:311; Redi (1688), 80; AvL (1702), 124; Andry (1700), 13.
6 Aldrovandi (1602), 641; Kircher (1658), 39; idem (1646), 834; Lemnius (1583), 465; Digby (1644), 215; Gassendi (1658), 2:260.
7 Paracelsus (1965–8), 5:59, 61; Burgersdijck (1642), 260–1; see also Goedaert (1700), 1:171.
8 Burgersdijck (1642), 255, 257; Helmont (1648), 36; Highmore (1651), 26–7; Descartes (1897–1910), 11:506; Regius (1646), 216–17; see also idem (1668), 162–3.

to the Royal Society on his own repeated observations in which a host of small insects arose from dying plants.[9]

Soon after midcentury, however, spontaneous generation became a *belle question* in the gatherings of savants cultivating an amateur interest in science and philosophy.[10] They now wrestled with contending philosophical conceptualizations of the process and looked more closely at purported instances. In 1668 the Italian naturalist Francesco Redi raised the stakes when he initiated a general campaign against the notion of generation from putrefaction with his *Esperienze intorno alla generazione degl' insetti,* which was translated into Latin three years later. Redi's attack rested on his own extensive observations and experiments, the most celebrated being those that demonstrated that maggots on putrefying flesh came from flies and not from the flesh itself, which produced no maggots at all if kept in a covered vessel. The impact of Redi's book soon became apparent in the Netherlands as elsewhere, touching off what rapidly developed into a sweeping repudiation of an assumption that had prevailed since antiquity. By the end of the century, leading naturalists were deeply committed to the denial of spontaneous generation.[11]

Both Swammerdam and Leeuwenhoek numbered among its most determined new opponents. Swammerdam found it deeply offensive and repeatedly attacked it in uncompromising terms. His indignant hostility winds like a pervasive thread through his natural history. Leeuwenhoek, in his turn, inveighed so often against spontaneous generation that, according to Hartsoeker, his observations might be taken in large part as having had no other purpose. Indeed, Leeuwenhoek's letters speak explicitly of his determination to free the world of its error in believing that a living creature could arise from putrefaction.[12]

Leeuwenhoek informed the Royal Society in 1687 of the opening offensive of his own campaign against that persisting belief. He had conducted a four-month study of the life cycle and sexual apparatus of the grain weevil (*Calandra granaria* Linnaeus) for no other reason, he wrote, than to demonstrate to the world that the pest came from a worm. The campaign thus begun was pursued with particular intensity during the following decade, when he undertook similar studies of the grain moth (*Tinaea granella* Linnaeus), the flea and the louse, aphids, and various flies.[13] He set out to expose how shellfish reproduced as well,[14] and wrestled in particular with the origin of eels.[15] As he wrote spe-

9 Digby (1644), 220; Goedaert (1662?–9), 1:104–5, 127–8, 2:274–83; Harvey (1651), 122, 179–80, 229; Birch (1756–7), 3:420.

10 Oldenburg (1965–), 1:259; see also Gassendi (1658), 2:263; Birch (1756–7), 1:213, 242, 258, 270, 294, 297, 418–19, 424, 444, 448, 501–2, 2:48–9.

11 Redi (1688), 10, 33–4. On Redi's impact in the Netherlands, see, for instance, Craanen (1689), 81–2, 85; Raven (1950), 375, 469, 471.

12 NH (1730), "Ext. crit.," 18; *AB* 8:176–7, 184–5, 276–9.

13 *AB* 7:4–37, 98–121, 8:276–91, 296–311, 304–7, 318–29; AvL (1694), 537–49 (see also AvL 1718, 375); AvL (1696), 28, 60, 77–99 (see also AvL to RS, 22 Sept. 1711, AvL Letters, fol. 151v–154r); AvL (1697), 197ff. According to F. J. Cole, Leeuwenhoek studied the reproduction and life cycles of some twenty-six animal types (Cole 1937, 7).

14 AvL (1694), 693–720; idem (1696), 119, 154.

15 *AB* 8:176–7, 184–7, 330–1; AvL (1693), 514–22.

cifically of his study of the grain moth, these were studies pursued not only to instruct himself but to turn the world from its errors.[16]

Both Leeuwenhoek and Swammerdam inevitably mobilized the microscope in the cause of their campaigns. The lens extended the reach of their researches into the mystifyingly small life cycles whose invisibility – often but not always due to the diminutive dimensions in which they played themselves out – provided spontaneous generation with some of its most resilient support. Malpighi resorted to the microscope to discover the insect eggs in plant galls, whose emergent flies misled even Redi, but Swammerdam went one step further: he compared the eggs in willow galls to those in the abdomen of the parasitic fly and cited their similarity as decisive evidence against what was advanced as an unassailable instance of spontaneous generation.[17] Having followed the life cycle of parasitic larvae he had found in the intestines of other larvae, he similarly concluded that nothing came from putrefaction. Leeuwenhoek likewise produced a tiny parasitic fly from maggots he found in the hollowed bodies of aphids, from which he also concluded that, the more one investigated the propagation of creatures, the more indisputable it became that no new creatures were made, by which he seems clearly to have meant that none were generated spontaneously.[18]

To underscore how the ostensibly "imperfect" creatures in fact generated no differently than other animals, Leeuwenhoek and Swammerdam turned the microscope to the bodily parts of these animals and, passing gradually from the visible to the invisible, revealed the elaborate structures and organs of sexual reproduction. In the 1660s, Swammerdam had already displayed the penis and testicles of the bee – the latter large enough to be easily seen with the naked eye, as Thévenot had remarked – and in the following years he astonished visitors with a microscopic look at the penis of a cheese skipper fly.[19] The issue of spontaneous generation was more in evidence, however, when Leeuwenhoek included an illustration of the magnified male sex organs in his study of the grain weevil. He provided the Royal Society with a similar drawing of the testicle of the flea (Fig. 20) and set its penis before a microscope for visitors, as he also did the male sex organs of the louse.[20] He wrote in addition of the swarming spermatozoa he had found in both the flea and the grain weevil and also in the mussel, sullied as it too was by the taint of spontaneous generation.[21]

16 *AB* 8:176, 184–5.

17 Adelmann (1966), 2:829 (see also Redi 1688, 92ff., 108–9; Farley 1977, 15); JS (1737–8), 2:732, 734–6; see also *AB* 6:48–71.

18 AvL (1697), 134–5; JS (1737–8), 2:713.

19 See Chapter 5, n. 23; F. Schrader (1681), 13.

20 *AB* 7:26–9 (figs. 3, 4 in "Table" I), 4:20–3; *Catalogus*, item 33; AvL (1697), 205–6 and plate opposite p. 204.

21 *AB* 3:324–5, 7:12–13; AvL (1696), 142–3, 145–6 (and regarding the background concern with spontaneous generation, see pp. 154–5).

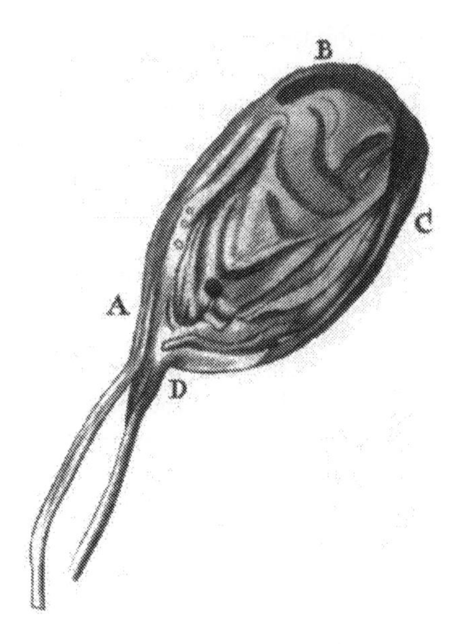

Figure 20. Leeuwenhoek's illustration of the dried testicle of a flea (*Philosophical Transactions,* no. 145 [1682/3], pl. 2). Photo courtesy of the Special Collections Department, University of Colorado at Boulder Libraries.

The female reproductive role had been more easily discovered. Andreas Libavius had written of finding the "womb" (*matricem*) filled with hundreds of eggs within the silkworm moth, and Thomas Moffett had even earlier described the eggs seen within the flea. Near the middle of the seventeenth century, the Italian anatomist Marco Aurelio Severino offered a small, highly stylized sketch of a "double-horned uterus" with eggs and tiny "fetuses" within a beetle, and Redi eventually wrote of the double ovaries crammed with eggs that he, apparently together with Steno, had found in flies.[22]

Malpighi's detailed microdissection of the sex organs of the female silkworm moth opened a new era, however. Swammerdam, having earlier discovered the large ovaries of the queen bee, would subsequently attempt to describe and illustrate the thousands of eggs (the oocytes) he now saw diminishing in size in the ovarioles until they escaped even the best of his microscopes (see Fig. 13).[23] The exploration, description, and illustration of the sex organs, male as well as female, became an essential part of his anatomies of insects, and even in his remarkable dissection of the louse, "only a point in size," he depicted the ovaries again with their diminishing eggs (Fig. 21). Leeuwenhoek too observed the

22 Libavius (1632), 401; Moffett et al. (1634), 276; Severino (1645), 346; Redi (1688), 38, 86.

23 Malpighi (1686–7), *Diss. epist. de bombyce,* 35–8, pl. XII (fig. 1); JS (1669), pt. 1:[105]; idem (1737–8), 2:471–2; *LJS* 75.

ovaries in a variety of similar animals, including the grain weevil, the flea, the fly of the parasitic maggot within the aphid, and the mite.[24] The microscope thus revealed an extravagantly elaborate sexual apparatus within such tiny, "imperfect" bodies.

Leeuwenhoek pressed his researches into the egg itself, observing the living larva of the flea still within its shell and even prematurely removing the louse from its egg (and once even from an egg still in the ovary). Within the translucent eggs of mites he watched developing areas of darkness give way to the more clearly defined parts of the later embryo, all culminating finally – and fatally – in the futile struggle of a newborn mite to exit the opened shell.[25] To close the reproductive circle, Leeuwenhoek even searched the contents of such diminutive eggs in a fruitless hunt for the initial spermatozoon.[26]

Aphids, however, harbored a startling surprise. Having failed to find their eggs in infested foliage, and recognizing this as a boon to the defenders of spontaneous generation, Leeuwenhoek turned to searching for the missing eggs in dissected aphids. To his great amazement, he found not eggs but tiny unborn young, numbering at times more than seventy in a single individual. The dissection of still smaller aphids revealed the development of the unborn, articulated bodies from translucent egg-shaped particles that diminished successively in size until they appeared no larger through the microscope, he wrote, than sand grains appeared to the naked eye.[27] Unexpected as the discovery was, nonetheless, the quantities of young within the bodies of the aphids offered striking confirmation that aphids came from aphids, not from putrefaction.[28]

The microscope's most memorable challenge to spontaneous generation, however, was its revelation of the minute complexity of insect anatomy. That intricacy, after all, was the focal point of Swammerdam's microscopic observations, the upshot of which, he stressed, rendered spontaneous generation absurd. How could putrefaction have produced the wisdom and inventiveness he encountered there? How could ingenious structure so narrowly circumscribed be reconciled with such an origin?[29] Leeuwenhoek drew the same lesson from his own observations. Like Swammerdam, he was also struck by the fabric of the insect wing: The abundance of vessels in its structure, the thousands of hairs that covered it, the smooth perfection of the once-folded wing unfolded, all argued that creatures adorned with such parts could not have arisen from putrefaction.

24 JS (1737–8), 1:83–4, pl. II (fig. VIII); *AB* 7:18–19; AvL (1694), 541; idem (1696), 134; idem (1697), 297; idem (1718), 108.

25 AvL (1694), 545; idem (1697), 201, 203; AvL to RS, 22 Sept. 1711, AvL Letters, fol. 149r–v, 153r. In early 1704, Leeuwenhoek also wrote of dissecting the "grains" of the cochineal dye in order to confirm that they were in fact the bodies of insects, and, repeatedly finding eggs within these grains, he also repeatedly discovered small, unborn insects within the eggs (AvL to RS, 21 March 1704, ibid., fol. 282r–287r).

26 Ruestow (1983), 201–2.

27 AvL (1696), 77–8, 81–2, 88, 95, 99–100; idem (1702), 281–2.

28 Steven Blanckaert, it should be noted, had already reported the viviparity of aphids before Leeuwenhoek (Blanckaert 1688b, 162).

29 JS (1737–8), 1:85, 346, 2:707–8, 713; idem (1675a), "Ernstige aanspraak," [11].

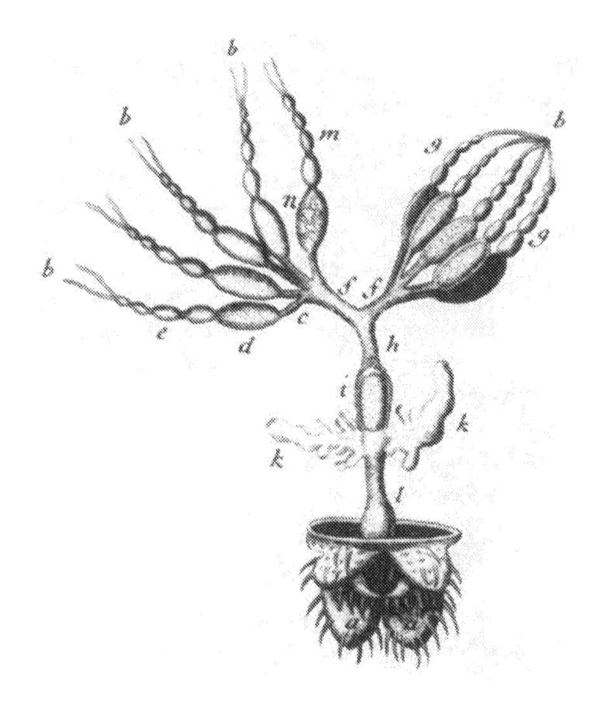

Figure 21. The ovaries of a louse by Swammerdam (JS 1737, 1:pl. 2). Photo courtesy of the Department of Special Collections, University of Chicago Library.

But even the hairs on a honeybee, or for that matter on a mite, displayed a precision and structure that denied such beginnings, and if a hair could bear such witness, a compound eye was eloquent indeed.[30]

Like Swammerdam, Leeuwenhoek not only insisted now on the "perfection," *volmaaktheid,* of the supposedly imperfect animals but repeatedly extolled them above their larger counterparts,[31] and the implications for spontaneous generation followed inescapably. Given the perfection and extraordinary orderliness in both its structure and its mating, argued Leeuwenhoek, even the mite could no more have come from putrefaction than could an elephant from a bit of dust; nor were sweat and putrefaction any more capable of engendering lice and fleas than was a manure pile a horse or ox. Indeed, if spontaneous generation were true, Swammerdam added, there was no reason why it should not produce a man as well, since animal bodies were equally amazing.[32] It was a

30 JS (1737–8), 2:432; AvL (1696), 61–2, 73, see also p. 4; AvL to RS, 22 Sept. 1711, AvL Letters, fols. 147r, 151;. AvL (1718), 88–9; idem (1694), 726, see also p. 702. Anton de Heide had also brought the microscope to bear on the "elegant structure" of the mussel (Heide 1684).

31 On their "perfection," see, for instance: *AB* 3:396–7, 7:18–19; AvL (1694), 569, 602; idem (1696), 7, 72, 91; idem (1697), 213, 308. On the comparison with larger animals: *AB* 8:324–5; AvL (1693), 466, 473–4, 531; idem (1694), 584; idem (1696), 39–40.

32 AvL (1694), 602; *AB* 6:64–5, see also 7:112–13; JS (1737–8), 2:669.

rhetorical gambit with a ring to it, and it continued to echo repeatedly, and sometimes ponderously, in the writings of succeeding naturalists.[33]

The microscope was not unfailingly hostile to spontaneous generation, however. For Kircher, it had rather revealed a pervasiveness of spontaneous generation that, had he not had frequent recourse to the instrument, he wrote, he would never have believed himself. The early literature on the microscope continued to include reported instances of spontaneous generation, while the committed defenders of spontaneous generation persisted, in their turn, in citing microscopic observations.[34] An account of the generation of a frog observed in a dewdrop may be dismissed as sheer silliness, but Kircher surely observed swarms of previously unseen nematodes and other forms of microscopic life. Leeuwenhoek later protested that, had Kircher had a good microscope, he would not have claimed so much that was untrue; yet Leeuwenhoek's own microscope revealed a world of living things that, even under his own sharpest lenses, seemed no more than moving particles of matter, "clouds of moving atoms" behind which, in the eighteenth century, spontaneous generation would in fact come storming back.[35]

Nor could the microscope's most distinctive arguments against spontaneous generation, the pervasiveness of reproductive generation and the intricate structure of formerly "imperfect" animals, make the case conclusively. In the first place, to explore the life cycle of every creature of doubtful origin was simply not possible. Leeuwenhoek early complained of being constantly challenged to explain away yet one more instance of spontaneous generation, as if he should be ready, he protested, to solve everything that was laid before him;[36] but a single case that he failed to refute was sufficient to sustain a whole system of belief.

The traditional lore was so flexible and accommodating, moreover, that the close observation of life cycles and reproductive anatomy could never in fact absolutely preclude the possibility of spontaneous generation even for any specific species, for sexual and spontaneous generation were not, after all, mutually exclusive. Thus caterpillars were said to come not only from dew and the leaves of plants but from butterfly eggs as well, and, according to Aldrovandi, the worms that gave rise to flies arose from either sex or putrefaction, just as wasps similarly derived from copulation as well as from rotting crocodile blood and horses. John Jonston, a Polish-born Scot who compiled a prominent survey of natural history a half-century later, likewise ascribed fleas not only to dust (especially when soaked with the urine of men or goats), putrefying moisture in the fur of

33 Malebranche (1958–70), 1:90–1, 3:347, 12:229–30; Ray (1693), 16; Réaumur (1734–1929), 1:30–1; H. Baker (1753), 229; Roger (1971), 233–8.

34 Kircher (1665), 2:347; see also idem (1658), 38, 40, 42–3; Borel (1656), 45; Zahn (1685–6), 3:102; Buonanni (1691), pt. 1:passim.

35 Griendel von Ach (1687), 28; Major (1932), 7; AvL (1718), 88; John Needham (1748), [636].

36 *AB* 8:176–7, 184–5.

dogs, and the sweat of slaves (at least in certain regions of the Indies), but to copulation too. Bacon also had noted that, of the living creatures that came from putrefaction, many afterward continued to propagate by mating.[37] The cumulative and unsystematic lore of spontaneous generation thus left plenty of room for sex, and the fact that the likes of mites, fleas, and lice had sex lives too was telling only to those already inclined to narrow nature's options. By contrast, the scholar Isaac Vossius chided those who taught that frogs came from frogs, mice from mice, and insects from insects but failed to point out that they came from slime and putridity as well.[38]

The intricacy now revealed in the construction of insects might well dispute their imperfection, but as an argument against spontaneous generation it too fell short, resting as it did on analogy and limits arbitrarily imposed on nature's creative capacities. Regius had not hesitated to describe how, in some quiet place in the soil, the earthy particles came together to produce not only a heart and blood vessels but bone, fibers, muscle, parenchyma, the liver, spleen, head, kidneys, bladder, and urethra, the makings of something more than an "imperfect" animal. In the preceding century, Paracelsus had boasted of being able to produce a homunculus in a flask, and in the next, Diderot proposed that "fermentation" if not putrefaction could indeed produce an elephant.[39] How much more conceivable, after all, was the formation of the complex body of an animal from the mixing fluids widely held to be at work within the womb? Consequently, the new revelations of minute anatomies were likely to prove persuasive only to those predisposed to be convinced.

The disenchantment with spontaneous generation reflected a shift in more general convictions, both philosophical and religious. On the threshold of the seventeenth century, spontaneous generation could fit very comfortably in the framework of ideas underlying the prevailing schools of philosophy. The scholastic natural philosophy taught in the universities revolved around basic concepts embodying the assumption that "generation" (and its converse, "corruption") pertained to the very essence of nature in general, animate or inanimate; the foremost rival philosophy at the time, the so-called chemical philosophy, emphasized generative processes in nature no less.[40] The educated European was thus shaped to assume that such processes were everywhere at work.

As the century progressed, however, the philosophic notion of "generation" as a universal process became increasingly problematic. "Form" was a central concept, and a philosophically technical one, in the Aristotelian logical analysis of natural phenomena; when joined to matter, Franco Burgersdijck had once taught at Leiden, the form of a thing established the nature of that thing and distinguished it from the other natural bodies of the world. Hence the emer-

37 Goedaert (1700), 1:159; Aldrovandi (1602), 205, 354; Jonston (1657), 92; Bacon (1877–89), 2:638.
38 Voorde & Heide (1680), 221–2n. 3, 532n. 4; Ray (1693), 15, 17; Vossius (1663), 61.
39 Regius (1646), 216–17; Paracelsus (1965–8), 5:62; Diderot (1875–7), 2:133.
40 Ruestow (1973), 18, 22; Pagel (1958), 115; idem (1944), 16–22.

gence (or actualization) of form was the process of generation. The precise origin and nature of form was historically a much exercised problem, however, and Jean-Baptiste Du Hamel, secretary to the recently founded Académie Royale des Sciences, still testified in the mid-seventeenth century that no other question in philosophy was more involved or more difficult to explicate. Indeed, it was a question that of late had only grown more intractable, and for Robert Boyle, who agreed that the origin of forms was one of the most noble but at the same time one of the most abstruse inquiries in natural philosophy, the very way the problem had traditionally been framed now seemed perverse. The assumptions that matter was inert and that motion endeavored only to persist – if it "endeavored" at all – undergirded a mechanical philosophy in terms of which the emergent "forms" of things were now in fact meaningless.[41]

Having undermined the traditional philosophies, however, the adherents of the aggressive new alternative began in time to insist as well on its inability to deal satisfactorily with what, short of the emergence of the rational soul, was the most dramatic instance of generation, the generation of living animals. The formation and organization of their bodies was "the greatest of all the particular Phaenomena" of the created world, insisted John Ray in 1691, but it was by now a commonplace among both friends and foes of mechanism that "blind, brute matter" or mechanical laws alone could not produce the "excellent Contrivances" that composed those bodies.[42] Despite the determinism suggested by its laws, moreover, the mechanical philosophy was often perceived as the heir of ancient atomism's identification with chance as well, and "Blind Chance" was hardly more acceptable that "Blind Necessity" as the parent of the elegant structures of living things. More was at issue than simply the rise of the mechanical philosophy, however; since Descartes, Regius, and others had earlier burdened it with no such limitations, mechanistic sensibilities themselves were shifting. Whereas Descartes had ultimately contrived an extended mechanistic account of the forming of the animal fetus in the womb, John Ray found such efforts so absurd as to "need no other Confutation than ha ha he."[43]

The philosophical difficulties were compounded by religious apprehensions as well. In the first half of the century, Johannes Baptista van Helmont had charged the scholastic doctrine of generation with atheism for deriving the forms of things from the powers and dispositions of matter. That generation was a natural process and hence appropriate to natural philosophy was a fundamental tenet of scholastic instruction, but van Helmont protested that generation was the cre-

41 Ruestow (1973), 19; Reif (1962), 139; Roger (1971), 326–9, 333; Du Hamel (1669), 137–8; Boyle (1699–1700), 1:29–36, 62. For an account of the general struggle with the concept of "form" in the sixteenth and seventeenth centuries that stresses rather the durability of the idea, see Emerton (1984), 60–75.

42 Ray (1691), 27–8, 33–4, 36–7, 217–18; NH (1722), 22, 216; Wilkins (1675), 83–4; Malebranche (1958–70), 12:228–9.

43 Wilkins (1675), 83–4; Leclerc (1704), 4:52; B. Nieuwentyt (1717), 579, 589; Descartes (1897–1910), 11:223–90; Ray (1691), 217; see also Garden (1690–1), 477; Roger (1971), 212, 219–24, 443.

ation of new forms from nothing and that the only power capable of doing that was the same Creator who had made the universe itself from nothing. A few decades later, however, it was the new mechanical philosophy that was stirring up religious concerns, even among the growing number of its adherents.[44]

To the prominent English divine Ralph Cudworth, the idea that animal bodies and other phenomena were formed only by "Material and Mechanical Necessity" or "the mere Fortuitous Motion of Matter" was not only irrational but atheistic. Cudworth, though, was ostensibly addressing himself to classical schools of thought, and while Descartes emphatically grounded his laws of nature on the character of divinity itself, contemporary atomists also pointedly rejected the "chance" of their classical predecessors and opted as well for a reliance on the providence of God.[45] Nonetheless, their insistence on the point betrayed their own sensitivity to the latent dangers in the philosophy they embraced, so that Cudworth's remonstrations spoke indeed to a widely felt uneasiness rooted in the philosophic drift of the times. There was an explicit reluctance as well, to be sure, to grant God too free a hand in the generation of his creatures;[46] but whatever the reason for restraining his activities, it left generation looking ominously like an independently creative capacity in nature, and spontaneous generation in particular began now to acquire an odor of impiety.[47]

A range of philosophical and religious issues thus came to bear on spontaneous generation. Few, perhaps none, were wholly new to the late seventeenth century, however, nor did they even now dictate convictions, even in intellectually up-to-date circles. The absurdity of Descartes's mechanistic explication of generation was not immediately or equally obvious to all. Hooke, for one, still embraced spontaneous generation in his *Micrographia,* betraying no awareness that piety might be in peril; to the contrary, he was ready to ascribe such generation unhesitatingly to an omnipotent Creator.[48] Nonetheless, the drift of ideas and changing sensibilities cast an ever more doubtful light on spontaneous generation, as Swammerdam bears witness.

Very early on, indeed, Swammerdam stigmatized the notion of spontaneous generation as profoundly irreligious. Repeatedly associating it with atheism, he denounced it as a Godless opinion sustained only by a heathen philosophy.[49] "Accidental" in nature, it belied the wisdom and providence on which Swammerdam insisted in the lives and design of insects and thus denied their witness to the Deity.[50] For Swammerdam, then, the issue of spontaneous generation was a choice between the Creator on the one hand and, on the other, "all the

44 Helmont (1648), 130, 132, 134; Reif (1962), 139–41.

45 Cudworth (1678), 147–8; Descartes (1897–1910), 8(pt. 1):61–2; Beeckman (1939–53), 3:43, 63; Goorle (1620), 247; Gassendi (1658), 1:311–26.

46 Cudworth (1678), 150; Ray (1691), 34–5; Descartes (1897–1910), 11:524; NH (1730), "Ext. crit.," 20.

47 Joblot (1718), pt. 2:44; Pluche (1732–50), 1:19, 4:526–30.

48 Du Hamel (1669), 219–20; Blanckaert (1688b), 2; Hooke (1665), 124–7, 130–1, 193–4.

49 JS (1737–8), 1:171, 2:394, 432, 669, 708, 712–13.

50 JS (1669), pt. 1:3, 23, pt. 2:40; idem (1675a), "Ernstige aanspraak," [11], 4; idem (1737–8), 1:85, 106, 171, 346, 2:394, 432–3, 669, 708.

atheists who say that the generation of these small animals occurs by accident."[51] But the natural order was also at stake, for to ascribe the birth of insects to accident and putrefaction, Swammerdam noted, was to render the firm order of nature itself wholly accidental.[52]

Swammerdam's protests suggest, however, that the implications of shifting outlooks were initially brought to bear only when they also bore on matters of more personal urgency. For one so plagued by spiritual self-doubts, preserving the religious significance of his life's preoccupation was no small concern, but the preservation of "order" struck a deep personal chord as well. Far from being a committed adherent of the mechanical philosophy, after all,[53] Swammerdam in fact showed little solicitude for the new laws of motion that underlay the seventeenth-century scientific revolution in mechanics and astronomy. Rather, order in nature had another meaning – or, rather, other meanings – for Swammerdam as a research naturalist. He perceived it, as we have seen, in the intricate construction of small creatures and in both the arrangement and the ingenious precision of the operation of their parts. At such moments, to be sure, order was not far removed from that purposeful complexity and design that bespoke the hand of God; but of no less importance to Swammerdam was the order he perceived in the unfolding lives of insects. Each animal had its ordained time, place, manner of life, and particular food, he wrote, and generation in particular followed an unalterable rule and order.[54]

That order too, however, was directly threatened by spontaneous generation. Even Hooke, in his *Micrographia,* had marveled that the same kind of insect could be engendered in several different ways, and Regius, decades before, had stressed that, given the almost infinite variety of particles and motions in the earth, the possible diversity of the offspring of spontaneous generation was virtually unlimited. No matter how many different animals we might imagine, he wrote, the countless shapes and arrangements of particles were capable of producing many more.[55] In effect, the naturalist could hardly know what to expect next. Hence, an essential first step in imposing order on insect life was to assert, as Swammerdam did, that all insects came only from eggs, "although most philosophers," he wrote, "maintained the opposite."[56]

But spontaneous generation intruded upon subsequent insect development as well, for metamorphosis too was widely conceived as the generation of one animal from the death and putrefaction of another.[57] It was an understanding bolstered indeed by striking – and presumably factual – accounts of caterpillars or chrysalides that produced not butterflies but quantities of flies (or sometimes worms that, pupating themselves, then engendered flies).[58] It was Harvey, however, who framed the issue for Swammerdam, for in 1651 the Englishman

51 JS (1737–8), 1:346. 52 JS (1669), pt. 1:3. 53 See Chapter 9, n. 90.
54 JS (1737–8), 2:710. 55 Hooke (1665), 193–4; Regius (1646), 219. 56 JS (1669), pt. 1:57.
57 See, for instance, Teresa of Avila (1979), 91–4; more prosaically, Goedaert (1700), 1:201.
58 Goedaert (1662?–9), 1:4, 42–5, 146–7, 2:239–41, and see also p. 136; Moffett et al. (1634), 192; Aldrovandi (1602), 253; Bacon (1877–89), 2:507.

had not only ascribed the emergence of the butterfly to the putrefaction of the caterpillar but had virtually identified a broadened conception of metamorphosis with spontaneous generation.[59]

For Swammerdam, predictably, Harvey's doctrine of metamorphosis pointed directly towards atheism,[60] but challenging as it did the orderliness of insect development as well, it too threatened the very meaningfulness and coherence of Swammerdam's effort as a naturalist. His *Historia insectorum generalis* was in fact a preliminary exposition of a new system of insect classification, one that ultimately provided the basis of the future great work Swammerdam intended. That system not only provided the unifying framework of his studies in natural history, consequently, but was in itself, to Swammerdam's mind, an original and important new contribution to the understanding of insect life.[61] Revolving as it did around four classifications ("orders") determined by four different life-cycle patterns, however, its viability rested on the inviolability of those patterns – on order, that is, in insect development. Swammerdam thus forged a framework of order that was central to his scientific enterprise, but it demanded an unyielding denial of spontaneous generation and the unpredictability to which it exposed the life cycle of insects through metamorphosis as well.

That systematic order he was determined to defend had several levels of significance for Swammerdam. More broadly, it confirmed that a reassuring order prevailed everywhere, for if the repeated generation, rearing, growth, and metamorphosis of insects were regulated and fixed, he asked, who would dare

59 Harvey (1651), 121–2, 188–9. 60 JS (1737–8), 2:669.

61 It has been argued that the four "orders" that constitute Swammerdam's system of classification were in fact "nothing more and nothing less than a means to arrange [*rangschikken*] his observations" (Visser 1981, 71). Swammerdam's sense of the ontological status of his orders is certainly not unproblematic, and he wrote at the beginning of the need, to be sure, to establish – literally "to make," *maaken* – beforehand some orders (*stellingen ende orderen*) of changes; these orders were to be understood as a means by which to grasp the different appearances (literally "exhibitions" or "spectacles," *vertooningen*) of insects and other "bloodless" animals (JS 1669, pt. 1:4–5; idem 1737–8, 1:3). But he continued in that passage, though not altogether clearly, that those orders would be like a brush with which the changing forms of these animals would be depicted "as if with their own colors" and displayed in all their splendor and elegance; and he shortly after elaborated the simile still further, declaring that his orders were the means by which, "as with the true varnish of nature itself," he intended to restore and brighten the dirtied painting of nature to its original beauty and splendor and thus to reveal more clearly the wisdom and goodness of God with respect to the growth, maintenance, and changing of the very smallest creatures (JS 1669, pt. 1:6, 56–7; idem 1737–8, 1:4, 38).

Although he persisted in writing of his four orders as indeed "our" or "my" orders (JS 1669, pt. 1:57; idem 1737–8, 1:4, 2:867), Swammerdam also wrote in the *Historia insectorum generalis* that the rules and foundations (*regelen ende gronden*) of his new system had been discovered nonetheless "in the nature of things" (JS 1669, pt. 1:5). He likewise early spoke of the orders (*regelen ende orderen,* now pertaining to the very relevant issue of the difference between chrysalides and pupae) as being the orders of the omniscient Creator and having been set immutably, again, "in the nature of things," so that, if we did not follow them closely, we would be constantly misled in our observations (ibid., pt. 1:21; idem 1737–8, 1:15). The assumption that these orders actually existed in nature is reflected as well in his more offhand remark that he was astonished that no one else before him had recognized his second order of change in nature (JS 1669, pt. 1:99; idem 1737–8, 1:219). For an overview of Swammerdam's orders, see Schierbeek (1947), 159–67, and *DSB,* s.v. "Swammerdam, Jan."

deny the fitting government of all the parts of the universe?[62] More narrowly, and with more immediate meaning for Swammerdam as an active scientist, the order reflected in his system gave coherence, direction, and meaning to his lifelong researches. The worth of years of invested time, labor, and skill depended on its viability, as did the prospect he foresaw of a vast new arena of significant and continuing discovery.[63]

The development of that system was indeed a creative and still-unfolding enterprise, a continuing discovery of order in – or imposition of order on – a multitudinous and at first sight chaotic realm of nature. Beyond the coherence, meaning, and purpose provided by Swammerdam's system, consequently, the discovery of order added excitement and a reaffirmation of developing skills – whereas its imposition offered the gratification of efficacy and control. Is it too much to suggest, then, that his system was tied as well to the joy that Swammerdam found in his research? We need not go so far, however, in order to conclude that Swammerdam's preoccupation with the order of rigidly preordained insect life cycles was intimately linked to the significance, both personal and scientific, of his lifelong work as a naturalist. Adapted to the distinctive circumstances of his life and temperament, the shifting conceptual inclinations of the age thus framed Swammerdam's hostility to spontaneous generation, and did so well before he turned his hand to microscopic dissection.[64]

Despite his enthusiasm for the microscopic anatomies he uncovered, therefore, Swammerdam's elaboration of their relevance to spontaneous generation was but an embellishment of a stand already rigidly fixed – so fixed, indeed, that strikingly contrary microscopic evidence failed to sway it. The microscopic anatomies themselves were ultimately also inconclusive, but there had been earlier observations, to be sure, of more decisive import. In his doctrine of metamorphosis, Harvey had insisted on the absence of a prior structure within the homogeneous "putrescent matter" of the former caterpillar from which the butterfly arose,[65] and hence Swammerdam's exhilaration when he discovered the butterfly and its parts hidden, to the contrary, within the caterpillar.[66] But when

62 JS (1669), pt. 2:40; idem (1737–8), 2:863.

63 Countless wonders were yet to be discovered about insect development, Swammerdam declared in the *Historia insectorum generalis* (pt. 1:131), and, although unclear as to whether he meant his four orders specifically as well as his basic conception of metamorphosis, he also wrote that, supported by his discoveries, these "important truths" could serve as a firm foundation for building a greater and more amazing structure and provide a torch, as it were, to lead the way to the discovery of innumerable other truths (dedication [3]–[4]). Indeed, he offered his own studies of the development of selected insects as examples to open the way to further discoveries (pt. 1:131, pt. 2:1). Having also spoken of the observations he was offering in the *Historia* as being founded upon or "anchored" in his orders (pt. 1:5), he also reminded his readers one last time in the concluding line to the text of the *Bybel der natuure* – as arranged, at least, by Boerhaave – that he had reduced all the insects he had studied to those orders, which rested indeed on the pupa or nymph as a foundation (2:873).

64 With a reference to "a constant and inviolate order in nature," though not explicitly with respect to generation, Swammerdam had rejected at least as early as 1666 the idea that insects were generated from plants (JS 1669, pt. 2:1–2).

65 Harvey (1651), 121–2.

66 See Chapter 5, n. 76.

Swammerdam later brought the microscope and his new skills in insect dissection to bear on the metamorphosis of the soldier fly *Stratiomyia furcata* (Fabricius), what he now confronted pointed toward a very different conclusion. The changes in the internal anatomy he observed as the grub became the fly struck him as miraculous, as if there had been a new creation, "or rather generation," or a resurrection in another body. While acknowledging that what he had now observed might indeed seem to warrant the conclusion that metamorphosis was a radical transformation after all, he nonetheless persisted in maintaining that the same animal endured throughout and that even the members of its body, though some were lost and others gained, remained nonetheless essentially the same.[67] The microscope, to be sure, provided the debate over spontaneous generation with some vivid and gripping new imagery, but the nature of Swammerdam's own commitment argues the deceptiveness of those trappings.

Leeuwenhoek's engagement with spontaneous generation unfolded very differently. His convictions emerged rather uncertainly amid a variety of microscopic observations, and while the compelling drive of conceptual and religious commitments is less in evidence, a background of interactive social contacts and associations is very much to the fore. In contrast to the early and decisive impact of Swammerdam's intellectual obsessions, consequently, Leeuwenhoek's repudiation of spontaneous generation took shape through a complex interplay of observations, on the one hand, and social interventions and influences, on the other. Nonetheless, as was the case with Swammerdam, very personal priorities appear also to have played their role throughout.

Leeuwenhoek's initial attitude toward spontaneous generation is unclear, and, even after he had emerged as an insistent opponent of the ancient doctrine, the roots of his hostility are far from obvious. Throughout the 1670s, his letters betray no evidence that he even recognized the subject as a significant one. He had seen a version of Redi's *Esperienze* by early 1675, but limited as he was in his knowledge of languages, he showed no awareness of its central theme.[68] On the few occasions that he brushed against the issue, not only the rarity of these occasions but his vagueness and ambiguity suggest if not indifference then at the least a surprising casualness. They seem to indicate that, although he was indeed inclined to look for signs of a more legitimate parentage, he was willing to accept spontaneous generation as fact. Struck in 1676 by the rapid increase in the numbers of microorganisms in a pepper infusion, he suggested that they "had been put together in a moment's time, so to speak"; but, declining to elaborate on what precisely he had in mind, he brushed the question aside as something he would leave to others.[69]

The following year, however, Leeuwenhoek appears to have shown more of an inclination to put certain supposed instances of spontaneous generation to the

67 JS (1737–8), 2:666, 680–1. Regarding the identification of this fly, see Chapter 5, n. 115.

68 *AB* 1:272–3. Leeuwenhoek commented very briefly only on his dissatisfaction with Redi's illustration of a louse.

69 *AB* 2:100–1; Ruestow (1984), 228–32.

test. Having removed the eggs from dissected lice and then unborn lice from the eggs, he turned to fleas as well, whose eggs, he said, he had never seen; nor, he added, had he ever noticed smaller fleas of progressively larger size. Consequently, having kept several fleas until they indeed produced eggs, he carried the eggs about in his pocket until they hatched and did likewise with the larvae until they spun cocoons and pupated, and within the pupa he at last recognized the flea again.[70] His interest in the eggs he had never seen and the apparent absence of growth (an absence he had also noted in the microorganisms of the preceding year) imply that the question of spontaneous generation was on his mind, and similar life-cycle studies (including the flea and the louse again) were to loom large in his later campaign against it. Nonetheless, he still made no explicit reference to what he would later condemn so insistently, and his attempt somewhat over a year later to explain the presence of liver flukes in sheep betrays a continuing readiness to let nature produce living things spontaneously.[71]

Leeuwenhoek had written of the fluke in February of 1679, but it was only a few months later that he took a stand that perhaps now entailed a general repudiation of spontaneous generation. In June he wrote Lambert van Velthuysen, philosopher and theologian at Utrecht, that he could not understand how any animal could be produced without propagation – *voortelingh* (or *voortteling*), the term he would consistently use in the future (though with inconsistent spelling) as the antithesis of spontaneous generation. His purpose in insisting on propagation in this instance appears to have been the rejection not of spontaneous generation, however, to which there is still no explicit reference, but of the claim that unborn mice already had within them the next generation of mice, a prospect that also conflicted with his advocacy by now of the role of the semen. Yet the continuation of the passage suggests how that advocacy was perhaps working also to undermine his tolerance of spontaneous generation. Reflecting on the origin of the microorganisms that appeared in rainwater, he now assumed that they derived from a "seed" or semen (*saet* or *zaad*) that had been carried aloft in rising droplets of water vapor. All that remained unresolved, he added – underscoring how the ideas were mingling in his mind – was whence came the seed of the spermatozoa within the testicles.[72]

In the interval since his initial reflections on the liver fluke, Leeuwenhoek's view of the spermatozoa had indeed undergone an enormous change, and his open opposition to spontaneous generation appears to have emerged nearly simultaneously with a new conception of the role of the spermatozoa. As far as can be established with certainty, Leeuwenhoek first confronted spermatozoa in 1677,[73] when a medical student at Leiden named Johan Ham brought him a flask of ostensible semen from a gonorrhea patient. In this fluid – very likely, in

70 *AB* 2:244–53.

71 *AB* 2:416–17. Regarding Leeuwenhoek's consciousness of the absence of detectable individual growth in microorganisms he observed, see also AvL (1696), 166.

72 *AB* 3:76–9.

73 In earlier observations of semen, Leeuwenhoek had discovered what he took then to be globules, which he doubtless expected to find (*AB* 2:290–1).

fact, rather the discharge symptomatic of the disease – Ham said he had discovered living animals, which Leeuwenhoek now observed as well. Oldenburg had earlier included semen among the subjects he had asked Leeuwenhoek to investigate with his microscopes; but Leeuwenhoek, personally repelled by the subject and even more by the prospect of having to talk about it, had quickly ended those observations. Ham's visit, however, spurred Leeuwenhoek to resume them (though of specimens from healthy individuals, beginning with himself),[74] initiating what became in time his celebrated, indeed notorious, association with spermatozoa.

Although his new observations aroused great interest, however – even calling Christiaan Huygens back to the microscope and leading to his collaboration with Hartsoeker[75] – Leeuwenhoek himself appears to have initially regarded the spermatozoa as but another variant of the microscopic life he was by now discovering in such abundance.[76] He did indeed now adopt an unorthodox insistence on the dominant role of the semen in reproduction, but that stance at this point rested rather on his puzzling perception of microscopic vessels in the semen that he seems to have taken for the rudiment of a vascular system.[77]

His thinking was soon jarred, however, by further unexpected discoveries. In his early observations of semen, Leeuwenhoek had relied on the spilled or excess residue of the act of coitus,[78] but in the spring of 1679 he wrote that, for various unexplained reasons, he had begun dissecting a male hare and, in so doing, cut the vas deferens, the duct conveying the semen from the testicles. Observing the matter that oozed out of the severed vessel, Leeuwenhoek found it filled with an unbelievable multitude of spermatozoa; turning then to the testicle itself, he discovered that the filaments of which it was composed contained spermatozoa as well. He does not appear to have expected to find them there, but his observations, soon confirmed in a widening variety of animals, convinced him that the spermatozoa did indeed originate in the testicles and that the testicles in fact served no other purpose.[79] His discovery entailed a dramatically altered understanding of the spermatozoa themselves, though he cloaked his own conversion in an impersonal assertion of vanquishment: Those who have maintained that the spermatozoa are produced by putrefaction and play no role

74 *AB* 2:290–1, 280–3, see also 3:74–5; AvL (1702), 65. Regarding the common identification of gonococcal discharge with semen, see, for instance, Tol (1674), cap. 1, §5; Craanen (1689), 750; Blanckaert (1683b), 232. Leeuwenhoek himself spoke of the putative semen Ham brought with him as that which was "spontaneously discharged" from the afflicted patient (*AB* 2:280–1, see also 3:74–5). I am grateful to Dr. F. Marc Laforce of the University of Colorado School of Medicine and Dr. Frank Judson of Denver General Hospital for reassurance that it is not unusual to find spermatozoa in such discharge.

75 See Chapter 1, nn. 95–6, 98–101. Regarding the rapid spread of a familiarity with the observation of spermatozoa in the Netherlands, see also: *LJS* 87; Bartholin (1673–80), 5:24. Johan Ham apparently continued his efforts too (F. Schrader 1681, 34–5).

76 *AB* 2:290–1.

77 See Chapter 9, n. 128.

78 *AB* 2:290–1, 328–9, 338–41, 3:10–11.

79 *AB* 3:8–19, esp. pp. 18–19, 202–9, 324–5 (see also pp. 38–9). Before these observations, Leeuwenhoek believed that the semen derived from the testicles but that the spermatozoa arose rather in the penis (*AB* 3:18–19).

in procreation must now succumb, he wrote. It was two months later that, having now posited a "seed" for microorganisms, he declared that the only remaining question was that of the origin of the seed of the spermatozoa themselves within the testicles.[80]

Perhaps inevitably Leeuwenhoek's new commitment to spermatozoa would impinge on his thinking about spontaneous generation – he seems, after all, to have previously numbered among those who also ascribed such generation to the spermatozoa themselves. But it was his discovery (while pursuing wide-ranging inquiries into various animal semens) of the spermatozoa in the semen of insects (Fig. 22) that soon elicited an explicit and unambiguous denial of spontaneous generation. It was his opinion, he wrote in November 1680, that we could now be assured that all animals, however small, come from propagation rather than putrefaction; for spermatozoa were to be found in the semen of the flea as well as of the larger animals, and why, then, would even the very smallest lack the same "perfection"?[81] An emerging campaign in behalf of the spermatozoa thus appears to have fathered a future campaign against spontaneous generation.

Hence, Leeuwenhoek's rejection of spontaneous generation seems indeed to have been determined by the microscope – but, again, how conclusive was the testimony? The logic of Leeuwenhoek's new stance at first sight seems clear enough. Now identified with sexual generation, the spermatozoa in the semen of animals thought subject to spontaneous generation indicated that these animals too propagated by reproduction, and the spermatozoa found in the smallest animals whose semen Leeuwenhoek could investigate also argued that the diminishing scale of living things posed no barrier to such reproduction. Clear as it might be, however, this reasoning fell short of being decisive.

The presence of spermatozoa was but an elaboration of the presence of reproductive organs, after all, and could come to grips no more effectively with the accommodating flexibility of the lore of spontaneous generation. In some respects, Leeuwenhoek's observations of the spermatozoa actually made the argument from analogous though ever-smaller sexual equipment more difficult to make. He had to disregard the fact that he had found no correlation between the size of animals and the size of their respective spermatozoa; indeed, he had been particularly struck by the large size of the spermatozoa of the flea.[82] To grant a similar "perfection" to microorganisms often smaller than the spermatozoa themselves was hence a less obvious step than it might at first have appeared. Microscopic observations that might seem to have encouraged such an inference in fact called into doubt the very consistency of diminishing scale and continuity of analogy it assumed. What besides observations and the logic of his reasoning helped, then, to transform Leeuwenhoek in the spring of 1679 into an opponent of spontaneous generation?

80 *AB* 3:18–21, 76–9.

81 *AB* 3:314–31, esp. pp. 328–9. F. J. Cole writes that Leeuwenhoek ultimately examined the spermatozoa of thirty different animal types, including eleven arthropods, two mollusks, seven fish, one amphibian, two birds, and seven mammals (Cole 1937, 8).

82 *AB* 3:324–5.

Figure 22. Leeuwenhoek's drawing of a spermatozoon of a dragonfly (*AB* 3:318). By permission of the President and Council of the Royal Society.

He was certainly at this time learning more about spontaneous generation as a topic of interest to the learned. Following up, as it were, on a letter from Nehemiah Grew in London the preceding year outlining the prevailing understanding of eggs and semen,[83] Leeuwenhoek's correspondence with van Velthuysen in early 1679 appears to have begun acquainting the Delft townsman with current discussions of yet broader issues concerning generation. Leeuwenhoek's June letter, to be sure, was written against the background of earlier correspondence between the two,[84] and we do not know what besides some papers of Leeuwenhoek's had already passed between them – nor hence the extent to which that prior correspondence had prompted Leeuwenhoek's remarks on propagation and the *saet* of microorganisms. However, that June letter, a relatively brief one, dealt essentially with matters pertaining to generation. Van Velthuysen, himself the author of an earlier treatise on generation that, having been revised, was soon to be republished,[85] responded immediately with an explication of recent theories so developed and so subtle, said Leeuwenhoek, that he at this point declared himself unfit to comment. A year later, however, Leeuwenhoek now wrote of his having become familiar with the various opinions about generation (*voortelinge*) and especially those of a "certain gentleman" who had written that meat and its juices in a tightly closed vessel engendered no living creatures.[86] Further testifying to his new awareness of Redi's experiments and their purpose, Leeuwenhoek in 1680 undertook his own parallel experiment with pepper infusions.[87]

To be sure, the results of the experiment failed to conform to Redi's: Microorganisms, though different, appeared in a sealed as well as an open glass tube. Yet if Leeuwenhoek was troubled by this, of which there is no evidence, any doubts had passed four months later when he wrote of the spermatozoa of the flea and their likelihood in the very smallest animals as well. The doubtful outcome of his experiment did not deter him, that is, from denying spontaneous generation soon after;[88] this only compounds the suspicion that something more – or other – than observations lay behind his new convictions. Nor do his letters yet betray the religious or philosophical sensibilities that might have brought his reticence and vagueness to an end and further induced him to take what was,

83 [Grew] 1678. To identify Grew as the author of this response, see *AB* 2:326–33.

84 *AB* 3:74–5.

85 Velthuysen (1657); idem (1680).

86 *AB* 3:86–7, 260–1.

87 *AB* 3:260–7.

88 Four years later, apparently, he turned again to an experiment with microorganisms that echoed Redi's, and now the outcome was very different and certainly reassuring (*AB* 6:62–3). But see also Fournier (1981a), 204–6, regarding Christiaan Huygens's similar experiments.

after all, a public stand. In the light, then, of the social sensibilities ascribed to Leeuwenhoek above, can we discount the possibility that his conversion was in fact triggered by his new awareness not only of the terms of the learned debate over spontaneous generation but also of the attention the issue commanded among the learned and the social alignments it implied?

The case for the critical influence of such social reference becomes only stronger as Leeuwenhoek's public stand emerges in fact as a public campaign, beginning with his extended study of the grain weevil in 1687 and continuing with particular intensity through the following decade. The very commitment to a campaign argues in itself that, for Leeuwenhoek, more was at stake than simply the veracity of a scientific fact. In the 1680s, indeed, he began to speak occasionally of the wisdom of God observed in the structure and order of his creatures, no matter how small,[89] and in 1687 he cited his first religious or metaphysical argument against spontaneous generation: that the emergence of an animal soul – an endowed capacity for movement – from unmoving things would be a miracle necessarily involving the omnipotent Creator.[90] (It would also violate the philosophical principle that nothing comes from nothing, he later added, further testifying thereby to his broadened education.)[91] Clearly, Leeuwenhoek had had philosophical inclinations – such as his mechanistic bent – prior to the education afforded by his new correspondents,[92] but his fuller awareness of learned opinion helped to crystallize the clearer and more developed convictions that now called for the renunciation of spontaneous generation.

Such appeals to religious and philosophical scruples still lacked the intensity of Swammerdam's pious protests, however, and, wholly absent in Leeuwenhoek's first denials of spontaneous generation, they remained rare even in his later letters. References to the "ordered" (or more literally, "regulated," *gereguleerde*) propagation of various animals as an argument against spontaneous generation also came only in the 1690s, and, even then, they lacked Swammerdam's level of anxiety over an endangered natural order.[93] The evidence of the impact of social interactions persists throughout, however, suggesting that the intensity of commitment that engendered and sustained his campaign stemmed rather from the bifurcated social milieu in which that campaign unfolded.

Leeuwenhoek surely responded to the fuller articulation and elaboration of ideas consistent with his own inclinations, but his susceptibility was inevitably heightened as well by the character of those from whom he learned them. Thus, his aspiration to the milieu of learned gentlemen also helped focus and sustain

89 *AB* 3:396–7, 6:306–7; AvL (1696), 62, 132; idem (1697), 235; idem (1702), 201. Further examples occur in the following years.

90 *AB* 7:34–5; see also *AB* 8:176, 184–5, 328–30; AvL (1702), 107.

91 AvL (1718), 88.

92 Leeuwenhoek had also assumed virtually from the beginning and with no detectable prompting that even microorganisms were endowed with the organized complexity of other animals (*AB* 2:340–1, 390–1, 3:54–5, 396–7).

93 AvL (1694), 725–6; idem (1696), 60, 119; see also AvL to RS, 22 Sept. 1711, AvL Letters, fol. 150v.

both his developing convictions about spontaneous generation and the researches he undertook to buttress them. Through his correspondence, Leeuwenhoek continued to direct his detailed studies of insect anatomy and life cycles to the "learned world,"[94] and it is not unlikely that he persisted in these efforts less in the expectation of converting the unconverted than of striking a responsive chord among the already convinced.[95]

But the dichotomy of the social arena in which Leeuwenhoek carried out these studies made for a decidedly tenser engagement. To be able to speak to the learned world with a unique authority it valued was deeply gratifying, but it was rather the citizenry of Delft that appears to have first provoked Leeuwenhoek's campaign, and it did so, and continued to do so, less by respectful interest than obstinate resistance. He had undertaken his initial study of the grain weevil because of those, apparently among the grain merchants, millers, and bakers of Delft, who had tried with all their might, he wrote, to convince him that the weevil and the grain moth both arose without propagation.[96] As the years passed, his fellow townsmen continued to greet his denials of spontaneous generation with obvious doubt or outright contradiction,[97] to the latter of which, at least, he had already acknowledged he did not take kindly. When he at last could show (or so he thought) that eels too produced their own kind, the prospect gave him exceptional pleasure in part because it would silence those who had been taunting him, he was convinced, behind his back – a suspicion, whether true or not, suggestive of antagonisms in the neighborhood rather than with correspondents and learned visitors.[98] "Prominent and learned persons" still often had to be convinced as well, it is true,[99] and, when in the mid-1690s he thought he had done enough, it was a revival of debate in learned circles that convinced him otherwise.[100] The streets of Delft, however, harbored the potential for direct and daily confrontation, and here Leeuwenhoek took issue with the beliefs of merchants, artisans, and workers in an ongoing exchange that was not without its moments of heat and resentment.[101]

94 *AB* 8:328–30; AvL (1696), 154.
95 See below regarding n. 103.
96 *AB* 7:4–9, 34–5, see also 8:176, 184–5, 306–11.
97 *AB* 7:34–5, 8:176–7, 184–5, 322–9; AvL (1694), 729; idem (1696), 28; AvL to RS, 22 Sept. 1711, AvL Letters, fols. 151v, 153v.
98 AvL (1693), 521–2. Leeuwenhoek's (mistaken) discovery of unborn young within eels also provided him great pleasure, he said, because of the relentless effort he had put into it. See also Bertin (1956), 5. I am grateful to Virginia Taylor for clearing up my own misunderstanding of the reproduction of eels.
99 AvL (1693), 514; *AB* 7:98–101; AvL (1696), 155.
100 AvL (1696), 30. Kircher's former pupil Filippo Buonanni in Rome had come to the defense of spontaneous generation in his *Observationes circa viventia, quae in rebus non viventibus reperiuntur* in 1691. Leeuwenhoek had apparently received a sketch of the contents from a correspondent (AvL 1694, 651–2, 693). In the 1690s and following, Leeuwenhoek would indeed give particular attention to the claims of both Buonanni and Kircher (AvL 1694, 587; idem 1697, 214–15; idem 1718, 87–8). Kircher's 1665 *Mundus subterraneus* was translated into Dutch in 1682 as *D'onder-aardse weereld,* and Leeuwenhoek cites it on at least two occasions (AvL 1694, 569, 571; idem 1718, 87).
101 *AB* 7:34–5; AvL (1694), 729; AvL to RS, 22 Sept. 1711, AvL Letters, fols. 151v, 153v.

Despite the resistance still also encountered among the learned, moreover, Leeuwenhoek could justifiably have identified the rejection of spontaneous generation with the community of progressive savants and perceived it as a hallmark of those "philosophical" minds for whom alone he said he wrote.[102] Although it would convince "all learned intellects" (*alle Geleerde verstanden*), he declared, his study of the grain weevil was unlikely to suffice for the grain merchants, millers, bakers, and "those who see no further than the end of their nose."[103] For Leeuwenhoek, consequently, his denial of spontaneous generation and the campaign he waged against it may have represented as well an assertion of the new social identity for which he was striving. If so, the intensity of that campaign derived not only from his combativeness and impatience with contradiction, but also from a determination to affirm his distance from his immediate social surroundings and the common man.

Shortly after Leeuwenhoek's death, Hartsoeker belittled his efforts by dismissing spontaneous generation itself as too absurd even to merit refutation.[104] Nonetheless, the researches Leeuwenhoek had undertaken had enhanced the stature of the microscope[105] and enriched the field of natural history. Always disposed to disparage his erstwhile rival, moreover, Hartsoeker could be so condescending only because the cause for which Leeuwenhoek had labored had triumphed: Spontaneous generation had indeed – for the moment – been rendered inane.[106] Hartsoeker's disparagement can be recast more sympathetically, however, as the question with which we have been wrestling: Why did the issue of spontaneous generation provoke such a sustained and intensive line of microscopic research? A cluster of preconceptions characteristic of the age undermined the older tradition, it is clear; but how did these ideas, otherwise curiously indifferent and in ways even discouraging to microscopic research, motivate Leeuwenhoek's aggressive assault? How did they acquire the emotional import capable of inspiring such a commitment? Mediated by certain social circles and contested by others, the upshot of these preconceptions acquired social implications that tugged at fretful personal sensibilities. Inevitably inconclusive about spontaneous generation itself, a remarkable program of microscopic observations emerged nonetheless from an entanglement of ideas and social circumstances that impinged upon Leeuwenhoek's self-perception.

102 See Chapter 6, n. 78. This is not to say, however, that even the learned circles with which Leeuwenhoek was in contact had wholly abandoned the idea of spontaneous generation. The Royal Society, for instance, was still entertaining occasional stories about spontaneous generation at least into the mid-1680s, when Thomas Birch's history of the society came to an end (Birch 1756–7, 4:282, 492; see also the ambiguous references on pp. 541–2).

103 *AB* 7:34–5.

104 NH (1730), "Ext. crit.," 18.

105 That the microscope had played a decisive role in bringing about the downfall of spontaneous generation, which presumably added to the instrument's prestige, was a common notion in the early eighteenth century: J. Belkmeer (1719), 53; Pluche (1732–50), 4:527.

106 Farley (1977), 8; Roger (1971), 333, 365.

CHAPTER NINE

Generation II: The search for first beginnings

Although also caught up in a warmly contested public campaign, and a more dissentient and solitary one at that, Leeuwenhoek's conception of the spermatozoa more clearly revealed the depth of his susceptibility to the content as well as the social implications of the preconceptions of the age. Coming to grips now with the generation of man as well, Leeuwenhoek set out to convince a largely skeptical world that the spermatozoa were the essential instrument of reproduction, an undertaking that both extended and diversified his studies of the spermatozoa.[1] His most searching observations were in pursuit, however, of a phantom framed by the same preconceptions that were challenging spontaneous generation, a phantom that, were it to have materialized, would not only have decisively confirmed Leeuwenhoek's beleaguered claims but also loomed as the preeminent biological discovery of the century.[2]

The underlying question and the preconceptions at issue linked Leeuwenhoek's effort to a broader series of celebrated observations and a wider arena of clashing speculative commitments, all still revolving around the origin of organic forms. Contending interpretations inevitably entailed diverse philosophic and religious outlooks that influenced the perception of not only uncertain microscopic images but the very promise of the instrument. Very personal priorities often imposed themselves as well, however, undermining the seemingly clearcut confrontation between speculative camps and spawning idiosyncratic variants. From the farrago of ideas and observations implications emerged that, even amid speculations about extravagantly elaborate unseen structures and the great secrets they concealed, often subtly discouraged recourse to the microscope.

1 Ruestow (1983), 200–2.

2 In later years, Leeuwenhoek resigned himself to the fact that the resistance to his claims would persist throughout his lifetime (AvL 1718, 166–8). Judging, indeed, from the disputations at Leiden toward the end of the century, his campaign had as yet had no more impact on the instruction at the nearby medical school than had his studies of the erythrocyte (Wiel 1686, cor. 11; Cloothack 1687, cor. 4; Cocquis 1688, "Theoretica," §40; Walker 1688, cor. 14; Mul 1694, ann. 14), and even those who rallied to the spermatozoa, including Huygens, Hartsoeker, and a wavering Boerhaave, still rejected Leeuwenhoek's denial of the mammalian egg (*OCCH* 13:526–7; NH 1708, 110, 113; idem 1722, 201–5; Boerhaave 1734, 327–8, 340–3). Cf. Roger (1971), 309–22.

The phenomena in question were those that pertained to the apparent formation of an animal within an egg or womb, a development in fact no more comprehensible to pious mechanists at the end of the seventeenth century than was spontaneous generation itself. Amsterdam's erudite Jean Leclerc took issue with those who maintained "that, from fluid alone, an animal is formed by mechanical laws and the heat of the uterus," for this would be as difficult to understand, he objected, as it would be if the heat of the fire caused molten metal to form the parts of a clock.[3] At the heart of the difficulty lay the purposeful austerity, again, of the otherwise easy imagery of mechanism.

It was a problem alien to the traditional schools of thought. Whereas the scholastic form negated the problem by encompassing all that made a specific animal what it was, other traditions ascribed early animal development to "forces" and other dynamic agents – "natures," "souls," "virtues," and the like – of which Harvey offered the century's most prominent and extended exposition. Concerned with the "causes" of the embryo, he identified within the chicken egg a "vegetative faculty" or "soul" – indeed, a "workman soul" (*anima opifex*) – and a "plastic force," "virtue," or "faculty" – or "forming faculty" (*facultas formatrix*) – that produced, prepared, and shaped the matter of the emerging chick.[4] Such active agencies were generally understood themselves, however, as basic, irreducible facts, susceptible to no further analysis. Although Harvey recognized the knowledge of causes as the foundation of perfect science,[5] the efficient causes he identified were or worked through forces of which little more could be known than when and whence they came and what effects they produced.

To the mechanists, this offered little in the way of understanding; hence, while Harvey was gathering the observations around which he built his treatise on generation, the *Exercitationes de generatione animalium,* Descartes was struggling rather to render the formation of the fetus comprehensible through the imagery of particles in motion. It was an undertaking he found discouragingly difficult,[6] and the final treatise that resulted was published only posthumously in 1664. Assuming a mix of male and female semens within the uterus, Descartes ascribed the initiation of embryonic development to the heating of the particles of the mixed semen – that is, their increased motion – by the ubiquitous, ultra-fine, and ultra-active particles of the first "element" of his universe. This heating caused a localized concentration of particles within the semen to expand, forming the beginning of the heart and forcing many of these particles out of the heart and into the surrounding mass of semen. Other particles then moved into the heart and, acquiring the agitated motion from particles encountered there, expanded in their turn, establishing the regular pulsing of the heart. That pulsing continued to drive what was now the blood in a developing system of veins and arteries throughout the semen, while the variously shaped particles of

3 Leclerc (1704), 4:120, see also p. 52.
4 Ruestow (1973), 19; Harvey (1651), 115–16, 120, 122, 139, 141–2, 146. See also, for instance: Jacchaeus (1653), 84–5; Deusing (1649), 129–30.
5 Harvey (1651), 141.
6 Descartes (1897–1910), 1:254, 5:261.

the blood, some irregular and easily entangled, others compact and very active, formed the various organs in their proper places within the emerging animal. And all this, as he said of the formation of the valves within the heart, "in accordance with the rules of mechanics."[7]

The appeal of such mechanistic explanation was far from universal, needless to say, and although Craanen took up and expounded Descartes's fantasy at Leiden, Ysbrand van Diemerbroeck at the medical school in Utrecht derided Regius's similar imaginings for explaining the obscure in terms of the more obscure.[8] Even the later mechanists of the century found it difficult to conceive how their laws of motion and their agitated particles could harbor the guidance needed to construct and faithfully reproduce the complex order and structure of an animal. Craanen notwithstanding, their philosophical commitment placed late-seventeenth-century mechanists in a distinctive impasse.

The chemical philosophers invested spirit-saturated "seeds" with "intelligence" or "innate knowledge" – the instructions, as it were, wrote van Helmont, for the things they were to do[9] – and, whereas Harvey's "vegetative faculty" produced the fetus by an "inborn disposition" (*connatum ingenium*), his "vegetative soul" was ruled in what it did by a binding command. Though denying knowledge or deliberation to that soul itself, Harvey ultimately resorted, indeed, to a supreme efficient cause that was endowed with foresight and intelligence and of which the other efficient causes, the souls and faculties, were only instruments. Like van Helmont and the scholastics as well, after all, the later mechanists could also heartily endorse a final recourse to God; but their mechanized nature could no more accommodate guiding faculties or souls than they could scholastic forms or, for that matter, tutored seeds. The most fundamental commitment of the mechanical philosophy denied its basic principles, matter and motion, any trace of cognizance or any embedded blueprints more complex than their own perseverance. Protesting that matter, no matter how finely divided, remained "a senseless and stupid Being still," John Ray compromised the mechanical philosophy by turning, as did Leclerc and Hartsoeker, to spiritual agencies.[10] But a stricter mechanism increasingly insisted that only existing corporeal structure could provide the framework and format for other structures yet to come.

Even Descartes had maintained that the animal and its parts were specifically predetermined by the particles of the semen, but classical thought and the church

7 Ibid. II:253–82; the final quotation is on p. 279.

8 NH (1730), "Ext. crit.," 45 (see also Craanen 1689, 712–13); Diemerbroeck (1685), *Anatomes*, 196.

9 Pagel (1958), 85; idem (1967), 241–2; Helmont (1648), 34.

10 Harvey (1651), 144, 146; Reif (1962), 193; Ray (1691), 33. (Ray remarks that there are many phenomena in nature that surpass or even partly contradict "Mechanick Powers" and need final causes or some vital principle; the greatest of such phenomena was indeed the formation and organization of bodies [ibid., 26–8].) Hartsoeker ultimately turned to "intelligences" that, subordinate to God, presided over essential physiological processes (see, e.g., NH 1712, 17–18), whereas Ray and Leclerc opted rather for Cudworth's "plastic nature" (Raven 1950, 376, 456; Roger 1971, 419–22).

fathers, notably Augustine, offered a notion of the actual rudimentary beginnings of the parts of the future animal within the semen. As with the chicken egg, wrote the Englishman Alexander Ross in 1652, "so the seed of other animals contains potentially the animall that shall be, with all its members; therefore the common opinion is, that seed is drawn from all parts of the body, because it contains in it all the parts." But the Church fathers had also proposed that a seed-like beginning for every creature that would ever appear had been created in the original Creation, and, together with its resolution of the mechanists's dilemma, the very extravagance of the prospect recommended it to those who sought not only to reconcile but to merge the new mechanical philosophy with a religious exaltation with pronounced biblical overtones.[11]

In 1674 in the first volume of his *Recherche de la vérité,* Malebranche, both a Cartesian and a priest with Augustinian leanings, began to elaborate the doctrine of the preexistence of all living creatures. Since the future flower could be discovered already in the tulip bulb, he wrote, it was reasonable to assume that the same was true of seeds and that the seeds of trees thus each contained a tree in miniature. Indeed, it was reasonable as well to posit an infinite number of trees within the seed (or, more precisely, within the *germe* of the seed), for it would contain not only the tree but the tree's own seeds as well, and these would enclose more trees and consequently further seeds, and so on to infinity. Thus a single seed contained the apple trees for countless centuries, which nature, in time, merely expanded or unfolded. What was said of plants, Malebranche continued, could be said of animals as well, so that the bodies of all men and animals destined ever to be born had perhaps been produced at the creation of the world and were all contained in the first females of their species.[12]

Despite its extraordinary challenge to the imagination, the idea of preexistent bodies successively encapsulated within a line of numberless generations proved increasingly compelling in the closing decades of the century. If true, it was indeed "something marvelous" to the younger Constantijn Huygens, yet his sober and otherwise tough-minded brother Christiaan found it not only reasonable but probable. By the early eighteenth century, it had won an extensive and influential following, from Leibniz to the French priest Noël-Antoine Pluche, one of the age's foremost popularizers of natural history. As opposed to the alternative of perpetual new creations, according to Pluche, encapsulated preexistence was more compatible with reason, experience, Scripture, and the power of God.[13]

Inevitably, however, many balked at the sweep and speculative presumption of the vision. In the Netherlands, Holland's own influential exponent of the

11 Descartes (1897–1910), 11:277; Ross (1652), 230; Adelmann (1966), 2:730–4, 749–50, 875–6; Roger (1971), 332–3.

12 Malebranche (1958–70), 1:81–3.

13 *OCCH* 9:355, 361. (In the judgment of those who participated in the International Huygens Symposium in Amsterdam in August 1979, Christiaan Huygens stood out among all the major figures of the Scientific Revolution for his marked lack of interest in philosophical matters [Hall & Van Helden 1980, 138].) Pluche (1732–50), 1:20; Roger (1971), 334–53, 364–84.

genre of popular natural theology, Bernard Nieuwentyt, affirmed in 1717 that, if there was anything regarding generation on which all the naturalists of the day agreed, it was that in the first beginnings of all animals their parts were already packed and rolled up together; but he reserved judgment, nonetheless, on preexistence since the initial Creation. Hartsoeker, uneasy as well over such difficulties as accounting for the birth of "monsters," found the infinite smallness necessarily presumed at the Creation to be absurd, and, without more certain proof, Pierre Lyonet also subsequently declined to embrace an hypothesis that so startled his imagination.[14]

Nonetheless, as Pluche attested, the doctrine of encapsulated preexistence brought to bear a powerful combination of philosophical and religious inducements. It continued to speak directly to a mechanistic understanding of nature in which mechanism itself had turned emphatically sterile. Although we may acknowledge that the laws of motion alone produced the variety and succession of beautiful things in the world, argued Malebranche, we cannot convince ourselves that those laws could ever, even through sexual union, form so complex and purposeful a machine as the body of an animal or plant. Those laws could suffice, however, to unfurl and enlarge (*pour développer*) the parts of a body that was already organized, and, to avoid recourse to a special providence to explain generation, he concluded in effect, we must believe in the existence of such preexistent miniatures.[15] Leibniz likewise affirmed that, since its formation did not seem explicable within the framework of nature, we had to assume the preexistence of the organic body. He wrote Hartsoeker in 1710 that everything that was material could and did develop mechanically through the communication of motion alone, but this was because of a divine preformation long ago of the machines that now changed and unfolded: "thus the miracle will only be in the origin of things, and all that follows with respect to material things will be mechanical." "Most philosophers today will not admit any true production of plants and animals," remarked Réaumur in 1734; "they recognize only unrollings [*développemens*]."[16]

Though unnerving, the breathtaking perspective of numberless centuries also had its compelling appeal, for it highlighted the doctrine's potential for religious gratification as well. Most immediately, it glorified divinity as the sole and direct source of all creative power and concentrated that power in the original Creation. It conformed, hence, to a flourishing theological emphasis on the radical sovereignty of God while also evoking the aura of a more primitive Scriptural traditionalism.[17] But the doctrine lent itself to the cultivation of religious

14 B. Nieuwentyt (1717), 294, 579–80; NH (1730), "Ext. crit.," 19–20; Hublard (1910), 48–9.

15 Malebranche (1958–70), 12:228–9, 252–3.

16 Leibniz (1875–90), 6:152, 619, 3:508. (Leibniz's mechanism and his doctrine of preexistence were subordinated, however, to his doctrine of monads: immaterial "atoms of substance" or "metaphysical points" that, in addition to identifying with Aristotle's substantial forms, he also described as unifying forces or "souls" of material bodies [ibid. 4:479, 482, 6:610].) Réaumur (1734–1929), 1:360.

17 See Deason (1986), 167–91; Roger (1971), 224–5, 331.

sensibilities in other ways as well. Stressing how incomprehensible it truly was that the first fly had contained all future flies, Malebranche warned nonetheless that those who balked at such thoughts denied God his due by confining his infinite power to the reach of their own finite imagination. Indeed, it is in the very incomprehensibility of infinite smallness, Malebranche suggested, that we confront the hallmark of divinity. With good reason we fear to probe too deeply into the works of God, he wrote, for we confront infinity everywhere, and the mind is dazzled, lost, and frightened before the smallest of those works; but the mind thus recognizes its weakness, and, perceiving the infinity in which it is lost, grasps the greatness of God.[18]

Leibniz also extolled the doctrine of preexistence as a radiant display of the perfections of God, and, though he seized on it in terms of his own distinctive metaphysics of innumerable "monads," it yielded religious rewards to less subtle intellects too. Descanting on the preexisting beginnings of humans, animals, and plants, Wijer Willem Muys asked who could ponder such a display of God's power without experiencing a certain holy awe (*sacer quidam horror*). Some pious critics of preexistence (such as Hartsoeker) recoiled from ascribing "monsters" and deformity directly to the Divinity, but, for others, preexistence indulged a religious exaltation that powerfully validated a philosophy of nature.[19]

Having posited yet another realm of incredibly elaborate but dramatically contracting structures, the doctrine of preexistence inevitably invited appeals to the microscope, despite the repeated – and often jubilant – insistence by proponents of the doctrine that its true foundations had nothing to do with the senses. Despite his own early allusion to the lens, indeed, Malebranche initially presented his vision of preexistence in order to underscore the limitations both of the senses and of the imagination, neither of which could grasp the diminishing dimensions of structure that reason disclosed. The mind's vision, he insisted, should not be limited by the vision of the eye, and his identification of preexistence with an exalted reliance on reason to overcome the limits of the imagination and the senses was echoed in the following century. The Swiss naturalist Charles Bonnet acclaimed encapsulated preexistence as one of the finest victories of pure understanding over the senses, and Henry Baker managed to make the point in rhyme:

> Amazing Thought! what Mortal can conceive
> Such wond'rous Smallness! – Yet, we must believe
> What Reason tells: for Reason's piercing Eye
> Discerns those Truths our Senses can't descry.[20]

18 Malebranche (1958–70), 9:1118–19, 1:81–3.

19 Leibniz (1875–90), 3:562 (see n. 16 above); Muys (1738), preface [32]–[3]; Roger (1971), 397–418.

20 Malebranche (1958–70), 1:80–3; Bonnet (1779–83), 5:84, 8:64–5, 15:222–4; H. Baker (1742), 252. See also: Andry (1700), 298; Blanckaert (1701), 1:199; idem (1683a), 28–9.

Microscopic observations continued to be called upon to bear witness nonetheless, and none served the cause of preexistence better than the oft-cited observations of seeds. That the plant seed was itself a product of sexual fertilization was not recognized until near the end of the seventeenth century, and, since the plant embryo within the seed was often visible even to the naked eye, it early suggested that a preformed miniature indeed preceded the future plant. The seed embryo was also an obvious subject for the early microscope[21] and thus inevitably linked the new instrument to the emerging doctrine of preexistence. Whether the microscope could reveal the rudiments and first beginnings of the plant within the seed was no trifling question among physicians and philosophers, Craanen reportedly said; those who claimed that it could, he added in acknowledgment of the inferences that were at stake, believed as well that the complete structure of a man lay likewise within the semen of his father. But since the part of the seed that became the first rudiment in generation was so very small, he demurred, "let us ponder, I pray," whether the microscope could really discover there the prepared and prearranged parts of the entire plant. Though otherwise sharing Craanen's Cartesian sympathies, however, Blanckaert was among the many who felt the issue had been settled: Within the heart of the seed was the complete delineation of the plant, he wrote, citing indeed the great assistance provided by the microscope.[22]

Malpighi, Grew, and Leeuwenhoek all in fact studied the plant seed closely with their varied instruments,[23] but the advocates of the preformed miniature in generation too often simply reduced such studies to a confirmation of older and cruder accounts while ignoring what these more recent microscopic studies failed to see. The anatomist Albinus was more scrupulous. A close examination of the seed revealed the leaves at the top of the embryo plant, he confirmed, and, joined to a little root, a more solid and homogeneous part that would become the trunk; but much still escaped even the lens, and whether there were as well flowers and fruits in the seed remained only speculation. Nonetheless, Charles Bonnet, champion of preexistence (though not necessarily encapsulated) that he was, still casually remarked a century later that the microscope would convince us that the organized body of the future plant with all its essential parts already existed within the seed.[24]

In asserting that the microscope would likewise vouch for the miniature animal preexisting in the egg, Bonnet's facile rhetoric of preexistence appropriated a history of complex and problematic observations even more heavy-

21 Ritterbush (1964), 88–108. Leeuwenhoek, for one, explicitly denied that plants interacted sexually (*AB* 5:232–5, 246–7, 6:120–1). The young Christiaan Huygens also showed how readily the plant embryo in the seed came to mind as a fitting subject for the early microscope (*OCCH* 1:321).

22 Craanen (1689), 713; Blanckaert (1701), 1:199; idem (1683a), 28–9; Luyendijk-Elshout (1975), 302.

23 Malpighi (1686–7), *Anatome plant.*, 75–82, and *Anat. plant. pars altera*, 1–16; Grew (1682), 203–12. Leeuwenhoek's studies of the seed are discussed later in this chapter.

24 B. S. Albinus (1719), 11; Bonnet (1779–83), 5:88.

handedly.[25] Adherents of virtually every shade of opinion had long cultivated an analogy between the plant seed and the initial rudiment of animal and human generation, whatever that rudiment might be, embryo, egg, or spermatozoon;[26] but what was observed in the case of animals was far more uncertain than the embryo plant in the seed. Ruysch described a preserved human embryo the size of a lettuce seed as a confused and unformed mass that, even when observed with a microscope – and, indeed, with some Leeuwenhoek microscopes at that, Ruysch attested – revealed no vestige of a head or limbs. So much for those, he said in effect, who claimed to have seen embryos with hands, feet, and eyes a few days after conception and indeed with a head and limbs even in the ovum (as the Amsterdam doctor Theodorus Kerckring had reported)! So likewise de Graaf, who Leeuwenhoek himself informs us had also used a microscope in such observations, had similarly remarked how little he could distinguish in a "shapeless" twelve-day-old embryo of a rabbit.[27]

Significant and systematic new observations pertaining to animals did accompany the speculative debates about embryonic development nonetheless, and they were observations that placed a dramatic realm of phenomena before the lens. Taking up an ancient Greek suggestion, sixteenth-century observers had already revived what became the most celebrated technique for studying early animal development: the successive opening of incubating chicken eggs, to which, in the seventeenth century, the microscope was now also applied.[28] It was to this line of observations that Bonnet was obviously alluding.

The chicken egg, indeed, provided easy access to progressive stages of embryonic development, although what the observer confronted was nonetheless still far more difficult to make sense of visually than was the embryo plant within the seed. The structures at issue were in fact seldom truly microscopic, but they were often obscured by pallidity and a lack of crisp definition as well as by smallness, and a clear reading of what precisely was taking place as shapes and colors appeared, changed, and perhaps disappeared did not come easily. The earliest detectable manifestations of embryonic development in the egg thus intimated different things to observers with different expectations. As introduced into the controversy over the nature of sexual reproduction, consequently, the relevance of microscopic observations ultimately rested on predispositions. In key instances, ideas shaped not only the observers' interpretations of observations but their judgments as well, it appears, as to what was actually seen. It re-

25 Bonnet (1779–83), 5:88.

26 Harvey (1651), 76, 120, 179–80; Highmore (1651), 81–4; JS (1669), pt. 1:64; Blanckaert (1701), 1:199; idem (1683a), 28–9; B. S. Albinus (1719), 23. Regarding Leeuwenhoek, see later in this chapter. Such analogies indeed had a long history; see also: Adelmann (1966), 2:901–2; Cole (1930), 39; Ritterbush (1964), 89.

27 Ruysch (1705b), 30–1; Graaf (1677), 406; *AB* 5:170–1. A slim but influential work published in 1671, Kerckring's *Anthropogeniae ichnographia* pictured an impossibly well-formed human embryo that he claimed to have found in the egg only three days after conception (Roger 1981, 230–1).

28 Adelmann (1966), 2:733–4, 747–8, 755–7, 779–80, 824; Kircher (1646), 834; Borel (1656), 15.

quired rigorous and conscious self-discipline to stave off the intrusion of preconceptions, but such restraint then resulted in observations that could offer little to resolve the basic question of generation as it was framed. When unchecked, however, those encroaching preconceptions also shaped attitudes pertinent to the very employment of the microscope.

Harvey's *Exercitationes de generatione animalium* was in large part structured around just such an extended study of the early happenings within the egg. Often aided by a magnifying glass (*perspicillum*), Harvey attended particularly closely to what he took to be the first beginnings of the chick, which no one before him, he claimed, had yet observed.[29] His account indeed reflects a peering scrutiny of the early phenomena of the developing embryo. Soon after the beginning of incubation, he noted a tiny pool of very bright and transparent fluid (the subgerminal fluid, measuring perhaps a half-centimeter across by the end of the first day of incubation) lying in the middle of a small white disk (the blastoderm) on the yolk. During the following days, that pool would increase in size, and, by the fourth day of incubation – or the late third, if a lens or bright light was brought to bear – Harvey now observed tiny red lines around the edges of the fluid and, in its middle, a red and pulsing point so small, he wrote, that it vanished when it contracted. What seemed to be a vein emanating from the point and branching through the clear fluid had also emerged, and, toward the end of the fourth day and the beginning of the fifth, the pulsing point grew into what now appeared to Harvey a small and very delicate vesicle. In some eggs, indeed, it already looked like a double vesicle whose reciprocal pulsing motion pumped the blood it apparently contained into the veins.[30]

In more advanced eggs, however, the view of the pulsing vesicle was now blurred as if by an intervening haze, Harvey noted, and with brighter light and magnifying glasses he discovered the cause of that indistinctness to be a little cloud that continuing observations revealed to be the rudiment of the body of the chick. Having carefully repeated these observations, he was uncertain for some time as to whether that rudimentary body had condensed out of the original clear fluid in which it floated – like clouds in the air, he wrote – or was produced from an effluvium exuded from the veins. "The beginnings of even the greatest things are very small," he explained, "and because of the littleness, very obscure." As the emerging rudiment of the body became denser, however, it seemed now to Harvey to be surrounding or attached to the vein, and the "testimony of reliable eyes" convinced him that the first threads and visible foundation of the body were in fact the veins, to which the later parts of the body, and the viscera in particular, were attached like fungus to the bark of a tree.[31]

The rudimentary body itself appeared to be of a mucous or gelatinous consistency in which, "in so far, at least, as we can see or ascertain by any reasoning,"

29 Harvey (1651), 45, 49, 54, 62.
30 Ibid., 47, 49, 51, 53; Adelmann (1966), 3:1031–2.
31 Harvey (1651), 54, 58–9, 64, 182.

he could discover no differentiated parts or substances. The shape of that rudiment as a whole began to appear as if divided into two major parts, however, the largest of which, the future head, soon divided further into four vesicles filled with a perfectly limpid water; in the middle of the two largest of those vesicles, according to Harvey, the future pupils of the eye now sparkled and gleamed like crystal. On the sixth day, he reported, the flesh of the heart began to grow upon the pulsing vesicle, and, soon after, the white rudiments of the liver, lungs, and other viscera appeared as well. The rudiments of the bones soon followed, emerging as milky lines resembling the thin threads of a spider's web.[32] To Harvey, this series of observations confirmed that the chick embryo was formed by the progressive emergence of its organs from what was initially a homogeneous and jellylike rudiment of the body, and, to designate this process of "unintelligible separation" (*divisio obscura*), he appropriated the word *epigenesis*.[33]

Despite the closeness and scrupulousness of Harvey's observations, however, they were embedded in an elaborate mesh of preconceived ideas that shaped not only how he interpreted what he saw but also, at times, how he saw it. Reflecting his fundamentally premechanistic understanding of nature, one in which a host of varied dynamic agencies expressed themselves, he interpreted the beginning of the chick as the breaking forth of the latent "plastic force" in the initial small white disk in the egg, and, because of the art, providence, and divine intelligence displayed in the structure of the chick, Harvey had no doubt that the cause that produced it also had to be a soul. For Harvey, souls were indeed the source of all the vital functions and the very life of an animal,[34] a conviction that further predisposed him to assume the progressive emergence of structure and differentiation in the forming of the embryo and that, abetted by other observations also colored by his animistic outlook, shaped how he attended to the phenomena he confronted in the egg.

Having underscored the divine supervision evident in the structure of the chick, Harvey had then set out to discover when and where in the egg the Creator's agent (*vicarium*), analogous to the substance of the stars and redolent of art and understanding, made its first appearance. Bolstered by citations from the Bible, numerous observations had persuaded Harvey that it was indeed in the blood that the soul driving generation first manifested itself, and that it was there that it lingered last in death as well. He had observed in frequent vivisections, he testified, that even after all other signs of life had ceased, including the beating of the heart, he could still detect a faint agitation or undulation in the blood itself. From many other observations he had also determined that the pulsing point that appeared in the initial small pool of clear fluid in the chicken egg was both sensitive to touch and capable of motion, so that he attributed to the blood a "sensitive" as well as a vegetative and also motile soul.[35]

32 Ibid., 56–8, 61–2, 125, 254.
33 Ibid., 121–2, 124–5, 255; Pagel (1967), 233.
34 Harvey (1651), 45 (see also p. 49), 131, 250.
35 Ibid., 52, 127–8, 149–51, 250.

Aristotle too had begun his account of the developing embryo with the pulsing point, and, although he granted the heart priority over the blood – a priority Harvey would emphatically reverse – he had also remarked that the beginnings of the heart looked like nothing so much as a throbbing speck of blood. Later Renaissance proponents of epigenesis (though before it had the name) were Aristotle's heirs, but they were also given to a more spiritualist cast of mind and honored the blood in particular as the first receptacle of the animating spirit. To be sure, Harvey's observations alone may well have been sufficient to convince him of the initial dynamism in the blood; but, at a minimum, it was certainly reassuring, perhaps even exhilarating, to find those observations conforming so closely to an ancient description that derived from Aristotle and, in the face of the threatening mechanism of Harvey's day, seemed to reaffirm the spiritual forces at work in nature.[36]

The exhilarating impact of the coincidence of observation and metaphysical inclination is reflected in Harvey's sometimes lyrical exaltation of the blood, characterized as not only the soul's first and principal seat in the body but as indeed the soul itself. Ascribing reason and understanding to the activity of the blood, he declared that, by reason of its powers, it could justifiably be considered "spirit" (*spiritus*). It was hence also "the hearth, the Vesta, the household divinity, the innate heat, the sun of the microcosm, [and] Plato's fire," and, as the instrument of God, its divine and wondrous faculties could never be adequately glorified.[37] These were stirring sentiments for a scientific treatise, and that he was predisposed to see that which would justify them is reflected in his detection of the "faint agitation" supposedly lingering in the blood of dead animals.[38]

A still fuller weave of mutually accommodating preconceptions conspired to shape what Harvey perceived in the egg, however. To Harvey's mind, the seat of the soul corresponded as well to the "generative particle" (*pars* or *particula genitalis*) that, in the generation of the "more perfect" animals (those endowed, that is, with blood), was the first particle of the body to exist, the place where life first began (and finally ended) and whence the other parts of the body all derived. Both Aristotle and the logic of epigenesis called for the existence of this particle, and, for Harvey, the initial appearance of the thin red threads and the pulsing point answered to that expectation.[39] As understood by Harvey, the very concept of the generative particle dictated a progressive development of structure and differentiation that had not existed before, and thus what he reported in the incubating egg reflected precisely what his metaphysical preconceptions required. Hence, we may surmise, his inclination ultimately to see the viscera growing like fungus on the veins (conforming as well again to Aristo-

36 *Hist. An.* 6.3.561^{a}10–12; cf. *Parts An.* 2.1.647^{b}4–6; Harvey (1651), 149–50, 153; Pagel (1967), 246–7, 341, see also pp. 243, 309, 336; R. Frank (1980), 38.
37 Harvey (1651), 151–3, 179, 247–8, 250.
38 Ibid., 151; see also Harvey (1628), 28.
39 Harvey (1651), 149, 121–2; *Gen. An.* 2.1.735^{a}15–25.

tle)[40] rather than an embryonic body condensing like a cloud – neither of which perceptions, to be sure, was right. Hence, too, his failure to notice the faint structural beginnings of the embryo that did in fact precede the more salient pulsing point of blood.

Harvey's preconceptions did not simply run roughshod over observations, however; rather, the gratifying correspondence of metaphysical commitment and features that were more readily apparent in what was observed biased the perception of other, more uncertain features. What was less distinctly visible and at the same time inconsistent with a scheme of things that seemed to have been so compellingly confirmed appears to have become more difficult to recognize.

The apparent agreement of appearances and metaphysics encouraged the assumption that, beyond what was already seen, there was little more of consequence to be sought in what was confronted through the lens. Harvey was fully aware of the possibility of invisible parts, and he began his account of his observations of the incubating egg by indeed affirming that, in their subtlety, the first threads of nature usually eluded the keenness of both the eye and the mind. He explicitly raised the question whether the rudimentary body of the chick, though still unseen because of its delicacy and transparency, might in fact have come into existence simultaneously with, rather than after, the blood and the pulsing point. Although his observations could not exclude that possibility, he rejected it on the basis of what he had said about the blood, the soul, the generative particle, and two other concepts inherited from the past: the "radical moisture" (*humidum radicale*) and "innate heat" (*calidum innatum*).[41]

The innate heat was accepted by everyone, said Harvey, as that which first initiated the process of generation (or, more precisely, the "vegetative operations," *operationes vegetabiles*), and he ultimately identified it too with the blood;[42] but the radical (or "primitive," *primigenium*) moisture represented a further elaboration in Harvey's thinking that had its own distinct application to what he saw in the chicken egg. Like the innate heat, it was widely cited but incorrectly understood, he asserted. Designating it the origin of even the generative particle, Harvey identified it not with the blood but with the original small pool of clear liquid from which he believed the blood and the pulsing point themselves had been made. Defined by Harvey as a supremely pure and simple body that was nonetheless the origin of all the future parts of the embryo, the concept of the radical moisture thus in itself embodied an initial assumption that differentiated structure arose from homogeneity.[43] It was one more strand in the complex weave of interrelated ideas that helped to persuade Harvey that the images he saw before him were sufficient, that where nothing more was seen, it was very likely because nothing more was there. Since what he perceived in the early development of the chick embryo answered so fully to his anticipations of how God and his active agencies operated within the

40 *Parts An.* 2.1.647^{b}1–4.

41 Harvey (1651), 42, 59; Pagel (1967), 257.

42 Harvey (1651), 116, 247–50, 252.

43 Ibid., 115, 125, 244, 252.

egg, it argued that he need seek no further to observe the essential processes of generation.

The conviction, on the other hand, that there were still prior, as-yet-unseen rudimentary structures arose from the desire to render the process of generation comprehensible, a commitment that found a fuller expression in mechanism. Harvey's was a mind of a very different stamp, however, subordinating the quest for such comprehensibility to an affirmation of the spiritual energy at work in nature. The souls he posited in living things were very much a part of the natural world, to be sure. Partly corporeal, they also acted "naturally" (*naturaliter*) and in the same way that heavy bodies fell and light bodies rose – in a way, that is, akin to Aristotle's principles of motion and rest, which Harvey continued to accept as the essence of what was natural in natural things.[44] Nonetheless, not only were those souls never *wholly* corporeal, their actions, though natural, were also emphatically inconceivable. The operations of the vegetative soul that gave rise to generation surpassed human comprehension no less than divinity surpassed man himself, proclaimed Harvey, and were so wondrous that human minds could not penetrate their inscrutable splendor.[45]

Even in the realm then of natural – though spiritual – efficient causes, Harvey was willing, perhaps eager, to renounce the aspiration to fuller philosophic understanding for the sake of the religious intimations of incomprehensibility. The embryonic parts in the egg simply arose "as if the whole chick were produced by a pronouncement or command of the divine Creator: 'Let there be a pallid, homogeneous mass [*massa similaris alba*], and let it divide into parts and grow; and while it grows, let the parts be separated and delineated; and let these certain parts be harder, denser, and paler, and these parts softer and more imbued with color.' And so it was." Together with the pinpointing of other efficient causes employed by God in the process, that was the essential explanation of generation: an evocation of divine fiat and Harvey's own echo of Genesis.[46] The sense of nature itself entailed in Harvey's more extended account was, in contrast to mechanism's commitment to structural definition, that of a patterned flow of energy; seemingly made manifest in his observations, it, too, gratified a familiar strain of piety. Its grip thus strengthened, it worked on the perception of uncertain images and, in denying the likelihood of significant structure yet to be discovered, diminished the promise of the microscope.

Only months after Harvey's *Exercitationes de generatione animalium,* a similar if less elaborate study of the early chick embryo appeared from the hand of Harvey's compatriot and acquaintance Nathaniel Highmore. Highmore also used what in his English text he explicitly called a "Microscope," but he also referred to it at times as simply "this Glasse," so that it may in fact have been but a single lens not unlike what Harvey himself had used.[47] What he saw with it proved quite different, however, and behind the different observations very different ideas were at work as well.

44 Ibid., 146, 250. 45 Ibid., 83–4, 250. 46 Ibid., 125, 143–5, 255.
47 Highmore (1651), 69, 71–4; Cole (1930), 40; R. Frank (1980), 26–8.

Highmore directed himself not to Harvey but to Sir Kenelm Digby, whose mechanistic approach had ascribed generation to the action of moisture and heat and, to Highmore's mind, had given "Fortune and Chance" the "preheminency."Despite his strictures upon Digby's views, however,[48] Highmore shared in part at least the mechanistic reluctance to let structure slip away, and, for an alternative, he turned to the doctrine of diverse and insensibly small particles that, drawn from the various organic parts of the body to which they still corresponded, were gathered in the semen as "seminal Atomes." It ultimately took a "specifick soul" to arrange them as the first beginning of the new animal, to be sure, but they represented hidden structure preceding generation nonetheless, structure of which the microscope might yet discover some further evidence.[49] Hence, and with such expectations, Highmore also undertook to scrutinize the incubating chicken egg.

On the third day, Highmore's report of what he saw began to deviate significantly from Harvey's. Having also noted the initial small pool of limpid fluid, Highmore now remarked a small body in its middle that, though clear itself, was nonetheless somewhat "obscurer" than the fluid around it, and within its own middle was a small white spot (presumably the nucleus of Pander, often obvious to the naked eye). Highmore's account is not without some obscurity of its own, but it was clearly the same small body he speaks of as "the white Circle and spot" later in the same day, when it appeared "by the Microscope" to be in effect the back and neck of the embryo (measuring very likely 6-7 mm as it lay in the area pellucida) arching around the heart (perhaps a millimeter in length) at the center.[50]

It was at this point or shortly after that, with the help of bright light or a lens, Harvey had first seen the pulsing red point, and on the fourth day, indeed, the heart now appeared to Highmore as well to be filled with blood and moving. Through Highmore's microscope or "Glasse," however, the heart appeared to be not a pulsing point of blood but a "perfectly fashioned" heart already provided with its auricles. (In fact, the heart would not have reached its full four-chambered development by then, but it was on its way, and its appearance could indeed suggest a more complex structure.) Highmore could also now discern the head, consisting of three "bubbles," and some "small obscure clouds" attached to the arching back of the embryo where the future wings and legs would be.[51]

48 Highmore (1651), 4, see also p. 83; for Highmore's other reservations about Digby's doctrine, see ibid., 12–14, 22, 34–9.

49 Ibid., 26–8, 42–6, 62–3, 66, 80–4.

50 Ibid., 68–70; Adelmann (1966), 2:943n, 3:1072. Emilio Parisano had already noted the nucleus of Pander in 1621 and likened it in size to a millet seed (ibid. 3:1035). Highmore referred to the back and neck of the embryo as the *carina,* literally "keel"; on the traditional use of this term, see ibid. 3:1087–91.

51 Highmore (1651), 69–72; Patten (1951), 196–200. The heart in fact begins pulsing, though weakly, after about only twenty-nine hours of incubation (ibid., 120), and Parisano had also already remarked that it began to beat before it contained visible blood (Adelmann 1966, 2: 760).

To Highmore, these and further observations revealed the daily development of seminal atoms that, floating in the small pool of clear fluid, had earlier been gathered and joined together. Believing, incorrectly, that the initial white spot became the heart and the white circle around it the curving back and neck of the embryo, Highmore asserted that the spot and circle never disappeared "but grow larger still, till they discover themselves what they are." He assumed as well that, "as Nature hath need of them," the other parts of the embryo also "grow up to their offices in their visible figures." Thus the seminal atoms formed the chick in the egg even before incubation, he wrote (having in fact observed the white spot with a surrounding circle also in the unincubated egg), just as the "germen" of the plant was perfectly formed in the seed though its parts remained concealed.[52]

His observations hardly sustained such conclusions, of course, but his anticipation of ever-receding structure may have helped him no less than his microscope to see the curving rudiment of the embryo before or at least as early as he saw the pulsing point and to see the latter, indeed, not as a point but a structured heart. Harvey had also seen the early white spot in the middle of the pool of fluid, after all, but still failed to perceive the arching structure around it. That he in fact suggested that, in some way, the white spot actually gave rise to the pulsing point underscores the spot's compatibility with his conceptualization of epigenesis;[53] a simultaneously emerging structure around it, however, was another matter. Yet this was precisely what Highmore, by contrast, was looking for.

Highmore's assumption of hidden structure would also have fostered a sharper attentiveness to what the lens in particular revealed, even though Harvey had been applying the instrument over a greater span of years.[54] What was of greatest interest to Highmore could only be discovered by the microscope, and, as opposed to a desire simply to clarify or elaborate on what the naked eye had seen, the expectation of diverse unseen structures would have inclined him to search for visual cues suggesting features that diverged from the way things seemed to the naked eye. He would have been more responsive, consequently, to what was different and distinctive in magnified images.

A dramatically more accurate description of the early embryonic development of the chick appeared twenty-one years later, in 1672, in two treatises addressed to the Royal Society by Malpighi.[55] Although he in fact would minimize the contribution of the microscope,[56] its role was often evident, and it was a meth-

52 Highmore (1651), 66, 68, 80–4 (esp. p. 82). The nucleus of Pander is often visible in the unincubated egg (Adelmann 1966, 3:1016).

53 Harvey (1651), 46, 48, 51.

54 Harvey had already written in 1628 of using the *perspicillum* in observing the hearts of flies and wasps and other small animals (Harvey 1628, 28). On the other hand, Kenelm Digby's *Two Treatises*, to which Highmore was responding and which hence perhaps first stimulated his observations, appeared only in 1644. In any event, Highmore was only fifteen years old in 1628.

55 *Dissertatio epistolica de formatione pulli in ovo* and *Appendix repetitas auctasque de ovo incubato observationes continens*, in Adelmann (1966), 2:932–1013, see also p. 824.

56 Ibid. 2:985, see also p. 829.

od for mounting the blastoderm on glass for microscopic observation that finally, though still only dimly, enabled Malpighi to discern the first "threads" (*stamina*) of the chick that he was seeking.[57] He was acutely conscious of defending his account of what he saw from the inroads of speculation, but he too had his theoretical leanings, after all, and they were not without their effect on his observations.

For the first time, Malpighi now reported the succession of emerging longitudinal structures – furrows, ridges, thickenings, and folds (measuring in the range of 2 mm and greater in length) – that begin to develop after the middle of the first day of incubation. He also believed, however, that he had begun his observations with an egg that, although fertilized, had not as yet begun incubation (though, clearly, it had, perhaps as the result of the warmth of the Italian summer on which Malpighi himself remarked).[58] Like Harvey and Highmore before him, Malpighi noted the small pool of clear fluid in this egg, but he also now reported what he already took to be the first filaments of the back of the embryo and the beginning of the head (probably the primitive streak and perhaps the node),[59] forms he had discovered when he peered at the apparently mounted blastoderm against the sunlight.[60]

After what according to Malpighi was twelve hours of incubation, he also detected two rows of vertebrae (the somites, which normally in fact begin to appear after about twenty hours of incubation) and, though vaguely, what he believed to be the beginnings of the brain.[61] Other forms followed, and after only a day he thought he even saw the motion of the heart. Six hours later he identified the heart with certainty and, after eight more hours, its pulse and structure, which he thought he might have vaguely glimpsed before. The heart, he correctly suggested – and illustrated more decisively – was formed from what was originally a bent and widened vessel; but he continued that, since everything was still so mucous, transparent, and white, the structure could not be clearly seen no matter what kind of instrument was used to assist the eye. It was only after forty hours by Malpighi's reckoning that the fluid being pumped through what he also now prematurely saw as the chambers of the heart acquired a dark tawny or rusty color that only gradually changed to a full blood red.[62]

Malpighi's study of the chick embryo continued through the third week to the threshold of hatching, but it was his observations of the initial stages of embryonic development that spoke most immediately to the now celebrated question of the first origins of animal form. Far more rigorously than either Harvey

57 Ibid. 2:833, 865–6, 867n. 5, 5:2236–7.
58 Ibid. 2:940–1, 941n. 3.
59 Ibid. 2:944–7, 944n. 8, 984–93, 986n. 9, see also 2:842, 3:1089, 1165; Patten (1951), 59, 70–1, 84–7. Like Highmore (see n. 50 above), Malpighi also at least initially used the term *carina* to refer to the embryo's back or, more specifically, its rudimentary spinal column; see Adelmann (1966), 3:1089–90.
60 Adelmann (1966), 2:944–5.
61 Ibid. 2:946–7; Bracegirdle & Freeman (1978), "A Reference Table of Chick Development," endpapers.
62 Adelmann (1966), 2:950–5, 954n. 3, 996–7, 1000–1, 3:1359–64.

or Highmore, Malpighi strove to keep his observations untainted by theorizing, and he was well aware of the particular danger of succumbing to speculation as observations approached the limits of what was visible. Having allowed himself at one point to reflect on the possible invisible existence of the embryo, he caught himself up short. Noting that the beginnings of the embryo were so minute, so intricate, and so concealed that the senses were easily deceived, he declared such conjecture "absolutely pointless" and returned to describing what he could clearly see.[63]

Nonetheless, Malpighi's inclination to assume prior but invisible structures surfaced in other passages as well. When he remarked that, because everything was so mucous, white, and transparent he could not detect the structure of the heart even with a lens, his point in fact had been that we could not assume its absence simply because it was not yet seen. We could at least "suspect" (*vereri possumus*) that the structure of the heart became visible only when it began to move, he urged, and that it had existed earlier, but in a state of invisible quiescence. He was less – but only slightly less – cautious in the second of his two treatises. It was now "probable" that the heart was already present even before the clear fluid that was to become the blood had begun to change its color; hence, he still "nourished the conjecture" (*conjecturam foveo*) that the fluid, vessels, and heart had all existed before they became visible.[64]

Taking the first structures that he saw in the middle of the early blastoderm for the embryo itself, and also believing that the egg in which he first saw them had not begun to incubate, Malpighi concluded more portentously that those first "threads" of the chick that had "once lain hidden" had a "deeper origin" and had existed beforehand in the egg. "We might surmise" (*suspicari possimus*), he wrote with a studied tentativeness once again, that the embryo with the beginnings of nearly all its parts lay floating and concealed in the initial clear pool of liquid.[65] Years later, looking back with apparent self-satisfaction on his treatises on the chick, he invited everyone to philosophize at their pleasure on the wonderful phenomena he had described; but he also now asserted his own opinion more confidently that those phenomena demonstrated the existence in the blastoderm of a compendium of the animal, or of a first outline (*delineationes*) of its principal parts, that in time became visible to the eye.[66]

Such was also the immediate and persisting impression of others who read Malpighi's treatises. The original recipients, the members of the Royal Society, perceived the first treatise as already "importing chiefly" that "the first rudiments of the principal parts of the chick" were found in the impregnated egg even before incubation – a reading Oldenburg was quick to communicate to such correspondents as Swammerdam and de Graaf. Hence, Blanckaert would soon write that the microscope "teaches us not only that the complete plant is contained in the seed but that the entire little animal is in the egg before incu-

63 Ibid. 2:819–21, 824–7, 956–7.
64 Ibid. 2:954–7, 996–7.
65 Ibid. 2:944–5, 956–7, 976–7.
66 Ibid. 2:865–7, 867 n. 5.

bation; and who, being guided by reason, would deny the same in the eggs of women?" To be sure, Haller cited Malpighi's account of the development of the chick's chambered heart from a crooked vessel among the reasons for his own temporary disenchantment with preexistence, but Malpighi was more likely to be seen as having enhanced rather than harmed the case for preexistence, whose proponents often cited him.[67]

To do so, however, was to misrepresent Malpighi's thinking – to say nothing of his observations – for he showed little interest in and no sympathy for the doctrine.[68] Moreover, although he believed indeed that he had discovered the first structures in the egg before incubation, it was explicitly after impregnation. When, in his later writings, he allowed himself more freedom to speculate, he also sketched a hypothetical process of emerging form that, although preceding the visible embryo, was hardly in the spirit of preexistence: He proposed that the first threads of the parts of the chick appeared in fact as small walls, as it were, that rose and outlined shaped cavities into which seeped the clear fluid of the surrounding pool and from which then came the spine, vertebrae, and vesicles of the head.[69] Though it is by no means clear, he may have thought of these walls as having also arisen only after impregnation.[70] He ascribed their rise to "nature" and infused the fluid they contained with an organizing "plastic spirit,"[71] but he granted the masculine semen a "plastic force" as well, although its role remains obscure.[72] Be that as it may, Malpighi's theorizing avoided any suggestion that the hidden structure preceding the embryo might have had a lengthy, much less cosmic, history.

Limited in scope as they were, however, and though purposefully barred from his descriptive accounts, Malpighi's theoretical inclinations were not without a consequential impact on his observations. Since the animal was discovered already largely formed, he wrote at the beginning of his first treatise, the effort to study generation in the egg in fact proved futile, "for, unable to attain to the first beginning, we are forced to wait for the successively emerging parts to reveal themselves."[73] Regardless of the measure of his success or failure, though, he did thus believe that there *had* been a "first beginning" (*primus ortus*) prior to what he could subsequently see and, hence, an earlier existence of embryonic parts that still remained unobserved. Together with his accumulated experience in researching fine structure, it was that belief that encouraged him to bring the microscope to bear on the blastoderm once more, to experiment with new techniques to expose what still lay hidden – indeed, to hold the mounted blastoderm up to the sun – and to seize the more readily on the faint hints of form thus revealed.

67 Birch (1756–7), 3:16, 30; Adelmann (1966), 1:378–9, 2:919–20; Graaf (1677), 465; Blanckaert (1686), 483–4; idem (1701), 1:257; Boerhaave & Haller (1740–4), 5 (pt. 2): 502n, cf. 498n. See also: Garden (1690–1), 475–6; Malebranche (1958–70), 1:82–3; Leibniz (1875–90), 4:480; Schiller (1978), 34.

68 Adelmann (1966), 2:885–6, 886n. 2, 909; in anticipation of what is to follow on Leeuwenhoek and the spermatozoa in the text, see also p. 868.

69 Ibid. 2:866, 867n. 5.

70 Ibid. 2:869, 885, 930, cf. 860, 862, 863n. 1.

71 Ibid. 2:866, 867n. 5.

72 Ibid. 2:861–2, 863n. 1.

73 Ibid. 2:934–7.

Nonetheless, the impact of Malpighi's microscopic study of the chick embryo remained a problematic one. Despite the frequent appeals to Malpighi in the continuing debate, he had left the basic questions pertaining to reproduction unresolved, as he himself underscored. Moreover, his legacy to microscopic practice was an uncertain one. To be sure, he had moved the observation of the chick embryo to another level, and his scrupulous distinction between observation and interpretation encouraged a more critical evaluation of the implications of microscopic images in themselves. Curiously but nonetheless tellingly, however, Malpighi himself evidenced an ambivalence toward the instrument, and, though having by its means discovered new embryonic beginnings, he in fact belittled its contribution.[74] Indeed, though henceforward a necessary accessory in studies of the chick embryo, the microscope remained incapable of conclusively deciding the fundamental question as it was posed throughout the following century as well: Was embryonic development truly the making or only the unveiling of form? On the other hand, preexistence doctrine or the presumption of preformed embryonic structure also proved surprisingly equivocal, after all, in its encouragement in turn of microscopic research.

Both Swammerdam and Leeuwenhoek also left their mark on the debates over preexistence. Although they echoed many of the commonplaces of the time, however, the social and emotional contexts that shaped their scientific commitment more generally also shaped their views of preexistence in strikingly distinctive, indeed radical, ways. In large part, this was mediated through the atypical orientation of their respective research preoccupations, for neither showed much interest in exploring the embryo in the chicken egg.[75] On the other hand, nor did either succeed in establishing an alternative tradition of continuing and systematic research into the origins of animal form. As great as their impact was and as close their identification with the microscope, their legacy in terms of the furtherance of microscopic research in the field was an ambivalent one.

Swammerdam was in fact the first to have proposed the idea of preexistence in print. Having announced in his *Historia insectorum generalis* that he had at last penetrated the darkness surrounding the generation of animals, he declared that, in truth, there was no generation at all but only a propagation and growth of parts. We could now explain the perfect fetus born to a parent lacking its own limbs, he wrote – for all the parts were contained within the egg (a clarification he offered three years later) – and grasp the meaning of the biblical assertion that Levi, in the loins of his father, had paid a tithe long before his birth. An unnamed savant with whom Swammerdam had shared his discoveries had even suggested that Swammerdam had revealed the basis of original sin as well, since

74 See n. 56 above.

75 Robert Hooke did turn Leeuwenhoek's attention to chicken eggs in late 1679 (*AB* 3:146–7), and, though that encouragement seems to have resulted in only a brief interest, Leeuwenhoek remarked a few years later on his efforts to discover the spermatozoa in the yolk of the chicken egg (*AB* 4:64–7). Also see n. 142 below.

everyone would have been similarly enclosed in the loins of their initial ancestors.[76]

Although Swammerdam implied that he would have more to say on the subject after further research, he added little throughout his life to those few explicit – and problematic – lines in the *Historia*. Three years later, in his *Miraculum naturae sive uteri muliebris fabrica,* he specified the female egg as the site of preexistence (a point that jarred with Levi's sojourn in the loins of his father, needless to say) but otherwise simply restated in Latin what he had earlier said in Dutch; these lines also remained virtually unaltered in the manuscript of the great work he intended.[77] Nonetheless, even if less explicitly, other passages throughout his writings appear as well to have reflected his adherence to preexistence (and his subordination, for that matter, of the male semen to the egg).[78]

Indeed, he had also written in the *Historia insectorum generalis* that, just as a plant grew from a seed that already contained leaves and a tiny sprout, so an insect emerged from an egg that likewise contained all its parts.[79] Insisting that the egg was the animal itself wrapped in a skin, he subsequently described the parts of the insect within the still unlaid egg as growing from invisible but nonetheless *wesentlyke* (*wezenlijke*) – either "real" or "essential" – parts that preceded them.[80] Indeed, he conceived of the egg in these terms even before it was fertilized. He reported in the *Historia* that he had discovered the beginnings of the eggs of the moth *Orgyia antiqua* (Linnaeus) already within the caterpillar and had observed their development through the chrysalis and into the ima-

76 JS (1669), pt. 1:51–2; idem (1672), 21. Daniel LeClerc and Jean Jacques Manget identified the savant, a "very learned gentleman," with whom Swammerdam had talked about these matters as Malebranche (LeClerc & Manget 1685, 1:497; see also Cole 1930, 42–3; cf. Schierbeek 1947, 155). Whereas Swammerdam does explicitly declare his adherence to Malebranche's theologically based epistemology at one point (JS 1737–8, 2:705), nowhere does he identify Malebranche with the doctrine of preexistence. Swammerdam was indeed acquainted with Malebranche (see, e.g., *LJS* 128, 154), and, along with the tulip in the bulb, the latter would refer to the parts of the future bee within the larva and to the butterfly within the caterpillar as part of his argument for preexistence (Malebranche 1958–70, 12:230, 253). But if Swammerdam had indeed carried out such insect "dissections" while in France in the 1660s (see Chapter 5, n. 31), where he would presumably have talked with Malebranche about preexistence if he did, it seems noteworthy that Malebranche did not mention such observations until 1688, years after they had been made public in Swammerdam's *Historia insectorum generalis,* which Malebranche cited. Earlier, to be sure, in the *Recherche de la vérité,* Malebranche had already included Swammerdam's assertion that the frog could be seen in the frog egg in his argument for preexistence, but then too he had cited Swammerdam's *Miraculum naturae* (Malebranche 1958–70, 1:83). I have also been unable as yet to discover any passages where Malebranche himself associated preexistence with the doctrine of original sin.

77 JS (1672), 21–2; idem (1737–8), 1:34.

78 Regarding Swammerdam's views on the contribution of semen to reproduction, see Ruestow (1985), 219–20.

79 JS (1669), pt. 1:64. Swammerdam speaks here of the insect "seed," *saat,* as well, but the context and other passages equating "seed" and insect "egg" make his meaning clear (see ibid., pt. 2:2, 18; idem 1737–8, 1:273).

80 JS (1737–8), 1:41, 2:568, 603–4, 728; idem (1669), pt. 1:60, 72, 86, pt. 2:22, 30, 42, see also pt. 1:100. He also speaks of the *weesentlijke* (*sic*) parts of the butterfly hidden under the skin of the caterpillar (ibid., pt. 2:43).

go.[81] Concluding again that the egg was the animal itself, Swammerdam took these observations as particularly revealing evidence that no essential change took place in the insect's parts.[82] The logic is obscure, but not the fact that the egg to which he was referring – visibly growing with its parent, as he also noted[83] – included the egg before its fertilization. Likewise, he reaffirmed in his manuscripts that the "entire animal" was perfected in the frog's ovary as well.[84]

Such passages interspersed throughout his observations do not connote a purposeful effort to demonstrate or confirm preexistence, however, and the congruity between observations and the belief in preexistence was precarious at best. Like the moth's eggs within the caterpillar, certain observations were very suggestive to a mind suitably inclined, and the microscope in particular surely provided some exciting moments, such as discovering the "eggs" already packed by the thousands in the ovaries of the future queen bee still maturing within her cell.[85] In truth, however, the microscopic images Swammerdam encountered had little else to offer in behalf of preexistence. Opening beetle eggs on the verge of being laid, he could apparently discover only a whitish fluid, and, although he was thrilled by the sight of tiny embryos developing in the eggs of the miraculous, crystalline snail, none could be seen in the smaller eggs even though they too had already left the ovary.[86]

81 JS (1669), pt. 2:30, see also pt. 1:136; idem (1737–8), 2:565; Schierbeek (1947), 267 (regarding pl. 33). I am grateful to L. Lavenseau of the Laboratory of Neuroendocrinology, the University of Bordeaux, for confirming in a letter dated 26 Sept. 1984 that Swammerdam may well have seen a suggestion of the future eggs in the caterpillar of *Orgyia antiqua*. Dr. Lavenseau writes: "Comme chez *Bombyx mori,* les ovaires de la chenille d' *Orgyia antiqua* sont bien visibles au dernier stade larvaire et se distinguent des testicules par la forme. Un test supplémentaire consiste à crever ces gonades. Chez le mâle les lobules testiculaires laissent échapper une sorte de 'purée' cellulaire. Chez les femelles l'enveloppe déchirée de l'ovaire libère les ovarioles bien individualisés qui renferment des taches plus opaques correspondant aux futurs ovocytes. A la fin du dernier stade larvaire, il n'est pas interdit de penser que Swammerdam, malgré les moyens optiques réduits dont il disposait ait reconnu ces détails."

Since insect eggs are also among the subjects to which Swammerdam applied the lens in the *Historia insectorum generalis,* it is not improbable that he used the instrument in this instance as well, even though his microscopic dissections lay in general in the future (see Chapter 5 regarding n. 31). It remains striking, however, that he never mentions the microscope in the context of the eggs in the caterpillar of *Orgyia antiqua* and speaks of them, indeed, as already visible, *al sigtbaar.* When I raised the question whether Swammerdam might have been able to recognize the beginnings of the eggs in the caterpillar even with the naked eye, Dr. Lavenseau responded (by a letter dated 18 March 1991) that – if I may paraphrase him – Swammerdam may have been able to detect the minute oocytes in the ovarioles at the very end of the last instar, but before then would seem very problematic. At earlier stages of development the content of the juvenile gonads looks quite transparent, and it becomes difficult even to distinguish a testicle from an ovary without a microscope and the appropriate illumination. See also Cole (1930), 81.

I might also note that, in contrast to the preceding observation, Swammerdam later observed the beginnings of the reproductive organs of a butterfly only during the sixth to eighth days of the chrysalis (JS 1737–8, 2:585).

82 JS (1737–8), 2:568.

83 JS (1669), pt. 2:30.

84 JS (1672), 21; idem (1737–8), 2:470, see also pp. 809–10.

85 JS (1737–8), 2:471–2, 477–8, see also p. 604.

86 Ibid. 1:175–6, 303.

Swammerdam could ascribe the invisibility of the smallest snail embryos to their transparency and fineness, but to disregard what the frog egg ultimately revealed to the microscope required a greater investment of faith and commitment. Having systematically observed the beginnings of the frog's development, Swammerdam described the newly laid egg as a small, black globule whose most pronounced feature by the second day was a deep furrow that ran around it and nearly divided it in two. He could at this point see no entrails with his microscope and no other contents than small, yellowish granules that made up the substance of the egg. Vague entrails subsequently began to appear, however, and by the fifteenth day a distinct intestine had taken shape; but it, like the rest of the emerging animal, seemed to Swammerdam to be loosely composed of the globular granules of the original egg.[87]

Reading this description, the critic of preexistence (in anticipation of later arguments) might well have insisted that the visibility of the granules before the larger parts they subsequently composed bore witness to the initial absence of those larger parts even in rudimentary form.[88] Swammerdam gave no hint of sensing such implications, however, and he persisted in asserting that the black spot in the frog's egg was the little animal, the tadpole in its initial vestment (*rok*), the frog itself. Although no longer describing it as complete with all its parts, he still maintained that the entire animal in the form of the egg reached its full perfection in the ovary.[89]

His persistence in such assertions despite such uncongenial observations underscores the strength of Swammerdam's own predisposition to preexistence; but the roots of that predisposition are difficult to discover, for Swammerdam was not sympathetic to important threads in the typical argument crafted in behalf of preexistence in the last decades of the century. He did not share the mechanistic inclination that now also increasingly entailed a denial of nature's own capacity to produce the complex, purposeful structure of living things,[90] nor did he look

87 Ibid. 2:789, 813–19.

88 Caspar Wolff (1759), 72; Adelmann (1966), 2:921–3.

89 JS (1737–8), 1:42–3, 2:470, 789, 812–15.

90 Indeed, Swammerdam's own mechanistic inclinations appear to have been quite stunted. He did at one point liken the egg of the miraculous, viviparous snail itself to a *horologie* (JS 1737–8, 1:178), and he had grasped and exploited the fruitfulness of a mechanistic outlook in physiological experimentation and explanation. In his *Tractatus . . . de respiratione,* he had argued through mechanical analogies and the use of instrumental apparatus that, in respiration, the air was inhaled as well as exhaled by its being pushed rather than by attraction or an imaginary avoidance of the vacuum. He also embraced, in this treatise, Descartes's doctrine of rarefaction and condensation, and often referred to a "subtle matter" – a hallmark of the Cartesian influence in medicine and physics (JS 1667b, 46, 49, 71, 72, 119). Nonetheless, the mechanistic emphasis of this early treatise is atypical of his other writings, just as the characterization of the snail egg as a watch was uncharacteristic imagery for Swammerdam. (The only other passage of which I am aware wherein he likens an animal body to a machine in his works on natural history is JS 1737–8, 2:859; there the animal body is the human body.) Even in the treatise on respiration, moreover, Swammerdam's mechanism pertains essentially to the macrostructure of the body rather than to the minute mechanical processes with which the mechanical philosophy was preoccupied. Though having cited Descartes himself against too great a reliance on pure speculation, Swammerdam very likely had the mechanistic imagery of the Cartesians in

with kindness on the related commitment to keeping nature comprehensible.[91] (Indeed, Swammerdam's sympathy for the religious teachings of Jacob Boehme would have required, at the least, a remarkable tolerance for a conception of nature that stressed an extravagantly irrational creative dynamism.)[92]

Like the mechanistic adherents of preexistence, to be sure, Swammerdam replaced "generation" with "growth," but it is still by no means clear what he himself understood to be happening as the parts of the preexistent animal developed. The way he spoke of "growth" does not suggest a clearly, much less mechanistically, conceived process, and in one instance in the Latin translation of Swammerdam's manuscripts, his compatriot Hieronymus Gaubius, one of Boerhaave's students, rendered one of Swammerdam's Dutch expressions for the growth of parts as "epigenesis." In contrast to his indignation over Harvey's doctrine of metamorphosis, indeed, Swammerdam seems to have had few if any reservations about the epigenetic process Harvey had finally graced with a name, and Swammerdam himself explicitly embraced it in another but related context.[93] Consequently, Swammerdam's own conception of preexistence may after all have entailed something akin to Harvey's obscure and nonmechanistic process of organic development. At a minimum, Swammerdam was not overly exercised about conveying the comprehensibility of whatever processes he had in mind, and there is good reason to believe that he never grasped them very clearly himself. Experience perpetually taught him, he ultimately wrote, that the true causes of the growth of animals and their parts were unsearchable to mortal man.[94]

If mechanism and its commitment to comprehensibility were not at issue, neither can Swammerdam's piety satisfactorily account for his attraction to preexistence. Although he insisted so often and so fervently on the direct derivation of all living things from the hand of God, he in fact alluded to the original creation of the preexistent plants and animals only indirectly and passingly with his reference to the savant who had spoken of original sin. If piety prompted his commitment to preexistence, moreover, why did Swammerdam completely ignore the possibility upon which van Helmont had insisted, that all generation

Footnote 90 *(cont.)*
mind when, in the *Historia insectorum generalis,* he scoffed at medical theory that, deriving from such speculation, proved incapable of producing any results (JS 1669, pt. 2:6, 7). Furthermore, though he himself continued to use the phrase "subtle matter" with respect to a possible substance moving through the nerves, its nature remained emphatically uncertain in Swammerdam's writings (JS 1737–8, 2:836–7, 840–1, 854–6, 860).

91 See Chapter 5 regarding nn. 147, 148; see also JS (1737–8), 2:867–8; cf. Swammerdam's reference to Malebranche on p. 705.

92 Regarding Swammerdam's familiarity with and appreciation of Boehme, see JS (1737–8), 1:332, 2:593, 596–7, 667. With respect to Boehme's view of nature, see Hvolbek (1984), esp. chap. 5. There were Boehmist groups in Amsterdam at least by 1651 (Stoudt 1968, 307–8).

93 JS (1737–8), 2:568; idem (1669), pt. 1:26, 43, 101, see also p. 85.

94 JS (1737–8), 2:785, see also p. 680; Ruestow (1985), 222–4, 228–9. Mary P. Winsor also notes that Swammerdam had no clear idea of the distinction between growth and differentiation (*DSB,* s.v. "Swammerdam, Jan").

was a new creative act of God?[95] Preserving God's creating hand as the origin of all living things was of no small moment to Swammerdam, as we have seen, but preexistence in fact kept it a distant hand. In a poetic vein, Swammerdam could write of God's pulling and tugging at the early human embryo, and he reported having observed, in the transformation of the tadpole, God's almighty arm revealed in splendor.[96] Although omnipotence loomed very large among the attributes he repeatedly ascribed to God,[97] Swammerdam's insistence on a preexisting animal nonetheless forestalled any new acts of divine creativity.

Swammerdam's explicit remarks on preexistence are far too brief to reveal much about the deeper significance it may ultimately have held for him, but it is possible to search for that significance indirectly through his recurrent, more developed, and often ardent passages on metamorphosis. Indeed, those passages frequently echo Swammerdam's phrases pertaining to preexistence, and, leaving aside the explicit parallel he at one point drew between metamorphosis and the development of the human embryo within the womb, he repeatedly likened the pupa to the insect egg.[98] A clear understanding of the pupa provided the foundation for an understanding of the egg as well, he early remarked,[99] and it was in similar terms and with similar emphasis that he continued to insist henceforward that the egg and the pupa were both the "animal itself."[100] In both cases, what occurred was no more than an unveiling of parts that had grown in concealment and then dried out from a state of fluidity to firmness.[101] (It was also the growth of parts preceding metamorphosis, however, that Swammerdam explicitly identified as epigenesis, and he subsequently described the causes of metamorphosis as no less inscrutable than the Divinity that made it happen.)[102]

The similarities in how he spoke of both preexistence and metamorphosis and the analogies he so often drew between the egg and pupa suggest that, for Swammerdam, preexistence and metamorphosis held some common or kindred meaning. A peripheral idea, in fact, that Swammerdam probably never clearly or fully worked out, preexistence appears in this light not as an accommodation of the power of God, on the one hand, and mechanism and comprehensibility, on

95 See Chapter 8 regarding n. 44.
96 JS (1675a), 421–2; idem (1737–8), 2:828–9.
97 See Chapter 5 regarding nn. 156, 159.
98 JS (1737–8), 2:792–3; idem (1669), pt. 1:148–9, 86, 103; idem (1737–8), 1:273, 2:603–4, 728. Quite apart from Swammerdam's own train of thought, subsequent writers on generation, including Haller, also frequently perceived his discovery of the butterfly in the caterpillar as an observation that at least suggested the notion of preexistence; see, for instance, Garden (1690–1), 476; Malebranche (1958–70), 12:230; Boerhaave & Haller (1740–4), 5(pt. 2):496–7; Adelmann (1966), 2:884; see also Schiller (1978), 34.
99 JS (1669), pt. 1:7.
100 JS (1669), pt. 1:33, 42, 72, 86, 145, pt. 2:30; idem (1737–8), 1:6, 28, 2:568, 603–4, 728.
101 JS (1669), pt. 1:15, 42–3, 55–6, 58–61, 72, 86, 143–4, 148, pt. 2:21; idem (1737–8), 1:28, 37, 41–2, 2:716, see also p. 565.
102 JS (1669), pt. 1:26, 38, 40, 43, 101; idem (1737–8), 2:680; Ruestow (1985), 228–9. Swammerdam dropped the explicit identification with epigenesis in the *Bybel der natuure,* however (see JS 1737–8, 1:272), although he did commit himself, with respect to metamorphosis, to precisely what Harvey meant by epigenesis, a successive growth of parts (ibid. 1:19, 28).

the other, but as a further expression of Swammerdam's distinctive insistence on a natural foundation for the rigid regulation of insect generation and development. It was a priority that gave his conception of preexistence itself a distinctive cast.

Indeed, Swammerdam first broached the idea of preexistence in the *Historia insectorum generalis* while discussing insect metamorphosis. Though he also wrote that it was the discovery of the eggs in human ovaries that initially prompted him to suspect that all animals had eggs,[103] the text of the *Historia* intimates rather that it was his studies in insect generation that evoked the more radical notion of the preexisting animal in the egg itself. He approached the idea in the *Historia* by declaring that the generation of insects was so clear that it might enable us to reach the true principles of animal generation, and in the immediate background were his discoveries of the eggs in the queen bee and the butterfly and, in particular, the eggs of *Orgyia antiqua* in the caterpillar. Judging from the text of the *Historia,* the latter was the key discovery;[104] but be that as it may, his studies of insect generation and metamorphosis would appear, at a minimum, to have provided the initial context for his formulation of the concept of preexistence. It is thus not unreasonable to assume that the concerns underlying those studies were at work in shaping the new concept as well.

Indeed, in first putting forward the notion, Swammerdam endowed it with a specific significance also essential to his reduction of metamorphosis to a simple unveiling of parts: Having reduced generation as well to but a propagation and growth of parts, preexistence too was explicitly a barrier against chance.[105] As such, preexistence was hence one more bulwark to the reliable consistency of insect life cycles; it added whatever persuasiveness it could muster to the repudiation of spontaneous generation and scotched other potential sources of irregularity as well. John Ray was among those who noted, for instance, that the Cartesian account of the formation of the fetus failed to provide "the least reason why an Animal of one Species might not be formed out of the Seed of another," and Hooke still proclaimed with his own mechanistic imagery that, not only could the same insect be produced in different ways, but it might itself produce different kinds of progeny.[106]

From his concern for reliable life cycles may also have followed Swammerdam's ambivalence toward divine omnipotence, a wariness also evident, to be

103 JS (1737–8), 1:305.

104 JS (1669), pt. 1:51; idem (1737–8), 1:34. In both the *Historia insectorum generalis* (JS 1669, pt. 2:30–1) and the *Bybel der natuure* (JS 1737–8, 2:565, 568), Swammerdam refers to the "infinite uses" of the discovery of the "eggs" in the caterpillar of *Orgyia antiqua* and then directs the reader to the location of his earlier statement on preexistence. See also Harcourt Brown (1934), 281.

105 JS (1669), pt. 1:51; idem (1672), 21; idem (1737–8), 134. That the egg, like the pupa, involved no more than a growth of parts, he asserted, destroyed accidental generation just as the similar nature of the pupa uprooted the idea of radical metamorphosis (JS 1669, pt. 1:103; idem 1737–8, 1:273–4). Because of these implications with respect to both metamorphosis and generation, Swammerdam spoke of the recognition of the common process of development of the egg and pupa as a discovery of "infinite use." Compare this with the preceding note.

106 Ray (1691), 28; Hooke (1665), 193.

sure, in his treatment of metamorphosis. In the remarkable transformation he observed during the metamorphosis of the soldier fly, Swammerdam also confronted what he said was the Deity moving in a visible demonstration of divinity;[107] but by insisting on the persisting insect nonetheless, Swammerdam in fact declined simply to leave what went on within the chrysalis to the pleasure of God – as others had done, he knew.[108] After all, divine omnipotence, if unrestrained, was as threatening to nature's dependable constancy as was chance.

Indeed, despite continued theological promptings that God had not only created but still governed all things, even those philosophers and naturalists of the late seventeenth century who were most solicitous of God's interests proved no more inclined than the scholastics to allow him to intervene directly in nature's daily operations. When reasons were adduced for this scruple, they were generally couched in terms of a protective concern for God's honor and image. It was demeaning for the Creator to have a hand in the drudgery of nature's everyday chores, wrote Cudworth in effect, while, with regard specifically to generation, the slow processes of nature were inconsistent with omnipotence. John Ray found the latter point particularly compelling, but he also added that, if granted a hand in all such matters, God would be saddled as well with the responsibility for the monsters and mistakes of generation.[109]

There is reason to suspect, however, that the resistance to an overly busy Divinity rested in large part, rather, on a seldom acknowledged solicitude for nature – or, in truth, science and philosophy. Wittingly or not, Cudworth also echoed the scholastic concern for nature's autonomy when he argued that the Creator's direct and continual intervention would nullify nature as something significant and distinct from the Deity, so that nothing would happen naturally. As conceived by the naturalists and natural philosophers, however, the unspoken concern was very likely to have been rather the threat an incomprehensible omnipotence posed to any graspable order and regularity in nature – a dilemma, after all, that had long troubled the theologians.[110] The implications were hardly less consequential for the intricate patterns of insect life, and although fending off chance was always in part a defense of the divine origin of living things, the preexistent animal, like the persisting insect in metamorphosis, nonetheless distanced the unfathomable power of God from the day-to-day processes of natural history as well.

If the parallel drawn between his views on metamorphosis and preexistence holds good, Swammerdam's concern was to assert the persistence of a physical core, the "animal itself," that served less to explain the origins of structure than

107 JS (1737–8), 2:666.
108 JS (1669), pt. 1:32.
109 Cudworth (1678), 150; Ray (1691), 34–5; see also Descartes (1897–1910), 11:524; NH (1730), "Ext. crit.," 20. Regarding the theological insistence on God's continuing governance of all things, see, for instance, the Confession of the Dutch Reformed Church, article 13 (Cochrane 1966, 197).
110 Cudworth (1678), 150; Oakley (1984), 50–65, 77–84.

to provide a natural vehicle for the enduring pattern of the life cycle of the species. The persistence of structure throughout the butterfly's metamorphosis was called upon, after all, to testify to the continuity of the animal, rather than the continuity of the animal to the persistence of structure. Passages suggest that, just as the parts characteristic of the butterfly grew or developed under the skin of the caterpillar before metamorphosis, so the parts of the embryo developed within the egg in the ovary. In the former case, at least, Swammerdam explicitly called that development epigenesis, the antithesis to the preexistence doctrine elaborated by later proponents more concerned with enduring structure.[111]

As was true of his doctrine of metamorphosis, consequently, the aspect of preexistence that Swammerdam appears to have seized upon was the prospect it offered of a firmer foundation for the predetermination of insect generation and life cycles. It thus further secured a natural order essential not only to Swammerdam's system but to his practice of natural history, a practice made all the more dear in turn, it has been argued, by the easement it offered his emotional ordeals. The conception of preexistence shaped by these priorities, however, remained ultimately indifferent to the microscope.

Abbreviated and cryptic as it was, Swammerdam's only explicit espousal of a general notion of preexistence dates from before his pioneering recourse to the lens, which then appears to have offered little to warrant a further elaboration of the theme. In what seem to have been the critical observations in *Orgyia antiqua,* even the "beginnings of the eggs" are described as already visible (*al sigtbaar* [*zichtbaar*]) in the caterpillar and occasioned no reference to the lens,[112] whereas more distinctively microscopic observations failed to elicit any fuller or clearer articulation of what Swammerdam had in mind. Indeed, those microscopic observations perhaps most directly relevant, his study of the embryonic development of the frog, offered decidedly negative evidence and a perverse encouragement, hence, to leave the issue obscure.

Nor does Swammerdam's commitment to preexistence appear to have stimulated any sustained line of microscopic research. He betrayed no inclination, for instance, to apply his impressive new skills in a search for preformed structures specifically to confirm it, and, given Swammerdam's distinctive conception of preexistence, the incentive to do so was limited. The persisting core of the animal that guaranteed invariable generation posited no vestige of the infinitely elaborate and endlessly diminishing structural complexity more generally identified with the doctrine of preexistence. For Swammerdam, continuity not structure was the issue, and at some point not long before it emerged in generation, the "animal itself" was still very likely physically ill-defined, perhaps no more structurally developed than the black spot in the frog's egg. Consequently, when

111 See n. 102 above. In the *Historia insectorum generalis* (JS 1669, pt. 2:26), Swammerdam also speaks of the parts of the insect within the egg as developing from a singular "internal principle" (*inwendige begintsel*).

112 See n. 81 above.

in the course of his systematic studies of life cycles Swammerdam failed to find the later animal fully prefigured in its beginnings, he had no reason to assume that the unseen parts were there nonetheless and might yet be discovered if he persevered with the microscope. Conceived as an adjunct to his practice of natural history, Swammerdam's distinctive notion of preexistence did little to promote an aggressive and continuing recourse to the high-powered lens.

Although he adhered even less than Swammerdam to the fully developed doctrine of preexistence, Leeuwenhoek, by contrast, pressed against the furthest limits of his microscopes in searching for the preformed animal that preceded generation. This inevitably followed in part from the great novelty in his approach to animal generation, his emphasis on the spermatozoa; but it followed too from preconceptions he did share with orthodox preexistence theory (that Swammerdam did not) on the presumed imperatives of structure in generation. Nonetheless, his ultimate conclusion was a pensively somber one now explicitly proclaiming the exhausted potential of the microscope.

Like Swammerdam after his provocative initial passage, Leeuwenhoek generally avoided addressing the full, sweeping prospect of preexistence directly; indeed, among the ideas that Velthuysen had communicated in 1679 on which Leeuwenhoek had politely declined to comment was "the contention of some learned people about the generation of plants and animals, that it did not take place anew."[113] Nonetheless, having subsequently remarked himself that no new creatures were made by God – or, indeed, were made "anew"[114] – Leeuwenhoek repeatedly stressed in later decades that living creatures derived (*afhangen, afkomstig zijn*) from the first Creation.[115] In 1687, while trying to impress upon the members of the Royal Society the unimaginable (and truly microscopic) "perfection" within the plant seed, he still declined to expand on his speculations on the derivation of all plants from those created in the beginning; but he now clearly had such speculations, and it was because they were of a kind that might prove offensive that he withheld them, he said. The following year he was more forthcoming, stating as a certainty that within the ripened grain seed already lay the beginnings of not only the plant but its future seeds. By 1690 the younger Constantijn Huygens was ascribing to Leeuwenhoek a full-blown doctrine of preexistence: "He seems to be of the opinion," he wrote Christiaan, "that the *semences* of plants and animals contain animals [and plants, he presumably meant to say] that in their own *semences* contain yet others *in in-*

113 *AB* 3:86–7. In the following few years, Leeuwenhoek was apparently also exposed to echoes of Malebranche, for Leeuwenhoek wrote in 1685 of having often been told that the flower itself could be seen in the tulip bulb (*AB* 5:302–3; and see n. 12 above). Indeed, Malebranche's *Recherche de la vérité* also appeared in Dutch translation in 1680–1 (Malebranche 1958–70, 1:xix).

114 *AB* 7:378, 34–5.

115 See, for instance: AvL (1693), 474; idem (1696), 135; idem (1702), 149–50; AvL to RS, 26 Feb. 1703, AvL Letters, fol. 238v.

finitum, and that there are no new creatures at all in the world, but only those that have already been made and only grow and expand [*augmenter*]. . . ."[116]

Leeuwenhoek's commitment to the spermatozoa had greatly complicated the difficulties already surrounding animal generation in particular, however, and his further reflections on the spermatozoa underscored, in fact, how far he really was from the doctrine Constantijn attributed to him. Attempts to explain the origin of the spermatozoa themselves varied widely, from spontaneous generation to Hartsoeker's initial suggestion that they came from the air we breathed and the food we ate and were separated out of the blood in the testicles (an idea he himself later derided as preposterous).[117] Having rejected – or apparently renounced – spontaneous generation, Leeuwenhoek came to assume that, being animals themselves,[118] the spermatozoa propagated in the testicles from a "seed," or semen (*saet*), or "seminal stuff" (*zaadelijke stoffe*) of their own. (Although he did believe he had in fact seen males and females among the spermatozoa, he tended to think that they did not really mate but rather reproduced along the lines of the parthenogenesis of aphids.)[119] At least by 1685, however, Leeuwenhoek not only posited the beginnings of other spermatozoa within the spermatozoon but lodged the preformed future animal or human there as well,[120] thus relocating the essential problem of animal generation to that spermatozoan seminal stuff.

Yet in the very same letter in 1685, Leeuwenhoek also asserted that "the first essential stuff [*eerste wesentlijke stoffe*], or the beginning from which a man is produced (which to us is incomprehensibly small), will remain hidden and inscrutable." Nor did he venture in subsequent letters to speculate on the nature of that seminal stuff – or of the *saet,* presumably the same – from which he was now convinced the spermatozoa had come. The generation of animals was thus quickly lost in mystery again, as was that of plants, after all, for Leeuwenhoek also declared the stuff that was the beginning of the embryo plant to be no less inaccessible.[121] Despite his occasionally suggestive rhetoric, consequently, Leeuwenhoek had a very uncertain grip on any meaningful notion of preexistence, and his later letters often make it clear that, in insisting on derivation from the Creation, he in fact meant (at least by then) lineal descent and not the persistence of the individual.[122]

116 *AB* 6:308–9, 7:384–7; *OCCH* 9:354–5. Leeuwenhoek had in fact undertaken to discover the future ear within the cereal grain; failing in this, he did succeed in finding the ear in the young dissected seedling and remained convinced that the future grains were already within that ear and that the beginning of those grains had hence already lain within the original ripened grain as well (*AB* 7:372–87).

117 *AB* 3:18–19; Overkamp (1686), 423; NH (1708), 107; idem (1722), 193.

118 From the first, Leeuwenhoek had indeed spoken of the spermatozoa as "little animals" – *dierkens* or *animalcula* – and in the mid-1680s he even explicitly insisted that each spermatozoon had a soul (*AB* 5:176–9, 248–9, 266–7). Leeuwenhoek identified the "living soul" of an animal with an endowed capacity for movement (see, e.g.: *AB* 5:246–7, 7:34–5).

119 Ruestow (1983), 214, 216–20.

120 *AB* 5:264–7, 236–7.

121 *AB* 5:266–7. Regarding seeds, see also *AB* 7:378–81.

122 See, for instance: *AB* 7:34–5; AvL (1696), 135; idem (1702), 111, 149–50.

Nonetheless, Leeuwenhoek did share perhaps the basic intuition underlying developed preexistence theory, Swammerdam notwithstanding: that organized structure could only arise from some preceding and comparable structure.[123] It was an intuition Leeuwenhoek believed he had in fact found confirmed in plant seeds, and now at a genuinely microscopic level even within the embryo. Beginning in 1685, he repeatedly described not only the embryonic "leaves" and their beginnings, but now also the "vessels" in the parts destined to become the trunk and root of the tree.[124] Thirty years later, in speaking of the twenty-five hundred vessels his observations indicated were enclosed in a pear seed, he reported as well what he took to be the valves he had long imagined in such vessels.[125]

But the intuition of a necessary continuity of structure was in the first place a predisposition that seized upon conveniently suggestive microscopic images and discounted the inconvenient. Though he had always been a bit tentative about the valves, to be sure, Leeuwenhoek left no doubt that, in what were in fact patterns of cells in the seed embryo, he truly saw the vessels of the later tree;[126] for if all the vessels of the tree were not in the seed, he asked, how could the seed produce the tree?[127] His readiness to indulge that predisposition emerged perhaps even more strikingly, though also more fleetingly, in his early observations of semen. Before taking up the cause of the spermatozoa, indeed, he had first insisted on the primacy of the semen in reproduction on the basis of a still puzzling perception of a microscopic network of vessels in the semen, a network he appears to have conceived as a very delicate, preembryonic vascular system

123 In his response to images of the cross section of the funiculus of seeds in particular, indeed, Leeuwenhoek reversed Swammerdam's priorities and clung to a continuity of comparable structure while seeming to let the persisting creature itself slip away (Ruestow 1983, 222–3). How such structural continuity could actually produce reproduction remained nonetheless a mystery; the minuteness of the parts of which plants and animals were made was no more comprehensible than the expanse of the universe, Leeuwenhoek wrote in 1692, and we could grasp even less how those parts were put together so that one living thing produced another (AvL 1693, 483–4).

Leeuwenhoek did believe, however, that oozing fluid and globules might produce such an elemental form of organized structure as vessels: *AB* 3:414–15 (cf. *AB* 5:208–9, 6:80–1); *AB* 6:50–3; see also AvL (1693), 504–7.

124 *AB* 5:208–11, 218–23, 230–1, 258–9, 6:252–309; AvL (1696), 44–5, 53; idem (1697), 280. Leeuwenhoek usually made no distinction between the cotyledons and ordinary leaves (*AB* 5:219 n. 4, 225n. 13), but see *AB* 5:258–9.

125 AvL (1718), 179; see also the following note.

126 Leeuwenhoek had vigorously and very early insisted (in disagreement with Nehemiah Grew) that patterns of wood cells were indeed vessels he had clearly "seen," but he continued to speak more tentatively of what he "called," "judged," or "imagined" to be valves in the larger vessels of the wood, even though he had observed indications of such structures and believed the hydraulics of the tree demanded them: *AB* 1:48–53, 2:6–9, 12–13, 32–7, 3:158–61, 170–1 (cf. 180–1); Baas (1982), 84, 87.

127 AvL (1696), 53; AvL to RS, 26 Feb. 1703, AvL Letters, fol. 234r; AvL (1718), 33, 176. Regarding Leeuwenhoek's insistence that later organic forms had to be within the immediately antecedent ones, see also: AvL (1702), 95–6, 103, 305–6, 450; AvL to RS, 26 Feb. 1703, AvL Letters, fol. 238v.

with even the beginnings of the vital organs.[128] In time, the vessels would vanish from his letters, and he would ultimately repudiate his early accounts, but not before he had forwarded to the Royal Society several of his own ink drawings of these vessels that betrayed at the time no hint of hesitancy or uncertainty (Fig. 23).[129]

Not only did this predisposition build upon whatever raw materials Leeuwenhoek's microscopic images provided, it also defied unyieldingly contrary observations. Few observations could have argued more forcefully against the preexistent structure of the animal than those that Leeuwenhoek, like Swammerdam, made on the embryonic development of the frog. (That they also led to the discovery of the capillary circulation, however, was no small compensation.) As he described these observations in 1688, the initial eggs appeared to consist of a great number of rather complex globules in a watery fluid, and he continued to see only globules as he opened the changing and in some cases already moving eggs from day to day during the following week. Even the dissected body of a young, free-swimming tadpole struck Leeuwenhoek as being composed of globules alone, with no intestines, nerves, or vessels.[130] Years later in 1701 he remarked how the parts within the transparent body of a newly hatched spider also appeared as if composed of globules similar to those he had earlier seen in spider eggs.[131]

Like Swammerdam as well, however, Leeuwenhoek failed to acknowledge the implications of his observations of the early frog embryo, and at the end of the century he still insisted that, although the form of the frog was not yet to be discovered in the tadpole, the frog lay enclosed there nonetheless. He struck a more uncertain note when he remarked in 1701 that, as the yolk of the egg changed rather late into the intestines of the chick, the same was true, he believed, of the newborn spider.[132] Yet he still wrote the following year that all the wonders of the silkworm moth's anatomy had to have been contained in

128 *AB* 2:292–7, 336–7; Ruestow (1983), 191–3, 196–9. Honoré Fabri had anticipated Leeuwenhoek in 1666 by positing a network of vessels in the semen of the rooster that he also supposed to be the vessels and nerves of the future animal (Adelmann 1966, 2:913), and several eighteenth-century observers would also report having observed filaments or threads in the semen (Buffon 1841–4, 3:58; John Needham 1748, [643]; Wrisberg 1765, 95–6; Castellani 1973, 40, 50, 62–3). Nonetheless, the vessels reported – and illustrated – in the semen by Leeuwenhoek have continued to baffle modern commentators (see *AB* 2:295n. 31; Castellani 1973, 40). J. Kremer has asserted that Leeuwenhoek's "vessels" are in fact "canals" with swimming spermatozoa between the coagulated parts of the semen directly after ejaculation (1977, 5). I have elsewhere taken issue with a passage in the English translation of Leeuwenhoek's 18 March 1678 letter to Grew in *AB* that has Leeuwenhoek implying that at one point the spermatozoa were in fact within the vessels he saw (Ruestow 1983, 192n. 29). It should be noted that Kremer's explanation of the vessels would be quite compatible with that earlier translation, which still seems to me, however, to go well beyond the original Dutch text.

129 *AB* 4:8–11, 2:364–7 ("Tables" 18, 19).

130 *AB* 8:10–15.

131 AvL (1702), 362–3.

132 Ibid., 92, 362–3. Regarding Leeuwenhoek's reference to the development of the chick embryo, see n. 142 below.

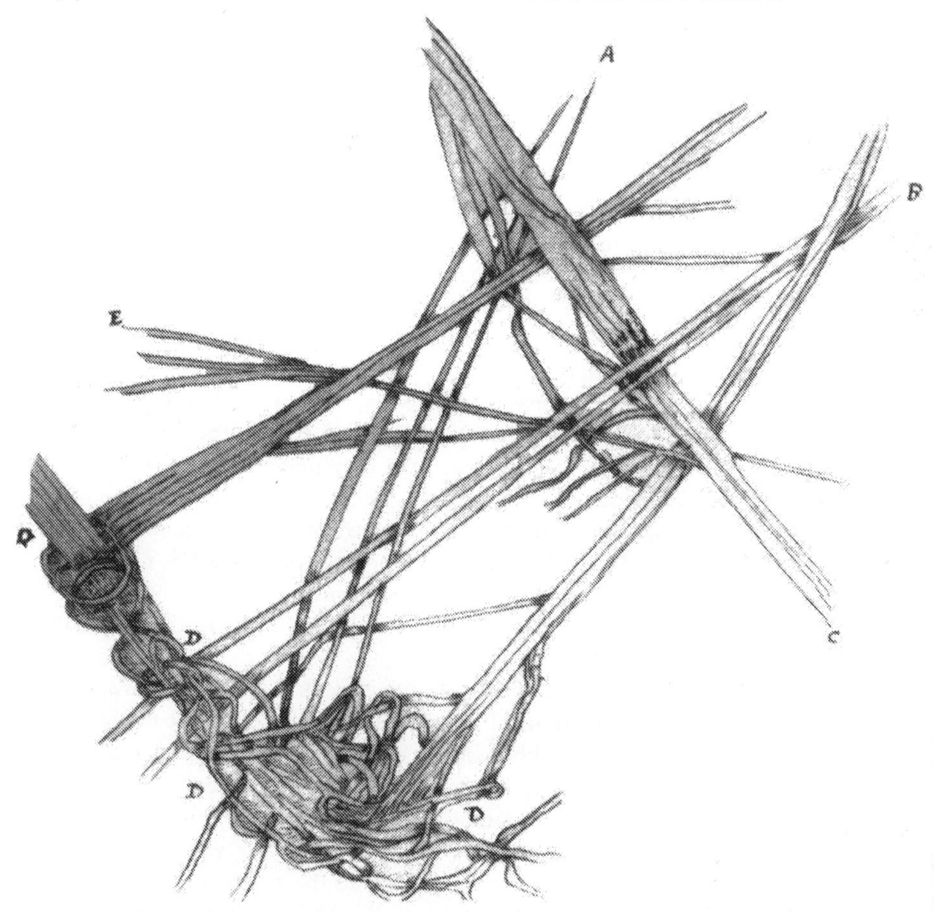

Figure 23. An ink drawing by Leeuwenhoek of the vessels he reported having observed in semen. By permission of the President and Council of the Royal Society.

the spermatozoon, and a decade later he would ask how the muscle fibers and even the membranes they were wrapped in could come from the spermatozoon were they not already present there.[133]

Despite the obscurity of the spermatozoan seminal stuff, consequently, Leeuwenhoek continued to insist not only that the seed harbored the perfection of the plant but also that the spermatozoa likewise contained "all the beginnings of the perfection" with which a developed creature was provided – indeed, the "entire man."[134] The structural continuity necessarily underlying generation

133 AvL (1702), 450; idem (1718), 22.

134 AvL (1696), 53; idem (1702), 96, 105. See also, regarding the spermatozoa: *AB* 7:384. Regarding the perfection or entirety of the tree being within the seed: AvL (1696), 57; idem (1697), 278, 280; idem (1702), 103; AvL to RS, 26 Feb. 1703, AvL Letters, fol. 238v.

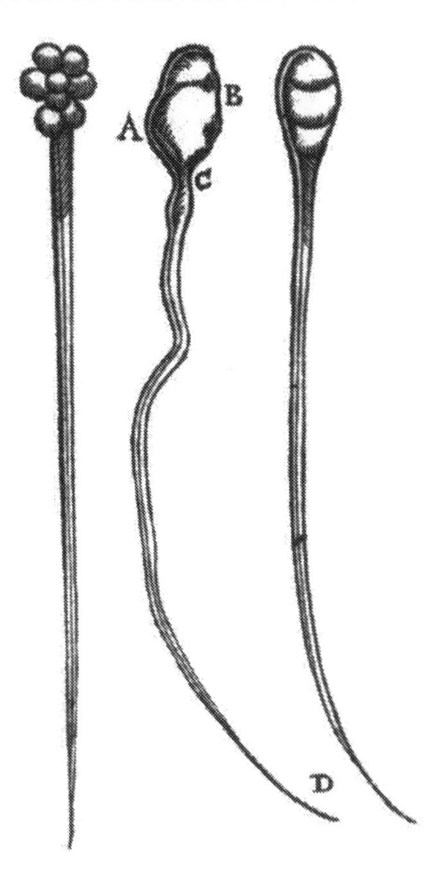

Figure 24. In the center is a human spermatozoon based on a drawing by Leeuwenhoek (*Philosophical Transactions,* no. 142 [1678], "Tab" 2). Included in an array of eight such figures in the original plate, those on the left and right are the remains of a human and a canine spermatozoon, respectively. Photo courtesy of the Special Collections Department, University of Colorado at Boulder Libraries.

(bolstered as well by his understanding of the spermatozoa itself as an animal with its own complex anatomy)[135] thus proved an enduring and compelling preconception, and, despite Leeuwenhoek's wariness of more extravagant claims, it posited the prospect of a remarkable discovery yet awaiting him in the spermatozoon. As a momentous truth pertaining indeed to the *beginselen* of things, the discovery of the preformed human – or animal – would have been the crowning achievement of his microscopy, and although he ultimately failed to reach it, it was not for a lack of trying.

135 *AB* 3:20–1; AvL (1718), 404–5.

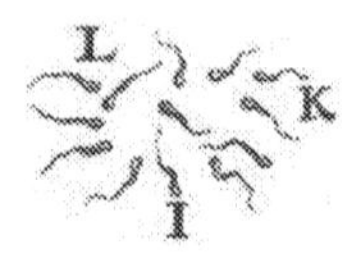

Figure 25. Spermatozoa as observed and drawn by one of Leeuwenhoek's draftsmen when left by Leeuwenhoek to draw them as he saw them (see n. 136). Photo courtesy of the National Library of Medicine, Bethesda, Md.

In addition to his fascination with the cloudlike multitudes of spermatozoa sweeping back and forth before his gaze, indeed, Leeuwenhoek had from the very beginning scrutinized the individual spermatozoon as well. In an early exchange of letters with Nehemiah Grew in 1678, he had already forwarded drawings of spermatozoa magnified some two thousand times, and among them was a figure of a human spermatozoon that was not surpassed in accuracy for another century and a half (Fig. 24; cf. Fig. 25).[136] He chafed nonetheless at the shortcomings of his techniques,[137] and, having initially observed the semen in capillary tubes,[138] he eventually turned to diluting a small particle of semen with a drop of water and spreading it as thinly as possible on a piece of glass. With the spermatozoa thus displayed – as if they were lying in a field on a bright day, he wrote – Leeuwenhoek achieved what he considered his clearest and most penetrating observations.[139]

Still nonetheless frustrated by how little he could see, he even attempted to strip the spermatozoa of their outer skin with a small moistened brush as they lay dried upon the glass;[140] but the effort proved futile, and repeated attempts to peer into their inner structure yielded little more than a suggestion of globular shapes.[141] There were moments, to be sure, when he suspected that he might have glimpsed something more, and he wrote at one point in 1685 that he thought he could almost say that "there lies the head and there as well the shoulders and there the hips." But the perception was fleeting and far too nebulous, and he hence resigned himself to the hope that he would one day find an animal with spermatozoa large enough to allow the animal within to be

136 *AB* 2:346–9 and "Table" XVI; Hughes (1955), 15. In contrast to Leeuwenhoek's own early mammoth depiction of spermatozoa, Fig. 25 represents spermatozoa as one of his draftsmen saw them years later through one of Leeuwenhoek's instruments (AvL 1718, 404, and fig. 5 in the plate facing p. 401). To be sure, the microscopes Leeuwenhoek used on the spermatozoa varied from "very good and very magnifying" instruments (*AB* 7:10–13) to "ordinary" (*gemeen*) ones (*AB* 3:108–9); yet Leeuwenhoek himself commented that his draftsmen saw objects smaller, indeed half the size, through the microscope than did he (AvL 1694, 546, 550–1; idem 1702, 56, 231; AvL to RS, 24 April 1705, AvL Letters, fol. 368v; AvL 1718, 31).

137 *AB* 2:348–9; AvL (1702), 99.

138 *AB* 2:362–3; AvL (1702), 321, 323.

139 AvL (1702), 99–100. At times, at least, Leeuwenhoek observed them by candlelight, with a small concave mirror to enhance the illumination (ibid.). For Hartsoeker's not dissimilar procedures (apart from the lighting), see NH (1730), 49.

140 AvL (1702), 325.

141 *AB* 5:190–3 and "Table" XX (figs. 33 and 34); AvL (1696), 146; idem (1702), 98–9, 301–2 (and figs. 2–6 in the plate facing p. 300), 322.

seen.[142] That hope, of course, was never realized, and what would have been his culminating triumph relentlessly eluded him.

Apart from Leeuwenhoek's personal frustration, moreover, the implications of that failure also cast a doubtful light on the future of the microscope. By the end of the century, indeed, Leeuwenhoek had accepted the barriers he confronted in both the spermatozoon and the seed as insurmountable, and, reflecting his own assessment of his instruments and skills, he concluded in effect that no one was likely to succeed where he had failed. What was still hidden within seeds and spermatozoa became the "great secret" to Leeuwenhoek, and by the mid-1690s his efforts had convinced him it could not be reached.[143] He was still assuring Boerhaave in 1717 of the enclosed though unseen lineaments of the future animal in the spermatozoon,[144] but they had been absorbed into a mystery he now believed would never be revealed.

The inadequacy of even his finest lenses[145] thus testified that the true workings of generation were concealed in the same impenetrable and stupefying smallness into which nature seemed everywhere to recede. He had already in 1685 stressed the incomprehensibility of "the smallness from which a man was produced from generation to generation,"[146] and in the following decade, his letters increasingly remarked the inconceivable minuteness of nature's parts.[147] It was not a wholly new theme in Leeuwenhoek's letters, to be sure,[148] and its greater prominence now doubtless reflected in part the new flourish in his rhetoric generally. However, it could not but reflect as well his experience of the unyielding impenetrability of life's definitive capacity.

Indeed, it was with the great secret in seeds and semen in mind that, in 1703, Leeuwenhoek recalled Christiaan Huygens's remark shortly before the latter's

142 *AB* 5:236–7. Despite the way he sometimes expressed himself, however, Leeuwenhoek did not in fact expect to find in the spermatozoon a miniature replica of the future animal. When in 1699 a celebrated hoax reported and illustrated the discovery of the lineaments of a man in spermatozoa (see Cole 1930, 68–72), Leeuwenhoek called it to the attention of the Royal Society in order to deny it, and his argument stressed the unlikelihood that the preexistent man would look like a man (AvL 1702, 92–5; the relevant plate, which has been placed in the body of the wrong letter, is found facing p. 68). He cited Malpighi concerning the developing chick embryo to make his point, but his own observations of the chick embryo years before appear to have already impressed him with how unlike the future chick the early embryo looked (*AB* 4:64–5). Leeuwenhoek had also acknowledged the diversity in the outward appearance of embryonic development in other animals, including the frog, in the context of which he also made reference to a pea-sized human embryo (*AB* 4:66–7). Nor indeed did the major proponents of preexistence expect the preformed embryo to be a replica of the future animal; see Malebranche (1958–70), 12:229, 253; Leibniz (1875–90), 3:579, see also 6:619–20; Bonnet (1779–83), 8:61–2; Haller (1758), 2:182–6.

143 AvL (1696), 53–4; idem (1702), 96, 306, see also pp. 103, 105–6; AvL to RS, 26 Feb. 1703, AvL Letters, fol. 240v.

144 AvL (1718), 405.

145 See n. 136 above.

146 *AB* 5:264–5.

147 AvL (1693), 473, 492; idem (1702), 106, 306, 360; AvL to RS, 24 April 1705, AvL Letters, fol. 369r; AvL to RS, 22 Sept. 1711, ibid., fol. 147v; AvL (1718), 62, 316.

148 Writing of water particles and the vessels in microscopic animals, Leeuwenhoek had already written to Christiaan Huygens, Sr., in 1679 of "the wonders of the dimensions" he imagined in nature (*AB* 3:66–7, see also pp. 396–7).

death that perhaps astronomical observation had reached its limit.[149] Although Leeuwenhoek never ventured such a sweeping judgment about the microscope, his emphasis on the futility of any further search into the nature of generation struck a repeated and resonating chord in his later letters. He might be wrong, he granted in 1703, but he doubted that anything more of the great secret could be discovered.[150]

This acquiescence in hopelessness derived from the same intuitive presumption that demanded a continuity of structure in the first place; for, convinced as he was of the necessary body within the spermatozoon, Leeuwenhoek's failure to find it could only bespeak then an enduring and unyielding concealment. In his observations of the capillary circulation, of the blood corpuscle, of the muscle fiber, and even, at an initial level of observation at least, of the seed embryo, the interplay between images and ideas forged a sensation of discovery and understanding. In the case of the spermatozoa, however, the absence of what Leeuwenhoek could construe as a reflection of his expectations in microscopic images proclaimed, rather, an impenetrable smallness in which the essential processes of generation would be forever hidden.

Leeuwenhoek's epochal studies of the spermatozoa were provoked and sustained by the social context in which they unfolded – by the encouragement of supporters, the proddings of skeptics, and the seeming obstinacy of a broader audience[151] – for Leeuwenhoek's campaign for the spermatozoa entailed very personal implications. When still committed to the vessels in the semen (though it was Grew's doubts that had kept him looking for those as well), Leeuwenhoek had called upon the members of the Royal Society to place their trust in him rather than in Harvey and de Graaf. Though he subsequently transferred his allegiance to the spermatozoa, his observations and his interpretations were still the crux, and his abilities, the significance of his achievement, and the worth of his judgment were still at issue.[152] The discovery of the preformed human or animal within the spermatozoa would thus not only have gratified his expectations about the nature of generation; confirming the role he championed for the spermatozoa, it would also have vindicated Leeuwenhoek in a public controversy in which his ego and self-perception were again at risk.

149 AvL to RS, 26 Feb. 1703, AvL Letters, fol. 240v.

150 Ibid.; see also AvL (1702), 96, 103. Even in very early letters, to be sure, Leeuwenhoek had expressed his recognition of the fact that structures related to those he was researching lay ultimately beyond the reach of the microscope (see, e.g., *AB* 1:308–11, but note the subtly shifted tone in pp. 316–17); however, he was readier and more emphatic in his later years in acknowledging that his specific researches had reached the limits of what was possible. In addition to his letters pertaining to the great secret in seeds and the spermatozoa, see also the following letters with respect to other lines of observation as well: AvL (1694), 590–1; idem (1718), 191, 340; AvL to RS, 9 Jan. 1720, AvL Letters, fol. 227r; AvL to RS, 24 Jan. 1721, ibid., fol. 246v; AvL to RS, 27 June 1721, ibid., fol. 264r; AvL to RS, 1 May 1722, ibid., fol. 276v; AvL to J. Jurin, [Aug. 1723], ibid. (vol. 4, letter no. 86), fols. 325v.–326r. (Regarding this last letter, see also Chapter 7, n. 88.)

Regarding Leeuwenhoek's more general statements as to the inscrutability as well as the incomprehensibility of nature, see Chapter 6 regarding nn. 95–100.

151 See Chapter 7 regarding nn. 102–4.

152 *AB* 2:334–9.

High personal stakes thus spurred on his studies of the spermatozoa through at least the 1680s (and he was still soliciting the Royal Society's advice in 1700 as to what animal he should turn to next),[153] while a characteristic intuition of the age fashioned the prospect of a triumphantly conclusive discovery. Leeuwenhoek soon acquiesced in the unattainability of that discovery, however, and emphasized instead a resignation that, without his personal goads – his combativeness, social aspiration, and self-doubt – and the social circumstances that made them fruitful, was unlikely to foster a continuing tradition of microscopic research. In the case of his researches on the spermatozoa, then, the interplay of ideas and microscopic images promoted an understanding of nature that further circumscribed the promise of the microscope.

153 AvL (1702), 222.

CHAPTER TEN

A new world

What was the discovery of the spermatozoa in fact a discovery of? Having failed to expose their hidden secret, Leeuwenhoek left the question unanswered, and, bereft of a conclusion, it remained in dispute throughout the eighteenth century.[1] But it was soon enveloped in a broader debate over the nature of microscopic life in general, which contending naturalists struggled to assimilate to sometimes radically different philosophies of nature. The bizarre and teeming populations Leeuwenhoek's microscopes had revealed quickly became the most celebrated of the new microscopic discoveries, and while firsthand observers mused on the infinite numbers that very likely still lay undetected,[2] poets and popular writers across Europe repeatedly evoked the prodigious multitudes thriving on a leaf, in a drop of water, and even in great stone cliffs.[3] Indeed, reechoing literary allusions to microscopic life attested that the microscope had drastically changed the understanding of the world; but the ultimate nature of that change remained uncertain, muddled by the puzzles posed by microscopic life itself, puzzles that it might well have seemed only the microscope could resolve.

What, again, had been discovered? The most popular metaphor set microscopic life apart as a separate "world,"[4] whereas evocations of the metaphysical hierarchy of the chain of being reduced it to a step in the descent toward nothingness.[5] To be sure, it was by no means wholly settled that it was life at all, for even Leeuwenhoek was unable to persuade all his visitors that the swarms in his pepper water were in fact alive, and doubts and cautions persisted throughout the eighteenth century. Be it from the ambition to discover something new, from illusions within the microscope or eye, or from simple inexperience or carelessness, the ease with which the nonliving in the microscopic realm

1 Cole (1930), 16–36.

2 AvL (1694), 587; Joblot (1718), pt. 2:41; Hill (1752a), 105; H. Baker (1753), 231; see also Müller (1786), xviii.

3 [Addison] (1712), 6:132; Fontenelle (1818), 2:43; Thomson (1866), 54–5; Martinet (1778–9), 3:9; Scheuchzer (1731–5), 1:17; Schatzberg (1973), 128, 140, 288, 305; Terrada Ferrandis (1969), 39.

4 See, for instance: B. Nieuwentyt (1717), 268, 738; [John Needham] (1745), 1; Müller (1786), i; Linnaeus & Roos (1767). As often as not, it was also specifically a "new world"; see, for instance: [Effen] (1731–5), 3:210, 301; P. Musschenbroek (1736), 543; Bonnet (1779–83), 1:xxviii–xxix; Gleichen-Russworm (1778), 1.

5 Lovejoy (1936), 60–1; Jones (1966), 29, 85, 106, 111, 148, 159, 177; Schatzberg (1973), 140–1, 325; [John Needham] (1745), 1, 3–4; cf. Hill (1752a), 105.

could be taken for the living was underscored by those both sympathetic and unsympathetic to the claims of microscopic discovery.[6]

Nonetheless, when all was said and done the pervasiveness of microscopic life was generally conceded; but what kind of life it was and how it impinged upon the other world of larger living things remained vexed questions, questions that, for the eighteenth century, pertained more broadly to the understanding of nature and life itself. Laden, hence, with import imposed on them by eighteenth-century observers, images of microscopic life came to bear upon portentous themes that embraced even the much-contested issue of generation. It was no inconsequential role that thus seemed to have been forged for the microscope.

The microorganisms most commonly observed – usually in infusions and hence loosely dubbed *infusoria*[7] – were generally taken to be animals, but animals that challenged the very notion of what an animal was. For no less astonishing than their minute scale were their bizarre forms and capacities that, as Henry Baker wrote, "serve to shew our Ignorance concerning the real Essence and Properties of what we term *Life*." Leeuwenhoek had been astounded not only by (what seemed) the rotifer's rotating wheellike organs – how a living animal could produce such a motion he could not imagine, he wrote[8] – but also by its revival after sometimes even years of desiccation. Whereas Hartsoeker simply would not believe it, Carl Linnaeus later in the century cited the rotifer's resurrection as one of nature's wonders; "who *would* believe it if they hadn't seen it?" But other bewildering surprises were soon to follow, for this was a world, remarked the Dane Otto Frederik Müller, the century's most systematic student of microscopic life, that, with its monsters of extraordinary shapes and unheard-of lives, abounded no less than the distant Indies with strange and marvelous things.[9]

Among the more appealing new "monsters" was volvox. A tiny globe composed of from five hundred to five hundred thousand cells, it is just visible to the naked eye and was described by Baker as no more than a moving point. Perpetually rolling through the water, its constant motion – now quickly, now slowly, left and right, up and down, wrote the German illustrator Rösel von

6 Dimelius (1685), §21; Leclerc (1714–30), 4:165–6; Buffon (1841–4), 3:57; Sellius (1733), 25–6; Linnaeus & Roos (1767), 11; Gleichen-Russworm (1777), 63–4; Wrisberg (1765), 89n; Müller (1786), xxii–xxiii. While Heinrich August Wrisberg alluded to the inexplicable movement of fine particles on the surface of almost every water drop – and his, of course, would not have been the only eighteenth-century allusion to Brownian motion – Otto Frederik Müller noted that the breath of the observer himself evoked a slow and tremulous motion in such particles, as did perhaps the movement of a passing carriage, too distant to be heard. Buffon, on the other hand, warned against the misleading appearances of currents within the drop as well as bubbles in the lens and dust on its surfaces. But see n. 39 below.

7 Rothschild (1989), 278.

8 AvL (1702), 405–6; see also *AB* 7:94–5; AvL to RS, 25 Dec. 1702, AvL Letters, fol. 206v; AvL to RS, 4 Nov. 1704, ibid., fol. 320r–321r, 326r; AvL (1718), 64, 67, 69.

9 AvL (1702), 401–13; AvL to RS, 3 Nov. 1703, AvL Letters, fol. 255r; AvL to RS, 4 Nov. 1704, ibid., fol. 323r–324v; NH (1730), "Ext. crit.," 57–8; Linnaeus & Roos (1767), 6 [emphasis added, to be sure]. (Linnaeus's *Mundum invisibilem* appeared as a disputation under the name of his student Johannes Carolus Roos, but Linnaeus considered himself the author of all such disputations undertaken under his direction [Smit 1980, 118–19].) Müller (1786), i.

Rosenhof – fascinated early observers, and all the more since, as Baker noted, it lacked any sign of head, tail, or fins. Leeuwenhoek, who had already encountered volvox, seems to have doubted that it was in fact alive, but he recognized that its movements would convince others otherwise.[10]

Stranger still was the amoeba. Believing it but a smaller variant of a creature Baker had earlier named *proteus* (later identified as a ciliate, not an amoeba), Rösel von Rosenhof at midcentury first described and illustrated the true amoeba, subsequently characterized by Linnaeus as a mucous point capable of assuming any figure at any moment. Pieter Boddaert in the Netherlands also remarked Rösel von Rosenhof's account of amoeboid fission; lacking parts for generation, noted Boddaert, proteus simply broke into pieces, each one of which was a young new animal. To Boddaert's compatriot Martinus Houttuyn, however, the oddest thing about proteus was that it seemed to be only a membrane or bladder filled with crystalline globules, so that, when the membrane burst, the globules dispersed and the proteus itself disappeared. "Now will you call such a thing an animal?" he asked. The microscope had broached a strange new world indeed.[11]

What did this new world have to do with the old world that encompassed it? How were these bizarre animals related to other, more familiar creatures around and often within which they thrived? Apart from the metaphysics of the chain of being and the metaphor of another world, the answer that came most readily to hand was an ominous one, for physicians were quick to associate microscopic animals with the "worms" of medical tradition.[12] The new observations thus fostered a growing conviction that all one ate or drank and even the air one breathed were filled with either invisible worms or the microscopic "insects" from which the worms (and the infusoria) came.[13] In 1658, reacting to a devastating outbreak of the plague in Naples and Rome, Kircher had also linked such microscopic worms to the commonplace notion of *seminaria* comprising imperceptibly small, self-propagating particles that infected the body with disease.[14]

10 H. Baker (1753), 323–4; Rösel von Rosenhof (1746–61), 3:618; AvL (1702), 156–61.

11 H. Baker (1753), 260; Rösel von Rosenhof (1746–61), 3:622–4 and pl. CI (figs. D, E, F, G, V); Linnaeus & Roos (1767), 18 (concerning Linnaeus's authorship, see n. 9 above); Boddaert (1778), 309–10; [Houttuyn] (1761–85), 18:199–200; Bütschli (1887–9), Bd. 1, Abt. 3, p. 1121; Cole (1926), 16; and see n. 4 above; Elkins (1992). Christiaan Huygens had already described the "entrails" he was able to observe in larger infusoria as only "bubbles" that sometimes seemed to fill their bodies and, in one instance, seemed to roll inside the animal as it sped along (*OCCH* 13:707, 711–12, 714, 716, 721–2, 727–8).

12 See, for instance: Kyper (1654), 197; Bidloo (1715c), 20–4; Paulitz (1696), c. 1, §§6–14; Lipstorp (1687), preface, §§13, 14, 21; Lange (1688), preface, 22.

13 [Craanen] (1685), pt. 1:32; Paulitz (1696), c. 2, §§5, 8, 12, 13; Commelin (1694), §8; Goedaert (1700), 3:118–19; *OCCH* 13:524–5; Gray (1696b), 284; Réaumur (1734–1929), 4:430–1, 433–5; Boerhaave & Haller (1740–4), 6:178–9. See also Wilson (1995), 154–75.

14 Kircher (1658), preface, 29–39, 48–50, 115, 141–7; Fracastoro (1584), 77r–84r, 170r, 177r–179v, 184r–184v. Fracastoro's influence is evident in many seventeenth-century discussions of contagion and its agencies, whether designated *semina*, *seminaria*, or *seminia;* from the Dutch medical literature in particular, see, for example: Diemerbroeck (1685), *T. de peste,* 16, 42, 46; Kyper (1654), 166–7; Jacchaeus (1653), 146–7; Blaes (1661), 156, 172; Dimelius (1685), preface; see also Nutton (1990) and Belloni (1985).

To be sure, the growing disenchantment with spontaneous generation soon began unknitting the nexus of Kircher's doctrine, the linkage of putridity, disease, and worms,[15] but the sense of microscopic danger everywhere was hardly lessened by a compensatory deluge of invisible eggs, "seeds," or "seed" that followed.[16] What amazed Johannes Paulitz of Gdansk as he concluded his medical studies at Leiden in 1696 was that anyone escaped such contagion. In the eighteenth century, while an anxious Hartsoeker steeped himself in the stench of tobacco to ward off the poisonous insects that caused the plague, Réaumur reflected no less somberly on the insects of prodigious smallness that might be ravaging one's insides. Such creatures had perhaps caused more carnage, Linnaeus proposed, than had all the wars.[17]

In forging this fearful perspective on microscopic life, however, the direct experience of microscopic images played only a secondary role. Kircher, it is true, appealed to his own observations with the microscope and especially his ostensible discovery of worms in the blood of fever victims (in fact pus cells, it has been suggested, or rouleaux of adhering erythrocytes),[18] but Leeuwenhoek himself, absorbed as was no other of his time in the most startling images of microorganisms, appears to have resisted the notion they might cause disease.[19] (Even Hartsoeker worried about leaping, stinging, biting, and gnawing microbes he never saw, rather than the infusoria he surely did.)[20] It is safe to say that the proliferation of even wholly derivative allusions to the new observa-

15 Girolamo Fracastoro had also associated putrefaction with contagion (Fracastoro 1584, 81r–81v), and the linkage of worms, putrefaction, and disease retained its compelling logic in the seventeenth century even for many who were otherwise unsympathetic to Kircher's doctrine. (In the Netherlands see, e.g., Diemerbroeck 1685, *T. de peste,* 35–6, 175; L. Jongh 1685, §§16–17; regarding Kircher's prominent German supporters August Hauptmann and Christian Lange, see also Cogrossi 1714, viii–ix.) Nonetheless, although he asserted that physicians indeed unanimously agreed that putridity produced both diseases and invisible worms, Kircher (1658, 57) himself rejected spontaneous generation strictly conceived and insisted, rather, that the worms arose from *seminaria* or *semina* of plant or animal origin that resided in all putrefying matter (ibid., 41, 48–9).

16 For an explicit rejection of spontaneous generation for insect eggs as the source of parasitic worms, see: [Craanen] (1685), pt. 1:31–2; Steenevelt (1697), 14; see also Andry (1700), 10–11, 18–19, 21; Craanen (1689), 83–4; Lipstorp (1687), §§12–14, 18; Mason (1700), 11–12; Paulitz (1696), c. 2, §§5, 12, 13; Steenevelt (1697), 13; Bidloo (1715c), 17–18; Massuet (1733), 104, 111, 114; Boerhaave (1734), 394–5.

17 Paulitz (1696), c. 2, §16, see also §13; NH (1722), 93–4, 328–9, 331; Réaumur (1734–1929), 4: 435–6; Linnaeus & Roos (1767), 22 (see n. 9 above re Linnaeus's authorship).

18 Kircher (1658), 141–2, see also pp. 37–8; Castiglioni (1950), 756n; Singer (1915), 338.

19 Late in Leeuwenhoek's life, Hans Sloane of the Royal Society asked him to investigate whether, as was claimed, there were tiny animals in the fluid exuded from the skin in smallpox and scabies. Although willing to look into scabies if the opportunity arose, Leeuwenhoek doubted the existence of such animals. (Since they were not to be found in the blood, how could they get into those fluids? he asked.) Presumably unaware that the scabies mite had recently been recognized as indeed the source of the affliction (Cogrossi 1714, ix–x), Leeuwenhoek ascribed the belief in an animal in the scabies pustule to the common man – having said which, he also added that "some" were of the opinion that the air in plague time was filled with little animals as well. It was a reference made only in passing and with little evident interest, much less sympathy (AvL to J. Jurin, 7 July 1722, AvL Letters, fol. 296r–297r; AvL to J. Jurin, 13 June 1722, ibid., fol. 285r).

20 NH (1722), 92–7, 329, 331; idem (1710), 59–60; Andry (1700), 345.

tions sufficed for the recasting of traditional notions already arresting – not to say fascinating – by virtue of their own singular frightfulness.

On rare occasions, however, linkages between the two worlds of the microscopic and the everyday were conceived very differently. There had already been those, to be sure, who had proposed that, by consuming dangerous putridity, worms engendered in the body in fact benefited their hosts; but in the mid-seventeenth century, Cyrano de Bergerac offered the far more audacious suggestion that our own flesh, blood, and spirits might be only a tissue (*tissure*) of small animals that, all together, produced the activity called life. Before the century was out, Leibniz was earnestly arguing that, although each living creature had a dominant soul, every part of its body and every drop of its fluid were full of other living things that had souls of their own.[21] But some muffled dissent, a bit of literary whimsy, and fragments of an eccentric metaphysics did not bespeak a notion that could claim a substantial following as yet, and, well into the eighteenth century, the anticipations of a reciprocally beneficent and indeed necessary intertwinement of visible and invisible life were neither as familiar nor as gripping as pernicious worms and morbid *seminaria*. Hence, through the first half of the century, observations of microscopic life seldom evoked thoughts of a shared well-being or mutual dependency.

In the most striking instance of their having done so, however, much derived directly indeed from the suggestiveness of microscopic images in themselves, which inspired some extraordinarily idiosyncratic elaborations. As something of a preliminary, Theodorus Kerckring had already announced in 1670 in Amsterdam that, through his microscope, he had observed various organs of the body to be teeming with minute animals. Since a home, though habitation wore it down, also gleamed most brightly when inhabited, he wrote, it was unclear whether the organs were being preserved or destroyed by the perpetual motion of that microscopic throng. Kerckring's passing thought was but a meager forerunner, however, of the imaginings published some forty years later by Gottfried Sellius of Gdansk.

Then also resident in the Netherlands, Sellius had turned his microscope to the shipworm (*Teredo navalis* Linnaeus), whose assault on the ships and dikes of the Republic had intensified alarmingly. Very likely observing the extensive ciliated surfaces within the worm (in fact a bivalve mollusk) together with the variety of small organisms that infest it both dead and alive, Sellius reported that all its internal organs consisted, indeed, of a dense accumulation of microscopic animals engaged in a variety of sometimes frenetic motions.[22] Having also no-

21 Aldrovandi (1602), 657 (see also Monconys 1665–6, pt. 1:178); Bergerac (1656–61), 127; Leibniz (1875–90), 6:618–19.

22 Kerckring (1670), 178; Sellius (1733), 18–24, 105, 107, 110, 112, 209. Regarding the concern in the Netherlands over the shipworm, see, for instance: Schama (1987), 607–8; C. Belkmeer (1733), 42, 49–51. (Belkmeer also provides another instance of the microscope's being brought to bear on the shipworm's anatomy.)

On the ciliated surfaces in the shipworm, see Lazier (1924), 458–9. Regarding parasites and symbionts within the organism, see Turner & Johnson (1971), 287–90. What has been dubbed the "gland of Deshayes" in shipworms also consists in fact largely of bacteria (Popham &

ticed, moreover, that not only those motions but the profusion and density of the animalcules varied from part to part within the worm, Sellius proposed that it was their number, arrangement, and activities that actually distinguished one organ from another.[23]

Assuming that the animalcules had a reason of their own for roaming in the worm, however, Sellius also inferred that, for their own nourishment, they were in search of the particles of wood the worm devoured – particles that, in their quantity, softness, and minute size, had in effect been prepared for the animalcules.[24] Since they would have also perished with the worm, however, the animalcules had to leave it the nourishment it too required, and nature may even have contrived not only to limit what they consumed, Sellius surmised, but to nourish the worm from what they excreted.[25]

Sellius had been eerily prescient up to this point about the shipworm's symbiotic nourishment,[26] but, apparently responding as well to images of the worm's developing eggs and larvae (the latter brooded in its gills), he elaborated his vision of interdependency still further. In places, he saw the animalcules gathered within numerous elliptical shapes, and among the more remarkable spectacles he thought he observed were animalcules whirling in clusters of small vortices (perhaps the motions of ciliated tissue). Sellius took these forms to be the developing eggs of the shipworm, which, each encompassing hundreds of animalcules, wandered throughout the body of the worm, their rotation and the numbers of animalcules they contained diminishing as the eggs matured (Fig. 26). Such illusory perceptions persuaded Sellius that the eggs of the worm were formed or at least perfected by the whirling motion of the animalcules, and that the animalcules within the egg absorbed or consumed each other until a single, remaining individual became the embryo itself.[27]

Footnote 22 *(cont.)*

Dickson 1973, 338–40), but Ruth D. Turner of Harvard University has advised me (by a letter dated 22 May 1989) that it is unlikely that Sellius managed to observe these bacteria, first identified by means of the transmission electron microscope. (Although still uncertain about the shape of the animalcules he saw, Sellius ventured that he had perhaps discerned an oblong figure smaller than spermatozoa but larger than the creatures – presumably the bacteria first observed by Leeuwenhoek – that were found in the matter adhering to one's teeth [Sellius 1733, 23, 24–5].) I am indeed most grateful to Prof. Turner for her comments on my abbreviated rendering of Sellius's account, and my text above continues to reflect her suggestions as to what Sellius may or may not have seen.

Kerckring may also have observed the microorganisms that flourish on disintegrating tissue or been misled by still active ciliated surfaces, but illusions associated with early microscopes perhaps played a role as well; see Belloni (1962), 65–73; Zanobio (1971), 29, 32–4.

23 Sellius (1733), 22, 107, 209.

24 Ibid., 104–5, 194.

25 Ibid., 108.

26 The bacteria composing the gland of Deshayes (see n. 22 above) appear indeed to fix nitrogen for the shipworm and to provide the enzymes that enable it to digest cellulose (Waterbury, Calloway, & Turner 1983). Nitrogen-fixing bacteria have also been found in the gut of the shipworm (Carpenter & Culliney 1975).

27 Sellius (1733), 18–23, 112, 140–3, 148.

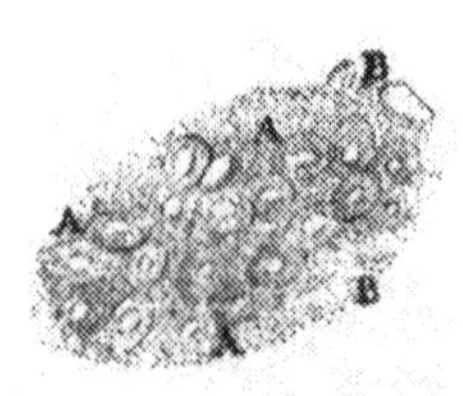

Figure 26. Sellius's depiction of the "vortices" of animalcules in the process of forming themselves into the shipworm's eggs (Sellius 1733, pl. 2, fig. 7). The image does very likely reflect Sellius's observation of the eggs of the shipworm at one point or another in their development (personal letter from Ruth D. Turner; see n. 22 of this chapter). Photo courtesy of the Department of Special Collections, University of Chicago Library.

Although it surpassed the reach of our imagination, Sellius declared, his observations had disclosed a new "end" or purpose of the worm: to provide a world, as it were, created particularly for the animalcules within it. Hence, the worm existed not only for its own sake but for its numberless denizens as well, providing them with domicile and sustenance. However, as the microscopic animalcules needed the worm to provide them food, he concluded, so also the worm needed the animalcules to propagate, and thus the one hand washed the other.[28]

The building of larger organisms from living microscopic particles was taken up in a grand way midcentury in sweeping speculative systems of which the most influential – and notorious – was that elaborated by the (eventual) comte de Buffon, presiding over the Jardin du Roi in Paris.[29] Incorporating the notion into a universal theory of generation, however, Buffon degraded the life in those particles and rejected the very thought-world of ends and purposes in nature. Determined to grant matter its own generative power and to eschew the recourse to divinity in natural science,[30] he elaborated a doctrine that bore the mark of the radical materialism of the Enlightenment, and, insisting that the mechanical philosophy be broadened to accommodate "penetrating forces" that produced organized bodies from matter,[31] he bent the observations of microscopic life to his purposes. Buffon's theory, however, was but the most elaborately developed instance of a more general inclination – more general even than that of

28 Ibid., 179, see also p. 194.

29 Shirley Roe indeed contends that nearly every commentator on biological matters for a generation after the appearance of Buffon's system shaped their views in direct response to his (S. Roe 1992, 439). In addition to Buffon's system, however, see also: Réaumur (1751), 2:363–4; Bonnet (1779–83), 5:207–8; Müller (1773–4), 21–2; Oken (1805), 22–3, 123, 194–5.

30 Roger (1971), 539, 541–2, 546, 556, 581, 584.

31 Buffon (1841–4), 3:17–18, 20–1.

materialism – to restore nature's dynamism, a restoration in which images of microscopic life were indeed deeply implicated.[32]

Buffon proposed that plants and animals were composed of minute and enduring "living organic parts"[33] composed themselves of yet simpler and indestructible "living organic molecules"[34] of an uncertain but wholly natural origin – perhaps the penetration of atoms of light into atoms of air, earth, or water, he at one point suggested.[35] Like the more complex "living organic parts," which were also gathered and reassembled in the process of reproduction, the molecules were particularly abundant in the *germes* of plant seeds and in the seminal fluids of animals, and in those latter fluids (produced ostensibly in both males and females) they clustered together to form spermatozoa.[36] Incapable, however, of producing anything or developing any further, the spermatozoa were themselves "living organic parts" akin to those that composed the substance of plants and animals and that appeared as infusoria in infusions where such substances decayed.[37]

Buffon had thus set out to argue that complex organisms could arise from simple corpuscles of matter; transforming the infusoria and other microscopic beings into early stages in the progress of organization, however, he diminished the quality of life and animality ascribed to them. Kindred to the spermatozoa, as Buffon maintained they were, the infusoria were only initial assemblages of organic molecules and hence too simple in structure to be animals themselves.[38] His own observations of the infusoria (and spermatozoa) he read accordingly, seeing the strange subtleties of microscopic movements as devoid of the willful activity that specifically betokened animal life.[39]

The infusoria were decidedly "living," nonetheless, and critical to Buffon's argument were also microscopic observations that pertained to their ostensible origins. Indeed, the materialists of the radical Enlightenment were eager to re-

32 Roger (1980), 275. Quite independently of the materialist and vitalist trends of the period, however, the ubiquitous activity and reproductive exuberance attributed to microscopic animalcules filled nature with another kind of dynamism as well. With reference to the presumed pervasiveness of microscopic life, a wide variety of authors continued to exclaim more or less with Fontenelle that "everything is alive, everything is animated": Fontenelle (1818), 2:43; Thomson (1866), 54–5; Martinet (1778–9), 3:9; Uilkens (1816), 21; Schatzberg (1973), 288. The tinier the size of the typical member of an animal species, it was also often maintained, the greater the fertility of the species: NH (1722), 334; Boerhaave & Haller (1740–4), 3:638; Swaving (1799), 59; Wilkinson (1984), 145.

33 Buffon (1841–4), 3:8, 14–15, 84–5, 99.

34 Ibid., 3:99–100, 120–1.

35 Ibid., 3:121–2; Roger (1971), 549–50.

36 Buffon (1841–4), 3:8, 20, 84–5, 87, 100.

37 Ibid. 3:20, 84–5, 87.

38 Ibid., 3:73, 84–5, 87, 99.

39 Ibid. 3:99; see also pp. 87–8 regarding the spermatozoa. Eighteenth-century observers repeatedly sought to distinguish the microscopic motions that signified life from those that did not, but it remained a very subjective judgment, and one that circumscribed life too narrowly as often, indeed, as it granted it too freely. See also: Verheyen (1710b), 68–9; [Houttuyn] (1761–85), 18:203–4, 19:97; Sellius (1733), 26; Gleichen-Russworm (1777), 30–1, 61; idem (1778), 7–8, 97–8; Miles (1741), 727–8.

vive the cause of spontaneous generation, and, committed to forces in matter that produced as many organized living bodies as possible, Buffon emerged as its most conspicuous new champion. Convinced that it was the most frequent and universal form of generation, he embraced it as fundamental to his general theory.[40] Hence, the formation of infusoria from organic molecules freed from decomposing plant or animal matter represented a conception of nature upon which he was determined to insist. Turning to microscopic observations of infusions, then, he found the ostensible evidence he was seeking.[41]

The awareness of the infusoria had helped shape Buffon's materialist bent into a developed theory, and observations offered seeming confirmation; but the seminal impact of the actual images of microscopic life was more pronounced in other observers. For assistance with the microscope and infusion experiments, Buffon had turned in 1748 to John Turberville Needham, an English priest who, having already acquired a reputation as a microscopist, was conveniently resident in Paris. Buffon laid out his theory for Needham at the outset, and, although the latter was emphatic in his rejection of materialism, significant affinities emerged in their thinking. Whatever their other differences, Needham affirmed, they agreed on a "real productive Force" in nature.[42] Needham, however, set about building a theory of generation of his own, and his close scrutiny of extraordinary microscopic images gave his speculations a different cast.

Following his collaboration with Buffon, Needham described – and illustrated – spontaneous generation in infusions as he himself saw and understood it. It was plain, he wrote, that, with the evaporation of salts and volatile parts, the organic substance in the infusion became softer, more divided, and attenuated. To the naked eye it appeared gelatinous, but the microscope revealed that when the substance was at its "highest Point of Exaltation" – just breaking into life, "as I may say" – it consisted of innumerable filaments that, swelling from active and productive internal forces, were "perfect Zoophytes" – living, that is, but definitively neither plant nor animal – "teeming with Life and Self-moving." Rays that moved with "extreme Vivacity" shot out from small particles and from extremities of the gelatinous substance that were "active beyond Expression, bringing forth, and parting continually with, moving progressive Particles of various Forms, spherical, oval, oblong, and cylindrical, which advanced in all Directions spontaneously. . . ." These were the "true microscopical Animals," he asserted, that naturalists had so often observed.[43] The microscope seemed to

40 Farley (1977), 28–9; Buffon (1841–4), 3:12–13, 17–18, 20–1, 112, 120; cf. Roger (1971), 557.

41 Buffon (1841–4), 3:84; Roger (1971), 543, 552, 556.

42 S. Roe (1983), 159–84, see esp. pp. 160, 171–7; John Needham (1748), [618–19], [633], [644–5], 665; Buffon (1841–4), 3:84. Questioning Leeuwenhoek's doctrine of the spermatozoa, Needham had already proposed three years before that, "if we make the Wisdom of the Almighty, as far as we can possibly trace, our Standard in this Particular," it was as reasonable to assert that the fetus was generated "from a Lifeless Point of Matter" as from an animalcule ([John Needham] 1745, 83). Needham and Buffon were at one as well in their firm rejection of preexistence theory: John Needham (1748), [618–30]; Buffon (1841–4), 3:157.

43 John Needham (1748), [644–7], [653–4].

have now revealed the very process of spontaneous generation, whose overthrow had been celebrated not long before as one of the conclusive achievements of seventeenth-century science.

To see such images in this way required a ready imagination, to be sure, but the images exercised an influence of their own in forging an articulated theory. Needham perceived what apparently included growing molds and their motile spores as a changing, amorphous mass of matter that "resolved into" or "shed" a variety of moving shapes and particles. He also spoke of the emerging particles or creatures as the product of "Vegetation" and ascribed the generation of infusoria to a "vegetating Force" residing in every microscopic point of animal and vegetable substances.[44] At a minimum, hence, his observations inspired a far greater emphasis on a specific vivifying force pervading an amorphous matter than on the accumulating corpuscular constructions that had been Buffon's point of departure. In Needham's own words, he opposed a "real Vegetation" to Buffon's "Evolution of organical Parts." Buffon doubtless had good reason to remark the influence of his ideas on Needham – "the ideas I gave him on this subject have borne more fruit in his hand than they would have done in mine," he wrote of Needham's infusion experiments – but microscopic images decisively shaped Needham's conceptions as well. The gifted Italian naturalist Lazzaro Spallanzani ascribed Needham's theory to his attachment to a hypothesis, to be sure, but also to illusions.[45]

Redolent of materialism, despite its author, and tainted as well by his collaboration with Buffon, Needham's theory found few sympathizers beyond those who, like Diderot, shared Buffon's materialistic inclinations.[46] Nonetheless, other observers were also now persuaded by confusing observations of both fungi and infusoria that some kinds of fundamental transmutations were indeed occurring in organic substances at the microscopic level. In 1766, Baron Otto von Münchausen described how the "blackish dust" (the spores) of grain smut and other fungi swelled when placed in warm water for several days and changed into agile little animals (very likely now indeed infusoria). In time, however, these animals seemed to form firmer masses with white veins through which "polyps" moved (a reading, perhaps, of moving nuclei or protoplasm in the hyphae), all of which constituted the beginning of a new fungus. Having called

44 Ibid., [644–6], [648–50], [653].

45 Ibid., [644]; S. Roe (1983), 164, 166, 168ff.; Buffon (1841–4), 3: 84; Spallanzani (1786), 2:36–8, 40–1, 95–6; see also idem (1934), 2:579–80. Whereas Buffon's own accounts did indeed at times echo Needham's descriptions (Buffon 1841–4, 3:100, 156), Needham noted that deciding between their two views – meaning presumably the possibility of confirming Buffon's corpuscular imaginings – probably lay well beyond the microscope (John Needham 1748, [645]).

46 S. Roe (1983), 182. Like Buffon's doctrine, Needham's accounts also easily provoked religious and philosophical biases. Despite what was at best in fact only a passing familiarity with those accounts, Voltaire did not hesitate to attack Needham bitterly as an advance man for atheism (S. Roe 1985), but Diderot took a contrary stand: In the dream that Diderot fabricated for him, d'Alembert contemplated an infusion through a microscope and exclaimed that Needham ("l'Anguillard," an allusion to Voltaire's hostile mockeries) was right: "I believe my eyes; I see them; how many there are! How they go! How they come! How they wriggle!" (Diderot 1875–7, 2:131). See also Farley (1977), 28–9.

attention to Münchausen's account, Linnaeus himself now also observed how the "seeds" of fungus moved about in the water like fish until, "by a law of nature thus far unheard of and surpassing all human understanding," they changed back from animalcules to the immobile beginnings of future fungi again.[47]

Such was the origin of the Linnean genus *Chaos* (in the class Vermes) that in addition to the supposed animals arising from smuts and other fungi – *Chaos ustilago* and *Chaos fungorum* – included the infusoria, which Linnaeus elsewhere suggested were the "living seeds" of mold. Confusing observations of microscopic organisms had thus lodged the infusoria in a realm of organic indeterminateness and mutability where the boundary between plant and animal life was not only problematic but often breached. It was a realm that, inspired by microscopic images, reflected and encouraged a revived readiness to believe in fundamental transmutations of organic substance.[48]

That readiness was also revealed in observations of infusions uncomplicated even by any obvious intrusions of fungi. The year before Münchausen's account, Heinrich August Wrisberg had also described what he took to be the process by which the decaying animal and vegetable matter in infusions produced infusoria. The first change to be seen in new infusions, he related, was the appearance of microscopic molecules or vesicles and air bubbles that collected at the surface. Identified by Wrisberg with similar entities (though now explicitly "*in*organic") composing plant and animal parts, the molecules separated from the vegetable and animal particles in the infusion; forming a skin over the infusion, they gathered into clumps that further motion and agitation then formed into what, because of their movement, were called infusion animalcules. Fully aware of the debate that was swirling around Needham and Buffon, Wrisberg was hence prepared and obviously willing to observe such a transmutation, but he insisted that his own conclusions were drawn from observations not hypotheses.[49] At a minimum, easily misconceived images facilitated the seeming confirmation of the thought.

Guided even if reluctantly by such accounts, even astute observers initially ill-disposed to organic mutations in infusions could also be persuaded by such images. Müller, the century's most systematic student of microscopic life, rejected both a (presumably Linnean) "chaotic" realm and Buffon's theory of the origin of infusoria, the former as the product of inadequate observation and the latter,

47 Linnaeus & Roos (1767), 2, 11–14 (re Linnaeus's authorship, see n. 9 above); Ainsworth (1976), 23. Ernst Almquist points out that the fungus of grain smut does not produce motile forms but that numerous infusoria and bacteria do soon appear in an impure culture of the spores (Almquist 1909, 162). Regarding the moving polyps in the hyphae, Münchausen had already likened the partially transparent substance of the spores to the substance of the hydra, the prototypical "polyp" of the age (Linnaeus & Roos 1767, 13; concerning the hydra, see later in this chapter).

48 Linnaeus & Roos (1767), 12 (see n. 9 above); Linnaeus (1766–8), 1:1326–7; Ainsworth (1976), 23–4. Although Gleichen-Russworm doubted Münchausen's account of spores becoming animals that then became fungus again (Gleichen-Russworm 1778, 101), he himself had earlier reported that pollen grains sealed for a day or more in water became infusoria (idem 1764, 30).

49 Wrisberg (1765), 82–9; see also Gleichen-Russworm (1778), 75.

of pure speculation. Nonetheless, he too in time noted a similarity between, on the one hand, the "crystalline globules" that derived from and then produced new fungi and, on the other, the animalcules moving about the film of vesicles produced by decomposing animal and vegetable substances in infusions. He had once rejected any identity between those animalcules and the similar looking but unmoving vesicles of the film itself; but prompted, he wrote, to look more closely through his microscope, he now saw what Needham and Wrisberg had seen before him: globular vesicles (or tiny points) separating from that film and, trembling, gradually beginning to move and run about. He now described, indeed, how animal and vegetable fragments decomposed into microscopic vesicles or globules that, like the crystalline globules of the fungi, loosened from the larger mass and produced the smaller and simpler infusoria. Nor were his broader speculations on the significance of these observations now without their echoes of Buffon.[50]

Though he had doubted them, the descriptions furnished by prior observers had prepared Müller for the recognition of what he had denied before, and antecedent ideas thus again imposed significance on uncertain and hence malleable microscopic images. However, if we may credit his own account, it was the images themselves that, studied more closely, ultimately persuaded him against his own initial inclinations that microscopic particles of decomposing matter were truly becoming animals after all. Indeed, Needham, Münchausen, Linnaeus, Wrisberg, and Müller apparently all remained convinced that they had accurately related what they had seen and that observations, indeed, not hypotheses had been decisive. Thus microscopic images were made to reveal a resurgent dynamism in nature that late-seventeenth-century mechanism had ultimately suppressed. By the same token, the instrument, the microscope, seemed to have at last fulfilled its promise and provided unique access to a profound and fundamental truth about nature and its processes.

The acceptance of microscopic transmutations and even the perception of an apparently manifest dynamism in these changes were abetted by the seemingly simple transparency of so many forms of microscopic life. Strikingly at odds with the earlier stress on the complexity of even the smallest animals – an emphasis paradoxically also associated with the microscope – this diaphanous simplicity was an aspect of the wondrousness of the newly discovered realm of life. Persistent and pervasive, however, the images of that simplicity became insistent, and other celebrated eighteenth-century studies in natural history encouraged microscopic observers to consider those images more seriously. These studies, moreover, entailed further associations with a truly extraordinary dynamism, and hence deepened the impact of such images on the shifting eighteenth-century understanding of nature at large.

In the 1730s, Leiden's influential champion of Newtonian physics Petrus van Musschenbroek still posited not only muscles, nerves, and blood vessels but even

50 Müller (1786), vi–vii, xxi–xxv; idem (1773–4), 1:20.

blood corpuscles in creatures a billionth the size of a grain of sand; but to increasing numbers of eighteenth-century observers, the transparency of the infusoria – at times like crystal, noted Buffon – seemed in fact to testify stubbornly to an absence of any meaningful internal structure. Pieter Boddaert included both proteus and volvox among microscopic animals that seemed to lack organic parts, and Linnaeus suggested that the bodies of the "chaotic" animalcules were only "marrow" (*medullaria*) with little or no organized construction. Müller defined the *Infusoria* proper (in contrast to the *Bullaria,* also largely microscopic but with internal as well as external parts) as gelatinous, homogeneous, and lacking any division of parts, and he characterized the smallest and simplest as no more than a gelatinous point.[51]

The willingness to credit the images of apparent simplicity – to abandon, that is, the assumption of a still unseen complexity – bore upon how microorganisms were conceived as forms of life. Buffon cited the simple appearance of some infusoria in arguing that, though living, they were nonetheless no more than primitive "organic parts," whereas Müller, conversely, distinguished the visibly more internally structured microorganisms (the *Bullaria*) from the simpler *Infusoria* by still ascribing the former to parental propagation.[52] In doing so, both Buffon and Müller attested the implicit significance they attached to the images of transparent homogeneity in microscopic observations; those images bespoke a simpler level of life and the attendant possibility – or conceivability – of radical transmutations.

The images of course had always been there – witness Leeuwenhoek's frustrated search for the innards of the spermatozoa – but their credibility as truthful images was on the rise. Doubtless, their unyielding persistence took its toll in time, but so also did the aggressive new advocacy of a dynamism in matter. That advocacy exploited those images, to be sure, but at the very least it in turn drove home the point that the assumption that life always necessarily entailed complexity could by no means be taken as self-evident.

A very distinct series of celebrated observations and discoveries in natural history also vested images of transparent protoplasm with powerful new associations. Indeed, the most notorious of the strange new creatures explored with the microscope in the eighteenth century was not in itself truly microscopic. Observed by Leeuwenhoek and others at the beginning of the century, the freshwater hydra was rediscovered in 1740 by Abraham Trembley, a Swiss tutor in Holland who now also disclosed properties of the hydra that were so startling – "so singular, so contrary to the generally accepted ideas about the nature of animals," Trembley himself attested – that they made his experiments among the

51 P. Musschenbroek (1736), 27 (and see, among other influential writers in the Dutch Republic in particular who continued to echo the assumption of the complex anatomy of microorganisms, Lyonet in Lesser 1745, 2:91n; B. Nieuwentyt 1717, 735; Massuet 1752, 1:165; Martinet 1778–9, 3:33); Buffon (1841–4), 3:99; Boddaert (1778), 309–10; Linnaeus & Roos (1767), 23 (re Linnaeus's authorship, see n. 9 above); Müller (1786), vii, 2; see also Wrisberg (1765), 5. Eighteenth-century observers did at times describe such structures as the contractile vacuole, the nucleus, and nucleolus (Hughes 1955, 4–5; Cole 1926, 16).

52 Buffon (1841–4), 3:99; Müller (1786), vii–viii.

most exciting in eighteenth-century science. Here is nature surpassing our fancies, remarked the *Histoire de l'Académie Royale des Sciences;* while, to Trembley's compatriot Bonnet, the creature Réaumur dubbed the *polyp* bordered on another universe.[53]

In addition to its ability to survive appalling distortions, the most astonishing property of the hydra was the regenerative power of its fragments. No other animal had ever so occupied the naturalists, declared the German microscopist Wilhelm Friedrich von Gleichen-Russworm in 1778. "We cut them with scissors and knives, turned them inside out like a stocking or glove, and stuck one within the other," and yet from their ruins still arose new offspring. "These are Truths," wrote Baker of the rise of new hydras from their parent's pieces (and of both the revival of desiccated rotifers and Needham's derivation of "eels" – nematodes – from supposed grains of smutty wheat), "the Belief whereof would have been looked upon some Years ago as only fit for *Bedlam*."[54] These "truths" fascinated not only naturalists, moreover, for they were seized upon as well by imaginative materialists intent on dramatizing nature's generative powers: Diderot fantasized in d'Alembert's fictitious delirium about human "polyps" on Jupiter or Saturn that split into an infinite number of "atomic men" that then metamorphosed and escaped like butterflies, so that a whole province could be peopled from the fragments of a single man.[55]

Wondering whether the imperceptible parts of an animal with such capacities would prove unusual as well, however, Trembley had also probed the bodily substance of the hydra with a microscope. (The specimens with which he worked, from the species *Hydra viridissima* Pallas, *H. vulgaris* Pallas, and *H. oligactis* Pallas, were themselves generally at least over a centimeter long and hence readily visible to the naked eye.) What he observed proved even more surprising than he anticipated, for the hydra's astonishing capacities resided in a bodily substance that appeared to be virtually devoid of any organized internal structure. Conceiving of its body as simply a single layer of "skin" rolled into a tube or intestine – all stomach, remarked Bonnet – Trembley studied that skin repeatedly under the lens to discover of what it was composed; but all he could detect were small granules in a glairy substance that helped hold the granules together, and very loosely at that, for he saw the granules dispersed by water currents. The granules had to be bound together by something more, he believed, but his search for that something more, which he expected to be vessels, remained unavailing.[56]

53 AvL to RS, 25 Dec. 1702, AvL Letters, fol. 207r–208v; Trembley (1744), 1, 3, 20, 150; Académie Royale des Sciences (1744), 34; Bonnet (1779–83), 8:181; J. Baker (1952), 36–7.

54 Gleichen-Russworm (1778), 88; H. Baker (1753), 256; John Needham (1743), 640–1; [idem] (1745), 85–6. Needham was in fact observing the nematode *Tylenchus tritici* (J. Baker 1952, 166–7).

55 Diderot (1875–7), 2:130.

56 Trembley (1744), xii, 7, 26, 53–8, 60, 131, and see pl. I (figs. 2, 3); Bonnet (1779–83), 7:112; J. Baker (1952), viii–ix, 13–17, 34, 54; Grayson (1970–1), 437–9, 441. The "granules" were apparently Zoochlorellae in *Hydra viridissima* Pallas, carotene granules in the brown species, and nematocysts on the tentacles (J. Baker 1952, 54, 56, 71).

Its lack of discernible internal structure did not prove as exciting as the hydra's other properties, to be sure, but it made an impact nonetheless. Though not ruling out the possibility of invisible vessels, Bonnet acknowledged that the structure of the hydra seemed to be the simplest of any animal known, and Buffon, who also noted that some animals appeared to be only coagulated glair, agreed that the organization of the hydra was of the very simplest kind. Haller, in denying Boerhaave's doctrine that all membranes were composed of vessels, cited the many small animals bereft of organs and offered the microscopic observations of the hydra as his example. Later, at the turn of the century, the French zoologist Baron Georges Cuvier described the body of the hydra as a homogeneous and gelatinous *pulpe* with no visible organization.[57]

In addition to its more notorious capacities, consequently, the hydra became an exemplar of active life inhering in a soft and radically simple body; as such, it rendered the apparent simplicity of many infusoria more credible as well. An affinity between the hydra and many infusoria (and, for that matter, fungus spores) was now indeed widely assumed, and, though their similar transparency, pliability, and apparent simplicity suggested it in the first place,[58] that assumed affinity reciprocally argued the reliability of those images in the infusoria. The body of the similar but larger analogue, after all, had been closely searched.

As an exemplar and analogue, moreover, the hydra also associated such living simplicity with both an irrepressible fecundity and tenacious vitality. A common simplicity in itself facilitated the acceptance of a transmutation of transparent, unorganized specks of decaying matter into transparent, unorganized specks of animal life, and, in the light of the hydra's celebrated capacities, who could confidently deny such viscid specks the possibility of extraordinary vital powers of their own?[59]

With the constraints on the imagination thus loosened, the expectations – and apprehensions? – of observers worked more freely on uncertain images of shifting shapes and agitated particles, more easily perceived now as vested with innate motion and twitching perhaps with the first hints of animate life. The interplay of images and ideas thus encouraged a growing number of prominent naturalists to accept the spontaneous generation of infusoria, and they continued to do so well into the following century.[60] More broadly, the consciousness of

57 Bonnet (1779–83), 7:112–13; Buffon (1841–4), 3:16, 86; Haller (1757), 1:24; Cuvier (1800–5), 2:362.

58 Cuvier (1800–5), 2:362; Gleichen-Russworm (1778), 88–91, 101; Bonnet (1779–83), 8:221–2; Wrisberg (1765), x–xi, 95; Linnaeus & Roos (1767), 13, see also p. 14.

59 There were those, indeed, quite ready to extend the hydra's generative dynamism far more broadly in the microscopic world. Münchausen proposed that polyplike creatures (*creaturae polypiformes*) provided the beginnings of all fungi and perhaps every "fermentation" (*fermentatio*) as well, a prospect Linnaeus was apparently fully prepared to take seriously (Linnaeus & Roos 1767, 14; re Linnaeus's authorship, see n. 9 above). At the beginning of the next century, Lorenz Oken's strange theory of generation identified the polyp with the "male principle" in nature in general (Oken 1805, 116, 155).

60 Farley (1977), chap. 3.

the ubiquity and inconceivable abundance of microscopic life – recognized as one of Leeuwenhoek's legacies[61] – turned those images into tokens of a more general dynamism underlying nature virtually everywhere.

It was a dynamism with great suggestive potential. At the beginning of the century, the imaginative French diplomat Benoît de Maillet had cited the emergence of microscopic life in grass infusions to bolster a theory of invisible and indestructible *semences* from which, in the retreating waters that had once covered the world, plants and animals had first similarly arisen. A century later, Erasmus Darwin rested his rendering of the theme on an altered understanding of microscopic life. Echoing a considerable history now of microscopic observations (and citing Müller's classifications *Monas, Vibrio, Proteus,* and *Vorticella,* as well as "mucor-stems" of mold and the droplet lens, the "crystal sphere," which Müller had also doubtless used),[62] Darwin proceeded to his own rhymed speculations on the rise of man:

> So, view'd through crystal spheres in drops saline,
> Quick-shooting salts in chemic forms combine;
> Or Mucor-stems, a vegetative tribe,
> Spread their fine roots, the tremulous wave imbibe.
> Next to our wondering eyes the focus brings
> Self-moving lines, and animated rings;
> First Monas moves, an unconnected point,
> Plays round the drop without a limb or joint;
> Then Vibrio waves, with capillary eels,
> And Vorticella whirls her living wheels;
> While insect Proteus sports with changeful form . . .[63]

Adhering to the spontaneous generation of infusoria, Darwin moved it to a primordial setting beneath the sea:

> Organic Life beneath the shoreless waves
> Was born and nurs'd in Ocean's pearly caves;
> First forms minute, unseen by spheric glass,
> Move on the mud, or pierce the watery mass;

whence more elaborate forms of life now evolved:

> These, as successive generations bloom,
> New powers acquire, and larger limbs assume;

61 Schatzberg (1973), 288.

62 Müller (1786), xviii. Wrisberg also apparently used a high-powered single lens for his more demanding observations (Wrisberg 1765, 17, cf. 8n), whereas Buffon and Needham used a compound instrument (Buffon 1841–4, 3:56–7), which Gleichen-Russworm cited as a source of their errors, at least regarding spermatozoa (Gleichen-Russworm 1764, 9).

63 Maillet (1748), 2:223–4; idem (1968), vii, 5, 15; Darwin (1803), 25–6, see also "Additional Notes," 9–10.

so that in time man himself arose

> . . . from rudiments of form and sense,
> An embryon point, or microscopic ens![64]

Such were the implications the eighteenth-century vested in microscopic life and the succession of observations that began in particular with Leeuwenhoek. Shifting and jostling throughout the century, contending versions of what that life in fact amounted to often in retrospect seem naïve and at times hopelessly misguided, but they pertained to enormous issues: the very conception of what an animal was, the interaction between different but overlapping dimensions of life, the origin of organized living things, and the autonomous dynamism of nature. These were impressive themes. Nonetheless, despite such implications – and the expectation, indeed, that so much was yet to be learned from the infusoria[65] – the eighteenth-century failed to develop the study of microscopic life into a sustained and integrated field of research. Its pursuit remained sporadic, and its contending interpretations produced little in the way of progressively developing new lines of investigation.

To be sure, new efforts and new discoveries were not lacking. The first treatise devoted specifically to microscopic life appeared in Paris in 1718 in Louis Joblot's *Descriptions et usages de plusieurs nouveaux microscopes*. As the century progressed, new internal structures were observed (Joblot first described the contractile vacuole and Trembley and Müller depicted the macronucleus in protozoa), new behavior reported (such as reproduction by fission and the formation and discharge of food vacuoles), and new species discovered.[66] The effort was also begun to classify them. In 1752, Hill made what was, by his own account, the first attempt to name the new world of microscopic animals and to incorporate them within a broader natural history of animal life. Linnaeus also included them within his taxonomic system, and, by his death in 1784, Müller had classified 379 species of "infusoria" (largely protozoa, but including as well a mix of bacteria, diatoms, nematodes, planarians, rotifers, and other sufficiently tiny odds and ends of life). "What labor! What patience!" exclaimed Bonnet of Müller's efforts.[67]

The new discoveries had come only sporadically, however, and although many of Leeuwenhoek's observations of protozoa were confirmed – often unwittingly, since his priority frequently went unrecognized – clear descriptions of bacteria became a rarity. Except for Müller's, the efforts at classification also

64 Darwin (1803), 26–7, 28.
65 Gleichen-Russworm (1778), 74.
66 Cole (1926), 15; J. Baker (1948–55), 90:99, 94:421–3; idem (1952), 155–62; Lechevalier & Solotorovsky (1974), 387; Corliss (1986), 477; see also idem (1978–9), 98:50–1.
67 Hill (1752b), preface; Linnaeus (1766–8), 1:1326–7; Müller (1786); Cole (1926), 18; Corliss (1986), 476; Bonnet & Haller (1983), 1135. (Bonnet, who in fact believed the classification of microscopic life to be virtually impossible [Bonnet 1779–83, 8:222–3], was still referring to Müller's earlier and less fully developed classification in Müller 1773–4.)

left much to be desired. Rejecting many earlier descriptions of microorganisms as fantasy, Hill reduced the number of different species discovered by mid-century to thirty-five, while Linnaeus's results were not only also very limited but particularly confused, reflecting little serious microscopic research of his own. Müller's achievement hence stands alone among eighteenth-century studies of microorganisms, and, since he lacked immediate heirs of significance, such efforts had come to a halt by the opening decades of the following century.[68] Even in the face of such exceptional forms of life, wonder – much less simple curiosity – had its limits as a stimulus to research.

Nor did apprehensions over microscopic life as pathological give rise to a sustained and focused effort. At the very end of the eighteenth century, a Dutch observer of the infusoria remarked, to be sure, that the numberless animals living in the diseased bodies of men and animals loomed as a wide field for research,[69] but any serious interest in such a field appears in fact to have faded during the course of the century. What had perhaps been the only proposal for a systematic, institutionalized study of microorganisms as a cause of disease appeared in an anonymous Parisian pamphlet in 1727 as part of a brazen hoax.[70]

The notion of a mutually beneficial intertwinement proved no more stimulating, even though fashionable clichés provided an edifying rationale for pursuing such a theme. That nothing was created uselessly and without its purpose was an eighteenth-century commonplace, as was the further elaboration that the individual ends of individual things were interwoven in a grand fabric that sustained the whole of creation. In the Netherlands in particular, influential spokesmen and popularizers of the new scientific trends called for the systematic study of the purposes and ends of things, a separate field of science that Petrus van Musschenbroek called *Teleologia*. All things are necessarily bound together, he said of its subject matter, and, since nothing is useless, whatever exists benefits the entire world.[71]

Sellius had indeed conceived his study of the shipworm and its microscopic denizens in just such terms. Harmful though the shipworm was to man, he wrote, it too had been created for the sake of certain ends intended for the preservation and perfection of the world,[72] ends that involved the animalcules that lived within the worm as well. The habitation and sustenance they found there constituted in itself, for Sellius, the discovery of yet another purpose of the worm,[73] while, beyond even their reciprocal role in sustaining their host, the animalcules also collaborated in fulfilling some of the worm's other ends: They

68 Dobell (1932), 375–82; Bulloch (1938), 29, cf. 78, 171–2; Hill (1752b), 1–12; Corliss (1986), 476; Almquist (1909), 169.

69 Swaving (1799), 82–3.

70 M.A.C.D. (1727), 9. This pamphlet followed another in a similar vein (idem 1726). Regarding the hoax and its author, see Belloni (1961), 582–6.

71 P. Musschenbroek (1762), 1:2–4; Bots (1972), 29–32; see also NH (1708), 123; Martinet (1778–9), 1:22; Lussingh (1783), 77; Porjeere (1783), 35; Schatzberg (1973), 326; Roger (1971), 242–9.

72 Sellius (1733), 173, 190, 207.

73 Ibid., 179, 215.

fed multitudes of little fish with the superabundance of eggs they made for the worm, and, though Sellius failed to make the connection explicit, they obviously joined in consuming the excess wood in the sea.[74] Tying the symbiosis within the worm to the broader well-being of the sea, Sellius demonstrated how the inclusion of microscopic life could add another dimension to nature's complex weave of integrated purposes.

Pious incentives plumped for such research as well. Understood almost universally (by those, that is, who acknowledged any purposes or ends in nature at all) as a further revelation of the wisdom and purposefulness of God,[75] the study of the ends of nature's creatures conformed indeed to a popular genre of natural theology,[76] and Sellius was again in step. Not only did the recognition of the shipworm's ends defend the universe from being ascribed to chance, he urged, but contemplating how the worm achieved its ends showed how the hand of God could be recognized even in the most detested animals – and in the smallest animals as well, he might have added.[77] Neither as pious instruction, however, nor as an insight into how nature was knit together did Sellius's effort – or any others like it, if such there were – give rise to a persisting line of study into how microscopic life interacted with the larger world in which it was embedded.[78] Where fear of pestilence had failed, was a taste for edification likely to do much better?

But if such promptings failed to motivate pious and more conservative naturalists to commit themselves to an enduring program of microscopic research, the prospect of reshaping the understanding of nature along materialist lines proved no more inspiring to the more radically inclined. Turning to the infusoria as a manifestation of matter's dymanism at work, eighteenth-century materialists and their perhaps reluctant allies recast spontaneous generation in the image – the eighteenth-century image – of modern science, with its aspiration to the generality and consistency of laws and its appeal to systematic observation and experiments. Appealing to Newtonian gravity ("the penetrating force of weight") as an analogy for the organizing forces he posited in nature, Buffon insisted that his theory revolved around emphatically "general" effects that themselves conformed to the proper meaning of "mechanical" principles. (In an

74 Ibid., 170, 177–80, 199–200, 210, 215.
75 See the citations, for instance, in n. 71 above.
76 Roger (1971), 227–31, 242–9; Bots (1972), 5–32.
77 Sellius (1733), 173, 194.
78 Gleichen-Russworm was much struck by his discovery of "infusion animals" living in the fluids of a number of earthworms, and he pursued these observations at some length; nonetheless, he never suggested the possibility of a symbiotic (or parasitic) relationship between these microscopic organisms and their host (Gleichen-Russworm 1777, 10, 59–61). Though alluding to the "stupendous prodigies" (*stupenda prodigia*) of animalcules living in the fluids of men and animals "as if in their own proper worlds," Muys did so only very much in passing and also without reflecting on their possible beneficial (or pathological) significance (Muys 1714, 56). To be sure, determining the ends of microorganisms and their utility to the rest of the world was acknowledged to be a particularly difficult problem (P. Musschenbroek 1762, 1:3–4; Bots 1972, 99).

intriguing echo of Descartes's constant quantity of motion in the universe, he also identified his organic molecules with an unchanging "quantity" of life.)[79] Further distancing itself from the old lore and its ties to popular belief, the new doctrine – or, more properly, doctrines – of spontaneous generation also appealed primarily to microscopic observations that presumed a purposeful and esoteric research effort.[80] Despite its new scientific ethos and its pertinence to an aggressive but warmly contested philosophy of nature, however, the eighteenth-century proponents of spontaneous generation initiated no persisting tradition of microscopic research. Buffon's own recourse to microscopic observation had proved a passing one, and it was a recourse he ultimately disparaged.[81]

The revived issue did provoke a string of experiments, to be sure; conceived roughly along the lines of Leeuwenhoek's earlier efforts, however, which in turn had echoed Redi's, they represented no unfolding, advancing, or expanding line of inquiry but rather a continuing – and continuously frustrated – quest for a conclusive answer to a single question: Were infusoria spontaneously generated or were they not? More often than not, moreover, these experiments were undertaken to challenge Needham's and Buffon's claims rather than to build upon them and explore the potential fruitfulness of materialist hypotheses.[82]

The wonder of a new realm of life that challenged basic understandings of the old; a pervasive threat to life and health; the prospect, on the one hand, of a deeper revelation of the extensiveness, intricacy, and wisdom of God's Creation, or, on the other, of a radically different science that would open a new era in the understanding of nature; the possibility, indeed, of decisive evidence pertaining to the major philosophical and religious issues of the day – all these were implicated in the study of microscopic life in the eighteenth century. Yet they failed to inspire a continuous new tradition of research. What more was needed?

79 Buffon (1841–4), 3:17–18, 120. Contrary to what Bonnet implied, Needham, Wrisberg, and Müller also pointedly rejected the older doctrine of "equivocal generation," and Needham maintained, as did Wrisberg less directly, that his own doctrine was explicitly concerned with "general laws" (Bonnet 1779–83, 8:223; S. Roe 1983, 171–2, 174; Wrisberg 1765, 101n; Müller 1786, xxiii).

80 Beyond the infusoria, to be sure, Buffon and others still insisted on the spontaneous generation of parasitic worms and even some insects (Buffon 1841–4, 3:112–19; Farley 1977, 31, 34–8), but only very rarely might the latter verge upon the realm of everyday observations.

81 J. Baker (1952), 166; S. Roe (1992), 443.

82 Joblot (1718), pt. 2:39–40, 44. (Writing before the issue was revived, however, and hence believing that spontaneous generation had already been satisfactorily disproved, Joblot seems to have been trying to determine, rather, whether the infusoria derived from eggs in the air or in the hay of the infusion.) John Needham (1748), [633], [637–40]; P. Musschenbroek (1762), 1:53–5; Hill (1752a), 92, 110, 117; Ellis (1769), 138–41 (see also Wright 1756, 553–6); Gleichen-Russworm (1778), 74–5, 77; Darwin (1803), "Additional Notes," 1–11; Roger (1971), 697; Bulloch (1938), 74–6, 78–9. Regarding analogous experiments on molds as well, see Lechevalier & Solotorovsky (1974), 343–4.

Conclusion

That Leeuwenhoek's observations of microscopic life failed to produce a more continuous and continuously advancing field of research proved all too typical of the eighteenth century. Despite the issues at stake and an enduring fascination with earlier discoveries, not one of the various lines of eighteenth-century microscopy achieved the accelerating rate of discovery now expected of a developing field of science.[1] To the contrary, the pace of discovery markedly slackened after the passing of the generation of Hooke, Malpighi, Swammerdam, Leeuwenhoek, and Nehemiah Grew.[2] The weave of circumstances that had elicited sustained research from Swammerdam and Leeuwenhoek, at least, suggests that an interplay of technology, ideas, social contexts, and individual temperaments also lay behind the waning of such commitment.

The desultory use of the microscope in the eighteenth century is indeed something of a puzzle. Eighteenth-century anatomists acknowledged its place in anatomy, to be sure, and Haller, at midcentury, reaffirmed the need for the lens in physiological research.[3] Haller himself knew Leeuwenhoek's published letters well (in the Latin editions) and cited them with respect to blood corpuscles, capillaries, and muscle fibers in particular. Nonetheless, eighteenth-century investigations of even the capillary vessels made but meager progress. Haller admitted that he had little to add to Leeuwenhoek's descriptions, and further efforts to grasp the mechanisms of capillary circulation ended in inconclusive results and controversy.[4]

The late eighteenth century did yield some striking new observations of the erythrocytes. In 1761, the Englishman F. H. Eyles Stiles and Giovanni Maria della Torre, observing together in Naples with the latter's minute droplet lenses, found the contrast between the raised edge and the concavity on either side of the human erythrocyte so pronounced that they now confidently described the

1 Regarding efforts to determine the characteristic growth patterns of modern science, see: Holton (1962), 388–95; D. Price (1986), 4–11; see also Crane (1972), 12–13.

2 Marian Fournier has recently shown how the publications dealing with microscopy fell precipitously from the mid-1680s (particularly if one excluded the works of a single individual, Leeuwenhoek) until a sudden revival in the 1740s that seems, in large part, to have been occasioned by Trembley's work on the polyp (Fournier 1991, chap. 2).

3 Haller (1757–66), 1:iv; see Chapter 4, n. 93.

4 Haller (1757–66), passim; Boerhaave & Haller (1740–4), passim in Haller's notes; *DSB*, s.v. "Lieberkühn, Johannes Nathanael"; Miller (1981), 270, 272, 277.

corpuscle as a circular ring with a hole in the middle! A little over a decade later, however, and after his own extensive study of the human erythrocyte, Stile's compatriot William Hewson denied the hole and asserted that the corpuscle was indeed disk-shaped, though, in his turn, he now mistook the dark appearance of the concavities for a smaller solid particle in the center of the disk. Nonetheless, Hewson's influential one-time colleague John Hunter still wrote near the end of the century of the red "globules" in the blood, and whereas Hewson had reemphasized the physiological importance of the red corpuscles, Hunter insisted that the "red part," meaning the corpuscles, was not after all a universal component of the blood and had received more attention than it deserved.[5]

Although the importance of the fibers of the body was perhaps universally accepted, the microscope also added little in the eighteenth century after Leeuwenhoek (and Georgius Baglivi) to the knowledge of the muscle fiber in particular and even less, of course, to speculations on the more abstract, fictitious entities of fiber theory.[6] Nor, for that matter, did the suggestive speculations about microscopic life that envisaged a living corpuscular structure for plants and animals make any impact on anatomical and physiological studies of human and animal bodies.

The notoriety of the spermatozoa also failed to keep a meaningful line of inquiry alive. Christiaan Huygens had ultimately judged the spermatozoa the most important of the microscope's discoveries, and Gleichen-Russworm still vouched that no microscopic discovery had made such a stir. Following Leeuwenhoek's death, nonetheless, little interest was shown in observing them first-hand again until the middle of the century, when Needham and Buffon did so in their effort to frame a new theory of generation. Their opponents responded with renewed observations of their own, but the results provided little that surpassed what Leeuwenhoek had reported decades before.[7]

Microscopic studies of the embryonic development of the chick fared somewhat better, but, despite significant new efforts and fundamental questions that remained in dispute, a persisting succession of researchers still failed to emerge. In 1758, Haller published an extended study of the chick embryo as his *Sur la formation du coeur dans le poulet,* a "masterpiece of toil and insight," according to one Dutch admirer,[8] that devoted 470 pages to 284 separate "observations" and a further 195 pages to an interpretation that still argued the case for preexistence. Within the year, however, Caspar Friedrich Wolff, a recent graduate of the University of Halle, included a series of similar observations in a forcefully argued theory of epigenetic development. There followed a respectful ten-year

5 Stiles (1765), 247, 249, 253–7; Hewson (1774), 303–4, 306–7, 309–17, 323; J. Hunter (1794), 40–2, and 42–3n.

6 Bastholm (1950), 179ff.; Berg (1942), 394ff.

7 *OCCH* 13:524–7; Gleichen-Russworm (1778), 3; Castellani (1973), 37–8, 47–8, 56; John Needham (1748), [632]–[3], [642]–[4], [653]–[4]; Buffon (1841–4), 3:58ff; Spallanzani (1934), 3:335ff.; Ledermüller (1756), 8 and passim.

8 Boddaert (1778), xxv.

exchange between Wolff and Haller,[9] but as probing – and still inconclusive – as their efforts were, they represented not a sustained tradition of research but a belated resumption of work undertaken a century before, a fitful thread of observations that was taken up meaningfully again only in the early nineteenth century.[10]

Despite a flourishing interest in entomology in the eighteenth century, insect dissection also continued nonetheless at a halting pace, highlighted though it also was by one stunning new contribution. The final appearance of the bulk of Swammerdam's own pioneering dissections, it is true, had been delayed for over half a century. Preparing his accumulated studies for publication had been a major preoccupations of his later years, but when he died in early 1680 that preparation, just short of completion, came with its author to a premature end.[11] He bequeathed the manuscripts, drawings, and engraved plates to Thévenot, but when Thévenot himself died twelve years later, they were sold, still unpublished, with his estate[12] – a sad epilogue to what had been years of continued and deeply valued support. The manuscripts remained in France until Boerhaave learned of their whereabouts; he purchased them in 1727 and ten years later published them in both Dutch and Latin as the two-volume *Bybel der natuure* (or *Biblia naturae*).

Apart from the *Ephemeri vita,* consequently, Swammerdam's microscopic dissections had long remained largely unknown; but so little had been done to further what he and Malpighi had begun that the *Bybel der natuure* proved hardly less astonishing when it finally appeared. To Bonnet at midcentury, Swammerdam's dissections were still masterpieces of their kind that, were it not for one later paragon, might have been taken for the last word in human *industrie.*[13] Yet they failed even now to inspire a continuing field of research into insect anatomies, and Boerhaave offered the curiously – and perhaps significantly – equivocal judgment that there would never by anything like the *Bybel der natuure* again.[14]

Nonetheless, Bonnet had gone on to remark that, despite his great ability, Swammerdam had indeed been surpassed by Lyonet,[15] whose *Traité anatomique de la chenille, qui ronge le bois de saule* had been published in The Hague in 1762. Devoting over six hundred pages and eighteen plates to the caterpillar

9 S. Roe (1981), 45–6.

10 Joseph Needham (1959), 185, 223–4.

11 On Swammerdam's labors to prepare his studies for publication, see, for instance: *LJS* 80–2, 95–6, 100–1, 157. In a letter apparently written in the winter of 1678–9, Swammerdam indeed informed Thévenot that "my Great Work" was at last in fact complete (*LJS* 141). Swammerdam's Dutch texts were still being translated into Latin, however (*LJS* 20).

12 *LJS* 20–2. Corresponding from France, Thévenot had in fact been aiding Swammerdam's efforts to prepare those manuscripts for the press. See *LJS* 161, see also pp. 114, 116.

13 Bonnet (1779–83), 7:129–30n. Although more interested in Swammerdam's preparation techniques, Rudolph Forsten referred to the *Bybel der natuure* in his 1776 survey of the contributions of Netherlanders to physiology as "stupendo illo opere" (Forsten 1776, 43).

14 Boerhaave (1962–4), pt. 2:344.

15 Bonnet (1779–83), 7:130–7n.

of a single species (the goat moth, *Cossus cossus* [Linnaeus]), the *Traité* was overwhelmingly preoccupied with its internal anatomy, and Lyonet, the quintessential empirical observer – and a meticulous engraver to boot – insisted on omitting nothing he could see and depict.[16] The plethora of parts was expected to astonish, and their crowded abundance, noted Lyonet himself, made his plates unlike any that had ever been done before.[17] To drive the point home, he counted some of the parts he had described and illustrated, including at least 4,041 muscles in the body (including many muscle fascicles he counted as separate muscles), 236 major vessels, 232 secondary vessels, and 1,336 still smaller branches in the tracheal system, to say nothing of the even tinier levels of ramifications he distinguished that ultimately diminished in prodigious numbers beyond the reach of his microscope. (He now distinguished as well three separate tunics in the tracheal wall, though he also now doubted the tracheal system's respiratory role.)[18]

No less than Swammerdam's *Bybel der natuure,* however, Lyonet's study of the goat moth caterpillar was also a very singular piece of work. Réaumur, to be sure, had encouraged naturalists to explore the concealed wonders and hidden mechanisms of insect anatomies, and he was not alone in having made sorties of his own in the field.[19] Nonetheless, Lyonet himself was conscious of the uniqueness of what he had undertaken. Openly and unashamedly in search of celebrity through the display of his remarkable skills – in engraving as well as dissection – he had chosen insect dissection precisely because it was so little practiced. Unlike human anatomy (which by comparison he also found loathsome and smelly), it was not so populated a field of research as to dim the prospects of surpassing what others had already achieved, he wrote.[20] Hence, he conceived of his treatise as less, if at all, a contribution to a developing field of science than an insular display of the wonders of both God's creations and his own mortal talents.

Nor was it perceived by others as pertaining to an integrated and expanding field of knowledge. In the extensive correspondence between Bonnet and Haller, the former a guiding hand for many naturalists and the latter a preeminent physiologist who professed a weakness for comparative anatomy,[21] they alluded rarely and briefly to Swammerdam's earlier insect dissections, and then only in order to compare his skill – not his discoveries – to Lyonet's. Although the latter was granted the palm, his accomplishment was treated no differently; it was admired as a "prodigy" of skill, but specific discoveries went unremarked, as did

16 Lyonet (1762), x–xi, 2–3.
17 Ibid., xxi, 585.
18 Ibid., 584, 411, 102–3, 79, see also pp. 99, 107–8, 237, 241. On Lyonet's muscle count, see: Cole (1951), 178; Hublard (1910), 71; cf. Freeman (1962), 178.
19 Réaumur (1734–1929), 1:15–17; see also Geer (1752–78), 1:7, 610–22, pls. 1, 2, 4, 2:pls. 12, 14, 23, 33, 36, 4:pl. 15, 7:pl. 3.
20 Lyonet (1762), xiv–xvii; Hublard (1910), 66.
21 Bonnet & Haller (1983), 253.

any continuing line of research to which they pertained.[22] After Lyonet, indeed, little more on insect anatomy appeared until the revival of the field early in the following century, and his treatise remained the only eighteenth-century effort in insect dissection that ranks with the earlier achievements of Swammerdam and Malpighi.[23]

There are few more striking instances of the long hiatus in microscopic research, however, than that in plant anatomy. Its nineteenth-century resumption would play a seminal role in the forging of cell theory, but to no small extent that resurgence simply picked up where seventeenth-century observers had left off.[24] Following Hooke's beginnings, Nehemiah Grew and Malpighi in particular had laid the foundations for systematic study in the field, and an interest in such microscopic observations continued to echo in both popular and academic literature.[25] Nonetheless, microscopic plant anatomy remained at a virtual standstill through the eighteenth and into the nineteenth century.[26]

At first glance, it seems not unreasonable to ascribe the faltering career of microscopic research in the eighteenth century to the widespread distrust of the instrument itself, for which there is ample evidence. Quite apart from the impatience with blatantly silly claims,[27] influential commentators had even Leeuwenhoek in mind when they cautioned against the great susceptibility to error in microscopic observations; the imagination, it was urged, and the longing to come up with something new too easily conjured what did not exist. Microscopic observers themselves readily granted the prevalence of deception. Motivated in part by jealousy, Hartsoeker not surprisingly cited Leeuwenhoek as an example of those who thought they saw a thousand things through the microscope that they never really saw at all. Sellius sounded a similar note, though with no mention of Leeuwenhoek, and remarked solemnly on the absurdities that the microscope had perhaps introduced into natural philosophy. No real partisan of the instrument in the first place, Buffon agreed in 1776 that the microscope had produced more error than truth.[28]

Buffon ascribed that error to the character of humankind rather than to the instrument, but even the devotees of the microscope called attention to its limitations. Baker complained of the distortion and indistinctness of the bead lenses

22 Ibid., 125, 126, 226, 968. Bonnet also wrote to John Turberville Needham of "the hand of the astonishing Lyonet" (Bonnet & Needham 1986, 305).

23 Freeman (1962), 175, 178, 180; cf. Rooseboom (1940), 33, 36–8. Lyonet himself had more to offer, but it was published only posthumously in the nineteenth century; see Lyonet (1832).

24 Fournier (1983), 61; Gasking (1967), 168; J. Baker (1948–55), 89:107–10, 121–3.

25 Christian Wolff (1962–), abt. 1, 6:601–13, 20.3:408–28; Terrada Ferrandis (1969), 64; Martinet (1778–9), 3:265; Schatzberg (1973), 74, 137, 161–2, 321. It has been stressed that Leeuwenhoek's observations in plant anatomy, as extensive as they were, were far less systematic than those of Malpighi and Grew (Arber 1940–1, 230), but for a fuller appreciation of Leeuwenhoek's achievement in this area, see Baas (1982).

26 Arber (1942–3), 14.

27 *Journal literaire* 18 (1731): 19–21; Boddaert (1778), xxxv.

28 Leclerc (1714–30), 4:165–6; NH (1710), 82; Sellius (1733), 25; Buffon (1885), 1:328.

in particular, while other experienced observers decried rather the distorted or blurred images of the compound microscopes.[29] When Haller remarked that he could see the spermatozoa clearly through none of the lenses of his Culpepper instrument (a widely used English make), the Dutch naturalist Martinus Houttuyn applauded him for an honesty few others had shown. Conscious of the flurry over Buffon's claim that the spermatozoa were only moving globules, another otherwise respectful critic concluded that the "great man," Buffon, had in fact never seen them even with Needham's microscope.[30] Many early instruments appear to have perversely compensated for what they obscured, moreover, by the variety of artifacts they added.[31] It has proved tempting, hence, to ascribe the decline of microscopic research not only to doubts about the microscope but to the real inadequacies of the instruments of the day.[32]

The modern era of light microscopy began indeed with the improved achromatic microscopes of the following century. The principle of reducing the chromatic aberration of lenses by combining components of different kinds of glass with different refraction indices had been successfully applied to low-power compound microscopes by the end of the eighteenth century, but the smaller, high-powered versions of such lenses were the most difficult to make,[33] and only in the late 1830s, consequently, did the new instruments begin to surpass the performance of the finer simple instruments. Nonetheless, the way was now open to the rapid and dramatic improvement of the light microscope, which approached the peak of its potential near the end of the nineteenth century.[34]

Although the new technology was essential to the subsequent achievements of nineteenth-century microscopy, however, neither the distrust nor the deficiencies of earlier instruments adequately accounts for the falling off of microscopic research in the preceding century. In botany, embryology, and physiology, such research was reviving even before the improved achromatic instruments,[35] and an older, simpler technology still had a significant contribution to make. In 1827, having finally discovered the true mammalian ovum in the follicle, Karl Ernst von Baer studied it with a simple microscope that showed the ovum so clearly "that a blind man would hardly deny it." Six years later, it was again a simple microscope that enabled Robert Brown to describe the nucleus in plant cells and to remark the cytoplasmic currents that passed between the nucleus and the

29 H. Baker (1766), 69; see Chapter 1, n. 50.

30 Boerhaave & Haller (1740–4), 5(pt. 1):354n. *e;* [Houttuyn] (1761–85), 1:284; Boddaert (1778), 342–3 (cf. [Houttuyn] 1761–85, 1:283). On the "flurry" over Buffon's treatment of the spermatozoa, see also, for instance, Gleichen-Russworm (1764), 9. For a recent reconsideration of the instrument Buffon and Needham may have used, however, see Sloan (1992).

31 Rooseboom (1967), 276; Belloni (1962), 65–73; Zanobio (1971), 29, 32–4.

32 See, for instance, Schlichting (1951), 3890; and Turner (1967), 177.

33 Daumas (1972), 152–60, 249ff; Cittert (1934b), 293; Turner (1980), 12; see also Hughes (1959), 7.

34 Cittert (1935), 57–60; idem (1934b), 293; Cittert & Cittert-Eymers (1951), 73; Turner (1967), 177–91; Cosslett (1966), 57.

35 Hughes (1959), 7, 10; Coleman (1977), 24; J. Baker (1948–55), 89:121–3, 93:161; Pickstone (1977), 33; Coleman (1988), 24; *DSB,* s.v. "Rudolphi, Karl Asmund."

cell wall.[36] While Brown's observations provided the point of departure for nineteenth-century cell theory,[37] Jacob Henle, one of the theory's early adherents, still affirmed in 1841 that Leeuwenhoek had often observed as much or more with his single lenses as one could see with the finest modern compound instruments.[38]

Shortly after, to be sure, one of the initial architects of early cell theory, Matthias Jacob Schleiden, remarked that, except for Brown's observations, the important discoveries of the past twenty years had been made with the compound microscope; yet, even for Schleiden, whether the simple or compound was in fact the superior instrument was still a legitimate question. Raising the prospect of the scientific immortality to be won in botany in particular, he went on to stress indeed how much was still to be discovered with the single lens.[39] The new achromatic compound microscopes clearly made possible the most consequential lines of subsequent nineteenth-century microscopy – of that there can be no question. Nonetheless, the limited optical potential of earlier instruments did not preclude a continuing, more vigorous pace of consequential discovery in the eighteenth century. The simple microscopes were capable of much more than had as yet been achieved.[40]

Nor was it either the awareness or the reality of microscopic deception, both of which continued to flourish throughout the revival of microscopic research as well. Though an admirer of the new microscopes, Schleiden in 1845 also underscored the distortions and deceptions still encountered in microscopic observation; indeed, his own account of cell formation, important as it was to the genesis of cell theory, subordinated his observations of cell division to the perception of a fictional process in which new cells were formed within the interior of the old. Those who insisted rather on cell division, he maintained, had been deceived.[41] In any event, it was certain that someone had.

Several mid-nineteenth-century observers now persuaded themselves as well that they had at last observed what Leeuwenhoek had so forbearingly denied himself, the visceral structures within the spermatozoa (though now assumed to be parasites),[42] while other microscopists of the period debated whether organs analogous to those in larger animals were also observed within protozoa.[43]

36 Hughes (1959), 7; Baer (1986), 218, 222; idem (1827), 12; R. Brown (1833), 710–13.

37 Schleiden (1841), 282.

38 Henle (1841), 134. Henle did stress here, to be sure, the long-recognized advantages of compound instruments, a broader field of view and brighter illumination. He also remarked, however, that most histological material was sufficiently distinct at ×300 and that more than ×400 was generally useless (ibid., 136), which was certainly within the range of older microscopes (see Chapter 1 regarding n. 40). As noted in the previous paragraph, simple microscopes were in fact superior to the compound in other areas of optical performance into the 1830s.

39 Schleiden (1849), 578, 580.

40 Ford (1985), 35, 118–19, 125, 128, 130, 150 (fig. 25); Hughes (1955), 5; Bracegirdle (1993), 305.

41 Schleiden (1849), 31–3, 577, 585–9, 591; Bechtel (1984), 323–4, 326. See also: Jacyna (1983), 78–9; Reiser (1978), 79–80.

42 Hughes (1959), 20–1; Farley (1982), 44.

43 Pouchet (1848), 516–18; Hughes (1955), 10–11; Cole (1926), 24–5.

A classical paper on muscle fibers noted that the new microscopes seemed indeed to have spawned only new differences of opinion, and further advances in techniques and technology later in the century continued to generate heated controversy over what was seen.[44] The debates only advertised the continuing difficulty of determining what precisely it was that the microscope revealed.

The importance of nineteenth-century improvements in the microscope is not of course in doubt. Together with advancing preparation techniques, they dramatically expanded the reach (and comfort) of scientific observation.[45] The point is that, since disputed observations and a recognized susceptibility to deception did not prevent the aggressive advance of microscopic research in the nineteenth century, they were unlikely to have done so a century earlier. Rather, such susceptibility and debate are characteristic of the continuing progress of instrumental science.[46] Consequently, that many early microscopes were liable to unreliable images that lent themselves to flights of fancy cannot explain the halting pursuit of microscopic research in the eighteenth century. That there were skeptics with reasonable doubts should not in itself have proved a barrier.[47] The decisive obstructions lay elsewhere, woven, it appears, into a fabric of ideas, social circumstances, and personal dynamics such as that which, in a previous age, had both resisted and prompted the microscopic discoveries of Swammerdam and Leeuwenhoek.

Of all the avenues to discovery, few seem as uncomplicated as a new technology that can detect what was not detectable before. Simply looking at whatever comes to hand, as Hartsoeker had done with his new droplet lens, would presumably yield something new.[48] Discovery, however, is not so simply won. At a minimum, an apt conceptual framework must be available to give meaning to what is encountered, and some minimal social involvement and like-mindedness

44 Hughes (1955), 10; Baxter & Farley (1979), 138–9; Farley (1982), 35, 160–1.

45 See, for instance, Baxter & Farley (1979), 139–40; Turner (1967), 191–2. Noting indeed that little of the microscopy involved in the development of cell theory in the 1830s entailed optical powers greater than those available in simple microscopes, R. H. Nuttall has argued that the critical importance of the new achromatic instruments to that development lay essentially in the greater physical comfort compound microscopes offered in the long observations that were necessary (Nuttall 1974, 77).

46 Indeed, with an enhanced capacity for artifacts (see, e.g., Franklin 1986, 165–91), modern instrumental technologies on the cutting edge of research entail a rich capacity to mislead, and the resulting "observations" are deeply colored by – indeed dependent upon – theory (Harold Brown 1987, 76, 200–1, 219). This has not prevented the aggressive application of these technologies, however. Regarding the electron microscope in particular, in addition to Franklin see Lynch (1985), 81–107, esp. p. 91.

47 In the twentieth century, the electron microscope, for example, also faced such doubts, which even the early pioneers of its development have not found unreasonable (see Marton 1968, vi, 19–20, 24; and Ruska 1987, 637).

48 Of the early years of X-ray crystallography, Dorothy Hodgkin has remarked that, for those who took up such research early on, "there was so much gold lying about that we couldn't help finding some of it" (Wolpert & Richards 1988, 72). Thomas F. Anderson also recalls that everything they looked at in the early days of electron microscopy was novel, so that two years' work produced thirty-one papers (Anderson 1975, 7–8).

are required if it is to be accepted as a contribution to scientific inquiry. A persisting line of continuing discovery demands even more of both ideas and social contexts: Ideas must shape the prospect of potentially open-ended discovery for instruments and techniques, and an enduring social framework must promote and sustain focused research.

In the eighteenth century, as had been the case in seventeenth-century Holland, ideas that simply construed an explorable world did not suffice. Indeed, eighteenth-century imaginations continued to cherish a seventeenth-century world fashioned, as it were, for the microscope. Hooke, to be sure, had not grasped the significance of the cell structure he had observed in plant materials, but, like Leeuwenhoek, he had gone on to posit hidden "Instruments and contrivances" in plant anatomy that in time, he proposed, might be discovered by better microscopes. In a popular "catechism" of natural theology a century later, Johannes Florentius Martinet, a Zutphen pastor, still echoed the vision of an intricate "fabric" (*weefsel*) in plants that included vessels, valves, sluices, and "trapdoors": "you would see a matchless embroidery, true masterpieces of mechanics, beautiful proofs of an astonishing omnipotence and wisdom."[49] The more he looked into the least of living things, Lyonet also wrote, the more orderliness (*arrangement*) and intelligence he found, and only the most sublime knowledge of hydraulics, chemistry, and mechanics could have built the machine he encountered there. Haller too stressed that the physiology of the human body as well ultimately rested on the operation of parts too small to be seen with the naked eye.[50] Routine as they were, such commonplace assumptions reaffirmed a pervasive potential for microscopic discovery; yet they proved a feeble stimulus to persisting research.

Lofty philosophical and religious implications also failed to inspire sustained research into such elaborate imagined structures. In the eighteenth century, the philosophical and religious overtones had, if anything, sharpened in the confrontation between preexistence and a doctrine of epigenesis that, in contrast to Harvey's, now pointedly rejected spiritualist affiliations and, as a result, smacked all the more of materialism. It was just that issue that had prompted Haller to undertake his own studies of the chick embryo, and his theoretical commitment entailed the prospect of a trove of preexistent structures with an exalted significance, for even Wolff acknowledged that the preformed embryo would have constituted an extraordinary proof of the existence of God.[51] But even when such researches were later resumed in the nineteenth century, it was Wolff's legacy rather than Haller's that proved the more stimulating.[52]

To some extent, of course, ideas conducive to sustained research had to correspond to what appeared in microscopic images, and many seventeenth- and eighteenth-century imaginings simply failed to materialize. By applying alcohol

49 Hooke (1665), 116; Martinet (1778–9), 3:265.
50 Lyonet in Lesser (1745), 2:91n–2n; Lyonet (1762), xv–xvi; Haller (1757–66), 1:iv.
51 S. Roe (1981), 26ff., esp pp. 36, 111.
52 Oppenheimer (1967), 63, 69, 133, 141, 296; cf. S. Roe (1981), 154–5.

and vinegar,[53] Haller succeeded indeed in revealing structures previously invisible in the viscous and transparent early embryo, a modest success that argued the possibility at least of still other unseen parts as well, but it was spotty and limited evidence at best. Moreover, as Haller himself pointed out, it demonstrated as well that the invisibility of such parts was not due only to their minuteness,[54] so that he too apparently saw little point in searching with stronger lenses.[55]

The confirmation microscopic images seemed on occasion to yield for other ideas often arose from circumstances too rare and fleeting or biases too idiosyncratic to endure in a tradition of continued observations. Leeuwenhoek surely considered the repetitive sexpartite construction of the red corpuscle and its constituent globules to be one of his great discoveries, and Boerhaave adapted its neatly descending scale of fluid particles to his immensely influential vascular physiology.[56] Not surprisingly, though, it is by no means clear that anyone else ever managed to see them again.

It is not simply a question of whether such ideas would in time prove right or wrong, however. The early decades of the research spawned by cell theory were framed by a wholly misconceived conception of cell formation.[57] At issue, rather, are the prospects for what at a given time and in a given place might be experienced as discovery, and interesting discovery at that, raising further alluring – often because uncertain – possibilities. The tidy wonder of those descending series of ever-similar but ever-smaller parts – fibers composed of fibers, vessels of vessels, and globules of globules – enjoyed an enduring influence in eighteenth-century thought, but their appeal was a conceptual neatness that anticipated few surprises and few new uncertainties to be explored.[58] After the limits to the series of muscle fibers had been determined, as Muys claimed to have done, what of interest remained to be revealed? Quite apart from being misconceived, fiber doctrine offered a meager research agenda that soon exhausted what it had to offer in the way of discovery.

In addition to some degree of concurrence with microscopic images, hence, a conceptual framework capable of sustaining microscopic research had to sustain

53 See, for instance, Haller (1758), 1:10–11, 161, 165, 221–2, 252, 287, 451, 2:66, 118.

54 Ibid. 1:215, 2:118–19, 176.

55 Haller spoke of the focal distance of his strongest lenses as being a *ligne* (ibid., 1:10). His units of measurement were in terms of the Bern inch, which he tells us was 10/11 of the Paris inch (ibid. 1:11), which would make his *ligne* 2.05 mm (see Chapter 1, nn. 40, 123). That Haller's *ligne* was a twelfth of an inch is confirmed by the proportion he cites between *lignes* and *centiemes* (Haller 1758, 1:205).

56 Boerhaave & Haller (1740–4), 2:296, 320–4, 3:638–9; cf. Boerhaave (1734), 127–8; see also Haller (1757–66), 1:114, 2:61–8.

57 *DSB*, s.vv. "Schleiden, Jacob Mathias" and "Schwann, Theodor Ambrose Hubert"; Coleman (1977), 28–9. Indeed, William Bechtel has more recently emphasized not only that the progress of cell theory in the nineteenth century was shaped more decisively by theoretical priorities than by empirical content (Bechtel 1984, 311, 324–5), but also that the successive stages in that progress all involved major tenets that were later rejected (ibid., 310).

58 Regarding in particular the appeal of the numerical and geometrical regularities of Leeuwenhoek's series of diminishing globules in the blood corpuscle, see, for instance, Brendel & Goesling (1747). According to Haller, indeed, Leeuwenhoek's sexpartite corpuscle had won adherents among eighteenth-century mathematicians as well as physicians (Haller 1757–66, 2:65).

as well an expectation of further encounters with things new and meaningful. Such a framework ultimately proved to be rather a mutable scaffolding of ideas that was (and were) constantly reworked in the light of ongoing observations themselves, and it emerged in the context of a persisting *social* framework for research as well.

Such an enduring social framework had been patently absent in the history of early microscopy. Indeed, the philosophical preoccupations of the rationalist Cartesians, the artistic commitments of the heirs of the tradition of miniature painting, and the alternative promise of the vascular physiology obstructed the development of microscopic research in the Netherlands most prejudicially by denying it access to secure institutional moorings, be it within the universities or craftsmen's workshops. What was lost ranged from the possibility of a sustaining career to an implicit, if narrow, social reassurance as to the value of the enterprise. In the absence of such support, the stimulus to what remained therefore an unusual pursuit could only come from personal drives that might well, and did, prove antithetical to establishing the very social framework needed to sustain what they had begun.

The social and temperamental circumstances that produced Swammerdam and Leeuwenhoek's commitment to microscopic research worked, indeed, to keep it out of an institutional setting. The very educational deficiencies that helped to focus Leeuwenhoek's commitment distanced him from the existing institutions of higher learning, and, though Swammerdam's training and demonstrated abilities were likely to have assured him an academic career like Malpighi's, it was an option he apparently never considered. The social discomfort that induced Swammerdam to withdraw into microscopic research precluded a more informal circle of followers as well, and though Leeuwenhoek was not similarly inclined to withdrawal, his temperament also predisposed him to eschew discipleship. Skeptical of both the ability and commitment of others and jealous, by his own account, of his independence, Leeuwenhoek explicitly and emphatically declined to teach his craft to others. Not one person in a thousand, he maintained, was fit for such studies, "for much time was needed and much money squandered." He brushed aside the medical students at Leiden in particular as concerned only with making money or acquiring a reputation for erudition, neither of which, he said, were to be achieved by means of the microscope.[59]

The exceptionality of his microscopic effort appears ultimately to have been more important to Leeuwenhoek than the continuing progress of microscopic discovery, for it was the uniqueness of his achievement, after all, that had as-

59 AvL (1718), 189, 168–9; see also Schlichting (1951), 3891. Leeuwenhoek may indeed have paid a price for having failed to cultivate any followers in the Netherlands. It is intriguing that, despite how widely his observations were cited in Europe, no one undertook to publish a last volume of his letters that he had prepared for publication and that, with plates and Latin translations, was available for purchase in his daughter's estate (*Catalogus*, 43).

sured him the attention of the learned elite. Not without reason, Leeuwenhoek saw himself and his practice of microscopic research as exceptional, and, shrouding prized methods and instruments in secrecy,[60] he saw little reason to alter that. Thus, for Leeuwenhoek as for Swammerdam, the very circumstances and sensibilities that provoked early microscopic discovery obstructed its institutionalization as a scientific discipline.[61]

Eighteenth-century Europe failed to provide the social setting that was lacking. Despite the endorsement of Haller and others, the medical schools did not generally respond to the prospect of microscopic research. Various universities, Leiden among them, had early included microscopes in their new collections of instruments, but the interest in the microscope in medical instruction appears, if anything, to have subsequently declined.[62] In the 1830s, one could still complete the standard classes in anatomy and physiology at Leiden without ever having seen a microscope.[63] Only in that decade, indeed, was the instrument finally beginning, in any systematic way, to penetrate the classrooms and curricula of the medical schools of Europe.[64]

Nor had new scientific societies provided a viable alternative in the interim. Though Leeuwenhoek became a corresponding member in 1699, the Parisian Académie Royale des Sciences showed little interest in microscopic research (or really in Leeuwenhoek himself),[65] and even the Royal Society's early support had dissipated rapidly. In addition to its ties to Leeuwenhoek, it is true, the London society had encouraged Hooke's preparation of the *Micrographia* and Grew's subsequent botanical studies that, under Hooke's stimulus, had likewise turned to the microscope. Speaking for the society, it was Henry Oldenburg who had also prompted Malpighi to undertake his classic study of the silkworm.[66] Still, the Royal Society evidenced little inclination and even less capacity to develop microscopic observation as a persisting field of research. Its active

60 In November 1681, the members of the Royal Society decided to request Leeuwenhoek to make known the novel instruments or methods – his "invention" – that he used in his microscopic observations, and Robert Hooke, as secretary, promised to convey that request in his next letter to Leeuwenhoek (Birch 1756–7, 4:104). If Hooke indeed carried out his charge, Leeuwenhoek declined to respond.

61 The little that we know of Anton de Heide, the third Netherlander after Swammerdam and Leeuwenhoek to have focused on the microscope in the late seventeenth century, also testifies to a life troubled by emotional difficulties. Apparently a very hard-working physician in Middelburg, he nonetheless ultimately repudiated medicine and denounced it as no more than deception and murder. He was also frustrated in his own quest for scientific recognition and died in a state of despair. A deeply religious man, he too, like Swammerdam, fell under the influence of Antoinette Bourignon (Man 1905, 22, 26, 40, 48, 51; see also Lindeboom 1983).

62 Hoog (1974), 255, 265; Rijnberk (1938), 2830; Warner (1982), 9.

63 Beukers (1983), 74, 76–8; see Chapter 4, n. 17. See also: Rooseboom (1967), 292n. 68; Terrada Ferrandis (1969), 24: Ackerknecht (1967), 125–6; Newman (1957), 107.

64 Beukers (1983), 69–70; Fournier (1983), 62; R. Kremer (1992), 84, 86; Lenoir (1988), 150; Tuchman (1993), 56–7, 76; *DSB*, s.v. "Siebold, Carl Theodor Ernst von."

65 Roger (1971), 183; Stroup (1990), 157–9, 165, 330n. 34, see also 117–18.

66 Birch (1756–7), 1:213 (and passim), 3:49; Arber (1940–1), 222, 225; M. Hunter (1981), 41; idem (1989), 265; Malpighi (1975), 1:355, 356n. 9, 374, 375n. 21.

interest in microscopy soon waned,[67] and its efforts to sponsor Grew's studies, though critical to their completion, in fact underscored how limited the society's capacity to support ongoing research truly was.[68] Like the growing number of new scientific societies that sprang up elsewhere in Europe in the eighteenth century, the Royal Society also lacked the commitment to training needed to sustain a research tradition.[69]

Although diverse naturalists did find positions in universities and other institutions, natural history remained nonetheless essentially an amateur enterprise,[70] as was reflected in the lives of those who showed a particular interest in microscopy. Spallanzani, to be sure, occupied a professorship in natural history at Pavia for thirty years, and Needham managed at least to end his career as the first director of the Académie de Bruxelles. In his last years, the German microscopist Martin Frobenius Ledermüller also added to his lawyer's livelihood a pension from the academy in Mannheim. Although characteristically members of a variety of such academies, however, most microscopic observers of note derived little if any support from those affiliations.

Trembley ended his long service as a private tutor when the generosity of his last charge freed him from any further need to work; Müller, also twenty years a tutor, achieved the same result through marriage. Hill was an apothecary (and, to some, a quack), and Gleichen-Russworm an ex-soldier who had retired to his inherited estates. A successful lawyer, Lyonet also made his way by deciphering codes for the government. In general, such eighteenth-century naturalists lacked an institutional framework within which, having perhaps focused their own microscopic efforts more sharply, they could recruit and train successors.

The enduring grip in anatomical practice and instruction of a rival research tradition underscored what microscopy was thus denied. Despite, indeed, the

67 Despite the determination of the Royal Society in July 1680 to pursue further studies of the infusoria, they almost immediately disappeared from the society's own activities (Birch 1756–7, 4:47, 49). Moreover, apart from Hooke's (and hence the society's) momentarily revived interest in muscle fibers two years later, again in relation to Leeuwenhoek's efforts (ibid., 140–2), the microscope itself only very rarely played a role in those activities for the remaining five years for which Thomas Birch published the minutes of the meetings (ibid., 178, 356, 443, 530–1, 546; cf. 305–6, 321). Leeuwenhoek's frequent letters continued to be read in meetings for a number of years, but even that show of interest wavered. Hartsoeker's suggestion that the members had begun to weary of those letters should perhaps be dismissed as spitefulness (NH 1730, "Ext. crit.," 60), but the society's responsiveness to its correspondent in Delft was in fact marred by significant periods of apparent indifference. Not unsympathetic to the suggestion that Edmond Halley in particular was disinclined to continue the correspondence with Leeuwenhoek, L. C. Palm remarks that Leeuwenhoek's relations with the Royal Society appear to have depended very much on the personal interest of the one or two society officials who handled the correspondence (Palm 1989b, 196–7, 199).

68 M. Hunter (1989), 261–78.

69 M. Hunter (1981), 38–44; McClellan (1985), 1–40, esp. p. 12. The Académie Royale des Sciences in Paris supported prominent individual scientists in their research with stipends and also included within its ranks a select number of "students," *élèves*, associated with those scientists; but in reality the Académie avoided involvement in scientific education (R. Hahn 1971, 16, 51, 53, 79). For an example of the limited role a local society could play in providing a context for microscopic research, see Fournier (1987), 171–4.

70 Roger (1980), 260–1; Porter (1988), 16–19.

ebbing fortunes of the vascular physiology in the mid-eighteenth century, the academic milieu at Leiden continued to accentuate injection technique through brilliant practice and imposing displays of collected preparations.[71] Other dynamics associated with institutionalization – and professionalization – were presumably at work as well: Established anatomists who had acquired the necessary skills of injection had a vested interest in sustaining its prestigious stature,[72] and ambitious beginners were sure to seize upon a research program with proven guidelines for winning professional recognition. So bolstered, the techniques of subtle anatomy in the Dutch medical schools remained virtually unchanged well into the nineteenth century,[73] coincidentally narrowing the elbowroom in which a new line of microscopic research could emerge.

The most visible manifestations of institutionalized research are facilities – buildings, instruments, and supporting equipment – and the gathering of practitioners; but institutionalization entails other circumstances also needed to transform a cluster of new discoveries by uncommon techniques into a continuing research program. In order to turn what Swammerdam and Leeuwenhoek had begun into a persisting endeavor, a compelling but less idiosyncratic motivation had to be instilled, uncommon skills and techniques passed on, and a framework of ideas cobbled together that could both raise and make good on the prospect of ongoing discovery. The confluence of these necessary circumstances was achieved within a social setting that promoted an enduring collective enterprise.[74]

Still an unconventional as well as difficult undertaking in the seventeenth century, the intense and persistent pursuit of microscopic research had then demanded an extraordinary and singularly personal motivation; but institutionalization in the nineteenth century introduced a more generic motivation identified with careers. Indeed, reputations won through research and skill with the microscope served very early to advance the careers of Jan Purkinje and Jacob Henle, both pioneers of instruction in microscopic research in the German uni-

71 Elshout (1952), chap. 4, esp. pp. 50–1, 56, 59, 73, 107, 115. Continuing to advance, indeed, the practice of injection reached its peak only in the nineteenth century (Beukers 1983, 66); cf. Cole (1921), 286–7, 340.

72 Shapin (1982), 164–5. For an expression of national pride as well in injection technique, see Forsten (1776), 12, 35–50.

73 Beukers (1983), 65, 69–70; Fournier (1983), 62.

74 On the recent historiography of "research schools" and their impact in the history of science in the nineteenth century and after, see Servos (1993). In their study of the emergence of radio astronomy, David O. Edge and Michael J. Mulkay have stressed how a new research technology not only frequently redefined initial problems but, through unexpected discoveries, opened up "new areas of ignorance" that became the sources of new problems and the foci of new research networks (Edge & Mulkay 1976, 354, 374, 388–91; cf. 376). Leeuwenhoek's observations of spermatozoa, capillaries, blood corpuscles, microorganisms, and the like were clearly potentially comparable beginnings; but Edge and Mulkay also point out the importance to the development of radio astronomy of the fact that, soon after its initial pioneering efforts, it was incorporated into social contexts that, among other things, allowed for the recruitment and training of future radio astronomers (ibid., 361–2, 364, 369–71, 373, 375, 380) – a point germane to the discussion that follows in the text.

versities.[75] In 1839, when Purkinje founded a modest physiological institute that was one of the earliest formal institutions to cultivate the use of the microscope, he had his eye in part, at least, on his status in the University of Breslau faculty.[76] Though surely not the sole motivation for the new breed of academic microscopists that followed, career goals and institutionalized competition would increasingly elicit intense application sustained through successive generations of researchers.[77] For those who entertained higher hopes, there was also the prospect of a persisting place, if not fame, in the collective memory of the profession.

A sustained and progressive tradition of microscopic research required as well a setting in which successive generations of researchers could be trained in the distinctive techniques of the field. Unfamiliar with Malpighi's methods of insect dissection, Swammerdam had discovered his own only through an exhausting ordeal; when many of Swammerdam's methods then were lost, all that remained was the impression of an extraordinary skill that even to Lyonet bordered on the prodigious. Yet Bonnet still wrote in turn of "the hand of the astonishing Lyonet" that few others could aspire to equal.[78] Haller's frustrated efforts to observe the erythrocytes of warm-blooded animals also exemplified the loss of what had been routine techniques to Leeuwenhoek.[79]

The use of instruments as well as manipulative techniques entail the subtleties of know-how. Even sympathizers, as we know, warned of the potential deceptiveness of microscopic observation, but a familiarity with commonplace distortions and illusions came only with experience. Recently identified by Robert Brown after it had misled so many eighteenth-century observers, Brownian motion was still difficult to characterize, testified Schleiden, and could only be grasped and recognized through frequent observation.[80] But such experience could be passed on to aspiring novices only through apprenticeship and guided practice, which called for just such a social setting as the microscope still lacked in the seventeenth and eighteenth centuries.[81]

By maintaining a collective quest for discovery, the institutionalization of research critically facilitates the fashioning of fruitful ideas as well. With persisting and ambitious researchers in pursuit of avenues to discovery, ideas are opportunistically tested, shaped, or conceived in the crucible of ongoing practice and

75 R. Kremer (1992), 89, 104, 109; Tuchman (1993), 55–8, 83.

76 R. Kremer (1992), 78–9, 81–3, 104.

77 For a developed emphasis on such factors in the history of physiological discovery more broadly in Germany in the nineteenth century, see Zloczower (1981).

78 Lesser (1745) 1: 41n; Bonnet & Needham (1986), 305.

79 Haller (1757–66), 2:50–1.

80 Schleiden (1849), 588.

81 Several observers of the processes of science have stressed the essential importance of such "craft" knowledge regarding instruments and techniques and have noted the learning by apprenticeship that it characteristically requires. See, for instance, Ravetz (1971), 77–8, 101, 103; Rouse (1987), 100, 108. For examples pertaining to the use of modern instruments, see also Hoch (1987), 486–7. Kathryn M. Olesko warns us against what she feels is an overemphasis on such "tacit knowledge," however (Olesko 1993).

forged to enhance the potential of the techniques and instruments at hand.[82] The institutional support for microscopy that began taking shape in the early decades of the nineteenth century, particularly in the German universities,[83] thus also yielded a conceptual framework capable of producing, sustaining, and indeed expanding microscopic discovery and, thereby, of engendering new traditions of microscopic research as well.

Grand philosophies of nature still flourished, and vitalist beliefs and a mechanist reaction against them formed an immediate and suggestive backdrop to early cell theory.[84] But a more tentative and fragmentary tissue of ideas shaped by institutionally nurtured microscopic practice was more immediately at issue in the long-term continuation and development of microscopic research. In an unpretentious and often circuitous interplay of ideas and practice, details and uncertain subtleties in microscopic images suggested their own narrowly focused questions that, against the background of research experience, intimated feasible opportunities for further discoveries. While cell theory may have indeed reflected more wide-ranging philosophic commitments, it emerged in the context of detailed descriptions and a multitude of questions raised by and judgments made about what was seen through the microscope.

Were the "granular coagulations" Schleiden observed within cells really the beginnings of cell nuclei?[85] Did the "small, sharply defined body" he saw within fully developed nucleus actually precede the nucleus and hence the cell (and was it, as it seemed – "judging from the shadows" – a thick ring or "thick-walled hollow globule")?[86] Were the shapes Theodor Schwann reported between older cells in cartilage truly bare nuclei with a nucleolus but devoid as yet of a surrounding cell? Was he correct in seeing and identifying slightly more complex patterns as nuclei with the beginnings of an expanding cell?[87] And with more immediate relevance to recently revived older lines of inquiry: Was the thin-walled vesicle observed within the yolk of the mammalian ovum another young cell – and in all likelihood then "the most essential rudiment" of the embryo – or was it the nucleus of the yolk that, like most nuclei (according to Schwann), was subsequently absorbed?[88]

82 Opposing it to the methodological notion of hypothesis testing, Karin D. Knorr has indeed stressed rather a process of "tinkering" in search of "opportunities for success" in modern laboratory practice (Knorr 1979, 362 ff.). "Ideas may be less tangible than research products," she later writes, "but they are no less circumstantially determined in the research process" (Knorr-Cetina 1981, 35, see also p. 4).

83 On the growing emphasis on and opportunities to work with the microscope in the German universities in the early nineteenth century, see, for instance, R. Kremer (1992), 79, 82, 88, 92; Tuchman (1993), 55–8, 76–7.

84 Schwann (1847), 186 ff., esp. p. 190; Coleman (1977), 11–13, 21, 23–6, 28; *DSB*, s.v. "Schwann, Theodor Ambrose Hubert"; cf. Lenoir (1982), chap. 3, esp. pp. 124–34, 142–55; Lenoir (1992), 54–6.

85 Schleiden (1841), 287.

86 Ibid., 284–5.

87 Schwann (1847), 18–19.

88 Ibid., 42–4, 48–9, see also pp. 22–3, 32, 39, 44, 62.

Often prompted by the more general thrust of cell theory, to be sure, and endowed by that theory with a broader significance, a rapidly accumulating body of very pointed questions specific to microscopic images was thus suggesting a multitude of further avenues for microscopic exploration. Robert Brown's account of his observations of the nucleus in plant cells had first opened for Schleiden what he himself called "a new path of inquiry,"[89] and Schleiden's own paper then led Schwann, in his own words as well, "to more extended researches in another direction."[90] Though Schleiden and Schwann's central concept of the formation of cells was in time discredited, the collective persistence and opportunism of nineteenth-century observers, both embracing and disputing Schleiden and Schwann's accounts, continued to forge further lines of inquiry from an expanding web of observations and attendant conjectures.[91]

Challenging even while building upon Schwann's account of the origin of cells, Carl Vogt turned to fish eggs to scrutinize that thin-walled vesicle and the "germinal flecks" he believed were cells destined ultimately to form the embryo. Recognizing its potential significance for pathology as well, Johannes Müller sought to confirm Schleiden and Schwann's theory in cancer tumors and embraced it as a powerful new theoretical framework.[92] Carl Theodor Ernst von Siebold also stimulated vital new lines of investigation when he distinguished a classification of single-celled organisms, *Protozoa,* likened their nuclei to those in animal cells, and proposed that those nuclei played a direct role in protozoan reproduction by fission.[93]

In striking contrast, indeed, to the pace of microscopic research in the preceding century, research revolving around cell theory unfolded and ramified astonishingly quickly. Following Brown's published account of his observations in 1833, Schleiden's paper appeared in 1838, and Schwann's treatise the very next year.[94] Only months later, Müller expanded cell theory into pathology, as did Jacob Henle, for that matter, in a paper on the formation of mucus and pus.[95] Vogt's observations and arguments appeared in 1842, and von Siebold's characterization of *Protozoa* three years later.

The rapidity of the succession of works building one upon the other reflected the emergence of new centers of microscopic research and, hence, a new community of microscopists in the German universities. It was under Johannes Müller's tutelage in his two-room personal laboratory at the University of Berlin that Schleiden and Schwann were brought together in the 1830s, when Henle was also present among Müller's protégés (as was Robert Remak, whose obser-

89 Schleiden (1849), 38; see also idem (1841), 282.

90 Schwann (1847), xiv, see also p. 8.

91 Coleman (1977), 28–30, 32, 35, 129–30.

92 Lenoir (1982), 137–8, 142–3.

93 Churchill (1989), 193–4, 213; see also Cole (1926), 10, 26–7; *DSB,* s.v. "Schultze, Max Johann Sigismund"; and Jacobs (1989), 228–9.

94 Brown's paper had in fact appeared two years earlier as a pamphlet for private distribution (*DSB,* s.v. "Brown, Robert"); Schleiden (1838); Schwann (1839).

95 Lenoir (1982), 143; Tuchman (1993), 59–60.

vations of cell division would soon pose a major challenge to Schleiden and Schwann's theory).[96] Though particularly influenced by his personal contact with Schleiden,[97] Schwann also drew, to be sure, upon the prior microscopic work of Müller and Henle.[98]

But the dramatic expansion of the ideas and observations associated with cell theory reflected as well a broader network of emerging centers of microscopic research, a network that would continue to expand in the German medical schools through at least the following two decades.[99] Indeed, beyond Müller's circle, Schwann's theory built upon the work of a wider and growing community of recent microscopists, including, among many, Purkinje,[100] Rudolph Wagner,[101] and Karl Ernst von Baer.[102] Von Baer pursued his critical research in microscopic embryology while professor of anatomy and then zoology at the University of Königsberg, his interest in those studies having been initially (though prematurely) aroused as a student at the University of Würzburg by Ignaz Döllinger, who had himself turned purposefully to microscopic research during the second decade of the century.[103] In addition to establishing an early center of microscopy at the University of Breslau, Purkinje too had stimulated von Baer as well as Schwann through his research;[104] and Wagner, at the universities of Erlangen and then Göttingen, was another vigorous new proponent of the microscope who helped arouse a succeeding generation of microscopists.[105] At the University of Bern in the late 1830s, Vogt had attended the anatomy and physiology courses of Gabriel Valentin, a microscopic anatomist who had studied under Purkinje; and von Siebold, having been inspired more informally (indeed inadvertently) by Döllinger and assisted by von Baer, in time succeeded Wagner at Erlangen and ultimately Purkinje at Breslau.[106]

Such a busily and closely – if not always happily – interactive community of scientist-academics had been notably absent in the earlier history of microscopy. The product of developing changes in the German universities, these researchers not only facilitated and accelerated the spread of the new ideas linked to cell theory but shaped and elaborated those ideas in critically distinctive ways. Indi-

96 *DSB,* s.vv. "Schleiden, Jacob Mathias," "Schwann, Theodor Ambrose Hubert," and "Remak, Robert"; Tuchman (1993), 55–6, 58–9, 63. On Müller's lab and its loose but very selective mode of operation, see Lenoir (1992), 42–5; see also Tuchman (1993), 57.

97 Schwann (1847), xiv, 2–3, 7.

98 Ibid., xi–xiv, 7–8, 10.

99 Richard Kremer points out that the three proposals submitted between 1836 and 1846 to the Cultural Ministry in Berlin for institutes for physiology all emphasized the microscope (R. Kremer 1992, 88, 104, 108). During this period, indeed, the instrument became a symbol of progressive medicine and medical instruction in the universities (see Tuchman 1993, 83).

100 Schwann (1847), 8, 42–3.

101 Ibid., 46–7.

102 Schwann's conception of cell formation was deeply influenced by von Baer's account of the developing ovum (Lenoir 1982, 116).

103 Ibid., 67–8, 70–3.

104 Ibid., 119; R. Kremer (1992), 82, 84.

105 *DSB,* s.v. "Wagner, Rudolph."

106 Lenoir (1982), 135; *DSB,* s.vv. "Vogt, Carl," "Valentin, Gabriel Gustav," and "Siebold, Carl Theodor Ernst von."

vidually ambitious,[107] they were all in the hunt for a new idea or an old idea newly twisted that could result in discovery.[108] Collectively persistent and interactive, they dramatically increased the likelihood that any such idea raised even in passing would be seized upon and given its chance to bear fruit. Given the brevity of Schwann's own career as an active researcher, virtually over by the end of the 1830s, even cell theory itself may serve as an example.[109]

In such an increasingly institutionalized – and professionalized – community of microscopists,[110] sustaining the research field implicitly became a major priority, so that ideas and achievements were measured in terms of their contribution to the collective effort. While Swammerdam's intended readership remains unclear – anatomists, naturalists, or spiritually benighted humanity in general? – Leeuwenhoek's "gentlemen philosophers" constituted an audience of broad and varied intellectual attainment. It was not unfitting, consequently, that reviewers both representing and addressing the literary world at large undertook to assess Leeuwenhoek's published letters, valued largely then for what was found "curious," "incredible," and "entertaining."[111] In the nineteenth century, however, a new preoccupation with careers and increasingly esoteric and technical lines of inquiry narrowed the intended audience, and microscopists directed their observations – and their ideas – at a select coterie of simultaneously collaborative and competitive fellow researchers who hailed what they judged useful to their vision of a common enterprise and inveighed against what they deemed obstructive, frivolous, or irrelevant.

Though a major force in the final rejection of Schwann's theory of cell formation, Rudolf Virchow nonetheless implicitly affirmed at midcentury the im-

107 As an acute expression of the ambition one might find in the circle around Johannes Müller, see Lenoir (1992), 51, regarding Émil du Bois-Reymond. On Henle's academic ambitions, see Tuchman (1993), 62, 83, 84–6.

108 Of the group of microscopists including Müller, Purkinje, Valentin, and Wagner – "to name only a few" – Arleen Tuchman writes that in the early 1840s they were all using the microscope in the hope of "raising" ("and perhaps answering") questions of physiological importance (Tuchman 1993, 86).

109 *DSB*, s.v. "Schwann, Theodor Ambrose Hubert."

110 Since "institutionalization" can as easily prove a constraining as an enabling force, my references focusing on its positive impact pertain specifically to the modest, rather loose forms that emerged in the early nineteenth-century academic context and that did so against the background of a virtual institutional vacuum with respect to earlier microscopy. Zloczower has stressed not only that the expansion of institutional support for research by no means necessarily ensures increasing discovery (Zloczower 1981, 74–5, 96, 121, 124–5) but that the development of institutional organization may indeed diminish discovery (ibid., 94, 98, 100, 113–14, 120, 125). For the kind of institutionalization I generally have in mind in the text above, Gerald Geison's definition of "research schools" suffices: "small groups of mature scientists pursuing a reasonably coherent programme of research side-by-side with advanced students in the same institutional context and engaging in direct, continuous social and intellectual interaction" (Geison 1981, 23). With his eye largely on contemporary science, David Hull asserts that, while it has become increasingly difficult for scientists to work in isolation (Hull 1988, 23, 286, 367, 395), the most productive scientific research still occurs in groups of from three to five (ibid., 286).

111 Bayle (1686), 4:819; Leclerc (1686–1718), 1:469; Rabus & Slaart (1692–1701), 2:6–13, 4:403–27, and passim.

portance of Schwann's "great discoveries" for the specific field of histology by lamenting the delayed recognition of their significance for pathology as well; indeed, even after half a century, he noted that the real importance of those discoveries, those "new facts," was still not fully appreciated.[112] A preeminent member of the profession thus acknowledged the impact of Schwann's discoveries in reshaping one specialized and ongoing field of research and predicted similar consequences for another; was this not the kind of acclaim for which microscopic researchers now yearned? The recognition by fellow professionals of an enduring contribution to persisting fields of research was the very measure of success.[113]

By the 1840s, consequently, a growing corps of researchers trained (no matter how haphazardly at times)[114] in microscopic techniques in an increasing number of university centers were now engaged in a collective and competitive search for further opportunities for microscopic discovery. In such a search, ideas are no less a tool than a consequence, and it is not unrealistic to assume that, just as instruments are redesigned to respond to new or resistant problems, ideas are similarly contrived and constantly refashioned to sustain the promise of continuing discovery with the instruments at hand.[115] The social framework of researchers and university centers in Germany (joined now by the Microscopical Society of London) sustained a perpetual context of active practice that thus shaped and engendered ideas attuned specifically to that practice, and the same social dynamics that spurred the search for discovery selected for ideas that could sustain it. Absent in the preceding centuries, a social context had taken shape that constituted a breeding ground for ideas that were likely to prove the most conducive to the fullest and continuing exploitation of the potential of the microscope.

Microscopy in the nineteenth century was therefore a different enterprise and a different experience than in the seventeenth; but there was a persisting dimension of that experience whose potential significance, veiled by the trappings of publicly staged careers and the professional disciplining of scientific literature, is

112 Virchow (1860), 2–3, see also p. 10.

113 While stressing the yearning for credit as a critical drive in science (Hull 1988, 281, 306, 309, 357, 376, 483), David Hull notes that modern scientists are primarily concerned with credit garnered from their professional peers and tend to disparage any recognition for their work they might receive from outside the profession (ibid., 306, 309, 394, 514). Success in science, according to Hull, is also indeed to be measured in terms not of the correctness of one's scientific results but of the extent to which they are built upon and incorporated into the work of those who follow (ibid., 283, 301).

114 Regarding the paucity of formal training in microscopy and anatomy Müller actually provided, see Lenoir (1992), 42–3.

115 From the perspective of his work in a very different realm of science, Werner Heisenberg has nonetheless written suggestively: "Experience teaches that it is usually not the consistency, the clarity of ideas, which makes them acceptable, but the hope that one can participate in their elaboration and verification. It is the wish for our own activity, the hope for results from our own efforts, that leads us on our way through science. This wish is stronger than our rational judgment about the merits of various theoretical ideas" (Heisenberg 1975, 225). See also Rouse (1987), 87–8, 100, 116; and n. 82 above.

more readily grasped in the scientific lives of Swammerdam and Leeuwenhoek. Career priorities did not wholly supplant more personal wellsprings of commitment,[116] nor did the absorption with careers and the collective enterprise strip the experience of discovery of more personal dimensions.[117] Conversely, however, even Swammerdam and Leeuwenhoek's profound personal experience was decisively shaped, indeed made possible, by social and cultural contexts.

That is not to deny those drives of obscure origin that compelled Swammerdam and Leeuwenhoek to manipulate and tinker in the first place,[118] to probe things, and to strive to assuage what Malpighi called the "human itch for understanding."[119] Whatever its roots, curiosity was undeniably for both Swammerdam and Leeuwenhoek an insistent passion.[120] But the social contexts that defined their opportunities and challenged their egos added a focusing intensity to curiosity – including the commitment to the microscope – and sustained their persistence, thus building the very experience of microscopic discovery.

Swammerdam and Leeuwenhoek were encompassed by different social circumstances, however, and they responded to those circumstances in very different ways. Apart from its having created sufficient wealth and leisure time and having fostered a general appreciation of learning, no consistent or specific influence can be readily ascribed to Dutch burgher society as such, within which Leeuwenhoek remained comfortably ensconced while Swammerdam struggled unhappily to escape. The social contexts that counted were more particular, and they counted not only because they pertained to the more immediate social environments in which these researchers found themselves, but because they also reflected Swammerdam and Leeuwenhoek's very different social sensibilities.

116 Timothy Lenoir remarks how microscopical anatomy seemed indeed for the moment, at least, the answer to the personal crisis of Émil du Bois-Reymond, one of the later lights of German physiological research, during his days as a student under Johannes Müller (who was himself manic-depressive) (Lenoir 1992, 51–2; *DSB*, s.v. "Müller, Johannes Peter"). There is even an echo of the traditional picture of Swammerdam's life in the biography of Schwann, the end of whose active years of research in the 1830s coincided with a turn to mysticism (*DSB*, s.v. "Schwann, Theodor Ambrose Hubert").

Individual accounts as well as more general studies of modern scientists suggest even more clearly, however – and with echoes particularly of Swammerdam – the depth of the personal concerns that can be entangled in scientific commitments; see, for instance, Keller (1983), passim; Eiduson (1962), 46–7, 50, 89–96, 114, 157, 162–3, 248, 255.

117 According to David Hull, scientific discovery is something that anyone who has experienced it wants to experience again as often as possible (Hull 1988, 305). For expressions of scientists' enthusiasm for discovery (and of very different kinds), see, for instance: Perutz (1991), 199; Austin (1978), 63–5, 100–1, 163–4, 189–90 (who subsumes discovery, as I read him, under "creativity"); Wolpert & Richards (1988), 144; Preston (1987), 168–70. "It is for moments of discovery that a creative scientist lives," writes Linus Pauling (1958, v); see also Eiduson (1962), 157, 161.

118 "I used to say, I think with my hands," remarked Dorothy Hodgkin when queried about her career in X-ray crystallography; "I just like manipulation" (Wolpert & Richards 1988, 71).

119 Arber (1942), 9–10.

120 Regarding Leeuwenhoek in particular, see *AB* 1:42–3; AvL to RS, 21 April 1722, AvL Letters, fol. 269r. For an overview of psychologists' efforts to understand the source and nature of exploratory behavior in recent decades in terms of "intrinsic motivation," see Deci & Ryan (1985), 11–40.

Increasingly suppressed in professionalized scientific writing, such sensibilities emerged even in Swammerdam and Leeuwenhoek's choice of literary forms for recounting their discoveries. Emphasizing, indeed advertising, his new social affiliations, Leeuwenhoek relied solely on a continuing series of personal letters, whereas Swammerdam aspired to a "great work" that by its very nature could best reconcile, on the one hand, his aversion to competition and social interaction and, on the other, his acute competitiveness and yearning for acclaim. Lending itself to long periods of isolated absorption, it would in time display abilities and achievement in a masterpiece whose very scope would preclude rivals.

These choices may in themselves have fostered different experiences of microscopic research. Leeuwenhoek's reliance on personal letters to diverse correspondents surely contributed to the wandering, disjointed manner in which he typically related his observations, and if, as it seems, he kept his research notes primarily for the sake of such letters,[121] that lack of order – or lack of a conscious pursuit of systematic coherence – may well have imbued much of his research as well.[122] Dedicated to a culminating great work, on the other hand, Swammerdam's research efforts very likely took shape in his mind as a vast but synthetic enterprise that was slowly but progressively unfolding.

In vesting their research with further meanings, however, their cultural (as opposed to their more narrowly social) milieu colored Swammerdam and Leeuwenhoek's experience of microscopic discovery still more vividly and, in doing so, further accentuated their divergence, for Swammerdam and Leeuwenhoek responded very differently to that milieu as well. Swammerdam, whose scientific writings betray a far greater responsiveness to other aspects of the high cul-

121 Leeuwenhoek himself testified to the nexus between his notes and his correspondence when he at times gave up taking notes because the Royal Society, he said, was not answering his letters (*AB* 8:177, 186–7; AvL to J. Chamberlayne, 17 May 1707, AvL Letters, fol. 27v).

122 The texts of Leeuwenhoek's letters relating his observations appear to have been largely transcribed directly, indeed, and sometimes years later, from the notes he had taken down as he observed (see, e.g., *AB* 6:18–19, 8:170–1, 236–7; AvL 1697, 185–6; idem 1702, 435; AvL to RS, 6 June 1710, AvL Letters, fol. 137v; AvL 1718, 69, 307–8). Already in 1676 he was referring to a journal, *Dag Register,* of his observations that he kept "now and then" (*AB* 2:160–1), and forty years later he spoke of writing down his observations and thoughts daily and in duplicate (AvL 1718, 231)! There was at least one instance in his later years, to be sure, when he also mentioned having written down his notes some time after having made the observations themselves; but, after all, he was reaching eighty-eight, he noted (AvL to RS, 9 Jan. 1720, AvL Letters, fol. 227v).

Leeuwenhoek's accumulating notes did not serve any effort toward synthesis or collation, however. Although he once, at least, had also kept copies of his letters (*AB* 3:82–3), in his later years he on occasion admitted to being unable to recall either what he had observed or what he had written, and, what is more, he did not particularly care and would not be bothered to check (AvL 1718, 236; AvL to RS, 20 Nov. 1720, AvL Letters, fol. 232r). In 1717, he also wrote of purposefully declining to look at his earlier notes on a subject he was researching again (AvL 1718, 321–2). Intriguing as well was his apparent decision to forward to the Royal Society a description of the rotifer from some older notes that had come to hand without updating it to conform to his more recent perception of the corona (see AvL 1718, 69 [cf. the description on p. 67 to AvL to RS, 4 Nov. 1704, AvL Letters, fol. 321r, 326r (figs. 3, 4); AvL to RS, 25 Dec. 1702, AvL Letters, fol. 206v; AvL 1702, 405–6]).

ture of his day – painting, poetry, philosophy, and religious reflection – imposed on the insect anatomies he scrutinized an elaborate overlay of aesthetic and religious significance. Encompassing the aesthetic, indeed, his religious sensibilities in particular powerfully shaped his response to microscopic research.

True to his age, Leeuwenhoek too embellished his accounts with religious allusions, and eighteenth-century observers as well still on occasion likened microscopic observation to perpetual worship.[123] Nonetheless, Swammerdam's temperament and personal trials endowed his religious response to microscopic discovery with a distinctive quality. The continuing revelation of a previously unseen subtlety and unexpectedness in nature's workings vested such discovery with a numinous aura and lit up an otherwise unhappy existence with moments of inspired exhilaration. For Swammerdam, that experience was a source of rare and extraordinary joy that at times transformed his religious anguish into a sensation of blessedness.

Leeuwenhoek's religious rhetoric often echoes Swammerdam's, but it bespeaks nonetheless a different range of religious feeling. He betrays little if any concern with his own spiritual state, and only very rarely, though it was such common coin, did he suggest that his microscopic researches might be looked upon as an act of piety. Nor is there any indication that Leeuwenhoek conceived his discoveries as gifts from God to a chosen recipient and hence as a sign of grace. Acute personal issues do not appear to have been at stake.

That impersonal piety appears as well to have stripped microscopic discovery of the presumed endlessness that for Swammerdam and others was part and parcel of its religious meaning.[124] The tenor of Leeuwenhoek's later letters tended to emphasize not what was still to be accomplished but rather the limits he had reached that were unlikely to be surpassed,[125] and his tempered religious outlook may indeed have denied him the awareness of working in an ongoing, open-ended enterprise. For Swammerdam as well, to be sure, the ultimate lesson of the microscope was the fundamental inscrutability of nature, but for Swammerdam that inscrutability was an intended lesson meant for spiritual instruction. Since that lesson was integral to the staging of humanity's spiritual drama, the research experience that rehearsed it was likely to prove perpetual. The very purpose of hidden microscopic secrets was indeed their eventual discovery. Reflecting on observations of muscle fibers in 1721, by contrast, Leeuwenhoek remarked rather on the many things the human species would not

123 Ledermüller (1762), preface [4]; Hill (1752a), 6.

124 Some of the more sweeping eighteenth-century assertions of the future continuation of microscopic discovery were also articulated in a religious context; see Muys (1738), preface [35]; H. Baker (1742), 310–11.

125 Leeuwenhoek still referred on occasion, it is true, to things yet to be discovered (AvL to RS, 9 Jan. 1720, AvL Letters, fol. 225r), but more characteristic were his allusions both to humanity's incapacity to search nature deeply and to hidden structures that he believed would remain forever beyond the microscope. See Chapter 9, n. 150.

discover.[126] His was indeed a finite enterprise, and he apparently believed that its most significant revelations were already running out.

As shaped in terms of his own personal priorities, nonetheless, microscopic discovery surely had an impact on Leeuwenhoek's sense of himself. Eighteenth-century proponents of the microscope remarked that microscopic observation enriched and developed the observer – it "improves the Faculties, and exalts our Comprehension," proclaimed John Hill[127] – and we cannot discount the possibility that for Swammerdam it also entailed what could well be called an existential dimension. An absorbing effort for which he seemed uniquely gifted, Swammerdam's microscopic dissections assumed a prominent place in a program of research that may have provided a desperately needed sense of personal meaningfulness and purpose. For Leeuwenhoek, his commitment to the microscope pertained to his measure of himself against a more patently social yardstick, but the lifetime of discovery that followed as a consequence reaped rich dividends nonetheless in a struggle against self-doubts.

In his later years, Leeuwenhoek's continuing discoveries were not only a source of joy – and perhaps all the more so because he did often believe them the limit of what could be achieved – but explicitly balm as well for a wounded ego.[128] Less explicitly, that balm transformed him. Anxious as he was to prove himself no common man, Leeuwenhoek's microscopic discoveries progressively enfolded him in an awareness not only of extraordinary achievement but also of extraordinary personal development.[129] Already by the early 1680s, the Delft draper who had once defensively urged his correspondents to keep in mind who he was – or rather, implicitly, who he was not – now anticipated that his discoveries would amaze the world, and by the last decades of his life he was conscious of having contemplated extraordinary ideas as well. Indeed, his ability to accommodate unprecedented microscopic images to the mechanistic drift of late-seventeenth-century thought forged what struck Leeuwenhoek as thrilling new insights into nature's workings, while his penchant for further elaborating his imaginings provided the stuff of what he himself regarded as exotic speculation. His life had thus proved one of singular achievements and experiences that were otherwise beyond the reach of even the thousandth man, as he himself might well have phrased it.[130]

Leeuwenhoek's self-consciousness throughout intimates yet another level in his life of discovery. When first introduced to the Royal Society in 1673, he could hardly have conceived what his abilities as well as his microscopes would prove capable of in the coming five decades of continuous research. For Leeu-

126 AvL to RS, 11 April 1721, AvL Letters, fol. 257v. As early as 1678, to be sure, while pondering the legs he attributed to various microscopic animals, Leeuwenhoek had already alluded to the inevitable superficiality of science (*AB* 2:390–1).

127 Hill (1752a), 6; see also H. Baker (1742), xiii, 310; Muys (1738), preface [35].

128 AvL to RS, 24 Jan. 1721, AvL Letters, fol. 246v.

129 See Chapter 6, regarding nn. 64, 129, 131. Recall that in those last decades he affirmed that it was also now only for the "philosophical" that he wrote (Chapter 6, n. 78).

130 See Chapter 6, n. 131.

wenhoek, consequently, that sustained and purposeful effort also entailed the exploration of new depths and new capacities within himself. Discovery in this instance entailed self-discovery as well.

Alien, indeed in ways inimical, to the research traditions of modern science, Leeuwenhoek and Swammerdam's commitment to the microscope nonetheless exposed a limitless new prospect for discovery. They dramatically enhanced the awareness of the unexpectedness of nature and of the unknown forms and phenomena that human ingenuity might aspire to bring to light. The lack of encouragement Leeuwenhoek offered his successors could not, after all, diminish the impact of his fifty years of pioneering observations. Together with the cultural patrimony upon which they drew, however, the social anxieties and circumstances that forged Leeuwenhoek and Swammerdam's commitment constructed in the first place a moving experience that touched their personal lives profoundly.

References

Primary sources

Manuscripts

Hoefnagel, Joris. I*gnis[:] Animalia rationalia et insecta.* Edith G. Rosenwald Collection; on deposit at the National Gallery of Art, Washington, D.C.

Hudde, Johan. Letter to Lambert van Velthuysen, 13 Oct. 1657. University of Amsterdam Library.

Leeuwenhoek, Antoni van. Letters. Royal Society Library, London. [AvL Letters]

Swammerdam, Jan. Papers. University of Göttingen Library, Göttingen. [JS Papers] (See also Swammerdam 1975.)

Published

Académie Royale des Sciences, Paris. 1744. *Histoire de l'Académie Royale des Sciences. Année M.DCCXLI.* Paris: Imprimerie Royale.

Adams, Archibald. 1710. A Letter . . . to Dr. Hans Sloane, R. S. Secr. Concerning the Manner of Making Microscopes, &c. *Philosophical Transactions,* no. 325, pp. 24–7.

[Addison, Joseph]. 1712. [The Pleasures of the Imagination, no. 10.] *The Spectator,* no. 420. Pp. 131–6 in vol. 6 of the 1st collected ed. London: S. Buckley & J. Tonson, 1712–15.

l'Admiral, Jacob. 1774. *Naauwkeurige waarneemingen omtrent de veranderingen van veele insekten of gekorvene diertjes.* Amsterdam: Johannes Sluyter.

Albinus, Bernard. 1702. *Oratio de ortu et progressu medicinae.* Leiden: apud Jordanum Luchtmans.

1711. *Oratio de incrementis et statu artis medicae seculi decimi septimi.* Leiden: apud Samuelem Luchtmans.

Albinus, Bernhard Siegfried. 1719. *Oratio inauguralis de anatome comparata.* Leiden: apud Henricum Mulhovium.

1721. *Oratio qua in veram viam, quae ad fabricae humani corporis cognitionem ducat, inquiritur.* Leiden: apud Henricum Mulhovium.

1754–68. *Academicae annotationes.* 8 vols. Leiden: J. & H. Verbeek.

Aldrovandi, Ulisse. 1602. *De animalibus insectis libri septem.* Bologna: apud Ioan. Bapt. Bellagambam.

Alsem, Arnoldus van. 1671. *Disputatio medica, inauguralis, de humoribus.* Leiden: apud viduam et heredes Joannis Elsevirii.

Andry, Nicolas. 1700. *De la génération des vers dans le corps de l'homme.* Paris: Laurent d'Houry.

Angel, Philips. 1642. *Lof der schilder-konst.* Reprint, Utrecht: Davaco, 1969.

Aristotle. 1984. *The Complete Works of Aristotle.* Edited by Jonathan Barnes. 2 vols. Princeton: Princeton University Press.

Aselli, Gaspare. 1627. *De lactibus sive lacteis venis quarto vasorum mesaraicorum genere novo invento.* Milan: apud Io. Baptãm Bidellium.

Back, Jacobus de. 1648. *Dissertatio de corde.* Rotterdam: ex officinâ Arnoldi Leers.

Bacon, Francis. 1877–89. *The Works of Francis Bacon.* Edited by James Spedding, Robert Leslie Ellis, and Douglas Denon Heath. Rev. ed. 14 vols. London: Longmans.

Baer, Karl Ernst von. 1827. *De ovi mammalium et hominis genesi epistolam.* Leipzig: sumptibus Leopoldi Vossii.

1986. *Autobiography of Dr. Karl Ernst von Baer.* Edited by Jane M. Oppenheimer; translated by H. Schneider. [Canton, Mass.]: Science History Publications USA.

Baglivi, Georgius. 1710. *Opera omnia medico-practica, et anatomica.* 7th ed. Lyon: sumptibus Anisson & Joannis Posuel.

Baker, Henry. 1740. An Account of Mr. Leeuwenhoek's Microscopes. *Philosophical Transactions,* no. 458, pp. 503–19.

1742. *The Microscope Made Easy.* London: R. Dodsley.

1753. *Employment for the Microscope.* London: R. Dodsley.

1766. A Report Concerning the Microscope-Glasses, Sent as a Present to the Royal Society, by Father di Torre of Naples, and Referred to the Examination of Mr. Baker, F.R.S. *Philosophical Transactions* 56:67–71.

Baker, Henry, [and Martinus Houttuyn]. 1778. *Het mikroskoop gemakkelyk gemaakt.* 3d ed. Amsterdam: de erven van F. Houttuyn.

Bartholin, Thomas. 1673–80. *Acta medica et philosophica Hafniensia.* 5 vols. in 4. Copenhagen: sumptibus Petri Hauboldi.

Basnage, Henri. 1687–1709. *Histoire des ouvrages des savans.* 24 vols. Amsterdam: Michel Charles le Cene, 1721.

Bayle, Pierre. 1686. *Nouvelles de la République des lettres.* 2d ed. 4 vols. Amsterdam: Henry Desbordes.

Beaumont, Elias Petrus de. 1698. *Dissertatio physica, de circulatione sanguinis in foetu.* Leiden: apud Abrahamum Elzevier.

Beeckman, Isaac. 1644. *Mathematico-physicarum meditationum, quaestionum, solutionum centuria.* Utrecht: apud Petrum Danielis Sloot.

1939–53. *Journal tenu par Isaac Beeckman de 1604 à 1634.* Edited by C. de Waard. 4 vols. The Hague: Martinus Nijhoff.

Belkmeer, Cornelius. 1733. *Natuurkundige verhandeling, of waarneminge; betreffende den houtuytraspende en doorboorende zee-worm.* Amsterdam: Erven van J. Ratelband.

Belkmeer, Jan. 1719. *Twee redenvoeringen.* Amsterdam: Joannes Loots.

Bergerac, Cyrano de. 1656–61. *L'Autre monde, ou les états et empires de la lune et du soleil.* Paris: Éditions sociales, 1968.

Bettini, Mario. 1659. *Recreationum mathematicarum apiaria novissima duodecim.* Bologna: sumptibus Joannis Babtistae Ferronii.

Bichat, Marie-François-Xavier. 1801. *Anatomie générale, appliquée a la physiologie et a la médecine.* Paris: Brosson & Gabon.

Bidloo, Govard. 1685. *Anatomia humani corporis.* Amsterdam: sumptibus viduae Joannis à Someren.

1715a. *Decades duae exercitationum anatomico-chirurgicarum.* In *Opera omnia anatomico-chirurgica.* Leiden: apud Samuelem Luchtmans.

1715b. *Dissertationes anatomico-physiologicae quatuor.* In *Opera omnia anatomico-chirurgica.* Leiden: apud Samuelem Luchtmans.

1715c. *Observatio, de animalculis, in ovino aliorumque animantium hepate detectis, ad virum celebrem Antonium van Leeuwenhoek.* In *Opera omnia anatomico-chirurgica.* Leiden: apud Samuelem Luchtmans.

1715d. *Vindiciae quarundum delineationum anatomicarum.* In *Opera omnia anatomico-chirurgica.* Leiden: apud Samuelem Luchtmans.

Bils, Lodewijk de. 1658. *Waarachtig gebruik der tot noch toe gemeende gijlbuis, beneffens de verrijzenis der lever, voorheen zoo lichtvaardig in 't graf gestooten.* Rotterdam: Joannes Naeranus.

1659a. *Aan alle ware liefhebbers der anatomie.* Rotterdam: Joannes Naeranus.

1659b. *Kopye van zekere ampele acte . . . rakende de wetenschap van de oprechte anatomije des menselijken lichaams.* Rotterdam: Joannes Naeranus.

Birch, Thomas. 1756–7. *The History of the Royal Society of London.* 4 vols. Reprint, Hildesheim: Georg Olms, 1968.

Blaes, Gerard. 1660. *Oratio inauguralis de iis quae homo naturae, quae arti, debeat.* Amsterdam: apud Petrum van den Bergh.

1661. *Medicina generalis.* Amsterdam: apud Petrum van den Berge.

1675. *Ontleeding des menschelyken lichaems.* Amsterdam: Abraham Wolfgangh.

Blanckaert, Steven. 1680–8. *Collectanea medico-physica, oft Hollands jaar-register der genees-en natuur-kundige aanmerkingen van gantsch Europa &c.* 4 vols. in 3. Amsterdam: Johan ten Hoorn.

1683a. *De kartesiaanse academie ofte institutie der medicyne.* Amsterdam: Johannes ten Hoorn.

1683b. *Lexicon medicum Graeco-Latinum.* Reprint, Hildesheim and New York: Georg Olms, 1973.

1685. *Praxeos medicae idea nova.* Amsterdam: ex officina Joannis ten Hoorn.

1686. *De nieuw hervormde anatomie, ofte ontleding des menschen lichaams.* 2d ed. Amsterdam: Jan ten Hoorn.

1688a. *Anatomia practica rationalis, sive rariorum cadaverum morbis denatorum anatomica inspectio.* Amsterdam: ex officina Corn. Blancardi.

1688b. *Schou-burg der rupsen, wormen, ma'den, en vliegende dierkens daar uit voortkomende.* Amsterdam: Jan ten Hoorn.

1701. *Opera medica, theoretica, practica et chirurgica.* 2 vols. Leiden: apud Cornelium Boutestein et Jordanum Lugtmans.

Bleyswijck, Dirck van. 1667. *Beschryvinge der stadt Delft.* Delft: Arnold Bon.

Boddaert, Pieter. 1778. *Natuurkundige beschouwing der dieren, in hun inwendig zamenstel, eigenschappen, huishouding enz.* Utrecht: J. van Driel.

Boerhaave, Herman. 1734. *Institutiones medicae.* 5th ed. Leiden: apud Theodorum Haak, Samuel Luchtmans, et Joh. et Herm. Verbeek; Rotterdam: apud Joan. Dan. Beman.

1738. Oratio quarta, sive sermo academicus, de comparando certo in physicis. Pp. 27–36 in *Opuscula omnia, quae hactenus in lucem prodierunt.* The Hague: J. Neaulme.

1751. *Methodus studii medici emaculata et accessionibus locupletata.* Edited by Albrecht von Haller. 2 vols. Amsterdam: sumptibus Jacobi a Wetstein.

1962–4. *Boerhaave's Correspondence.* Edited by G. A. Lindeboom. 2 vols. Leiden: E. J. Brill.

Boerhaave, Herman, and Albrecht von Haller. 1740–4. *Praelectiones academicae in proprias Institutiones rei medicae edidit.* 2d ed. 6 vols. in 4. Göttingen: Abram Vandenhoeck.

Boerhaave, Herman, and Frederik Ruysch. 1722. *Opusculum anatomicum de fabrica glandularum in corpore humano.* Leiden: P. van der Aa.

1751. *Opusculum anatomicum de fabrica glandularum in corpore humano.* Leiden: apud Cornelium Haak. This edition appears in Ruysch (1725–51), vol. 4.

Bohl, Johann Christoph. 1744. Dissertatio epistolica ad virum clarissimum, Fredericum Ruyschium . . . de usu novarum cavae propaginum in systemate chylopoeo, ut atque de corticis cerebri textura. Amsterdam: apud Janssonio–Waesbergios. This edition is included in Ruysch (1725–51), vol. 4.

[Boitet, Reinier]. 1729. *Beschryving der stadt Delft.* Delft: R. Boitet.

Bonnet, Charles. 1779–83. *Oeuvres d'histoire naturelle et de philosophie.* 18 vols. Neuchatel: Samuel Fauche.

Bonnet, Charles, and Albrecht von Haller. 1983. *The Correspondence between Albrecht von Haller and Charles Bonnet.* Edited by Otto Sonntag. Bern, etc.: Hans Huber.

Bonnet, Charles, and John Turberville Needham. 1986. *Science against the Unbelievers: The Correspondence of Bonnet and Needham, 1760–1780*. Edited by Renato G. Mazzolini and Shirley A. Roe. *Studies on Voltaire and the Eighteenth Century*, vol. 243. Oxford: Voltaire Foundation.

Bontekoe, Cornelis. 1684. *Korte verhandeling van 's menschen leven, gesondheid, siekte en dood*. The Hague: Pieter Hagen.

1689. *Alle de philosophische, medicinale en chymische werken*. 2 vols. in 1. Amsterdam: Jan ten Hoorn.

Borel, Pierre. 1655. *De vero telescopii inventore*. The Hague: ex typographia Adriani Vlacq.

1656. *Observationum microcospicarum centuria*. The Hague: ex officina Adriani Vlacq. Bound with *De vero telescopii inventore*.

Boyle, Robert. 1699–1700. *The Works of the Honourable Robert Boyle, Esq. Epitomiz'd*. Edited by Richard Boulton. 3 vols. London: J. Phillips & J. Taylor.

Bredius, Abraham, ed. 1915–22. *Künstler-inventare Urkunden zur Geschichte der Holländischen Kunst des XVI[ten], XVII[ten] und XVIII[ten] Jahrhunderts*. 8 vols. The Hague: Martinus Nijhoff.

Brendel, Johann Gottfried, and Johann Arnold Goesling. 1747. *Globulorum sanguinis Leeuwenhoekianorum rationes sextuplas*. Göttingen: typis Abrami Vandenhoek.

Brown, Robert. 1833. On the Organs and Mode of Fecundation in Orchideae and Asclepiadeae. *Transactions of the Linnean Society of London* 16: 685–745.

Browne, Thomas. 1964. *The Works of Sir Thomas Browne*. Edited by Geoffrey Keynes. 2d ed. 4 vols. London: Faber & Faber.

Buffon, Georges-Louis Leclerc, comte de. 1841–4. *Oeuvres complètes de Buffon*. 6 vols. Paris: Furne.

1885. *Correspondance générale*. Edited by H. Nadault de Buffon. 2 vols. Reprint, Geneva: Slatkine, 1971.

Buonanni, Filippo. 1691. *Observationes circa viventia, quae in rebus non viventibus reperiuntur*. Rome: typis Dominici Antonij Herculis.

Burger, Jacobus. 1694. *Disputatio medica inauguralis de sanguine*. Leiden: apud Abrahamum Elzevier.

Burgersdijck, Franco. 1627. *Idea philosophiae naturalis*. 2d ed. Leiden: ex officinâ Bonavent. et Abrahami Elzevir.

1642. *Collegium physicum*. 2d ed. Leiden: ex officinâ Elziviriorum.

Butterfield, Michael. 1678. Extract of a Letter from Mr. Butterfield, Mathematique Instrument-Maker to the French King, about the Making of Microscopes with Very Small and Single Glasses. *Philosophical Transactions*, no. 141, pp. 1026–7.

Calvin, Jean. 1559. *Institutes of the Christian Religion*. Edited by John T. McNeill; translated by Ford Lewis Battles. *The Library of Christian Classics*, vols. 20, 21. 2 vols. Philadelphia: Westminster Press, 1960.

Campdomercus, Johannes Jacobus. 1725. *Epistola anatomica, problematica quarta*. Amsterdam: apud JanssonioWaesbergios. This edition appears in Ruysch (1725–51), vol. 2.

Catalogus van het vermaarde cabinet van vergrootglasen, met zeer veel moeite, en kosten in veele jaren geïnventeert, gemaakt, en nagelaten door wylen den heer Anthony van Leeuwenhoek. 1747. Delft: Reinier Boitet.

Cats, Jacob. 1726. *Alle de Wercken van den Heere Jacob Cats*. 2 vols. in 1. Amsterdam: Johannes Ratelband, etc.; The Hague: Pieter van Thol en Pieter Husson.

Cloothack, Marcus. 1687. *Disputatio medica inauguralis de catalepsi*. Leiden: apud Abrahamum Elzevier.

Cluyt, Auger. 1634. *Opuscula duo singularia. I. De nuce medica. II. De hemerobio sive ephemero insecto, et majali verme*. Amsterdam: typis Jacobi Charpentier.

Cluyt, Dirck Outgerszoon. 1597. *Vande Byen*. Amsterdam: Broer Janiz, 1648.

Cochrane, Arthur C., ed. 1966. *Reformed Confessions of the Sixteenth Century*. Philadelphia: Westminster Press.

Cocquis, Petrus. 1688. *Disputatio medica inauguralis, continens selectiores quasdam positiones medicas*. Leiden: apud Abrahamum Elzevier.

Cogrossi, Carlo Francesco. 1714. *Nuova idea del male contagioso de' buoi*. Reprint with English translation by Dorothy M. Schullian. Rome: Sezione Lombarda della Società Italiana di Microbiologia, 1953.

Collegium Privatum, Amsterdam. 1673. *Observationum anatomicarum collegii privati Amstelodamensis, pars altera*. In *Observationes anatomicae selectiores Amstelodamensium 1667–1673*. Edited by F. J. Cole. [Reading, England]: University of Reading, 1938.

Commelin, Casparus. 1694. *Disputatio medica inauguralis de lumbricis*. Leiden: apud Abrahamum Elzevier.

[Craanen, Theodoor]. 1685. *Oeconomia animalis ad circulationem sanguinis breviter delineata*. Gouda: ex officina Guilhelmi vander Hoeve.

1689. *Tractatus physico-medicus de homine, in quô status ejus tam naturalis, quam praeternaturalis, quoad theoriam rationalem mechanicè demonstratur*. Edited by Theodorus Schoon. Leiden: apud Petrum vander Aa.

Cudworth, Ralph. 1678. *The True Intellectual System of the Universe*. London: Richard Royston.

Cuvier, Georges. 1800–5. *Leçons d'anatomie comparée*. 5 vols. Paris: Baudouin.

Darwin, Erasmus. 1803. *The Temple of Nature; or, the Origin of Society*. Reprint (with *The Golden Age*). New York and London: Garland, 1978.

Descartes, René. 1897–1910. *Oeuvres de Descartes*. Edited by Charles Adam and Paul Tannery. 12 vols. Paris: Léopold Cerf.

1972. *Treatise of Man*. Edited and translated by Thomas Steele Hall. Cambridge, Mass.: Harvard University Press.

Deusing, Anton. 1649. *Synopsis medicinae universalis seu compendium institutionum medicarum*. Groningen: typis Joannis Nicolai.

1651. *Anatome parvorum naturalium seu exercitationes anatomicae ac physiologicae, de partibus humani corporis conservationi specierum inservientibus*. Groningen: ex officina Johannis Sas.

1655. *De motu cordis et sanguinis itemque de lacte ac nutrimento foetus in utero, dissertationes*. Groningen: apud Franciscum Bronckhorst.

1659. *Idea fabricae corporis humani; seu institutiones anatomicae ad circulationem sanguinis, aliaque recentiorum inventa, accommodatae*. Groningen: typis Francisci Bronchorstii.

1660. *Erercitationes [sic] physico-anatomicae, de nutrimenti in corpore elaboratione*. Groningen: typis Francisci Bronchorsti.

1661. *Exercitationes physico-anatomicae de nutrimento animalium ultimo*. Groningen: typis Francisci Bronchorstii.

1662. *Resurrectio hepatis asserta*. Groningen: typis Francisci Bronchorstii.

1664. *Sylva-Caedua cadens: seu disquisitiones anti-Sylvianae de alimenti assumpti elaboratione et distributione*. Groningen: typis Francisci Bronchorstii.

1665. *Sylva-Caedua jacens: seu disquisitiones anti-Sylvianae ulteriores*. Groningen: typis Johannis Cülleni.

Diderot, Denis. 1875–7. *Oeuvres complètes de Diderot*. Edited by J. Assézat. 20 vols. Paris: Garnier Frères.

Diemerbroeck, Ysbrand van. 1672. *Anatome corporis humani*. Utrecht: sumptibus et typis Meinardi a Dreunen.

1685. *Opera omnia*. Edited by Timannus van Diemerbroeck. Utrecht: apud Meinardum à Dreunen et Guilielmum à Walcheren.

Digby, Kenelm. 1644. *Two Treatises*. Paris: Gilles Blaizot.

Dimelius, Johannes Wilhelmus. 1685. *Disputatio medica inauguralis de morbis contagiosis*. Leiden: apud Abrahamum Elzevier.

Dionis, Pierre. 1729. *L'Anatomie de l'homme*. 6th ed. Paris: la veuve d'Houry.

Drake, Roger. 1647. *Theses de circulatione naturali, seu cordis et sanguinis motu circulari*. In *Recentiorum disceptationes de motu cordis, sanguinis, et chyli, in animalibus*. Leiden: ex officina Ioannis Maire.

Drebbel, Cornelis. 1732? *Kort begrip der hoofdstoffelyke natuurkunde*. Amsterdam: P. G. Geysbeek & L. Groenewoud.

Drelincourt, Charles. 1680. *Praeludium anatomicum*. 3d ed. Leiden: apud Danielem à Gaesbeeck.

1684. *De foeminarum ovis*. Leiden: apud Danielem à Gaesbeeck.

Du Hamel, Jean-Baptiste. 1669. *De consensu veteris et novae philosophiae*. Oxford: W. Hall & Joh. Crosley.

1701. *Regiae Scientiarum Academia historia*. 2d ed. Paris: apud Joannem-Baptistam Delespine.

[Effen, Justus van]. 1731–5. *De Hollandsche Spectator*. 12 vols. Amsterdam: Hermanus Uytwerf.

Ellis, John. 1769. Observations on a Particular Manner of Increase in the Animalcula of Vegetable Infusions. *Philosophical Transactions* 59:138–52.

Elsholt, J. S. 1678–9. De microscopiis globularibus. *Miscellanea curiosa* 9–10:280–1.

Ettmüller, Michael Ernst. 1728. *Epistola anatomica, problematica duodecima*. Amsterdam: apud JanssonioWaesbergios. This edition appears in Ruysch (1725–51), vol. 2.

Everaerts, Anthony. 1661. *Novus et genuinus hominis brutique animalis exortus*. Middelburg: ex officinâ Francisci Kroock.

Evertze, Petrus. 1706. *Memoria demonstrationis organi chylificationis*. Sub praesidio Godefridi Bidloo. Leiden: apud Abrahamum Elzevier.

Feylingius, Johannes. 1665. *De macrocosmus, en microcosmus, ofte de wonderen van de groote en kleyne werelt, met allerhanden historien verligt*. Amsterdam: Samuel Imbrechts.

Folkes, Martin. 1723. Some Account of Mr. Leeuwenhoek's Curious Microscopes, Lately Presented to the Royal Society. *Philosophical Transactions*, no. 380, pp. 446–53.

Fontana, Francesco. 1646. *Novae coelestium, terrestriumque rerum observationes*. Naples: apud Gaffarum.

Fontenelle, Bernard Le Bouyer de. 1818. *Oeuvres de Fontenelle*. 3 vols. Paris: A. Belin.

Forsten, Rudolph. 1776. *Oratio de belgarum meritis in oeconomia corporis humani extricanda*. Harderwijk: apud Ioannem Mooien.

Fracastoro, Girolamo. 1584. *Opera omnia*. 3d ed. Venice: apud Juntas.

Fuhrmann, Josua. 1676. *Disputatio medica de catarrhis in genere*. Sub praesidio Theodori Craanen. Leiden: apud viduam et heredes Johannis Elsevirii.

Garden, George. 1690–1. A Discourse Concerning the Modern Theory of Generation. *Philosophical Transactions*, no. 192, pp. 474–83.

Gassendi, Pierre. 1658. *Opera omnia*. 6 vols. Reprint, Stuttgart and Bad-Cannstatt: Friedrich Frommann, 1964.

Gaubius, Hieronymus David. 1725. *Specimen inaugurale medicum exhibens ideam generalem solidarum corporis humani partium*. Leiden: apud Conradum Wishoff.

Geer, Charles de. 1752–78. *Memoires pour servir a l'histoire des insectes*. 7 vols. Stockholm: L. L. Grefing.

Gerstmann, Bartoldus Florianus. 1687. *Disputatio medica inauguralis de peste*. Leiden: apud Abrahamum Elzevier.

Geulincx, Arnold. N.d. *Physica vera quae versatur circa hunc mundum*. N.p.

Gleichen-Russworm, Wilhelm Friedrich von. 1764. *Das Neueste aus dem Reiche der Pflanzen*. [Nuremberg]: Christian de Launoy seel. Erben.

1777. *Auserlesene mikroskopische Entdeckungen bey den Pflanzen, Blumen und Blüthen, Insekten und andern Merkwürdigkeiten*. Nuremberg: Adam Wolfgang Winterschmidt.

1778. *Abhandlung über die Saamen- und Infusionsthierchen, und über die Erzeugung*. Nuremberg: Adam Wolfgang Winterschmidt.

Glisson, Francis. 1654. *Anatomia hepatis.* London: typis Du-Gardianis, impensis Octaviani Pullein.

Goedaert, Johannes. 1662?–9. *Metamorphosis naturalis, ofte Historische beschryvinghe vanden oirspronk, aerd, eygenschappen ende vreemde veranderinghen der wormen, rupsen, maedan, vliegen, witjens, byen, motten ende dierghelijcke dierkens meer.* With additions by Johannes de Mey. 3 vols. Middelburg: Jaques Fierens.

1700. *Metamorphoses naturelles ou histoire des insectes.* With additions by Johannes de Mey. [Translated from the Dutch.] 3 vols. Amsterdam: George Gallet.

Goorle, David van. 1620. *Exercitationes philosophicae quibus universa fere discutitur philosophia theoretica.* N.p.: sumptibus viduae Ioannis Comelini.

Gorter, Johannes de. 1735–7. *Medicinae compendium.* 2 vols. Leiden: apud Janssonios vander Aa, Balduinum vander Aa et Petrum vander Aa.

Graaf, Regnier de. 1668. *De virorum organis generationi inservientibus.* Leiden and Rotterdam: ex officina Hackiana.

1672. *De mulierum organis generationi inservientibus tractatus novus.* Leiden: ex officina Hackiana.

1677. *Opera omnia.* Leiden: ex officina Hackiana.

Gray, Stephen. 1696a. A Letter from Mr. Stephen Gray, Giving a Further Account of His Water Microscope. *Philosophical Transactions,* no. 223, pp. 353–6.

1696b. Several Microscopical Observations and Experiments. *Philosophical Transactions,* no. 221, pp. 281–2.

1697. A Letter from Mr. Stephen Gray, from Canterbury, May the 12th 1697, Concerning Making Water Subservient to the Viewing Both Near and Distant Objects. *Philosophical Transactions,* no. 228, pp. 539–42.

[Grew, Nehemiah.] 1678. Auctoris ad observatorem responsum. *Philosophical Transactions,* no. 142, p. 1043.

1682. *The Anatomy of Plants.* Reprint, New York and London: Johnson Reprint Corp., 1965.

Griendel von Ach, Johann Franz. 1687. *Micrographia nova.* Nuremberg: in Verlegung Johann Ziegers.

Guenellon, Pieter. 1680. *Epistolica dissertatio, de genuinâ medicinam instituendi ratione.* Amsterdam: apud Adrianum à Gaasbeek.

Haberkorn, Johannes Christianus. 1693. *Disputatio medica inauguralis, de medico physico.* Leiden: apud Abrahamum Elzevier.

Hahn, Ioh. Carolus. 1728. *De anatomes subtilioris utilitate.* Praeside Laurentio Heistero. Helmstedt: ex officina Pavli Dieterici Schnorrii.

Haller, Albrecht von. 1757–66. *Elementa physiologiae corporis humani.* 8 vols. Lausanne: Marci-Michael Bousquet.

1758. *Sur la formation du coeur dans le poulet.* 2 vols. Lausanne: Marci-Michael Bousquet.

1774–7. *Bibliotheca anatomica.* 2 vols. Zürich: apud Orell, Gessner, Fuessli, et Socc.

Hartsoeker, Nicolaas [NH]. 1694. *Essay de dioptrique.* Paris: Jean Anisson.

1696. *Principes de physique.* Paris: Jean Anisson.

1706. *Conjectures physiques.* Amsterdam: Henri Desbordes.

1708. *Suite des Conjectures physiques.* Amsterdam: Henri Desbordes.

1710. *Eclaircissemens sur les Conjectures physiques.* Amsterdam: Pierre Humbert.

1712. *Seconde partie de la Suite des Conjectures physiques.* Amsterdam: Nicolas Viollet.

1722. *Recueil de plusieurs pieces de physique.* Utrecht: la veuve de G. Broedelet et fils.

1730. *Cours de physique.* The Hague: Jean Swart.

Hartsoeker, Nicolaas, [and Christiaan Huygens]. 1678. Extrait d'une lettre de M. Nicolas Hartsoker écrite à l'auteur du journal touchant la maniere de faire les nouveaux microscopes, dont il a esté parlé dans le journal il y a quelques jours. *Journal des Sçavans* 6:371–3.

Harvey, William. 1628. *Exercitatio anatomica de motu cordis et sanguinis in animalibus*. Frankfurt: sumptibus Guilielmi Fitzeri.

1651. *Exercitationes de generatione animalium*. London: typis Du-Gardianis.

Heereboord, Adriaan. 1654. *Meletemata philosophica, maximam partem, metaphysica*. Leiden: ex officinâ Francisci Moyardi.

1663. *Philosophia naturalis, cum commentariis peripateticis antehac edita*. Leiden: ex officinâ Cornelii Driehuysen.

Heide, Anton de. 1684. *Anatome mytuli, belgicè mossel, structuram elegantem ejusque motum mirandum exponens*. Amsterdam: apud JanssonioWaesbergios.

1686. *Experimenta circa sanguinis missionem, fibras motrices, urticam marinam etc.* 2d ed. Amsterdam: apud JanssonioWaesbergios.

Helmont, Johannes Baptista van. 1648. *Ortus medicinae*. Edited by Franciscus Mercurius van Helmont. Amsterdam: apud Ludovicum Elzevirium.

Henle, Jacob. 1841. *Allgemeine Anatomie*. Leipzig: Leopold Voss.

Hevelius, Johannes. 1647. *Selenographia*. Gdansk: typis Hünefeldianis.

Hewson, William. 1774. On the Figure and Composition of the Red Particles of the Blood, Commonly Called the Red Globules. *Philosophical Transactions* 63(pt. 2): 303–23.

Hexham, Hendrick. 1648. *Het groot woorden-boeck*. Rotterdam: Arnovt Leers.

Highmore, Nathaniel. 1651. *The History of Generation*. London: John Martin.

Hill, John. 1752a. *Essays in Natural History and Philosophy*. London: J. Whiston, B. White, P. Vaillant, & L. Davis.

1752b. *An History of Animals*. London: Thomas Osborne.

Hind, Arthur M. 1915–32. *Catalogue of Drawings by Dutch and Flemish Artists Preserved in the Department of Prints and Drawings in the British Museum*. 5 vols. London: British Museum.

Hoefnagel, Jacob. 1592. *Archetypa studiaque patris Georgii Hoefnagelii Jacobus f: genio duce ab ipso scalpta*. Frankfurt a. Main: n.p.

1630. *Diversae insectarum volatilium icones ad vivum accuratissimè depictae*. N.p.: typisque mandatae a Nicolao Ioannis Visscher.

Hoefnagel, Joris, and Georg Bocskay. 1992. *Mira calligraphiae monumenta*. Edited by Lee Hendrix and Thea Vignau-Wilberg. Malibu: J. Paul Getty Museum.

Hoeve, Jacobus vander. 1690. *Disputatio medica inauguralis, continens ideam medicinae brevem*. Leiden: apud Abrahamum Elzevier.

Hollstein, F. W. H. 1949–. *Dutch and Flemish Etchings, Engravings, and Woodcuts, ca. 1450–1700*. 40 vols. to date. Amsterdam: Menno Hertzberger, etc.

Holwarda, Jan Fokkens. 1651. *Philosophia naturalis, seu physica vetus-nova*. Franeker: Idzardus Alberti.

Hoogstraten, Samuel van. 1678. *Inleyding tot de hooge schoole der schilderkonst: anders de zichtbaere werelt*. Rotterdam: Fransois van Hoogstraeten.

Hoogvliet, Arnold. 1738. *Mengeldichten*. 2 vols. Delft: Pieter vander Kloot.

Hooke, Robert. 1665. *Micrographia*. Reprint, New York: Dover, 1961.

1678. *Lectures and Collections*. London: J. Martyn.

1726. *Philosophical Experiments and Observations of the Late Eminent Dr. Robert Hooke*. Reprint, London: Frank Cass, 1967.

1935. *The Diary of Robert Hooke*. Edited by Henry W. Robinson and Walter Adams. London: Taylor & Francis.

Horne, Johannes van. 1652. *Novus ductus chyliferus*. Leiden: e typographeo Francisci Hackii.

1660a. *ΜΙΚΡΟΚΟΣΜΟΣ seu, brevis manuductio ad historiam corporis humani*. Leiden: ex officina Jacobi Chouët.

1660b. *Waerschouwinge, aen alle lieff-hebbers der anatomie*. Leiden: Daniel ende Abraham van Gaasbeeck.

1668. *Suarum circa partes generationis in utroque sexu observationum prodromus: ad celeberrimum virum D. Guernerum Rolfinckium, anatomicum veteranum exercitatissimum*. Leiden: apud Gaasbekios.

Horne, Johannes van, and Archibaldus Stephanides. 1660. *Dissertationis anatomicae de ductu chylifero, pars prior*. Leiden: apud Johannem Elsevirium.

Hortensius, Martinus. 1631. *Responsio ad additiunculam D. Ioannis Kepleri, Caesarei Mathematici, praefixam ephemeridi eius in annum 1624*. Leiden: ex officinâ Ioannis Maire.

Houbraken, Arnold. 1753. *De groote schouburgh der nederlantsche konstschilders en schilderessen*. 2d ed. 3 vols. Reprint, Amsterdam: B. M. Israë, 1976.

[Houttuyn, Martinus.] 1761–85. *Natuurlyke historie of uitvoerige beschryving der dieren, planten en mineraalen, volgens het samenstel van den Heer Linnaeus*. 37 vols. Amsterdam: F. Houttuyn, etc.

Hunter, John. 1794. *A Treatise on the Blood, Inflammation, and Gun-Shot Wounds*. London: George Nicol.

Huygens, Christiaan [ChrH]. 1678. Extrait d'une lettre de M. Huguens de l'Acad. R. des Sciences à l'auteur du journal, touchant une nouvelle maniere de microscope qu'il a apporté de Hollande. *Journal des Sçavans* 6:345–7.

1888–1950. *Oeuvres complètes de Christiaan Huygens*. 22 vols. in 23. The Hague: Martinus Nijhoff. [*OCCH*]

Huygens, Constantijn, Jr. [ConHj]. 1881. *Journaal van Constantijn Huygens, den zoon, gedurende de veldtochten der jaren 1673, 1675, 1676, 1677 en 1678*. Edited by the Historisch Genootschap te Utrecht. Utrecht: Kemink & Zoon.

Huygens, Constantijn, Sr. [ConHs]. 1892–9. *De gedichten van Constantijn Huygens*. Edited by J. A. Worp. 9 vols. Groningen: J. B. Wolters.

1897. Fragment eener autobiographie van Constantijn Huygens. Edited by J. A. Worp. *Bijdragen en mededeelingen, Historisch Genootschap te Utrecht* 18:1–122.

1911–17. *De briefwisseling van Constantijn Huygens (1608–1687)*. Edited by J. A. Worp. 6 vols. The Hague: Martinus Nijhoff.

J.H.S.M.F. 1694. *Wonderen der natuyr, of een beschryvingh van de wonderlijke geschapenheyt der natuyr*. The Hague: Meyndert Uytwerf.

Jacchaeus, Gilbertus. 1653. *Institutiones medicae*. 3d ed. Leiden: ex officinâ Joannis Maire.

Jacobaeus, Holger. 1910. *Holger Jacobaeus' Rejsebog (1671–1692)*. Edited by Vilhelm Maar. Copenhagen: Gyldendalske Boghandel, Nordisk Forlag.

Joblot, Louis. 1718. *Descriptions et usages de plusieurs nouveaux microscopes, tant simples que composez*. Paris: Jacques Collombat.

Jongh, Laurentius de. 1685. *Disputatio medica inauguralis de peste*. Leiden: apud Abrahamum Elzevier.

Jonston, John. 1657. *Historia naturalis de insectis*. Bound with *Historiae naturalis de quadrupedibus libri I*. Amsterdam: apud Ioannem Iacobi Fil. Schipper.

Journal des Sçavans. 1665–. Amsterdam: Pierre le Grand, etc.

Journal literaire. 1713–37. The Hague: T. Johnson, etc.

Kepler, Johannes. 1937–. *Gesammelte Werke*. Edited by Walther von Dyck, Max Caspar, and Franz Hammer. 18 vols. to date. Munich: C. H. Beck.

Kerckring, Theodorus. 1670. *Spicilegium anatomicum*. Amsterdam: sumptibus Andreae Frisii.

1671. *Anthropogeniae ichnographia*. Amsterdam: sumptibus Andreae Frisii.

King, Edmund. 1666. Some Considerations concerning the Parenchymous Parts of the Body. *Philosophical Transactions*, no. 18, pp. 316–20.

Kircher, Athanasius. 1646. *Ars magna lucis et umbrae in decem libros digesta*. Rome: sumptibus Hermanni Scheus.

1658. *Scrutinium physico-medicum contagiosae luis, quae pestis dicitur*. Rome: typis Mascardi.

1665. *Mundus subterraneus*. 2 vols. Amsterdam: apud Joannem Janssonium et Elizeum Weyerstraten.

1682. *D'onder-aardse weereld.* Translated from the Latin. 2 vols. in 1. Amsterdam: Erfgenamen van wylen J. Janssonius van Wassberge.

Kyper, Albert. 1650. *Anthropologia, corporis humani, contentorum, et animae naturam et virtutes secundum circularem sanguinis motum explicans.* Leiden: ex officinâ Adriani Wijngaerden.

1654. *Institutiones medicae.* Amsterdam: apud Joannem Janssonium.

Lairesse, Gerard de. 1707. *Het groot schilderboek.* 2 vols. Amsterdam: Erfgenaamen van Willem de Coup.

La Mettrie, Julien Offray de. 1751. *Oeuvres philosophiques.* 2 vols. Reprint, Hildesheim and New York: Olms, 1970.

Lange, Christian. 1688. *Pathologia animata.* In *Opera omnia.* Frankfurt a. Main: sumptibus Georgii Henrici Oehrlingii.

LeClerc, Daniel, and Jean Jacques Manget. 1685. *Bibliotheca anatomica sive recens in anatomia inventorum thesaurus locupletissimus.* 2 vols. Geneva: sumptibus Joannis Anthonii Chouët.

Leclerc, Jean. 1686–1718. *Bibliotheque universelle et historique.* 26 vols. Amsterdam: Wolfgang, etc.

1696. *Physica sive de rebus corporeis libri quinque.* Amsterdam: apud Georgium Gallet.

1704. *Opera philosophica.* 3d ed. 4 vols. in 2. Amsterdam: Joan. Ludov. de Lorme.

1714–30. *Bibliotheque ancienne et moderne.* 29 vols. Amsterdam: David Mortier.

Ledermüller, Martin Frobenius. 1756. *Physicalische Beobachtungen derer Saamenthiergens.* Nuremberg: George Peter Monath.

1762. *Nachleese seiner mikroskopischen Gemüths- und Augen-Ergtzung.* Nuremberg: Christian de Launoy.

1776. *Mikroskoopische vermaaklykheden.* Translated from the German. 4 vols. in 2. Amsterdam: erven van F. Houttuyn.

Leeuwenhoek, Antoni van [AvL]. 1693. *Derde vervolg der brieven.* Delft: Henrik van Kroonevelt.

1694. *Vierde vervolg der brieven.* Delft: Henrik van Kroonevelt.

1696. *Vijfde vervolg der brieven.* Delft: Henrik van Krooneveld.

1697. *Sesde vervolg der brieven.* Delft: Henrik van Krooneveld.

1702. *Sevende vervolg der brieven.* Delft: Henrik van Krooneveld.

1718. *Send-brieven.* Delft: Adriaan Beman.

1939–. *Alle de brieven van Antoni van Leeuwenhoek.* 12 vols. to date. Amsterdam: Swets & Zeitlinger. [*AB*]

Leibniz, Gottfried Wilhelm. 1875–90. *Die philosophischen Schriften.* Edited by C. I. Gerhardt. 7 vols. Reprint, Hildesheim: Georg Olms, 1960–1.

Lemnius, Livinus. 1583. *De miraculis occultis naturae.* Cologne: apud Theodorum Baumium.

Lesser, Friedrich Christian. 1745. *Theologie des insectes, ou démonstration des perfection de Dieu dans tout ce qui concerne les insectes.* Translated and annotated by Pierre Lyonet. 2 vols. Paris: Hugues-Daniel Chaubert & Laurent Durand.

Libavius, Andreas. 1632. *Observatio bombycum, historia singularis.* Pp. 377–405 in John Jonston, *Thaumatographia naturalis, in decem classes distincta.* Amsterdam: apud Guilielmum Blaeu.

Lieberkühn, Johannes Nathanael. 1745. *Dissertatio anatomico-physiologica de fabrica et actione villorum intestinorum tenuium hominis.* Leiden: Conrad et Georg. Jac. Wishof.

Linden, Johannes Antonides van der. 1653. *Medicina physiologica.* Amsterdam: apud Ioannem à Ravestein.

Linnaeus, Carl. 1766–8. *Systema naturae per regna tria naturae, secundum classes, ordines, genera, species, cum characteribus, differentiis, synonymis, locis.* 12th ed. 3 vols. in 4. Stockholm: impensis direct. Laurentii Salvii.

Linnaeus, Carl, and Johan Carl Roos. 1767. *Dissertatio academica, mundum invisibilem, breviter delineatura.* Uppsala: n.p.

Lipstorp, Gustavus Daniel. 1687. *Disputatio medica inauguralis, de animalculis in humano corpore genitis*. Leiden: apud Abrahamum Elzevier.

Luiken, Jan, and Kaspar Luiken. 1694. *Spiegel van het menselyk bedryf, vertoonende honderd verscheiden ambachten, konstig afgebeeld en met godlyke spreuken en stichtelyke verzen verrykt*. Leiden: A. W. Sijthoff.

Lussingh, Hendrik. 1783. Gods wijsheid in zijne werken. Pp. 68–102 in *Tael- en dichtlievende oefeningen, van het Genootschap ter spreuke voerende: Kunst wordt door arbeid verkreegen*, vol. 4. Leiden: C. van Hoogeveen, Junior, & C. Heyligert.

Lyonet, Pierre. 1757. Beschryving van een microscoopstel, geschikt tot het ontleden van kleine dieren, mitsgaders eenige aanmerkingen over het vermogen der vergrootende glazen. *Verhandelingen, Hollandse maatschappy der weetenschappen, Haarlem* 3:378–413.

1762. *Traité anatomique de la chenille, qui ronge le bois de saule*. The Hague: n.p.

1832. *Recherches sur l'anatomie et les métamorphoses de différentes espèces d'insectes*. Published by M. W. de Haan. Paris and London: J.-B. Baillière.

M.A.C.D. 1726. *Systême d'un medecin anglois sur la cause de toutes les especes de maladies*. Paris: Alexis-Xavier-Rene' Mesnier.

1727. *Suite du Systême d'un medecin anglois*. Paris: Alexis-Xavier-Rene' Mesnier.

Maillet, Benoît de. 1748. *Telliamed, ou, entretiens d'un philosophe Indien avec un missionnaire François sur la diminution de la mer, la formation de la terre, l'origine de l'homme, &c*. Edited by J. A. G. 2 vols. Amsterdam: L'Honoré et fils.

1968. *Telliamed, or Conversations between an Indian Philosopher and a French Missionary on the Diminution of the Sea*. Edited and translated by Albert V. Carozzi. Urbana, etc.: University of Illinois Press.

Malebranche, Nicolas. 1958–70. *Oeuvres complètes*. Edited by André Robinet. 20 vols. Paris: J. Vrin.

Malpighi, Marcello. 1661a. *De pulmonibus observatio anatomica*. [Bologna: typis Io. Baptistae Ferronii.]

1661b. *De pulmonibus epistola altera*. [Bologna: typis Io. Baptistae Ferronii.]

1665a. *De externo tactus organo anatomica observatio*. Naples: apud Aegidium Longum.

1665b. *Epistolae anatomicae de cerebro, ac lingua . . . quibus anonymi accessit exercitatio de omento, pinguedine, et adiposis ductibus*. Bologna: typis Antonij Pisarrii.

1666. *De viscerum structura exercitatio anatomica*. Bologna: ex typographia Iacobi Montii.

1669. *Dissertatio epistolica de bombyce*. London: apud Joannem Martyn et Jacobum Allestry.

1686–7. *Opera omnia*. 2 vols. London: apud Robertum Littlebury.

1697a. *De structura glandularum conglobatarum consimiliumque partium epistola*. London: apud Richardum Chiswell. This edition appeared bound with (or within) *Opera posthuma*.

1697b. *Opera posthuma*. London: A. & J. Churchill.

1975. *The Correspondence of Marcello Malpighi*. Edited by Howard B. Adelmann. 5 vols. Ithaca and London: Cornell University Press.

Mander, Karel van. 1604a. *Dutch and Flemish Painters*. Translated by Constant van de Wall. New York: McFarlane, Warde, McFarlane, 1936.

1604b. *Het schilder-boeck waer in voor eerst de leerlustighe iueght den grondt der edel vry schilderconst in verscheyden deelen wort voorghedraghen*. Reprint, Utrecht: Davaco, 1969.

Martinet, Johannes Florentius. 1778–9. *Katechismus der natuur*. 4 vols. Amsterdam: Johannes Allart.

Mason, Jacobus. 1700. *Dissertatio medica inauguralis de lumbricis*. Leiden: apud Abrahamum Elzevier.

Massuet, Pierre. 1733. *Recherches interessantes sur l'origine, la formation, le developement, la structure, &c. des diverses especes de vers à tuyau, qui infestent les vaisseaux, les digues, &c. de quelques-unes des Provinces-Unies*. Amsterdam: François Changuion.

1752. *Elemens de la philosophie moderne*. 2 vols. Amsterdam: Z. Chatelain et fils.

Merian, Maria Sibylla. 1730. *De Europische Insecten.* Amsterdam: J. F. Bernard.

N.d. *Metamorphosis insectorum surinamensium.* Amsterdam: Gerard Valck.

Meteren, Emanuel van. 1614. *Historie der Neder-landscher ende haerder na-buren oorlogen ende geschiedenissen, tot den iare M.VI.XII.* The Hague: Hillebrant Iacobssz.

Mey, Johannes de. 1742. *Al de godgeleerde en natuurkundige wercken.* 2 vols. Amsterdam: Johannes Rotterdam.

Miles, Henry. 1741. Some Remarks concerning the Circulation of the Blood, As Seen in the Tail of a Water-Eft, through a Solar Microscope. *Philosophical Transactions,* no. 460, pp. 725–9.

Moffett, Thomas, Edward Wotton, Konrad Gesner, and Thomas Penny. 1634. *Insectorum sive minimorum animalium theatrum.* London: Thom. Cotes.

Molyneux, William. 1685. A Letter . . . concerning the Circulation of the Blood As Seen, by the Help of a Microscope, in the Lacerta Aquatica. *Philosophical Transactions,* no. 177, pp. 1236–8.

1692. *Dioptrica Nova.* London: Benj. Tooke.

Monconys, Balthasar de. 1665–6. *Journal des voyages de Monsieur de Monconys.* 3 parts in 2 vols. Lyon: Horace Boissat & George Remeus.

Mort, Jacob le. 1696. *Chymia medico-physica, ratione et experientia nobilitata.* Leiden: apud Cornelium Boutesteyn [et] Fredericum Haaring.

Mul, Johannes. 1694. *Disputatio medica inauguralis, de fluore albo.* Leiden: apud Abrahamum Elzevier.

Müller, Otto Frederik. 1773–4. *Vermium terrestrium et fluviatilium, seu animalium infusorium, helminthicorum et testaceorum, non marinorum, succincta historia.* 2 vols. in 3. Copenhagen and Leipzig: Heineck, etc.

1786. *Animalcula infusoria fluviatilia et marina.* Copenhagen: typis Nicolai Mölleri.

Munniks, Wijnoldus. 1771. *Oratio inauguralis de summis quas anatome habet deliciis.* Groningen: apud Hajonem Spandaw.

Musschenbroek, Johan van. 1736. Lyst der natuurkundige, wiskundige, anatomische, en chirurgische instrumenten, welke by Jan van Musschenbroek, te vinden zyn te Leyden. In Petrus van Musschenbroek (1736).

Musschenbroek, Petrus van. 1736. *Beginselen der natuurkunde.* Leiden: Samuel Luchtmans.

1762. *Introductio ad philosophiam naturalem.* 2 vols. Leiden: Sam. & Joh. Luchtmans.

Muys, Wijer Willem. 1714. *Oratio inauguralis de theoriae medicae usu atque recta illam excolendi ratione.* Franeker: Franciscus Halma.

1738. *Investigatio fabricae, quae in partibus musculos componentibus extat.* Leiden: Joh. Arnold Langerak.

Needham, John Turberville. 1743. A Letter from Mr. Turbevil Needham, to the President. *Philosophical Transactions,* no. 471, pp. 634–41.

1745. *An Account of Some New Microscopical Discoveries Founded on an Examination of the Calamary, and Its Wonderful Milt-Vessels.* London: F. Needham.

1748. A Summary of Some Late Observations upon the Generation, Composition, and Decomposition of Animal and Vegetable Substances. *Philosophical Transactions* 45:[615]–66.

Needham, Walter. 1667. *Disquisitio anatomica de formato foetu.* London: typis Gulielmi Godbid.

Nieuwentyt, Bernard. 1717. *Het regt gebruik der werelt beschouwingen.* Amsterdam: de Wed. J. Wolters en J. Pauli.

Nieuwentyt, Eleazar. 1686. *Disputatio medica inauguralis de sanguine.* Leiden: apud Abrahamum Elzevier.

Nuck, Anton. 1685. *De ductu salivali novo, saliva, ductibus oculorum aquosis, et humore oculi aqueo.* Leiden: apud Petrum vander Aa.

1691a. *Adenographia curiosa et uteri foeminei anatome nova.* Leiden: apud Jordanum Luchtmans.

1691b. *Defensio ductuum aquosorum.* Leiden: apud Jordanum Luchtmans.

Nylant, P., and J. van Hextor. 1672. *Het schouw-toneel der aertsche schepselen.* Amsterdam: Marcus Willemsz. Doornick.

Oken, Lorenz. 1805. *Die Zeugung.* Bamberg and Würzburg: Joseph Anton Goebhardt.

Oldenburg, Henry. 1965–. *The Correspondence of Henry Oldenburg.* Edited and translated by A. Rupert Hall and Marie Boas Hall. 9 vols. to date. Madison: University of Wisconsin Press.

Oudaen, Joachim. 1712. *Poëzy.* 3 vols. Amsterdam: de Wed. P. Arentz en K. vander Sys.

Overkamp, Heidentryk. 1686. *Nieuwe beginselen tot de genees- en heel-konst.* 2d ed. Amsterdam: Timotheus ten Hoorn.

Paracelsus, Theophrastus. 1965–8. *Werke.* Edited by Will-Erich Peuckert. 5 vols. Basel and Stuttgart: Schwabe.

Pascal, Blaise. 1977. *Pensées.* Edited by Michel Le Guern. 2 vols. [Paris]: Gallimard.

Paulitz, Johannes Theodorus à. 1696. *Disputatio medica inauguralis de morbis animatis.* Leiden: apud Abrahamum Elzevier.

Pecquet, Jean. 1651. *Experimenta nova anatomica, quibus incognitum hactenus chyli receptaculum, & ab eo per thoracem in ramos usque subclavios vasa lactea deteguntur.* Paris: apud Sebastianum et Gabrielem Cramoisy.

Philosophical Transactions. 1666–. London: Royal Society.

[Plantade, François de.] 1699. Extrait d'une lettre de M. Dalenpatius à l'auteur de ces Nouvelles, contenant une découverte curieuse, faite par le moyen du microscope. *Nouvelles de la République des lettres* 12:552–4.

Plemp, Vopiscus Fortunatus. 1630. *Verhandelingh der spieren.* Amsterdam: Iacob Aertsz. Colom.

1632. *Ophthalmographia, sive tractatio de oculi fabricâ, actione, et usu praeter vulgatas hactenus philosophorum ac medicorum opiniones.* Amsterdam: sumptibus Henrici Laurentii.

1659. *Ophthalmographia sive tractatio de oculo.* 3d ed. Louvain: typis ac sumptibus Hieronymi Nempaei.

Pliny. 1938–63. *Natural History [Historia naturalis].* With an English translation by H. Rackham, W. H. S. Jones, and D. E. Eichholz. 10 vols. Cambridge: Harvard University Press / London: William Heinemann. [Pliny *HN*]

Pluche, Noël Antoine. 1732–50. *Le Spectacle de la nature.* 8 vols. in 9. Paris: la veuve Estienne et Jean Desaint.

Porjeere, Olivier. 1783. *Gods wijsheid in zijne werken.* Pp. 23–67 in *Tael-en dichtlievende oefeningen, van het Genootschap ter spreuke voerende: Kunst wordt door arbeid verkreegen,* vol. 4. Leiden: C. van Hoogeveen, Junior, & C. Heyligert.

Porras, Manuel de. 1716. *Anatomia galenico-moderna.* Madrid: Bernardo Peralta.

Porta, Giovan Battista della. 1962. *De Telescopio.* Edited by Vasco Ronchi and Maria Amalia Naldoni. Florence: Leo S. Olschki.

Pouchet, F. 1848. Note sur les organes digestifs et circulatoires des animaux infusoires. *Comptes rendus hebdomadaires sur séances de l'Académie des Sciences* 27:516–18.

Power, Henry. 1664. *Experimental Philosophy.* London: John Martin & James Allestry.

1934. In Commendation of ye Microscope. In "Dr Henry Power's Poem on the Microscope," by Thomas Cowles. *Isis* 21:71–80.

Rabus, Pieter, and Pieter vander Slaart. 1692–1701. *De boekzaal van Europe.* 19 vols. Rotterdam: Pieter vander Slaart.

Raey, Joannes de. 1654. *Clavis philosophiae naturalis, seu introductio ad naturae contemplatione, aristotelico-cartesiana.* Leiden: ex officinâ Johannis et Danielis Elsevier.

Ramon y Cajal, Santiago. 1920. *Reglas y consejos sobre investigacion cientifica.* 5th ed. Madrid: Nicolas Moya, Garcilaso, y Carretas.

Rau, Johannes Jacobus. 1694. *Disputatio anatomico-medica inauguralis de ortu et regeneratione dentium.* Leiden: apud Abrahamum Elzevier.

1713. *Oratio inauguralis de methodo anatomen docendi et discendi.* Leiden: apud Samuelem Luchtmans.

Ray, John. 1691. *The Wisdom of God Manifested in the Works of the Creation.* Reprint, New York and London: Garland, 1979.

1693. *Synopsis methodica animalium quadrupedum et serpentini generis.* London: S. Smith & B. Walford.

Réaumur, René-Antoine Ferchault de. 1734–1929. *Mémoires pour servir à l'histoire des insectes.* 7 vols. Paris: Imprimerie Royale & R. Lechevalier.

1751. *Art de faire éclorre et d'élever en toute saison des oiseaux domestiques de toutes especes, soit par le moyen de la chaleur du fumier, soit par le moyen de celle de feu ordinaire.* 2d ed. 2 vols. Paris: Imprimerie Royale.

Redi, Francesco. 1668. *Esperienze intorno alla generazione degl' insetti.* Florence: all'Insegna della Stella.

1688. *Experiments on the Generation of Insects.* Translated by Mab Bigelow. Reprint, New York: Kraus, 1969.

Regius, Henricus. 1646. *Fundamenta physices.* Amsterdam: apud Ludovicum Elzevirium.

1647. *Spongia, qua eluuntur sordes animadversionum, quas Iacobus Primirosius, Doctor Medicus, adversus theses pro circulatione sanguinis in Academia Vltrajectina disputatas nuper edidit.* In *Recentiorum disceptationes de motu cordis, sanguinis, et chyli, in animalibus.* Leiden: ex officina Ioannis Maire.

1668. *Medicina, et praxis medica, medicationum exemplis demonstrata.* 3d ed. Utrecht: ex officinâ Theodori ab Ackersdijck.

Reneri, Henricus. 1634. *Oratio inauguralis de lectionibus ac exercitiis philosophicis, in Illustris Gymnasii Ultraiectini inauguratio una cum orationibus inauguralibus.* Utrecht: ex officinâ Abrahami ab Herwiick et Hermanni Ribbii.

Ridley, Humphrey. 1695. *The Anatomy of the Brain.* London: Sam. Smith & Benj. Walford.

Rieger, Joannes Christophorus. 1763. *Introductio in notitiam rerum naturalium et arte factarum, quarum in communi vita, sed praecipue in medicina usus est.* 2 vols. in 4. The Hague: apud Petrum Gosse.

Rösel von Rosenhof, August Johann. 1746–61. *Der monatlich-herausgegebenen Insecten-Belustigung.* 4 vols. Nuremberg: Johann Joseph Fleischmann & C. F. C. Kleeman.

Ross, Alexander. 1652. *Arcana Microcosmi; or, the Hid Secrets of Man's Body Discovered.* London: Tho. Newcomb.

Ruysch, Frederik. 1665. *Dilucidatio valvularum in vasis lymphaticis, et lacteis.* Amsterdam: apud Janssonio Waesbergios, 1727. This edition appears in Ruysch (1725–51), vol. 1.

1701. *Thesaurus anatomicus primus.* Amsterdam: apud Joannem Wolters.

1702. *Thesaurus anatomicus secundus.* Amsterdam: apud Joannem Wolters.

1703. *Thesaurus anatomicus tertius.* Amsterdam: apud Joannem Wolters.

1704. *Thesaurus anatomicus quartus.* Amsterdam: apud Joannem Wolters.

1705a. *Thesaurus anatomicus quintus.* Amsterdam: apud Joannem Wolters.

1705b. *Thesaurus anatomicus sextus.* Amsterdam: apud Joannem Wolters.

1725. *Ontleedkundige verhandelingen over de vinding van een spier in de grond des baar-moeders.* Amsterdam: de Janssoons van Waasberge.

1725–51. *Opera omnia anatomico-medico-chirurgica, huc usque edita.* 4 vols. Amsterdam: apud Janssonio–Waesbergios.

1726. *Adversariorum anatomico-medico-chirurgicorum decas secunda.* Amsterdam: apud Janssonio–Waesbergios. This edition appears in Ruysch (1725–51), vol. 2.

1729. *Thesaurus magnus et regius qui est decimus Thesaurorum anatomicorum.* Amsterdam: apud Janssonio–Waesbergios. This edition appears in Ruysch (1725–51), vol. 4.

1733. *Adversariorum anatomico-medico-chirurgicorum decas tertia.* Amsterdam: apud Janssonio–Waesbergios. This edition appears in Ruysch (1725–51), vol. 2.

1739a. *Musaeum anatomicum Ruyschianum, sive catalogus rariorum, quae in authoris aedibus asservantur.* Amsterdam: apud Janssonio–Waesbergios. This edition appears in Ruysch (1725–51), vol. 1.

1739b. *Thesaurus anatomicus primus*. Amsterdam: apud Janssonio–Waesbergios. This edition appears in Ruysch (1725–51), vol. 3.

1744a. *Thesaurus anatomicus nonus*. Amsterdam: apud Janssonio–Waesbergios. This edition appears in Ruysch (1725–51), vol. 3.

1744b. *Thesaurus anatomicus sextus*. Amsterdam: apud Janssonio–Waesbergios. This edition appears in Ruysch (1725–51), vol. 3.

Schacht, Herman Oosterdijk. 1723. *Orationes duae, quarum prima est de firmitate artis medicae . . . altera de medico exercitato*. Leiden: apud Samuelem Luchtmans.

Scheuchzer, Johann Jacob. 1731–5. *Physica sacra*. 4 vols. Augsburg and Ulm: procurante Johanne Andrea Pfeffel.

Schleiden, Matthias Jacob. 1838. Beiträge zur Phytogenesis. *Archiv für Anatomie, Physiologie und wissenschaftliche Medicin*, pp. 137–76.

1841. Contributions to Our Knowledge of Phytogenesis. *Scientific Memoirs* 2:281–312. Reprint, New York and London: Johnson Reprint Corp., 1966.

1849. *Principles of Scientific Botany*. Translated by Edwin Lankester. London: Longman, Brown, Green, & Longmans.

Schott, Gaspar. 1657–77. *Magia universalis naturae et artis*. 4 parts in 2 vols. Würzburg and Bamberg: sumptibus haeredum Joannis Godefridi Schnwetteri [et] Joh. Martini Schnwetteri.

Schrader, Friedrich. 1679. *Disputatio medica inauguralis de venenis et antidotis*. Leiden: apud viduam et heredes Johannis Elsevirii.

1681. *Dissertatio epistolica de microscopiorum usu in naturali scientia et anatome*. Göttingen: sumptibus Bartholdi Fuhrmanns, typis Johannis Christophori Hampii.

Schrader, Justus. 1674. *Observationes et historiae omnes et singulae è Guiljelmi Harvei libello De generatione animalium excerptae, et in accuratissimum ordinem redactae*. Amsterdam: typis Abrahami Wolfgang.

Schreiber, Johann Friedrich. 1732. *Historia vitae et meritorum Frederici Ruysch*. Amsterdam: apud Janssonio–Waesbergios. This edition appears in Ruysch (1725–51), vol. 1.

Schwann, Theodor. 1839. *Mikroskopische Untersuchungen über die Uebereinstimmung in der Struktur und dem Wachsthum der Thiere und Pflanzen*. Berlin: G. E. Reimer.

1847. *Microscopical Researches into the Accordance in the Structure and Growth of Animals and Plants*. Translated by Henry Smith. London: Sydenham Society.

Schwencke, Thomas. 1743. *Haematologia, sive sanguinis historia, experimentis passim superstructa*. The Hague: Joh. Mart. Husson.

Sellius, Gottfried. 1733. *Historia naturalis teredinis seu xylophagi marini, tubulo-conchoidis speciatim belgici*. Utrecht: apud Hermannum Besseling.

Sénac, Jean Baptiste. 1749. *Traité de la structure du coeur, de son action, et de ses maladies*. 2 vols. Paris: Jacques Vincent.

Senguerdius, Wolferdus. 1685. *Philosophia naturalis*. 2d ed. Leiden: apud Danielem à Gaesbeeck.

Sepp, Jan Christiaan, and Christiaan Sepp. 1762-1860. *Beschouwing der wonderen Gods, in de minstgeachte schepzelen*. 8 vols. Amsterdam: J. C. Sepp.

Severino, Marco Aurelio. 1645. *Zootomia Democritaea*. Nuremberg: Literis Endterianis.

Spallanzani, Lazzaro. 1786. Observations et expériences faites sur les animalcules des infusions. [Translated by Jean Senebier.] 2 vols. in 1. Paris: Gauthier-Villars, 1920.

1934. *Le Opere di Lazzaro Spallanzani*. 5 vols. in 6. Milan: Ulrico Hoepli.

Steenevelt, Christianus à. 1697. *Dissertatio de ulcere verminoso*. Leiden: apud Jordanum Lugtmans.

Steno, Nicolaus. 1664. *De musculis et glandulis observationum specimen*. Amsterdam: apud Petrum le Grand.

1667. *Elementorum myologiae specimen, seu musculi descriptio geometrica*. Florence: ex typographia sub signo Stellae.

Sterre, Dionysius van der. 1687. *Tractatus novus de generatione ex ovo*. Amsterdam: apud Cornelium Blancardum.

Stiles, F. H. Eyles. 1765. Extracts of Three Letters . . . to Daniel Wray, Esq: F.R.S. concerning Some New Microscopes Made at Naples, and Their Use in Viewing the Smallest Objects. *Philosophical Transactions* 55:246–70.

Swammerdam, Jan [JS]. 1667a. *Disputatio medica inauguralis, continens selectas de respiratione positiones.* Leiden: apud viduam et haeredes Joannis Elsevirii.

1667b. *Tractatus physico-anatomico-medicus de respiratione usuque pulmonum.* Leiden: apud Danielem, Abraham. et Adrian. à Gaasbeeck.

1669. *Historia insectorum generalis, ofte algemeene verhandeling van de bloedeloose dierkens.* Utrecht: Meinardus van Dreunen.

1672. *Miraculum naturae sive uteri muliebris fabrica.* Leiden:apud Severinum Matthaei.

1675a. *Ephemeri vita, of afbeeldingh van 's menschen leven, vertoont in de wonderbaarelijcke en nooyt gehoorde historie van het vliegent ende een-dagh-levent haft of oever-aas.* Amsterdam: Abraham Wolfgang.

1675b. An Extract of a Letter, Lately Written to the Publisher by Dr. Swammerdam, of an Unusual Rupture of the Mesentery. *Philosophical Transactions,* no. 112, pp. 273–4.

1737–8. *Bybel der natuure.* Edited by Herman Boerhaave and translated into Latin by Hieronymus David Gaubius. 2 vols. Leiden: Isaak Severinus, Boudewyn Vander Aa, & Pieter Vander Aa.

1975. *The Letters of Jan Swammerdam to Melchisedec Thévenot.* Edited and translated by G. A. Lindeboom. Amsterdam: Swets & Zeitlinger. [*LJS*]

Swaving, Abraham Coenraad. 1799. Verhandeling over de infusie-diertjes. *Natuurkundige verhandelingen, Bataafsche maatschappy der wetenschappen, Haarlem* 1 (pt. 1): 49–84.

Sylvius, Franciscus dele Boë. 1679. *Opera medica.* Amsterdam: apud Danielem Elsevirium et Abrahamum Wolfgang.

Teresa of Avila. 1979. *The Interior Castle.* Translated by Kieran Kavanaugh and Otilio Rodriguez. New York, etc.: Paulist Press.

Terwen, Joannes. 1676. *Disputatio medica de cephalalgia, ex intemperie frigida.* Sub praesidio Theodori Craanen. Leiden: apud viduam et heredes Johannis Elsevirii.

Thomas à Kempis. 1892. *De imitatione Christi.* With an English translation. London: Kegan Paul, Trench, & Trübner.

Thomson, James. 1866. *The Poetical Works of James Thomson.* New York: American News Co.

Tol, Gysbertus van. 1674. *Disputatio medica inauguralis, de gonorrhoea virulenta.* Leiden: apud viduam et haeredes Johannis Elsevirii.

Torre, Giovanni Maria della. 1776. *Nuove osservazioni microscopiche.* Naples: n.p.

Trembley, Abraham. 1744. *Mémoires, pour servir à l'histoire d'un genre de polypes d'eau douce, à bras en forme de cornes.* Leiden: Jean et Herman Verbeek.

Tulp, Nicolaes. 1641. *Observationum medicarum, libri tres.* Amsterdam: apud Ludovicum Elzevirium.

Uffenbach, Zacharias Konrad von. 1753–4. *Merkwürdige Reisen durch Niedersachsen, Holland und Engelland.* 3 vols. Ulm and Memmingen: J. F. Gaum.

Uilkens, Jacob Albert. 1816. *Het nut en voordeel der insekten, bijzonder der zoodanige, die schadelijk genoemd worden, en van dewelken vele menschen eenen natuurlijken afkeer schijnt te hebben.* Amsterdam: Cornelis de Vries, etc. In *Verhandelingen, Maatschappij: Tot nut van 't algemeen,* vol. 16.

Vater, Abraham. 1727. *Epistola gratulatoria ad virum vere illustrem, Dominum celeberrimum Fredericum Ruyschium.* Amsterdam: apud Janssonio–Waesbergios. This edition appears in Ruysch (1725–51), vol. 4.

Velthuysen, Lambert van. 1657. *Tractatus duo medico-physici, unus de liene, alter de generatione.* Utrecht: typis Theodori ab Ackersdijck et Gisberti à Zyll.

1680. *Opera omnia.* 2 vols. Rotterdam: typis Reineri Leers.

Verheyen, Philippe. 1710a. *Corporis humani anatomiae.* 2d. ed. 2 vols. in 1. Brussels: Fratres t'Serstevens.

1710b. *Supplementum anatomicum sive anatomiae corporis humanis liber secundus*. Brussels: apud Fratres t'Serstevens.

Vieussens, Raymond. 1705. *Novum vasorum corporis humani systema*. Amsterdam: apud Paulum Marret.

Virchow, Rudolf. 1860. *Cellular Pathology*. Translated by Frank Chance. Ann Arbor: Edwards Brothers, 1940.

Volder, Burchardus de. 1695. *Exercitationes academicae, quibus Ren. Cartesii philosophia defenditur adversus Petri Danielus Huetii Episcopi Suessionensis Censuram philosophiae cartesianae*. Amsterdam: apud Arnoldum van Ravestein.

Vondel, Joost van den. 1820–4. *Dichterlijke werken van Joost van den Vondel*. 21 vols. Amsterdam: M. Westerman.

Voorde, Cornelis vande, and Anton de Heide. 1680. *Nieuw lichtende fakkel der chirurgie of hedendaagze heel-konst*. Rev. ed. Middelburg: Wilhelmus Goeree.

Vossius, Isaac. 1662. *De lucis natura et proprietate*. Amsterdam: apud Ludovicum et Danielem Elzevirios.

1663. *Responsum ad objecta Joh. de Bruyn professoris trajectini: et Petri Petiti medici parisiensis*. The Hague: ex officina Adriani Vlacq.

1666. *De Nili et aliorum fluminum origine*. The Hague: ex typographia Adriani Vlacq.

Wale, Johannes de. 1647. *Epistola prima de motu chyli et sanguinis: ad Thomam Bartholinum, Casp. filium*. In *Recentiorum disceptationes de motu cordis, sanguinis, et chyli, in animalibus*. Leiden: ex officina Ioannis Maire.

Walker, Thomas. 1688. *Disputatio medica inauguralis, de hydrope intercute, seu anasarca*. Leiden, apud Abrahamum Elzevier.

Wedelius, Christianus. 1737. *Epistola anatomica, problematica tertia et decima*. Amsterdam: apud JanssonioWaesbergios. This edition appears in Ruysch (1725–51), vol. 2.

Westerbaen, Jacob. 1672. *Ockenburgh*. In *Gedichten*, vol. 1. The Hague: Johannes Tongerloo.

Wiel, Petrus Stalpart vander. 1686. *Dissertatio medica inauguralis de nutritione foetus*. Leiden: apud Abrahamum Elzevier.

Wilkins, John. 1675. *Of the Principles and Duties of Natural Religion*. London: T. Basset, etc.

Wolff, Caspar Friedrich. 1759. *Theoria generationis*. Reprint, Hildesheim: Georg Olms, 1966.

Wolff, Christian. 1962–. *Gesammelte Werke*. Edited by J. École, H. W. Arndt, C. A. Corr, J. E. Hofmann, and M. Thomann. To date, 3 *Abteilungen* of 81 vols. in 99. Hildesheim, Zürich and New York: Georg Olms.

Wright, Edward. 1756. Microscopical Observations. *Philosophical Transactions* 49:553–8.

Wrisberg, Heinrich August. 1765. *Observationum de animalculis infusoriis satura*. Göttingen: apud viduam B. Vandenhoeck.

Zahn, Joannes. 1685–6. *Oculus artificialis teledioptricus sive telescopium*. 3 vols. in 1. Würzburg: sumptibus Quirini Heyl.

Secondary sources

Ackerknecht, Erwin H. 1967. *Medicine at the Paris Hospital 1794–1848*. Baltimore: Johns Hopkins Press.

Ackerman, James S. 1961. Science and Visual Art. Pp. 63–90 in *Seventeenth Century Science and the Arts*, edited by Hedley Howell Rhys. Princeton: Princeton University Press.

Adelmann, Howard B. 1966. *Marcello Malpighi and the Evolution of Embryology*. 5 vols. Ithaca: Cornell University Press.

Ainsworth, G. C. 1976. *Introduction to the History of Mycology*. Cambridge, etc.: Cambridge University Press.

Almquist, Ernst. 1909. Linné und die Mikroorganismen. *Zeitschrift für Hygiene und Infektionskrankheiten* 63:151–76.

Alpers, Svetlana. 1983. *The Art of Describing: Dutch Art in the Seventeenth Century*. Chicago: University of Chicago Press.

Anderson, Thomas F. 1975. Some Personal Memories of Research. *Annual Review of Microbiology* 29:1–18.

Arber, Agnes. 1940–1. Nehemiah Grew and Marcello Malpighi. *Proceedings, Linnean Society of London,* 153d Session, pp. 218–38.

1942–3. Nehemiah Grew (1641–1712) and Marcello Malpighi (1628–1694): An Essay in Comparison. *Isis* 34:7–16.

1953. From Medieval Herbalism to the Birth of Modern Botany. Pp. 317–36 in *Science, Medicine and History: Essays on the Evolution of Scientific Thought and Medical Practice Written in Honour of Charles Singer,* edited by E. Ashworth Underwood. 2 vols. London, etc.: Oxford University Press.

Arnheim, Rudolf. 1974. *Art and Visual Perception: A Psychology of the Creative Eye*. Rev. ed. Berkeley, etc.: University of California Press.

Austin, James H. 1978. *Chase, Chance, and Creativity: The Lucky Art of Novelty*. New York: Columbia University Press.

Baas, P. 1982. Leeuwenhoek's Contributions to Wood Anatomy and His Ideas on Sap Transport in Plants. Pp. 79–107 in *Antoni van Leeuwenhoek 1632–1723,* edited by L. C. Palm and H. A. M. Snelders. Amsterdam: Rodopi.

Bachrach, A. G. H. 1980. The Role of the Huygens Family in Seventeenth-Century Dutch Culture. Pp. 27–52 in *Studies on Christiaan Huygens: Invited Papers from the Symposium on the Life and Work of Christiaan Huygens, Amsterdam, 22–25 August 1979,* edited by H. J. M. Bos, M. J. S. Rudwick, H. A. M. Snelders, and R. P. W. Visser. Lisse: Swets & Zeitlinger.

Baker, John R. 1945. *The Discovery of the Uses of Colouring Agents in Biological Micro-Technique*. London: Williams & Norgate.

1948–55. The Cell-Theory: A Restatement, History and Critique. Parts 1–5. *Quarterly Journal of Microscopical Science, 3d ser.,* 89:103–25; 90:87–108; 93:157–90; 94:407–40; 96:449–81.

1952. *Abraham Trembley of Geneva, Scientist and Philosopher 1710–1784*. London: Edward Arnold.

Barber, Bernard, and Renée C. Fox. 1958–9. The Case of the Floppy-Eared Rabbits: An Instance of Serendipity Gained and Serendipity Lost. *American Journal of Sociology,* 64:128–36.

Bastholm, E. 1950. *The History of Muscle Physiology*. Copenhagen: Ejnar Munksgaard.

Baumann, E. D. 1949. *François dele Boe Sylvius*. Leiden: E. J. Brill.

Baxter, Alice, and John Farley. 1979. Mendel and Meiosis. *Journal of the History of Biology* 12:137–73.

Bechtel, William. 1984. The Evolution of Our Understanding of the Cell: A Study in the Dynamics of Scientific Progress. *Studies in History and Philosophy of Science,* 15: 309–56.

Bedini, Silvio A. 1963. Seventeenth Century Italian Compound Microscopes. *Physis* 5:383–422.

1966. Lens Making for Scientific Instrumentation in the Seventeenth Century. *Applied Optics* 5:687–94.

1967. The Instruments of Galileo Galilei. Pp. 256–92 in *Galileo, Man of Science,* edited by Ernan McMullin. New York and London: Basic Books.

Belloni, Luigi. 1961. Charlatans et *contagium vivum* au déclin de la première période de splendeur de la microscopie. Pp. 579–87 in *Comptes rendus du Congrès des sociétés savantes de Paris et des départements tenu a Chambéry-Annecy en 1960. Section des sciences*. Paris: Gauthier-Villars.

1962. Micrografia illusoria e «animalcula». *Physis* 4:65–73.

1963. Leeuwenhoek, Boerhaave a Bleyswyk sugli spermatozoi. *Physis* 5:327–32.
1966a. Appunti per una storia pre-Leeuwenhoekiana degli «animalcula». *Gesnerus* 23: 13–22.
1966b. Auf dem Wege zur Elementardrüse als Sekretionsmaschine: Forschungen des Kreises um Borelli (Auberius, Bellini-Zambeccari, Malpighi). Pp. 11–29 in *Medizingeschichte im Spektrum, Festschrift zum fünfundsechzigsten Geburtstag von Johannes Steudel,* edited by Gernot Rath and Heinrich Schipperges. *Sudhoffs Archiv,* suppl., vol. 7. Wiesbaden: Franz Steiner.
1966c. Essais d'anatomie de texture au XVIe siècle. Pp. 104–10 in *Current Problems in History of Medicine, Proceedings of the XIXth International Congress for the History of Medicine, Basel, 7–11 September 1964,* edited by R. Blaser and H. Buess. Basel and New York: S. Karger.
1967. Die Entstehungsgeschichte der mikroskopischen Anatomie. Pp. 269–96 in *Frühe Anatomie, Eine Anthologie,* edited by Robert Herrlinger and Fridolf Kudlien. Stuttgart: Wissenschaftliche Verlagsgesellschaft.
1968. Stensen-Andenken in Italien. Pp. 171–80 in *Steno and Brain Research in the Seventeenth Century,* edited by Gustav Scherz. Oxford, etc.: Pergamon Press.
1971. De la théorie atomistico-mécaniste à l'anatomie subtile (de Borelli à Malpighi) et de l'anatomie subtile à l'anatomie pathologique. *Clio medica* 6:99–107.
1985. Athanasius Kircher: Seine Mikroskopie die Animalcula und die Pestwürmer. *Medizinhistorisches Journal* 20:58–65.
Berg, Alexander. 1942. Die Lehre von der Faser als Form- un Funktionselement des Organismus. Die Geschichte des biologisch-medizinischen Grundproblems vom kleinsten Bauelement des Körpers bis zur Begründung der Zellenlehre. *Virchows Archiv für pathologische Anatomie und Physiologie und für klinische Medizin* 309:333–460.
Bergström, Ingvar. 1955. Disguised Symbolism in "Madonna" Pictures and Still Life. *Burlington Magazine* 97:342–9.
1956. *Dutch Still-Life Painting in the Seventeenth Century.* Translated by Christina Hedström and Gerald Taylor. London: Faber & Faber.
1963. Georg Hoefnagel, le dernier des grands miniaturistes flamands. *L'Oeil,* no. 101, pp. 2–9, 66.
1970a. De Gheyn as a Vanitas Painter. *Oud Holland* 85:143–57.
1970b. Vanité et moralité. *L'Oeil,* no. 190, pp. 12–17.
1974. Marseus, peintre de fleurs, papillons et serpents. *L'Oeil,* no. 233, pp. 24–9, 65.
Berkel, K. van. 1982. Intellectuals against Leeuwenhoek: Controversies about the Methods and Style of a Self-Taught Scientist. Pp. 187–209 in *Antoni van Leeuwenhoek 1632–1723,* edited by L. C. Palm and H. A. M. Snelders. Amsterdam: Rodopi.
Bertin, Léon. 1956. *Eels, a Biological Study.* London: Cleaver-Hume.
Beukers, H. 1983. De beginjaren van de microscopie aan de geneeskundige faculteiten te Utrecht en Leiden. *Tijdschrift voor de geschiedenis der geneeskunde, natuurwetenschappen, wiskunde en techniek* 6:65–81.
Biographie Nationale [Belgique]. 1866–1986. Académie royale des sciences, des lettres et des beaux-arts, Brussels. 44 vols. Brussels: H. Thiry–Van Buggenhoudt.
Blunt, Anthony. 1940. *Artistic Theory in Italy 1450–1600.* Oxford: Clarendon Press.
Bol, L. J. 1959. Een Middelburgse Brueghel-groep. IX. Johannes Goedaert, Schilder-Entomoloog. *Oud Holland* 74:1–19.
1960. *The Bosschaert Dynasty, Painters of Flowers and Fruit.* Leigh-on-Sea: F. Lewis.
1980. "Goede onbekenden." IV: Schilders van bloemen met klein gedierte als bijwerk. *Tableau* 3(1):368–73.
1984–5. Johannes Goedaert, schilder-entomoloog. Parts 1–3. *Tableau* 7(2):65–70; 7(3): 64–9; 7(4):48–54.
N.d. *Bekoring van het kleine.* N.p.: Stichting Openbaar Kunstbezit.
Bots, J. 1972. *Tussen Descartes en Darwin: Geloof en natuurwetenschap in de achttiende eeuw in Nederland.* Assen: Van Gorcum, H. J. Prakke, & H. M. G. Prakke.

Bracegirdle, Brian. 1978. The Performance of Seventeenth- and Eighteenth-Century Microscopes. *Medical History* 22:187–95.

1993. Seventeenth-Century Simple Microscopes. Pp. 295–305 in *Making Instruments Count: Essays on Historical Scientific Instruments Presented to Gerard L'Estrange Turner,* edited by R. G. W. Anderson, J. A. Bennett, and W. F. Ryan. Aldershot and Brookfield, Vt.: Variorum.

Bracegirdle, Brian, and W. H. Freeman. 1978. *An Atlas of Embryology.* 3d ed. London: Heinemann Educational Books.

Bradbury, S. 1967a. *The Evolution of the Microscope.* Oxford, etc.: Pergamon.

1967b. The Quality of the Image Produced by the Compound Microscope: 1700–1840. *Proceedings, Royal Microscopical Society* 2:151–73.

Brenninkmeyer-de Rooij, Beatrijs. 1984. Theories of Art. Pp. 60–70 in *The Golden Age: Dutch Painters of the Seventeenth Century,* by Bob Haak. Translated and edited by Elizabeth Willems-Treeman. New York: Harry N. Abrams.

Brown, Harcourt. 1934. *Scientific Organizations in Seventeenth Century France (1620–1680).* Baltimore: Williams & Wilkins.

Brown, Harold I. 1987. *Observation and Objectivity.* New York and Oxford: Oxford University Press.

Bulloch, William. 1938. *The History of Bacteriology.* London, etc.: Oxford University Press.

Bütschli, O. 1887–9. *Infusoria und System der Radiolaria. Abtheilung* 3 of *Band* 1 in *Dr. H. G. Bronn's Klassen und Ordnungen des Thier-Reichs.* Leipzig: C. F. Winter.

Carpenter, Edward J., and John L. Culliney. 1975. Nitrogen Fixation in Marine Shipworms. *Science* 187:551–2.

Castellani, Carlo. 1973. Spermatozoan Biology from Leeuwenhoek to Spallanzani. *Journal of the History of Biology* 6:37–68.

Castiglioni, Arturo. 1950. Gerolamo Fracastoro and the Doctrine of 'Contagium Vivum.' *Scientia Medica Italica* 1:747–59.

Chapman, A. Chaston. 1931. The Yeast Cell: What Did Leeuwenhoeck See? *Journal of the Institute of Brewing,* n.s., 28:433–6.

Chmelarz, Eduard. 1896. Georg und Jakob Hoefnagel. *Jahrbuch der kunsthistorischen Sammlungen des allerhöchsten Kaiserhauses* 17:275–90.

Churchill, Frederick B. 1989. The Guts of the Matter. Infusoria from Ehrenberg to Bütschli 1838–1876. *Journal of the History of Biology* 22:189–213.

Cittert, P. H. van. 1934a. *Descriptive Catalogue of the Collection of Microscopes in Charge of the Utrecht University Museum with an Introductory Historical Survey of the Resolving Power of the Microscope.* Groningen: P. Noordhoff.

1934b. The Optical Properties of the "van Leeuwenhoek" Microscope in Possession of the University of Utrecht. *Proceedings, Koninklijke Akademie van Wetenschappen, Amsterdam, Afdeeling natuurkunde* 37:290–4.

1935. Enkele opmerkingen betreffende de historische ontwikkeling van het oplossend vermogen van het microscoop. *Nederlands tijdschrift voor natuurkunde* 2(pt. 2):51–62.

1954. On the Use of Glass Globules as Microscope-Lenses. *Proceedings, Koninklijke Akademie van Wetenschappen, Amsterdam, ser. B, Physical Sciences,* 57:103–11.

Cittert, P. H. van, and J. G. van Cittert-Eymers. 1951. Some Remarks on the Development of the Compound Microscopes in the 19th Century. *Proceedings, Koninklijke Akademie van Wetenschappen, Amsterdam, ser. B, Physical Sciences,* 54:73–80.

Clarke, Desmond M. 1982. *Descartes' Philosophy of Science.* University Park: Pennsylvania State University Press.

Clarke, Edwin. 1968. The Doctrine of the Hollow Nerve in the Seventeenth and Eighteenth Centuries. Pp. 123–41 in *Medicine, Science and Culture: Historical Essays in Honor of Owsei Temkin,* edited by Lloyd G. Stevenson and Robert P. Multhauf. Baltimore: Johns Hopkins Press.

1978. The Neural Circulation. The Use of Analogy in Medicine. *Medical History* 22: 291–307.

Clarke, Edwin, and J. G. Bearn. 1968. The Brain "Glands" of Malpighi Elucidated by Practical History. *Journal of the History of Medicine and Allied Sciences* 23:309–30.

Cohen, Barnett. 1937. On Leeuwenhoek's Method of Seeing Bacteria. *Journal of Bacteriology* 34:343–6.

Cohen, I. B. 1980. *The Newtonian Revolution*. Cambridge, etc.: Cambridge University Press.

Cole, F. J. 1921. The History of Anatomical Injections. Pp. 285–343 in *Studies in the History and Method of Science*, vol. 2, edited by Charles Singer. Oxford: Clarendon Press.

1926. *The History of Protozoology*. London: University of London Press.

1930. *Early Theories of Sexual Generation*. Oxford: Clarendon Press.

1937. Leeuwenhoek's Zoological Researches. *Annals of Science* 2:1–46, 185–235.

1944. *A History of Comparative Anatomy from Aristotle to the Eighteenth Century*. London: MacMillan.

1951. History of Micro-Dissection. *Proceedings, Royal Society of London*, ser. B, 138:159–87.

Coleman, William. 1977. *Biology in the Nineteenth Century: Problems of Form, Function, and Transformation*. Cambridge, etc.: Cambridge University Press.

1988. Prussian Pedagogy: Purkyně at Breslau, 1823–1839. Pp. 15–64 in *The Investigative Enterprise: Experimental Physiology in Nineteenth-Century Medicine*, edited by William Coleman and Frederic L. Holmes. Berkeley, etc.: University of California Press.

Corliss, John O. 1978–9. A Salute to Fifty-Four Great Microscopists of the Past: A Pictorial Footnote to the History of Protozoology. Parts 1–2. *Transactions, American Microscopical Society* 97:419–58; 98:26–58.

1986. The 200th Anniversary of 'O.F.M., 1786': A Tribute to the First Comprehensive Taxonomic Treatment of the Protozoa. *Journal of Protozoology* 33:475–78.

Cosslett, V. E. 1966. *Modern Microscopy, or Seeing the Very Small*. Ithaca: Cornell University Press.

Crane, Diana. 1972. *Invisible Colleges: Diffusion of Knowledge in Scientific Communities*. Chicago and London: University of Chicago Press.

Crombie, A. C. 1967. The Mechanistic Hypothesis and the Scientific Study of Vision: Some Optical Ideas as a Background to the Invention of the Microscope. Pp. 3–112 in *Historical Aspects of Microscopy*, edited by S. Bradbury and G. L'E. Turner. Cambridge: W. Heffer & Sons.

Csikszentmihalyi, Mihaly. 1975. *Beyond Boredom and Anxiety*. San Francisco, etc.: Jossey-Bass.

Cudmore, L. L. Larison. 1978. *The Center of Life: A Natural History of the Cell*. New York: Quadrangle.

Damsteegt, B. C. 1976. Syntaktische verschijnselen in de taal van Antoni van Leeuwenhoek. Pp. 381–414 in Leeuwenhoek (1939–), vol. 9.

1982. Language and Leeuwenhoek. Pp. 13–28 in *Antoni van Leeuwenhoek 1632–1723*, edited by L. C. Palm and H. A. M. Snelders. Amsterdam: Rodopi.

Daumas, Maurice. 1972. *Scientific Instruments of the Seventeenth and Eighteenth Centuries*. Edited and translated by Mary Holbrook. New York and Washington: Praeger.

Davison, Alvin. 1923. *Mammalian Anatomy with Special Reference to the Cat*. 4th ed. Revised by Frank A. Stromsten. Philadelphia: P. Blakiston's Son.

Deason, Gary B. 1986. Reformation Theology and the Mechanistic Conception of Nature. Pp. 167–91 in *God and Nature: Historical Essays on the Encounter between Christianity and Science*, edited by David C. Lindberg and Ronald L. Numbers. Berkeley, etc.: University of California Press.

Deci, Edward L., and Richard M. Ryan. 1985. *Intrinsic Motivation and Self-Determination in Human Behavior*. New York and London: Plenum Press.

Delaissé, L. M. J. 1959. *Le Siècle d'or de la miniature flamande: Le Mécénat de Philippe le Bon. Exposition organisée à l'occasion du 400e anniversaire de la fondation de la Bibliothèque royale de Philippe II à Bruxelles, le 12 avril 1559. Bruxelles: Palais des Beaux-Arts; Amsterdam: Rijksmuseum, 26 juin–13 septembre 1959.*

1968. *A Century of Dutch Manuscript Illumination.* Berkely and Los Angeles: University of California Press.

Delbrück, Max. 1970. A Physicist's Renewed Look at Biology: Twenty Years Later. *Science* 168:1312–15.

Dictionary of Scientific Biography. 1970–80. Edited by Charles Coulston Gillispie. 16 vols. New York: Charles Scribner's Sons. [*DSB*]

Dijksterhuis, E. J. 1948. Mathematics in Leeuwenhoeck's Letters. Pp. 443–53 in Leeuwenhoek (1939–), vol. 3.

Dillard, Annie. 1974. *Pilgrim at Tinker Creek.* New York: Harper & Row.

Disney, Alfred N. 1928. *Origin and Development of the Microscope.* London: Royal Microscopical Society.

Dobell, Clifford. 1923. A Protozoological Bicentenary: Antony van Leeuwenhoek (1632–1723) and Louis Joblot (1645–1723). *Parasitology* 15:308–19.

1932. *Antony van Leeuwenhoek and His "Little Animals."* Reprint, New York: Dover Publications, 1960.

Dowey, Edward A., Jr. 1952. *The Knowledge of God in Calvin's Theology.* New York: Columbia University Press.

Drake, Stillman. 1970. Galileo and the Telescope. Pp. 140–58 in *Galileo Studies: Personality, Tradition, and Revolution.* Ann Arbor: University of Michigan Press.

Duchesneau, François. 1975. Malpighi, Descartes, and the Epistemological Problems of Iatromechanism. Pp. 111–30 in *Reason, Experiment, and Mysticism,* edited by M. L. Righini Bonelli and William R. Shea. New York: Science History Publications.

Durrieu, Paul. 1921. *La Miniature flamande au temps de la cour de Bourgogne (1415–1530).* Brussels and Paris: Librairie Nationale d'Art et d'Histoire & G. van Oest.

Eales, Nellie B. 1974. The History of the Lymphatic System, with Special Reference to the Hunter–Monro Controversy. *Journal of the History of Medicine and Allied Sciences* 29:280–94.

Eamon, William. 1985. Books of Secrets in Medieval and Early Modern Science. *Sudhoffs Archiv für Geschichte der Medizin und der Naturwissenschaften* 69:26–49.

Edge, David O., and Michael J. Mulkay. 1976. *Astronomy Transformed: The Emergence of Radio Astronomy in Britain.* New York, etc.: John Wiley & Sons.

Eiduson, Bernice T. 1962. *Scientists: Their Psychological World.* New York: Basic Books.

Elkana, Yehuda, and June Goodfield. 1968. Harvey and the Problem of the "Capillaries." *Isis* 59:61–73.

Elkins, James. 1992. On Visual Desperation and the Bodies of Protozoa. *Representations,* no. 40 (Fall), pp. 33–56.

Elshout, Antonie M. 1952. *Het Leidse kabinet der anatomie uit de achttiende eeuw: De betekenis van een wetenschappelijke collectie als cultuur historisch monument.* Leiden: Universitaire Pers Leiden.

Emerton, Norma E. 1984. *The Scientific Reinterpretation of Form.* Ithaca and London: Cornell University Press.

Engel, H. 1950. Records on Jan Swammerdam in the Amsterdam Archives. *Centaurus* 1:143–55.

Evans, Joan. 1931. *Pattern: A Study of Ornament in Western Europe from 1180 to 1900.* 2 vols. Oxford: Clarendon Press.

Evans, R. J. W. 1973. *Rudolf II and His World: A Study in Intellectual History 1576–1612.* Oxford: Clarendon Press.

Faber, J. A., et al. 1965. Population Changes and Economic Developments in the Netherlands: A Historical Survey. *A. A. G. Bijdragen* 12:47–110.

Faller, Adolf. 1986. Was erfahren wir über Jan Swammerdam (1637–1685) aus dem Briefwechsel Niels Stensens. *Gesnerus* 43:241–7.

Farley, John. 1977. *The Spontaneous Generation Controversy from Descartes to Oparin*. Baltimore and London: Johns Hopkins University Press.

1982. *Gametes and Spores: Ideas about Sexual Reproduction 1750–1914*. Baltimore and London: Johns Hopkins University Press.

Fétis, Édouard. 1857–65. *Les Artistes belges à l'étranger. Études biographiques, historiques et critiques*. 2 vols. Brussels: M. Hayez & T.-J.-I. Arnold.

Folter, R. J. de. 1978. A Newly Discovered *Oeconomia Animalis*, by Pieter Muis of Rotterdam (c. 1645–1721). *Janus* 65:183–204.

Ford, Brian J. 1973. *The Revealing Lens: Mankind and the Microscope*. London: George G. Harrap.

1985. *Single Lens: The Story of the Simple Microscope*. New York, etc.: Harper & Row.

1991. *The Leeuwenhoek Legacy*. [Bristol and London]: Biopress & Farrand Press.

Foster, Michael. 1901. *Lectures on the History of Physiology during the Sixteenth, Seventeenth and Eighteenth Centuries*. Reprint, New York: Dover, 1970.

Fournier, Marian. 1981a. Huygens' Microscopical Researches. *Janus* 68:199–209.

1981b. Jan Swammerdam en de 17e eeuwse microscopie. *Tijdschrift voor de geschiedenis der geneeskunde, natuurwetenschappen, wiskunde en techniek* 4:74–86.

1983. Hoofdlijnen van de microscopie in de negentiende eeuw. *Tijdschrift voor de geschiedenis der geneeskunde, natuurwetenschappen, wiskunde en techniek*, 6:59–64.

1985. De microscopische anatomie in Bidloo's *Anatomia humani corporis* (1685). *Tijdschrift voor de geschiedenis der geneeskunde, natuurwetenschappen, wiskunde en techniek* 8: 187–208.

1987. Drie Zeeuwse microscopisten uit de 18e eeuw. *Archief: Mededelingen van het Koninklijk Zeeuwsch Genootschap der Wetenschappen*, pp. 171–81.

1989. Huygens' Designs for a Simple Microscope. *Annals of Science* 46:575–96.

1991. *The Fabric of Life: The Rise and Decline of Seventeenth-Century Microscopy*. Doctoral diss., Twente University of Technology.

Frank, Mortimer. 1916. The History of the Discovery of the Secretory Glands and Their Function. *Johns Hopkins Hospital Bulletin* 27:302–9.

Frank, Robert G., Jr. 1980. *Harvey and the Oxford Physiologists: Scientific Ideas and Social Interaction*. Berkeley, etc.: University of California Press.

Franklin, Allan. 1986. *The Neglect of Experiment*. Cambridge, etc.: Cambridge University Press.

Freeman, R. B. 1962. Illustrations of Insect Anatomy from the Beginning to the Time of Cuvier. *Medical and Biological Illustration* 12:174–83.

Fretter, Vera, and Alastair Graham. 1962. *British Prosobranch Molluscs: Their Functional Anatomy and Ecology*. London: The Ray Society.

Frison, Edward. 1948. A Leeuwenhoek Microscope. *Microscope and Entomological Monthly* 6:281–7.

1963. De Microscoop van Dokter Somme, en de bruikbaarheid van de achttiend- en vroeg-negentiendeeuwse samengestelde microscopen voor wetenschappelijk onderzoek. *Scientiarum Historia* 5:1–10.

1965. Bij de driehonderdste verjaring van Robert Hooke's «Micrographia» (1665–1965). De opgang van de microscopie in Engeland in de tweede helft der 17de eeuw. *Scientiarum Historia* 7:92–104.

Fulton, J. F. 1938. The Influence of Boerhaave's *Institutiones Medicae* on Modern Physiology. *Nederlandsch tijdschrift voor geneeskunde* 82:4860–6.

Gasking, Elizabeth B. 1967. *Investigations into Generation 1651–1828*. Baltimore: Johns Hopkins Press.

Geison, Gerald L. 1981. Scientific Change, Emerging Specialties, and Research Schools. *History of Science* 19:20–40.

Gelder, J. G. van. 1976. The Animal and His "Lettres de Noblesse." *Apollo* 104(177): 346–53.

Geyl, Pieter. 1964. *The Netherlands in the Seventeenth Century: Part Two 1648–1715*. New York: Barnes & Noble / London: Ernest Benn.

Geymonat, Ludovico. 1965. *Galileo Galilei: A Biography and Inquiry into His Philosophy of Science*. Translated by Stillman Drake. New York, etc.: McGraw-Hill.

Gibson, Eleanor J. 1969. *Principles of Perceptual Learning and Development*. New York: Appleton-Century-Crofts.

Gibson, Robin. 1976. *Flower Painting*. Oxford: Phaidon / New York: E. P. Dutton.

Giehlow, Karl. 1915. Die Hieroglyphenkunde des Humanismus in der Allegorie der Renaissance, besonders der Ehrenpforte Kaisers Maximilian I. *Jahrbuch der kunsthistorischen Sammlungen des allerhöchsten Kaiserhauses* 32:1–232.

Gombrich, E. H. 1948. Icones Symbolicae: The Visual Image in Neo-Platonic Thought. *Journal of the Warburg and Courtauld Institutes* 11:163–92.

1961. *Art and Illusion: A Study in the Psychology of Pictorial Representation*. 2d ed. New York: Bollingen Foundation.

1976. *The Heritage of Apelles: Studies in the Art of the Renaissance*. Ithaca: Cornell University Press.

1982. *The Image and the Eye: Further Studies in the Psychology of Pictorial Representation*. Oxford: Phaidon.

1983. Review of *The Art of Describing*, by Svetlana Alpers. *New York Review of Books*, 10 November, pp. 13–17.

Grayson, Robin F. 1970–1. The Freshwater Hydras of Europe. 1. A Review of the European Species. *Archiv für Hydrobiologie* 68:436–49.

Grmek, M. D. 1970. La Notion de fibre vivante chez les médecins de l'école iatrophysique. *Clio medica* 5:297–318.

Haaxman, P. J. 1875. *Antony van Leeuwenhoek*. Leiden: S. C. van Doesburgh.

Haber, Ralph Norman, and Maurice Hershenson. 1973. *The Psychology of Visual Perception*. New York, etc.: Holt, Rinehart & Winston.

Hacking, Ian. 1983. *Representing and Intervening: Introductory Topics in the Philosophy of Natural Science*. Cambridge, etc.: Cambridge University Press.

Hahn, Roger. 1971. *The Anatomy of a Scientific Institution: The Paris Academy of Sciences, 1666–1803*. Berkeley, etc.: University of California Press.

Halbertsma, H. J. 1862. Johan Ham van Arnhem, de ontdekker der spermatozoïden. *Verslagen en mededeelingen, Koninklijke Akademie van Wetenschappen, Amsterdam, Afdeeling natuurkunde* 13:342–7.

Hall, A. Rupert, and Albert Van Helden. 1980. International Huygens Symposium, Amsterdam, August 22–25, 1979. *Isis* 71:138–9.

Hall, Marie Boas. 1965. *Robert Boyle on Natural Philosophy: An Essay with Selections from His Writings*. Bloomington: Indiana University Press.

Harting, Pieter. 1846. *Bijdragen tot de geschiedenis der mikroskopen in ons vaderland*. Utrecht: van Paddenburg.

1848–54. *Het mikroskoop, deszelfs gebruik, geschiedenis en tegenwoordige toestand*. 2 vols. Utrecht: van Paddenburg.

1867. Oude optische werktuigen, toegeschreven aan Zacharias Janssen en eene beroemde lens van Christiaan Huygens teruggevonden. *Album der Nature*, pp. 257–81.

Hauser, Arnold. 1986. *Mannerism: The Crisis of the Renaissance and the Origin of Modern Art*. Translated by Eric Mosbacher. Cambridge, Mass., and London: Belknap Press of Harvard University Press.

Haverkamp-Begemann, E. 1976. The Appearance of Reality, Dutch Draughtsmen of the Golden Age. *Apollo*, 104(177):354–63.

Heisenberg, Werner. 1975. Tradition in Science. Pp. 219–36 in *The Nature of Scientific Discovery: A Symposium Commemorating the 500th Anniversary of the Birth of Nicolaus Copernicus*. Edited by Owen Gingerich. Washington: Smithsonian Institution Press.

Hendrix, Marjorie Lee. 1984. *Joris Hoefnagel and the "Four Elements": A Study in Sixteenth-Century Nature Painting.* Ph.D. diss., Dept. of Art and Archaeology, Princeton University.

1985. Joris Hoefnagel's *The Four Elements. FMR America,* no. 9, pp. 78–85.

Hoch, Paul K. 1987. Institutional versus Intellectual Migrations in the Nucleation of New Scientific Specialties. *Studies in History and Philosophy of Science* 18:481–500.

Holton, Gerald. 1962. Scientific Research and Scholarship: Notes toward the Design of Proper Scales. *Daedalus* 91:362–99.

Hoog, Adriaan Cornelis de. 1974. *Some Currents of Thought in Dutch Natural Philosophy: 1675–1720.* Ph.D. diss., University of Oxford.

Hooykaas, R. 1972. *Religion and the Rise of Modern Science.* Grand Rapids, Mich.: William B. Eerdmans.

Hublard, Émile. 1910. *Le Naturaliste hollandais Pierre Lyonet, sa vie et ses oeuvres (1706–1789), d'après des lettres inédites.* Mons: Dequesne-Masquillier et fils.

Hughes, Arthur. 1955. Studies in the History of Microscopy. 1. The Influence of Achromatism. *Journal of the Royal Microscopical Society,* 3d ser., 75:1–22.

1959. *A History of Cytology.* London and New York: Abelard-Schuman.

Hull, David L. 1974. *Philosophy of Biological Science.* Englewood Cliffs, N.J.: Prentice-Hall.

1988. *Science as a Process: An Evolutionary Account of the Social and Conceptual Development of Science.* Chicago and London: University of Chicago Press.

Humbert, Pierre. 1951. Peiresc et le microscope. *Revue d'histoire des sciences et de leurs applications* 4:154–8.

Hunter, Michael. 1981. *Science and Society in Restoration England.* Cambridge, etc.: Cambridge University Press.

1989. *Establishing the New Science: The Experience of the Early Royal Society.* Woodbridge: Boydell.

Hutchinson, G. Evelyn. 1974. Aposematic Insects and the Master of the Brussels Initials. *American Scientist* 62:161–71.

Hvolbek, Russell H. 1984. *Seventeenth-Century Dialogues: Jacob Boehme and the New Sciences.* Ph.D. diss., University of Chicago.

Jacobs, Natasha X. 1989. From Unity to Unity: Protozoology, Cell Theory, and the New Concept of Life. *Journal of the History of Biology* 22:215–42.

Jacyna, L. S. 1983. John Goodsir and the Making of Cellular Reality. *Journal of the History of Biology* 16:75–99.

Jaeger, F. M. 1922. *Cornelis Drebbel en zijne tijdgenooten.* Groningen: P. Noordhoff.

Jones, William Powell. 1966. *The Rhetoric of Science: A Study of Scientific Ideas and Imagery in Eighteenth-Century English Poetry.* London: Routledge & Kegan Paul.

Jongejan, E. 1940. Van Leeuwenhoek's brieven en de nederlandse schrijftaal in de zeventiende eeuw. *Nieuwe taalgids* 34:300–7.

Jongh, E. de. 1982. *Still-Life in the Age of Rembrandt.* N.p.: Auckland City Art Gallery.

Judson, Richard J. 1973. *The Drawings of Jacob de Gheyn II.* New York: Grossman.

Keller, Evelyn Fox. 1983. *A Feeling for the Organism: The Life and Work of Barbara McClintock.* New York and San Francisco: W. H. Freeman.

Kessel, Erwin. 1933. Über die Schale von *Viviparus viviparus* L. und *Viviparus fasciatus* Müll., ein Beitrag zum Strukturproblem der Gastropedenschale. *Zeitschrift für Morphologie und Ökologie der Tiere* 27:129–98.

Kingma Boltjes, T. Y. 1941. Some Experiments with Blown Glasses. *Antonie van Leeuwenhoek* 7:61–76.

Knorr, Karin D. 1979. Tinkering toward Success: Prelude to a Theory of Scientific Practice. *Theory and Society* 8:347–76.

Knorr-Cetina, Karin D. 1981. *The Manufacture of Knowledge: An Essay on the Constructivist and Contextual Nature of Science.* Oxford, etc.: Pergamon Press.

Köhler, Wolfgang. 1969. *The Task of Gestalt Psychology.* Princeton, N.J.: Princeton University Press.

Koelbing, H. M. 1968. Ocular Physiology in the Seventeenth Century and Its Acceptance by the Medical Profession. Pp. 219–24 in *Steno and Brain Research in the Seventeenth Century,* edited by Gustav Scherz. Oxford, etc.: Pergamon Press.

Kremer, J. 1977. Antoni van Leeuwenhoek, the Founder of "Spermatology." In *Immunological Influence on Human Fertility: Proceedings of the Workshop on Fertility in Human Reproduction, Department of Biological Science, University of Newcastle, Australia July 11–13, 1977,* edited by Barry Boettcher. Sydney, etc.: Academic Press.

Kremer, Richard L. 1992. Building Institutes for Physiology in Prussia, 1836–1846: Contexts, Interests and Rhetoric. Pp. 72–109 in *The Laboratory Revolution in Medicine,* edited by Andrew Cunningham and Perry Williams. Cambridge, etc.: Cambridge University Press.

Kris, Ernst. 1927. Georg Hoefnagel und der wissenschaftliche Naturalismus. Pp. 243–53 in *Festschrift für Julius Schlosser zum 60. Geburtstage,* edited by Arpad Weixlgärtner and Leo Planiscig. Zürich, etc.: Amalthea.

Langer, Susanne K. 1957. *Problems of Art.* London: Routledge & Kegan Paul.

1967–82. *Mind: An Essay on Human Feeling.* 3 vols. Baltimore: Johns Hopkins Press.

Lazier, Edgar Locke. 1924. Morphology of the Digestive Tract of Teredo Navalis. *University of California Publications in Zoology* 22(14):455–74.

Lechevalier, Hubert A., and Morris Solotorovsky. 1974. *Three Centuries of Microbiology.* New York: Dover.

Lenoir, Timothy. 1982. *The Strategy of Life: Teleology and Mechanics in Nineteenth Century German Biology.* Dordrecht, etc.: D. Reidel.

1988. Science for the Clinic: Science Policy and the Formation of Carl Ludwig's Institute in Leipzig. Pp. 139–78 in *The Investigative Enterprise: Experimental Physiology in Nineteenth-Century Medicine,* edited by William Coleman and Frederic L. Holmes. Berkeley, etc.: University of California Press.

1992. Laboratories, Medicine and Public Life in Germany 1830–1849. Pp. 14–71 in *The Laboratory Revolution in Medicine,* edited by Andrew Cunningham and Perry Williams. Cambridge, etc.: Cambridge University Press.

Lievense-Pelser, E. 1977. De Remonstranten en de sekte van Antoinette Bourignon. *Nederlands archief voor kerkgeschiedenis, n.s.,* 57:210–21.

Lindberg, David C. 1976. *Theories of Vision from Al-Kindi to Kepler.* Chicago and London: University of Chicago Press.

Lindberg, David C., and Nicholas H. Steneck. 1972. The Sense of Vision and the Origins of Modern Science. Pp. 29–45 in vol. 1 of *Science, Medicine and Society in the Renaissance: Essays to Honor Walter Pagel,* edited by Allen G. Debus. 2 vols. New York: Science History Publications.

Lindeboom, G. A. 1959. *Bibliographia Boerhaaviana: List of Publications Written or Provided by H. Boerhaave or Based upon His Works and Teaching.* Leiden: E. J. Brill.

1968. *Herman Boerhaave: The Man and His Work.* London: Methuen.

1973. *Reinier de Graaf: Leven en werken 30-7-1641/17-8-1673.* Delft: Elmar.

1974. Antoinette Bourignon's First Letter to Jan Swammerdam: A Contribution to His Biography. *Janus* 61:183–99.

1975. A Short Biography of Jan Swammerdam (1637–1680). Pp. 1–27 in *The Letters of Jan Swammerdam to Melchisedec Thévenot.* Edited and translated by G. A. Lindeboom. Amsterdam: Swets & Zeitlinger.

1981. Jan Swammerdam als microscopist. *Tijdschrift voor de geschiedenis der geneeskunde, natuurwetenschappen, wiskunde en techniek* 4:87–110.

1982. Jan Swammerdam (1637–1680) and His *Biblia Naturae. Clio medica* 17:113–31.

1983. Anton de Heide als proefondervindelijk onderzoeker. *Tijdschrift voor de geschiedenis der geneeskunde, natuurwetenschappen, wiskunde en techniek* 6:121–34.

Lloyd, G. E. R. 1973. *Greek Science after Aristotle.* New York: W. W. Norton.

Longrigg, James. 1981. Superlative Achievement and Comparative Neglect: Alexandrian Medical Science and Modern Historical Research. *History of Science* 19:155–200.

Lovejoy, Arthur O. 1936. *The Great Chain of Being: A Study of the History of an Idea.* Cambridge: Harvard University Press.

Lumsden, Ernest A. 1980. Problems of Magnification and Minification: An Explanation of the Distortions of Distance, Slant, Shape, and Velocity. Pp. 91–135 in *The Perception of Pictures,* edited by Margaret A. Hagen. 2 vols. New York, etc.: Academic Press.

Lynch, Michael. 1985. *Art and Artifact in Laboratory Science: A Study of Shop Work and Shop Talk in a Research Laboratory.* London, etc.: Routledge & Kegan Paul.

Luyendijk-Elshout, Antonie M. 1975. *Oeconomia Animalis,* Pores and Particles: The Rise and Fall of the Mechanical Philosophical School of Theodoor Craanen (1621–1690). Pp. 294–307 in *Leiden University in the Seventeenth Century: An Exchange of Learning,* edited by Th. H. Lunsingh Scheurleer and G. H. M. Posthumus Meyjes. Leiden: Universitaire Pers Leiden & E. J. Brill.

McClellan, James E., III. 1985. *Science Reorganized: Scientific Societies in the Eighteenth Century.* New York: Columbia University Press.

McClelland, David C. 1962. On the Psychodynamics of Creative Physical Scientists. Pp. 141–74 in *Contemporary Approaches to Creative Thinking,* edited by Howard E. Gruber, Glenn Terrell, and Michael Wertheimer. New York: Atherton Press.

Major, Ralph H. 1932. *Classic Descriptions of Disease.* Springfield and Baltimore: Charles C. Thomas.

Man, J. C. de. 1905. *Antonius De Heide.* Middelburg: D. G. Kröler, Jr.

Marrow, James H. 1986. Symbol and Meaning in Northern European Art of the Late Middle Ages and the Early Renaissance. *Simiolus* 16:150–69.

Martin, John Rupert. 1977. *Baroque.* New York, etc.: Harper & Row.

Marton, L. 1968. *Early History of the Electron Microscope.* San Francisco: San Francisco Press.

Mendels, J. I. H. 1952. Leeuwenhoek's Taal. Pp. 314–20 in Leeuwenhoek (1939–), vol. 4.

Miller, Genevieve. 1968. Leeuwenhoek's Observations on the Blood and Capillary Vessels. Pp. 114–22 in *Medicine, Science and Culture: Historical Essays in Honor of Owsei Temkin.* Baltimore: Johns Hopkins Press.

1981. Early Concepts of the Microvascular System: William Harvey to Marshall Hall, 1628–1831. Pp. 257–78 in *The Analytic Spirit: Essays in the History of Science in Honor of Henry Guerlac,* edited by Harry Woolf. Ithaca and London: Cornell University Press.

Mills, A. A., and M. L. Jones. 1989. Three Lenses by Constantine Huygens in the Possession of the Royal Society of London. *Annals of Science* 46:173–82.

Möller, L. 1959. Anatomia-Memento mori. *Nederlands kunsthistorisch jaarboek* 10:71–98.

Moes, E. W. 1896. Een verzameling familieportretten der Huygensen in 1785. *Oud-Holland* 14:176–84.

Mols, Roger. 1954–6. *Introduction à la démographie historique des villes d'Europe du XIVe au XVIIIe siècle.* 3 vols. Louvain: J. Duculot.

Montias, John Michael. 1982. *Artists and Artisans in Delft: A Socio-Economic Study of the Seventeenth Century.* Princeton: Princeton University Press.

1989. *Vermeer and His Milieu: A Web of Social History.* Princeton: Princeton University Press.

Morge, Günter. 1973. Entomology in the Western World in Antiquity and in Medieval Times. Pp. 37–80 in *History of Entomology,* edited by Ray F. Smith, Thomas E. Mittler, and Carroll N. Smith. Palo Alto: Annual Reviews.

Napjus, J. W. 1939–40. De hoogleeraren in de geneeskunde aan de Hoogeschool en het Athenaeum te Franeker (1585–1843). XV–[XVIII]. Wijer Willem Muijs. *Bijdragen tot de geschiedenis der geneeskunde* 19:243–9; 20:49–53, 73–9, 97–103.

Needham, Joseph. 1959. *A History of Embryology.* 2d ed. Cambridge: Cambridge University Press.

Neisser, Ulric. 1976. *Cognition and Reality: Principles and Implications of Cognitive Psychology*. San Francisco: W. H. Freeman & Co.

Newman, Charles. 1957. *The Evolution of Medical Education in the Nineteenth Century*. London, etc.: Oxford University Press.

Nicolson, Marjorie. 1956. *Science and Imagination*. Ithaca: Cornell University Press / London: Oxford University Press.

Niesel, Wilhelm. 1956. *The Theology of Calvin*. Translated by Harold Knight. Philadelphia: Westminster Press.

Nordenskiöld, Erik. 1928. *The History of Biology: A Survey*. Translated by Leonard Bucknall Eyre. New York and London: Alfred A. Knopf.

Nordström, Johan. 1954–5. Swammerdamiana: Excerpts from the Travel Journal of Olaus Borrichius and Two Letters from Swammerdam to Thévenot. *Lychnos* 21–65.

Nuttall, R. H. 1974. The Achromatic Microscope in the History of Nineteenth Century Science. *Philosophical Journal* 9:71–88.

Nutton, Vivian. 1990. The Reception of Fracastoro's Theory of Contagion: The Seed That Fell among Thorns? *Osiris* 6:196–234.

Oakley, Francis. 1984. *Omnipotence, Covenant, and Order: An Excursion in the History of Ideas from Abelard to Leibniz*. Ithaca and London: Cornell University Press.

Olesko, Kathryn M. 1993. Tacit Knowledge and School Formation. *Osiris* 8:16–29.

Oppenheimer, Jane M. 1967. *Essays in the History of Embryology and Biology*. Cambridge, Mass., and London: MIT Press.

Osselton, N. E. 1973. *The Dumb Linguists: A Study of the Earliest English and Dutch Dictionaries*. Leiden: University Press / London: Oxford University Press.

Otto, Rudolf. 1931. *The Idea of the Holy: An Inquiry into the Non-Rational Factor in the Idea of the Divine and Its Relation to the Rational*. Translated by John W. Harvey. Rev. ed. London, etc.: Humphrey Milford & Oxford University Press.

Pächt, Otto. 1948. *The Master of Mary of Burgundy*. London: Faber & Faber.

1950. Early Italian Nature Studies and the Early Calendar Landscape. *Journal of the Warburg and Courtauld Institutes* 13:13–47.

Pagel, Walter. 1944. *The Religious and Philosophical Aspects of van Helmont's Science and Medicine. Bulletin of the History of Medicine*, suppl. no. 2. Baltimore: Johns Hopkins Press.

1958. *Paracelsus: An Introduction to Philosophical Medicine in the Era of the Renaissance*. Basel and New York: S. Karger.

1967. *William Harvey's Biological Ideas, Selected Aspects and Historical Background*. New York: Hafner.

Palm, L. C. 1978. Antoni van Leeuwenhoek en de ontdekking der haarvaten. *Tijdschrift voor de geschiedenis der geneeskunde, natuurwetenschappen, wiskunde en techniek* 1:170–7.

1989a. Italian Influences on Antoni van Leeuwenhoek. Pp. 147–63 in *Italian Scientists in the Low Countries in the XVIIth and XVIIIth Centuries*, edited by C. S. Maffioli and L. C. Palm. Amsterdam and Atlanta: Rodopi.

1989b. Leeuwenhoek and Other Dutch Correspondents of the Royal Society. *Notes and Records, Royal Society of London* 43:191–207.

Pas, Peter W. van der. 1975. Leeuwenhoeck's Correspondence at the Royal Society. *Aere Perennius* 18:3–11.

Patten, Bradley M. 1951. *Early Embryology of the Chick*. 4th ed. New York and Toronto: Blakiston.

Pauling, Linus. 1958. Foreword to *Moments of Discovery*, edited by George Schwartz and Philip W. Bishop. New York: Basic Books.

Pazzini, Adalberto. 1959. Evangelista Torricelli nella storia del microscopio. Pp. 99–102 in *Convegno di studi torricelliani in occasione del 350° anniversario della nascita di Evangelista Torricelli (19–20 ottobre 1658)*, by the Società Torricelliana di Scienze e Lettere, Faenza. Faenza: Fratelli Lega.

Pelikan, Jaroslav. 1961. Cosmos and Creation: Science and Theology in Reformation Thought. *Proceedings, American Philosophical Society* 105:464–9.

Perkins, D. N. 1982. The Perceiver as Organizer and Geometer. Pp. 73–93 in *Organization and Representation in Perception,* edited by Jacob Beck. Hillsdale, N.J., and London: Lawrence Erlbaum.

Perutz, Max. 1991. *Is Science Necessary? Essays on Science and Scientists.* Oxford and New York: Oxford University Press.

Pickstone, J. V. 1977. Absorption and Osmosis: French Physiology and Physics in the Early Nineteenth Century. *Physiologist* 20:30–7.

Pighetti, Clelia. 1961. Giovan Battista Odierna e il suo discorso su *L'occhio della mosca. Physis* 3:309–35.

Ploeg, Willem. 1934. *Constantijn Huygens en de natuurwetenschappen.* Rotterdam: Nijgh & van Ditmar.

Polanyi, Michael. 1969. *Knowing and Being.* Edited by Marjorie Grene. Chicago: University of Chicago Press.

Popham, J. D., and M. R. Dickson. 1973. Bacterial Associations in the Teredo *Bankia australis* (Lamellibranchia: Mollusca). *Marine Biology* 19:338–40.

Porter, Keith R., and Mary A. Bonneville. 1973. *Fine Structure of Cells and Tissues.* 4th ed. Philadelphia: Lea & Febiger.

Porter, Roy. 1988. The New Taste for Nature in the Eighteenth Century. *Linnean* 4: 14–30.

Portmann, Adolf. 1952. *Animal Forms and Patterns: A Study of the Appearance of Animals.* Translated by Hella Czech. New York: Schocken Books, 1967.

Preston, Richard. 1987. *First Light: The Search for the Edge of the Universe.* New York: Atlantic Monthly Press.

Price, Derek J. de Solla. 1986. *Little Science, Big Science . . . and Beyond.* New York: Columbia University Press.

Price, J. L. 1974. *Culture and Society in the Dutch Republic during the 17th Century.* London: B. T. Batsford.

Puyvelde, Leo van. 1944. *The Dutch Drawings in the Collection of His Majesty the King at Windsor Castle.* London: Phaidon Press / New York: Oxford University Press.

Randall, Lilian M. C. 1966. *Images in the Margins of Gothic Manuscripts.* Berkeley and Los Angeles: University of California Press.

Raven, Charles E. 1950. *John Ray, Naturalist, His Life and Works.* 2d ed. Cambridge: University Press.

Ravetz, Jerome R. 1971. *Scientific Knowledge and Its Social Problems.* Oxford: Clarendon Press.

Regteren Altena, Iohan Quirijn van. 1935. *Jacques de Gheyn: An Introduction to the Study of His Drawings.* Amsterdam: Swets & Zeitlinger.

1983. *Jacques de Gheyn, Three Generations.* 3 vols. The Hague, etc.: Martinus Nijhoff.

Reif, Sister Mary Richard. 1962. *Natural Philosophy in Some Early Seventeenth Century Scholastic Textbooks.* Ph.D. diss., Saint Louis University.

Reiser, Stanley Joel. 1978. *Medicine and the Reign of Technology.* Cambridge, etc.: Cambridge University Press.

Reznicek, E. K. J. 1961. *Die Zeichnungen von Hendrick Goltzius.* 2 vols. Utrecht: Haentjens Dekker & Gumbert.

Rijnberk, G. van. 1938. De geschiedenis der wetenschappen in Zweden. *Nederlandsch tijdschrift voor geneeskunde* 82:2828–30.

Ritterbush, Philip C. 1964. *Overtures to Biology: The Speculations of Eighteenth-Century Naturalists.* New Haven and London: Yale University Press.

1969. Art and Science as Influences on the Early Development of Natural History Collections. *Proceedings, Biological Society of Washington* 82:561–78.

Roe, Anne. 1953. *The Making of a Scientist.* New York: Dodd, Mead & Co.

Roe, Shirley A. 1981. *Matter, Life, and Generation: Eighteenth-Century Embryology and the Haller–Wolff Debate.* Cambridge, etc.: Cambridge University Press.

1983. John Turberville Needham and the Generation of Living Organisms. *Isis* 74: 159–84.

1985. Voltaire versus Needham: Atheism, Materialism, and the Generation of Life. *Journal of the History of Ideas* 46:65–87.

1992. Buffon and Needham: Diverging Views on Life and Matter. Pp. 439–50 in *Buffon 88: Actes du Colloque international pour le bicentenaire de la mort de Buffon (Paris, Montbard, Dijon, 14–22 juin 1988),* edited by Jean-Claude Beaune, Serge Benoit, Jean Gayon, Jacques Roger, and Doris Woronoff. Paris: Vrin / Lyon: Institut Interdisciplinaire d'Études Epistémologiques.

Roger, Jacques. 1971. *Les Sciences de la vie dans la pensée française du XVIIIe siècle: La génération des animaux de Descartes a l'Encyclopédie.* 2d ed. Paris: Armand Colin.

1980. The Living World. Pp. 255–83 in *The Ferment of Knowledge: Studies in the Historiography of Eighteenth-Century Science,* edited by G. S. Rousseau and Roy Porter. Cambridge, etc.: Cambridge University Press.

1981. Two Scientific Discoveries: Their Genesis and Destiny. Pp. 229–37 in *On Scientific Discovery; The Erice Lectures 1977,* edited by Mirko Drazen Grmek, Robert S. Cohen, and Guido Cimino. Dordrecht, etc.: D. Reidel.

Ronchi, Vasco. 1967. The Influence of the Early Development of Optics on Science and Philosophy. Pp. 195–206 in *Galileo, Man of Science,* edited by Ernan McMullin. New York: Basic Books.

1970. *The Nature of Light, An Historical Survey.* Translated by V. Barocas. Cambridge: Harvard University Press.

Rooseboom, Maria. 1939. Concerning the Optical Qualities of Some Microscopes Made by Leeuwenhoek. *Journal of the Royal Microscopical Society,* 3d ser., 59:177–83.

1940. Some Notes upon the Life and Work of Certain Netherlands Artificers of Microscopic Preparations at the End of the XVIIIth Century and the Beginning of the XIXth. *Janus* 44:24–44.

1959. Christiaan Huygens et la microscopie. *Archives neerlandaises de zoologie,* suppl. no. 1, pp. 59–73.

1967. The History of the Microscope. *Proceedings, Royal Microscopical Society* 2:266–93.

Rosenberg, Jakob, Seymour Slive, and E. H. ter Kuile. 1966. *Dutch Art and Architecture: 1600 to 1800.* Harmondsworth, etc.: Penguin Books.

Rothschild, Lynn J. 1989. Protozoa, Protista, Protoctista: What's in a Name? *Journal of the History of Biology* 22:277–305.

Rothschuh, Karl E. 1973. *History of Physiology.* Edited and translated by Guenter B. Risse. Huntington, N.Y.: Robert E. Krieger.

Rouse, Joseph. 1987. *Knowledge and Power: Toward a Political Philosophy of Science.* Ithaca and London: Cornell University Press.

Ruë, Pieter de la. 1741. *Geletterd Zeeland.* 2d ed. Middelburg: M. en A. Callenfels.

Ruestow, Edward G. 1973. *Physics at Seventeenth and Eighteenth-Century Leiden: Philosophy and the New Science in the University.* The Hague: Martinus Nijhoff.

1980. The Rise of the Doctrine of Vascular Secretion in the Netherlands. *Journal of the History of Medicine and Allied Sciences* 35:265–87.

1983. Images and Ideas: Leeuwenhoek's Perception of the Spermatozoa. *Journal of the History of Biology* 16:185–224.

1984. Leeuwenhoek and the Campaign against Spontaneous Generation. *Journal of the History of Biology* 17:225–48.

1985. Piety and the Defense of Natural Order: Swammerdam on Generation. Pp. 217–41 in *Religion, Science, and Worldview: Essays in Honor of Richard S. Westfall,* edited by Margaret J. Osler and Paul Lawrence Farber. Cambridge, etc.: Cambridge University Press.

Ruska, Ernst. 1987. The Development of the Electron Microscope and of Electron Microscopy. *Reviews of Modern Physics* 59:627–38.

Saine, Thomas P. 1976. Natural Science and the Ideology of Nature in the German Enlightenment. *Lessing Yearbook* 8:61–88.

Sassen, Ferdinand. 1941. Henricus Renerius, de eerste 'cartesiaansche' hoogleeraar te Utrecht. *Mededeelingen, Koninklijke Akademie van Wetenschappen, Amsterdam, Afdeeling letterkunde,* n.s., 4:853–902.

Schama, Simon. 1984. The Dutch Masters' Revenge. *New Republic* 190(19):25–31.

1987. *The Embarrassment of Riches: An Interpretation of Dutch Culture in the Golden Age.* New York: Alfred A. Knopf.

Schatzberg, Walter. 1973. *Scientific Themes in the Popular Literature and the Poetry of the German Enlightenment, 1720–1760.* Berne: Herbert Lang.

Scheltema, Pieter. 1886. *Het Leven van Frederik Ruijsch.* Sliedrecht: Gebroeders Luijt.

Schierbeek, A. 1939. Leeuwenhoeck en zijn globulentheorie. *Natuurwetenschappelijk tijdschrift* 21:185–9.

1947. *Jan Swammerdam (12 Februari 1637–17 Februari 1680), zijn leven en zijn werken.* Lochem: De Tijdstroom.

1950–1. *Antoni van Leeuwenhoek, zijn leven en zijn werken.* 2 vols. Lochem: De Tijdstroom.

Schiller, Joseph. 1978. *La Notion d'organisation dans l'histoire de la biologie.* Paris: Maloine.

Schlichting, Th. H. 1951. Waarom vond van Leeuwenhoek geen navolging? *Nederlandsch tijdschrift voor geneeskunde* 95:3887–92.

Schon, Donald A. 1963. *Displacement of Concepts.* London: Tavistock.

Seaver, Paul S. 1985. *Wallington's World: A Puritan Artisan in Seventeenth-Century London.* Stanford, Calif.: Stanford University Press.

Servos, John W. 1993. Research Schools and Their Histories. *Osiris,* n.s., 8:3–15.

Seters, W. H. van. 1933. Leeuwenhoecks microscopen, praepareer- en observatiemethodes. *Nederlandsch tijdschrift voor geneeskunde* 77:4571–89.

Shapin, Steven. 1982. History of Science and Its Sociological Reconstructions. *History of Science* 20:157–211.

Shapin, Steven, and Simon Schaffer. 1985. *Leviathan and the Air-Pump: Hobbes, Boyle, and the Experimental Life.* Princeton: Princeton University Press.

Singer, Charles. 1914. Notes on the Early History of Microscopy. *Proceedings, Royal Society of Medicine* 7:247–79, in the "Section of the History of Medicine."

1915. The Dawn of Microscopical Discovery. *Journal of the Royal Microscopical Society* 317–40.

1953. The Earliest Figures of Microscopic Objects. *Endeavour* 12:197–201.

Sinia, Rinse. 1878. *Johannes Swammerdam in de lijst van zijn tijd.* Hoorn: P. Geerts.

Slive, Seymour. 1953. *Rembrandt and His Critics 1630–1730.* The Hague: Martinus Nijhoff.

Sloan, Phillip R. 1992. Organic Molecules Revisited. Pp. 415–38 in *Buffon 88: Actes du Colloque international pour le bicentenaire de la mort de Buffon (Paris, Montbard, Dijon, 14–22 juin 1988),* edited by Jean-Claude Beaune, Serge Benoit, Jean Gayon, Jacques Roger, and Doris Woronoff. Paris: Vrin / Lyon: Institut Interdisciplinaire d'Études Epistémologiques.

Smit, Pieter. 1980. The Zoological Dissertations of Linnaeus. Pp. 118–36 in *Linnaeus: Progress and Prospects in Linnaean Research,* edited by Gunnar Broberg. Stockholm: Almqvist & Wiksell International / Pittsburgh: Hunt Institute for Botanical Documentation.

Snelders, H. A. M. 1982. Antoni van Leeuwenhoek's Mechanistic View of the World. Pp. 57–78 in *Antoni van Leeuwenhoek 1632–1723,* edited by L. C. Palm and H. A. M. Snelders. Amsterdam: Rodopi.

Stannard, Jerry. 1978. Natural History. Pp. 429–60 in *Science in the Middle Ages,* edited by David C. Lindberg. Chicago and London: University of Chicago Press.

Star, P. van der. 1953. *Descriptive Catalogue of the Simple Microscopes in the Rijksmuseum voor de Geschiedenis der Natuurwetenschappen (National Museum of the History of Science) at Leyden.* Communication no. 87 of the Rijksmuseum voor de Geschiedenis der Natuurwetenschappen, Leiden.

Sterling, Charles. 1952. *La Nature morte de l'antiquité à nos jours.* Paris: Pierre Tisné.

Stoudt, John Joseph. 1968. *Jacob Boehme: His Life and Thought.* New York: Seabury Press.

Stroup, Alice. 1990. *A Company of Scientists: Botany, Patronage, and Community at the Seventeenth-Century Parisian Royal Academy of Sciences.* Berkeley, etc.: University of California Press.

Stuldreher-Nienhuis, J. 1944. *Verborgen paradijzen: Het leven en de werken van Maria Sibylla Merian 1647–1717.* Arnhem: Van Loghum Slaterus.

Terrada Ferrandis, Maria Luz. 1969. *La anatomia microscopica en España: La doctrina de la fibra y la utilizacion del microscopico en España durante el Barroco y la Ilustracion.* Salamanca: Universidad de Salamanca.

Tierie, Gerrit. 1932. *Cornelis Drebbel (1572–1633).* Amsterdam: H.J. Paris.

Tuchman, Arleen M. 1993. *Science, Medicine, and the State in Germany: The Case of Baden, 1815–1871.* New York and Oxford: Oxford University Press.

Turner, G. L'E. 1967. The Microscope as a Technical Frontier in Science. *Proceedings, Royal Microscopical Society* 2(pt. 1):175–97.

1969. The History of Optical Instruments: A Brief Survey of Sources and Modern Studies. *History of Science* 8:53–93.

1980. *Essays on the History of the Microscope.* Oxford: Senecio.

Turner, R. D., and A. C. Johnson. 1971. Biology of Marine Wood-Boring Molluscs. Pp. 259–301 in *Workshop on Preservation of Wood in the Marine Environment, Portsmouth College of Technology, 1968: Marine Borers, Fungi and Fouling Organisms in Wood,* edited by E. B. Gareth Jones and S. K. Eltringham. Paris: OECD.

Van Helden, Albert. 1974a. "Annulo Cingitur": The Solution of the Problem of Saturn. *Journal for the History of Astronomy* 5:155–74.

1974b. Saturn and His Anses. *Journal for the History of Astronomy* 5:105–21.

1974c. The Telescope in the Seventeenth Century. *Isis* 65:38–58.

1977. *The Invention of the Telescope.* Philadelphia: American Philosophical Society.

1980. Huygens and the Astronomers. Pp. 147–65 in *Studies on Christiaan Huygens: Invited Papers from the Symposium on the Life and Work of Christiaan Huygens, Amsterdam, 22–25 August 1979,* edited by H. J. M. Bos, M. J. S. Rudwick, H. A. M. Snelders, and R. P. W. Visser. Lisse: Swets & Zeitlinger.

1983a. The Birth of the Modern Scientific Instrument, 1550–1700. Pp. 49–84 in *The Uses of Science in the Age of Newton,* edited by John G. Burke. Berkeley, etc.: University of California Press.

1983b. Roemer's Speed of Light. *Journal for the History of Astronomy* 14:137–41.

1985. *Measuring the Universe: Cosmic Dimensions from Aristarchus to Halley.* Chicago and London: University of Chicago Press.

Veen, Pieter Arie Ferdinand van. 1960. *De soeticheydt des buyten-levens, vergheselschapt met de boucken: Het hofdicht als tak van een georgische litteratuur.* The Hague: Van Goor Zonen.

Verbeek, Theo. 1992. *Descartes and the Dutch: Early Reactions to Cartesian Philosophy, 1637–1650.* Carbondale and Edwardsville: Southern Illinois University Press.

Verwey, Herman de la Fontaine. 1962. The Netherlands Book. Pp. 3–70 in *Copy and Print in the Netherlands: An Atlas of Historical Bibliography,* by Wytze Gs Hellinga. Amsterdam: North-Holland, etc.

Visser, R. P. W. 1981. Theorie en praktijk van Swammerdams wetenschappelijke methode in zijn entomologie. *Tijdschrift voor de geschiedenis der geneeskunde, natuurwetenschappen, wiskunde en techniek* 4:63–73.

Vries, Gerardus de. N.d. *Exercitatio physica gemina: altera de lumine; altera de lunicolis.* N.p.

Vries, Jan de. 1976. *Economy of Europe in an Age of Crisis, 1600–1750.* Cambridge, etc.: Cambridge University Press.

Vugs, J. G. 1975. Steno in Leiden. *Janus* 62:157–68.

Waard, Cornelis de. 1935. Le Manuscrit perdu de Snellius sur la réfraction. *Janus* 39:51–73.

Waard, Cornelis de, Jr. 1906. *De uitvinding der verrekijkers.* The Hague: H. L. Smits.

Warner, John Harley. 1982. "Exploring the Inner Labyrinths of Creation": Popular Microscopy in Nineteenth-Century America. *Journal of the History of Medicine and Allied Sciences* 37:7–33.

Watabe, Norimitsu, V. R. Meenakshi, Patricia L. Blackwelder, Elaine M. Kurtz, and Dana G. Dunkelberger. 1976. Calcareous Spherules in the Gastropod, *Pomacea paludosa.* Pp. 283–308 in *The Mechanisms of Mineralization in the Invertebrates and Plants,* edited by Norimitsu Watabe and Karl M. Wilbur. Columbia: University of South Carolina Press.

Waterbury, John B., C. Bradford Calloway, and Ruth D. Turner. 1983. A Cellulolytic Nitrogen-Fixing Bacterium Cultured from the Gland of Deshayes in Shipworms (Bivalvia: Teredinidae). *Science* 221:1401–3.

Westfall, Richard S. 1971. *Force in Newton's Physics: The Science of Dynamics in the Seventeenth Century.* London: Macdonald / New York: American Elsevier.

Wheelock, Arthur K., Jr. 1981. *Jan Vermeer.* New York: Harry N. Abrams.

Whitaker, Ewan A. 1978. Galileo's Lunar Observations and the Dating of the Composition of "Sidereus Nuncius." *Journal for the History of Astronomy* 9:155–69.

Wilberg Vignau-Schuurman, T. A. G. 1969. *Die emblematischen Elemente im Werke Joris Hoefnagels.* 2 vols. [Leiden]: Universitaire Pers Leiden.

Wilkinson, Lise. 1984. Rinderpest and Mainstream Infectious Disease Concepts in the Eighteenth Century. *Medical History* 28:129–50.

Wilson, Catherine. 1995. *The Invisible World: Early Modern Philosophy and the Invention of the Microscope.* Princeton: Princeton University Press.

Wilson, L. G. 1960. The Development of the Knowledge of Kidney Function in Relation to Structure – Malpighi to Bowman. *Bulletin of the History of Medicine* 34:175–81.

Winner, Ellen. 1982. *Invented Worlds: The Psychology of the Arts.* Cambridge, Mass., and London: Harvard University Press.

Wolfe, David E. 1961. Sydenham and Locke on the Limits of Anatomy. *Bulletin of the History of Medicine* 35:193–220.

Wolpert, Lewis, and Alison Richards. 1988. *A Passion for Science.* Oxford, etc.: Oxford University Press.

Yoder, Joella G. 1988. *Unrolling Time: Christiaan Huygens and the Mathematization of Nature.* Cambridge, etc.: Cambridge University Press.

Zanobio, Bruno. 1971. Micrographie illusoire et théories sur la structure de la matière vivante. *Clio medica* 6:25–40.

Ziggelaar, Augustine. 1980. How Did the Wave Theory of Light Take Shape in the Mind of Christiaan Huygens? *Annals of Science* 37:179–87.

Zloczower, A. 1981. *Career Opportunities and the Growth of Scientific Discovery in 19th Century Germany.* New York: Arno.

Zumthor, Paul. 1962. *Daily Life in Rembrandt's Holland.* Translated by Simon Watson Taylor. London: Weidenfeld & Nicolson.

Zupko, Ronald Edward. 1978. *French Weights and Measures Before the Revolution.* Bloomington: Indiana University Press.

Zuylen, J. van. 1981. On the Microscopes of Antoni van Leeuwenhoek. *Janus* 6:159–98.

1982. The Microscopes of Antoni van Leeuwenhoek. Pp. 29–55 in *Antoni van Leeuwenhoek 1632–1723,* edited by L. C. Palm and H. A. M. Snelders. Amsterdam: Rodopi.

Index

aberration
  chromatic, 17–19, 29, 285
  spherical, 17–18, 29
Académie de Bruxelles, 292
Académie Royale des Sciences, 25, 26, 199, 210, 291, 292n69
  Histoire de l'Académie Royale des Sciences, 273
Accademia dei Lincei, 38, 39n18, 58
Albinus, Bernard, 83, 83n17, 84n18, 86, 170n115
Albinus, Bernhard Siegfried, 83n17, 86, 93, 100, 170n115, 229
Aldrovandi, Ulisse, 48, 58, 107n10, 201, 208
Almquist, Ernst, 270n47
Alpers, Svetlana, 56n118, 77n64
amoeba, 262, 272, 275
Amsterdam, 123, 127, 131, 134, 147
  Athenaeum in, 83
anatomy, 85, 283, 291
  Leeuwenhoek and, 55, 165, 171–2, 217
  and the microscope, 64, 81–4, 96, 100, 103, 280
  see also comparative anatomy; insect anatomy; subtle anatomy
Anderson, Thomas F., 287n48
Angel, Philips, 73n51
aphid, 156, 203–4, 206, 251
Aristotle, 40, 105n4, 106, 145n180, 227n16, 233–5
Aselli, Gaspare, 42, 84
  De lactibus sive lacteis venis quarto vasorum mesaraicorum genere novo invento, 42
Ashton, Francis, 168
atomism, 33, 37, 39, 40, 106, 210
Auberius, L. C., 47n69
Augustine, St., 58n133, 226
Augustus, King of Poland, 161n66

Bacon, Francis, 37, 163, 209
bacterium, 31, 179–80, 264–5n22, 265n26, 276
Baer, Karl Ernst von, 285, 297
Baglivi, Georgius, 15, 281
Baker, Henry, 14–15, 28–30, 228, 261–2, 273, 284
Bartholin, Caspar, 125n77
Bayle, Pierre, 41
bee, 38, 48, 58, 127n87, 201, 204, 207
  anatomy, 110, 147
  drone, 109, 111n31, 114
  queen, 111n31, 205, 243, 247
  see also Swammerdam, and the bee
Beeckman, Isaac, 7n13
Belkmeer, Cornelius, 264n22
Belkmeer, Jan, 30
Bening, Simon, 69
Bergerac, Cyrano de, 264
Berlin, University of, 296
Bern, University of, 297
Bettini, Mario, 21
Beuningen, Koenraad van, 127n87
Bichat, Marie-François-Xavier, 92
Bils, Lodewijk de, 44–6, 85–6
Blaes, Gerard, 82–3
Blanckaert, Steven, 34–5, 84n18, 87, 89, 91, 93, 101, 108, 139, 171n119, 206n28, 229, 239
Bleyswijk, Abraham van, 171
Bleyswijk, Hendrik van, 149n11
Boddaert, Pieter, 262, 272
Boehme, Jacob, 245
Boerhaave, Herman, 92–3, 95, 98, 111n31, 123–5, 125n77, 130, 149n11, 152n28, 165, 170, 195, 223n2, 257, 282
  and the microscope, 95, 98–9
  and the vascular physiology, 92–6, 98, 274, 289
  Institutiones medicae, 92
Bois-Reymond, Émil du, 298n107, 300n116
Bonnet, Charles, 228–30, 273–4, 276, 279n79, 282–3, 294
Bontekoe, Cornelis, 65–6, 90–1, 93–4, 96, 170
Boreel, Willem, 7
Borel, Pierre, 37–8, 59, 81, 188
  Observationum microcospicarum centuria, 38

Borrichius, Olaus, 111n31, 123n74, 125n77
Bosch, Lodewyck Jans van den, 52
Bosschaert, Ambrosius, the Elder, 74
Bourignon, Antoinette, 127, 129, 291n61
Boyle, Robert, 26, 39, 106, 158–9, 169, 210
Breslau, University of, 294, 297
Brown, Robert, 285–6, 294, 296
Brownian motion, 261n6, 294
Buffon, Georges-Louis Leclerc, comte de, 17, 261n6, 266–72, 274, 275n62, 278–9, 281, 284–5
Buonanni, Filippo, 24n93, 54, 68
  Observationes circa viventia, quae in rebus non viventibus reperiuntur, 221n100
Burgersdijck, Franco, 202, 209
butterfly, 208, 212–14, 247, 249; *see also* Swammerdam, and the butterfly

Calandra granaria Linnaeus, *see* grain weevil
Calvin, Jean, 57–9, 77–9, 118
Cartesianism, 39–41, 62–6, 91, 183, 290
  and generation, 247
  Leeuwenhoek and, 183–4
  and mechanism, 39–41, 62–3, 67–8, 183–4, 189, 210–11
  and the microscope, 61–8, 79, 103
  and rationalism, 62–3, 65–9, 93, 184
Cats, Jacob, 57, 59
cell theory, 284–6, 289, 295–8
Cesi, Federico, 58
Charles II, King of England, 161n66
cheese skipper fly, 114, 204
circulation of the blood, 4, 41–3, 89–90, 93, 96, 185, 196
  Leeuwenhoek and, 94–5, 151, 155–6, 175–6, 178–9, 181–2, 184–5, 189, 199
Cluyt, Auger, 53
Cluyt, Dirck Outgerszoon, 109n22
cochineal, 206n25
Colvius, Andreas, 37
comparative anatomy, 108
  Leeuwenhoek and, 109, 180; and microscope, 108
  Lyonet and, 108n17
  Swammderdam and, 108n14, 109, 111–12, 124n75, 132, 144, 180
Cossus cossus (Linnaeus), *see* goat moth
Craanen, Theodoor, 63–7, 85, 170, 225, 229
  Oeconomia animalis, 63
  Tractatus physico-medicus de homine, 63, 66
Cudworth, Ralph, 211, 225n10, 248
Cuvier, Georges, 274

Darwin, Erasmus, 275
De Boekzaal van Europe, 150
Delbrück, Max, 130n100
Delft, 6, 46–7, 147, 155–7, 160, 163, 171, 182n35, 221
Descartes, René, 34–5, 39–41, 45, 61, 62, 66–7, 90, 92, 168, 183–4, 202, 210–11, 279
  and dioptrics, 18, 34–5
  and embryonic development, 224–5
  and generation, 210–11, 225
  and the microscope, 9, 18, 37, 62
  Discours de la méthode, 40
  La Dioptrique, 34
  Meditationes de prima philosophia, 40
  Principia philosophiae, 40
  Traité de l'homme, 41
Deusing, Anton, 45
Diderot, Denis, 269, 273
Diemerbroeck, Ysbrand van, 33, 43n38, 45, 225
Digby, Kenelm, 236
  Two Treatises, 237n54
dioptrics, 18, 32–5
Döllinger, Ignaz, 297
Drebbel, Cornelis, 6–8, 12, 38, 59, 60, 61
Drelincourt, Charles
  De foeminarum ovis, 171n120
Du Hamel, Jean-Baptiste, 210
Dutch Reformed Church, 57, 59, 78

East India Company, 156
eel, 151, 155, 176, 184, 194, 203, 221
embryonic development, 224, 230, 241, 295–6
  chick, 38, 65, 224, 230–41, 253, 257n142, 281–2, 288–9
  Descartes and, 224–5
  frog, 243–4, 246, 249, 253, 257n142
  Harvey and, 224–5, 231–8, 245, 246n102
  Highmore and, 235–9
  Leeuwenhoek and, 142, 175, 199, 241, 253 257
  Malpighi and, 237–41, 257n142
  plant seed, 229, 230, 251–2, 258
  Swammerdam and, 241, 244–6, 249
epigenesis, 232–3, 237, 245–6, 249, 288
Erlangen, University of, 297
erythrocyte, 30, 64, 188–9, 194–6, 263, 272, 280–1, 294
  Leeuwenhoek and, 64, 153–4, 175–6, 179, 182, 188–9, 190–1, 192–8, 258, 280, 289, 289n58, 293n74
  Swammerdam and, 188n51, 188, 189, 194
Eustachi, Bartolomeo, 42n31, 95n88

Fabrici, Girolamo, 101n115
Fabri, Honoré, 253n128
Feylingius, Johannes, 57
fiber, 92, 102
muscle, 82, 101–2, 254, 258, 281, 287, 289, 292n67; *see also* Leeuwenhoek, and muscle fiber,
flea, 21, 38, 76, 153, 156, 203–6, 208–9, 216, 218–19
anatomy, 107
Florence, 21
flower painting, 51–3, 56, 70–2
fluke, 216
cercariae, 139n142
"flying water scorpion," 136
Folkes, Martin, 30, 152n28
Fontana, Francesco, 7n11, 38
Fontenelle, Bernard Le Bouyer de, 87, 88, 267n32
Forsten, Rudolph, 97n93, 282n13
Fracastoro, Girolamo, 262n14, 263n15
Franeker, University of, 100, 106
Frederick I, of Prussia, 161n66
frog, 175, 178, 185, 188, 199; *see also* embryonic development, frog; tadpole
fungus, 269–71, 274, 274n59, 275, 279n82

Galen, 42, 43, 79, 89
Galileo Galilei, 1, 3, 6–7, 8n18, 12, 32, 36, 38, 40
Sidereus nuncius, 1, 36
Gassendi, Pierre, 39, 106
Gaubius, Hieronymus David, 136n130, 137n131, 245
generation (doctrine), 4, 38, 61, 65, 131, 200, 205, 223–5, 227, 229–31, 233–7, 240–1, 248, 251, 261, 266–8, 281; *see also* embryonic development; preexistence; preformation
animal, 109, 112, 122, 124
Cartesianism and, 247
Descartes and, 10–11, 225
Leeuwenhoek and, 168, 171–2, 216–17, 219, 229n21, 251–2, 257–8
plant seed, 229
spontaneous, 106, 201–3, 208–9, 211–13, 222, 223–4, 251; Hooke and, 202, 211–12; Leeuwenhoek and, 201, 203–4, 206–8, 215–22, 251; and the microscope, 38, 68n29, 124, 201, 204, 206–9, 214–15, 218, 222, 263, 266–71, 274–5, 278–9; Swammerdam and, 53, 201, 203–4, 206–7, 211–15, 220, 247
Swammerdam and, 109n21, 124, 241, 245, 247
George I, King of England, 161n66
Gheyn, Jacques de, 51n92, 51–2, 61, 70n38, 70–1, 73–4, 75n59, 76
gland, 30, 41, 43–4, 46, 65, 82, 89–90, 97–8, 108
Gleichen-Russworm, Wilhelm Friedrich von, 3, 270n48, 273, 278n78, 281, 292
goat moth, 283
Goedaert, Johannes, 53–9, 68, 70, 134, 202
Metamorphosis naturalis, 53, 54, 134
Goltzius, Hendrik, 70n40, 75n59
Goorle, David van, 33
Gorter, Johannes de, 92
Göttingen, University of, 297
Graaf, Regnier de, 46–8, 82, 86, 89, 90, 149, 171, 230, 239
and injection, 85–7
and Leeuwenhoek, 147–9, 170–1, 172n123, 258
and the microscope, 85
and Swammerdam, 46–8, 86, 89, 120, 120n64, 122, 125, 147
grain moth, 203–4, 221
grain weevil, 203–4, 206, 220–2
Grew, Nehemiah, 15, 149, 154, 157, 162n73, 171, 198, 200n102, 219, 229, 252n126, 256, 258, 280, 284, 291, 292
Groningen, University of, 45
Guenellon, Pieter, 91
Guericke, Otto von, 23n90

Haberkorn, Johannes, 196
Haller, Albrecht von, 30, 92, 95, 194, 240, 274, 280–3, 285, 288–9, 291, 294
Sur la formation du coeur dans le poulet, 281
Halle, University of, 281
Halley, Edmond, 292n67
Ham, Johan, 23n91, 216–17, 217n75
Hartsoeker, Nicolaas, 18–19, 23–5, 27, 34n162, 189, 199, 217, 223n2, 225, 227–8, 251, 256n139, 261, 263, 287
and Leeuwenhoek, 23, 108, 152–3, 160, 163, 164, 165, 203, 222, 223n2, 284, 292n67
microscopes, 13–16, 19, 22–25, 27, 29, 30, 36
microscopic discovery, the potential for (attitude), 173
Essay de dioptrique, 14n40
Harvey, William, 41–3, 89, 171, 202, 233, 258, 288

Harvey, William *(cont.)*
and embryonic development, 224–5, 231–8, 245, 246n102
and metamorphosis, 212–14
and the microscope, 107
Exercitatio anatomica de motu cordis et sanguinis in animalibus, 42
Exercitationes de generatione animalium, 224, 231, 235
Hauptmann, August, 263n15
Heere, Lucas de, 74n57
Heide, Anton de, 81–2, 84, 94, 207n30, 291n61
Heinsius, Antonie, 149n11, 163
Heisenberg, Werner, 299n115
Heister, Lorenz, 42n34
Helmont, Johannes Baptista van, 202, 210, 225, 245
Henle, Jacob, 286, 293, 296–7
Henshaw, Thomas, 199n95
Hessen-Kassel, Landgrave of, 161n66, 163
Heurnius, Otto, 43n38
Hewson, William, 27n120, 195, 281
Highmore, Nathaniel, 42n30, 202, 235
and embryonic development, 235–9
Hill, John, 16, 30n138, 276–7, 292, 303
Hire, Philippe de la, 26n111
Hodgkin, Dorothy, 287n48, 300n118
Hoefnagel, Jacob, 51n92, 52–4, 68, 70, 77n64
Hoefnagel, Jan, 52n101
Hoefnagel, Joris, 50–2, 54, 56, 69–70, 70n38, 72, 74
Hoeve, Jacobus vander, 91n61, 92n63
Holbein, Hans, 74n58
hollow nerve, 45, 65–6, 85, 92, 102, 198
Holwarda, Jan Fokkens, 106
Hoogstraten, Samuel van, 73, 74n57
Hoogvliet, Arnold, 168
Hooke, Robert, 2, 8, 12, 15, 22, 24, 24n96, 31, 59, 76, 111, 125n77, 157, 158, 160, 168, 185, 190, 247, 280, 284, 288, 291, 292n67
and insect dissection, 107, 114
and Leeuwenhoek, 26, 148n3, 149, 154, 158, 162n73, 164, 169, 199, 241n75, 291n60, 292n67
and spontaneous generation, 202, 211–12
Lectures and Collections, 162n73
Micrographia, 8, 22, 23n90, 24, 31, 76, 148n3, 158, 212, 291
Horne, Johannes van, 43, 45–7, 109, 111n31, 120n64, 121–4
Houbraken, Arnold, 74nn57–8
Houttuyn, Martinus, 262, 285
Hudde, Johan, 22n89, 22–3, 30–1, 61, 127n87, 143, 148
Hull, David, 170n113, 298n110, 299n113, 300n117
Hunter, John, 281
Huygens, Christiaan, 13, 18, 22n89, 23n91, 24–5, 27, 35, 38, 66, 76, 120, 140, 149n11, 153, 160, 184, 188n54, 199, 217, 219n88, 223n2, 229n21, 226, 250, 257, 262n11, 281
and astronomy, 1–4
and Leeuwenhoek, 2, 30, 159, 162–3, 168–9, 199
microscopes, 12–13, 15, 23, 25–30, 37
Discours de la cause de la pesanteur, 168
Traité de la lumière, 168
Huygens, Constantijn, Jr., 25, 28, 29n130, 160 163, 226, 250–1
Huygens, Constantijn, Sr., 24n96, 57, 59, 61, 71, 76
and Leeuwenhoek, 149, 154, 160, 162, 164 170, 173
and the microscope, 8, 138
hydra, 10, 270n47, 272–4
Hydra oligactis Pallas, 273
Hydra viridissima Pallas, 273
Hydra vulgaris Pallas, 273

illusionism (in art), 48–50, 52, 72–6
illustration, scientific, 48, 50–3, 54, 69–70, 72, 76, 83, 134
injection (technique), 84–9, 93–8, 100–1, 293
Graaf and, 85–7
and the microscope, 85, 98–100, 102–3
Ruysch and, 85n24, 86–9, 94–8, 100
Swammerdam and, 85–9, 98, 110, 117, 120n64
"insect" (term), 105–6
insect anatomy, 21, 29, 38–9, 79, 106–9, 112, 114, 134, 153, 206, 209, 283–4
Leeuwenhoek and, 108, 206–7
and the microscope, 106–7
Swammerdam and, 29, 107, 110–11, 113–14, 116–17, 132, 135–7, 140–5, 206–7, 214, 302
see also specific insect; insect dissection
insect dissection, 10, 107–9, 114, 129, 153, 205, 282–4, 294
Hooke and, 107, 114
Leeuwenhoek and, 94, 108, 113n38, 151, 153, 204–6, 216, 221
Malpighi and, 108, 112–13, 116, 121–2, 130, 132, 145, 205

Swammerdam and, 91, 105, 109–14, 116–17, 121, 126, 129–32, 135, 138, 142–3, 145, 167, 204–5, 214–15, 282–4
institutionalization of microscopic research, 288, 290–301

James II, King of England, 161n66, 164
Janssen, Sacharias, 7
Jardin du Roi, Paris, 17, 266
Joblot, Louis, 276, 279n82
Descriptions et usages de plusieurs nouveaux microscopes, 276
Jonston, John, 208
Journal des Sçavans, 26–7
Jurin, James, 197n88

Kepler, Johannes, 32, 34
Ad Vitellionem paralipomena, 33n151
Dioptrice, 32
Kerckring, Theodorus, 97, 230, 264, 265n22
Anthropogeniae ichnographia, 230n27
Kircher, Athanasius, 3, 21, 22n85, 37–8, 58, 83, 188, 201, 208, 262–3
D'onder-aardse weereld, 221n100
Mundus subterraneus, 221n100
Knorr, Karin D., 295n82
Königsberg, University of, 297

lacteal vessel, 42–4, 84, 89
l'Admiral, Jacob, 77n64
l'Admiral, Jan, 77n64
Lairesse, Gerard de, 56
La Mettrie, Julien Offray de, 93n74
Lange, Christian, 263n15
Langly, Willem, 47n71
Lavenseau, L., 243n81
LeClerc, Daniel, 242n76
Leclerc, Jean, 224–5
L'Écluse, Charles de, 72
Ledermüller, Martin Frobenius, 23n90, 292
Leeuwenhoek, Antoni van, 12, 23, 47, 64, 67, 79, 83, 97n93, 147, 157–8, 173, 175, 272, 280, 287, 290, 293, 300–1, 304
aggressiveness, 169–72, 174, 259
and anatomy, 155, 165, 171–2, 217; *see also* Leeuwenhoek, and comparative anatomy
and artists, 155, 156, 182n35, 256n136
and the capillaries, 83n15, 94, 95, 155, 156, 175, 176, 179, 181, 182, 184, 185, 197, 199, 253, 258, 280, 293n74
and Cartesianism, 183–4
and the circulation of the blood, 94–5, 151, 155–6, 175–6, 178–9, 181–2, 184–5, 189, 199
collection (microscopic preparations), 150–2
and the common man, 161, 163, 165, 174, 222, 263n19, 303
and comparative anatomy, 108, 109, 180
credibility, the problem of, 152–5, 284
and embryonic development, 142, 175, 199, 241, 253, 257
and the erythrocyte, 64, 153–4, 175–6, 179, 182, 188–9, 190–1, 192–8, 258, 280, 289, 289n58, 293n74
and experimental science, 158–60, 219
and generation (doctrine), 168, 171–2, 216–7, 219, 229n21, 251–2, 257–8; *see also* Leeuwenhoek, and embryonic development; ~ and preexistence; ~ and preformation; ~ and spermatozoa
globules, doctrine of, 64, 154, 188, 190, 192, 196
and Graaf, 147–9, 170–1, 172n123, 258
and Hartsoeker, 23, 108, 152–3, 160, 163, 164, 165, 203, 222, 223n2, 284, 292n67
and Hooke, 26, 148n3, 149, 154, 158, 162n73, 164, 169, 199, 241n75, 291n60, 292n67
and Huygens, Christiaan, 2, 30, 159, 162–3, 168–9, 199
and Huygens, Constantijn, Sr., 149, 154, 160, 162, 164, 170, 173
and insect anatomy, 108, 206–7
and insect dissection, 94, 108, 113n38, 151, 153, 204–6, 216, 221
and the learned world, 83, 161–2, 164–5, 165n83, 167–74, 183, 219–22, 291; *see also* Leeuwenhoek, and the medical profession; ~ and the Royal Society
and the leukocyte, 196
and mathematics, 35, 158, 160, 180–1, 184
and mechanism (doctrine), 67, 182–7, 197, 220, 288, 303
and the medical profession, 160, 170–2, 290
microscopes, 10–11, 14–16, 19, 21, 23n90, 24, 26, 30–2, 100, 146–8, 151, 153, 185, 230, 256n136, 286
microscopic discovery, potential for (attitude) 250, 257–9, 302–4
and microscopic life, 24, 26, 64, 67, 154, 159, 175, 178–81, 198, 208, 215–16, 218-19, 220n92, 260, 261, 262, 263, 275, 276 279, 280, 293n74
microscopic techniques, 16, 22, 151–4, 256, 291, 294
and muscle fiber, 151, 152n31, 158, 165, 185–7, 192n73, 197–8, 198n90, 280–1, 292n67. 302

Leeuwenhoek, Antoni van *(cont.)*
and observing, pleasure in, 152, 156–7, 175–6, 178–9, 182, 185, 191, 200
personal relations, 146, 149, 157, 162–3, 171, 290
and plant anatomy, 284n25
and the plant seed, 229, 251–2, 254, 257–8
and preexistence (doctrine), 241, 250–2
and preformation (doctrine), 217, 223, 250–9
publication, 150, 162, 170n110
and religion, 146, 166–7, 181, 219, 220, 302
and the Royal Society, 147–51, 154, 158, 160–2, 164, 169–73, 181, 195n81, 198–200, 203–4, 217, 250, 253, 257n142, 258–9, 263n19, 291, 291n60, 292n67, 301nn121–2, 303
and sciences (nonmicroscopical), 158–60, 168
secrecy, 19, 152–4, 157, 291
self-perception, 147, 157, 160–2, 166, 169, 172–4, 222, 258, 259, 291, 303–4
sensitivity (personal), 169–70, 174
skills, manual, penchant for, 148, 152–3, 158, 172, 186, 188, 197
social context, of his microscopy, 146–52, 155–7, 160–74, 195, 198–200, 220–2, 258–9, 290, 298, 300–1, 303; *see also* Leeuwenhoek, and the common man; ~ and the learned world; ~ social deficiencies
social deficiencies (personal), 162–6, 168–9, 290
and the spermatozoa, 23n91, 24, 26, 151, 155–6, 159, 172, 175, 178–9, 199–200, 204, 216–19, 223, 250–2, 254–60, 268n42, 272, 281, 286, 293n74
and spontaneous generation, 201, 203–4, 206–8, 215–22, 251
and Swammerdam, 30–1, 152, 163–4, 169–70
Send-brieven, 170n110
Leeuwenhoek, Maria van, 146, 151, 155n40, 156, 169
Leibniz, Gottfried Wilhelm, 67, 149, 164, 226–8, 264
Leiden, 56n120, 147
Leiden, University of, 34n163, 40, 63, 70, 72, 77, 202, 209, 271, 291
medical school, 41, 46, 63, 83, 87, 91–2, 109, 120, 189, 216, 223, 263, 290–1, 293; anatomical research, 41, 45–7
lens
achromatic, 19, 285
aspherical, 18–19
bead, 14n36, 19, 21–31, 95, 102, 148, 152, 275, 280, 284
Libavius, Andreas, 107, 205
Lieberkühn, Johannes Nathanael, 100
light, nature of, 34–5
Linnaeus, Carl, 261–3, 270–2, 274n59, 276–7
Mundum invisibilem, 261n9
Locke, John, 26
louse, 38, 147, 158–9, 203, 206–7, 209, 216
anatomy, 107, 111, 114, 116, 137, 140–1, 143, 204–5
lymphatic system, 42–4, 46, 84–6, 89
Lyonet, Pierre, 10, 15–16, 54, 113, 227, 282–4, 288, 292, 294
and comparative anatomy, 108n17
Traité anatomique de la chenille, qui ronge le bois de saule, 282–3

McClelland, David C., n84, 126
Magliabechi, Antonio, 164
Maillet, Benoît de, 275
Malebranche, Nicolas, 167, 181, 226–8, 242n76, 250n113
Recherche de la vérité, 167, 226, 242n76, 250n113
Malpighi, Marcello, 15, 41, 81–5, 90, 97, 105, 112, 114, 117, 119n62, 121–2, 124, 125n77, 138, 157, 167–8, 188, 204, 229, 280, 282, 284, 290–1, 294, 300
and comparative anatomy, 108, 109
and embryonic development, 237–41, 257n142
and insect dissection, 108, 112–13, 116, 121–2, 130, 132, 145, 205
and the microscope, promise of, 241
and preexistence, 240
and the silkworm, 112
Appendix repetitas auctasque de ovo incubato observationes continens, 237n55
Dissertatio epistolica de bombyce, 112–13, 124
Dissertatio epistolica de formatione pulli in ovo, 237n55
Mander, Karel van, 51n92, 52, 56, 56n118, 71 71n45, 74n58
Manget, Jean Jacques, 242n76
mannerism, 70n40, 71
Mannheim, 292
manuscript illumination, 48–52, 71n45, 72
Marsilius, Otto, *see* Schriek, Otto Marseus van
Martinet, Johannes Florentius, 288
Mary, Queen of England, 161n66
Massuet, Pierre, 30n139
Master of Mary of Burgundy, 49, 51, 69, 76
materialism (doctrine), 266–9, 273, 278–9, 288
Maurice, Count of Nassau (future Prince of Orange), 6, 8

mayfly, 56, 112–14, 117, 120, 140
anatomy, 109, 137, 140–1, 145
mechanism (doctrine), 34–5, 40, 62–3, 89, 93, 106, 185, 195, 202, 210, 224–5, 227, 233, 235–6, 244–7, 266, 271, 278, 288
Cartesianism and, 39–41, 62–3, 67–8, 183–4, 189, 210–11
Leeuwenhoek and, 67, 182–7, 197, 220, 288, 303
and the microscope, 62–3, 67–8, 79, 103
Swammerdam and, 144n176, 212, 244
medical profession, Leeuwenhoek and, 160, 170–2, 290
medical school, *see specific university*
and the microscope, 291, 293–7
Meerman, Jan, 149n11
Merian, Maria Sibylla, 77n64
metamorphosis, 54, 56, 124, 212–14; *see also* Swammerdam, and metamorphosis
microscope
and academic contexts, 62, 83, 86; *see also* microscope, and the medical schools
achromatic, 285–7
anatomy and, 64, 81–4, 96, 100, 103, 280
Cartesianism and, 61–8, 79, 103
compound, 7–8, 12–13, 15–16, 18, 24–5, 28, 30, 38, 275n62, 285–6
distrust of, 97–8, 284–7
electron, 287nn46–7
injection and, 85, 98–100, 102–3
insect anatomy and, 106–7
mechanism and, 62–3, 67–8, 79, 103
and the medical schools, 291, 293–7
miniature painting and, 54, 68–9, 75–7, 79, 103, 290
nature as revelation and, 58–9, 77–9, 103, 138
potential of, 173, 234, 288–9, 302
preexistence and, 228–30, 241, 249, 250
preformation and, 229–30, 241, 257–9, 288
Royal Society and, 291–2
simple, 8–10, 13, 15–16, 22, 24, 26–8, 231, 235, 237, 285–6
and spontaneous generation, 38, 68n29, 124, 201, 204, 206–9, 214–15, 218, 222, 263, 266–71, 274–5, 278–9
subtle anatomy and, 79, 80–2, 85–6, 97, 99, 101, 103
the vascular physiology and, 91, 93–5, 98–9, 101–3, 289–90
see also specific researcher
Microscopical Society of London, 299
microscopic illustration, 38, 112
microscopic life, 24, 38, 181, 198–9, 208, 260–2, 268, 271–2, 276–7, 279
and the dynamism of nature, 266–7, 269, 271–2, 274–6, 278
"infusoria," 261–2, 267–72, 274–9
as mutable, 269–72
as organic building blocks, 264–7, 281
as pathogenic, 262–4, 277, 278n78
purpose of in nature, 266, 277–8
as symbiotic, 264–6, 277, 278n78
see also fungus; Leeuwenhoek, and microscopic life; protozoa
microscopic research, institutionalization of, 288, 290–301
microscopic technique, 63, 84, 98, 104, 112, 237–8, 240, 287–9, 294
of Leeuwenhoek, 16, 22, 151–4, 256, 291, 294
of Swammerdam, 16, 98, 112–14, 129, 294
of Trembley, 16
Middelburg, 6–7, 53, 72, 74, 291n61
miniature painting, 48, 50–3, 54, 61, 69–71, 71n46, 74, 76
and the microscope, 54, 68–9, 75–7, 79, 103, 290
Swammerdam and, 132
mite, 8, 38–9, 58, 106–7, 153, 156, 167, 206–7, 209, 263n19
Moffett, Thomas, 8n17, 58, 107n10, 205
Molyneux, Thomas, 164–5
Müller, Johannes, 296–7, 298n108, 299n114, 300n116
Müller, Otto Frederik, 261, 261n6, 270–2, 275–7, 279n79, 292
Münchausen, Otto, Baron von, 269–71, 274n59
Musschenbroek, Johan van, 28
Musschenbroek, Petrus van, 271, 277
Musschenbroek, Samuel, 24
Musschenbroek shop, 188n54, 194
Muys, Wijer Willem, 30, 100–3, 194, 228, 278n78, 289
Investigatio fabricae, quae in partibus musculos componentibus extat, 101

Naples, 262, 280
natural history, 48, 50, 52–3, 54, 69–70, 132, 208, 222, 226, 271, 276, 292; *see also* Swammerdam, and natural history
nature as revelation, 54, 57–9, 61, 77–9, 106, 117, 228, 278–9, 283, 288
and the microscope, 58–9, 77–9, 103, 138
Swammderdam and, 88, 105, 116–19, 137–44, 211–12, 246, 302

Needham, John Turberville, 268–71, 273, 275n62, 279, 279n79, 281, 285, 292
Nepa cinerea Linnaeus, *see* "flying water scorpion"
Newton, Isaac, 18, 40, 93, 184
Principia, 184
Nieuwentyt, Bernard, 227
Nuck, Anton, 85, 87–91, 95, 97–8

Odierna, Giovan Battista, 21, 38, 39n18, 107, 108
Oken, Lorenz, 274n59
Oldenburg, Henry, 2–3, 124n75, 147–9, 154, 158–61, 166, 217, 239, 291
Oort, Johan, 116, 126, 128
Orgyia antiqua (Linnaeus), 242–3, 247, 249
Overkamp, Heidentryk, 92n63

Palingenia longicauda (Olivier), *see* mayfly; *see also* Swammerdam, *Ephemeri vita of afbeeldingh van's Menschen Leven*
Paludina vivipara, 138n138; *see also* Swammerdam, and the miraculous . . . snail
Paracelsus, Theophrastus, 202, 209
Paris, 268
Parisano, Emilio, 236nn50–1
Pascal, Blaise, 39, 167
Paul, St., 57, 142
Paulitz, Johannes, 263
Pecquet, Jean, 42n31
Peiresc, Nicolas-Claude Fabri de, 7, 38, 58
Pepys, Samuel, 73n53
Peter I, Czar, 161n66
Philosophical Transactions, 26, 148–50, 159, 162n73, 166, 170, 190
physiology, 41–3, 280–1, 285, 288, 291, 294; *see also* vascular physiology
Picard, Jean, 26n111
Piophila càsei (Linnaeus), *see* cheese skipper fly
plant anatomy, 284–5, 288, 296
Leeuwenhoek and, 284n25
Plemp, Vopiscus Fortunatus, 32–4, 79
Ophthalmographia, 32
Pliny, 58n133, 59, 73n53, 106
Pluche, Noël-Antoine, 226–7
polyp, 269, 273, 274n59, 280n2; *see also* hydra
Porta, Giovan Battista della, 12n26
Power, Henry, 2–3, 18, 37, 59, 107n12
preexistence (doctrine), 226–9, 240–1, 244, 246, 250, 257n142, 268n42, 281, 288
Leeuwenhoek and, 241, 250–2
Malpighi and, 240
and the microscope, 228–30, 241, 249, 250
Swammerdam and, 241–50, 252n123
preformation (doctrine), 225–6, 229, 237, 239-40, 288
Leeuwenhoek and, 217, 223, 250–9
and the microscope, 229–30, 241, 257–9, 288
Swammerdam and, 242–4, 246
proteus, 262, 275; *see also* amoeba
protozoa, 276, 286, 296
Purkinje, Jan, 293–4, 297, 298n108

Rabus, Pieter, 199n100
Raey, Joannes de, 41, 62n2, 62–4
Ramon y Cajal, Santiago, 185
Rau, Johannes Jacobus, 83n17, 84n18
Ray, John, 210, 225, 247–8
Razet, Jaques (=Jacob?), 51n92
Réaumur, René-Antoine Ferchault de, 54, 105n4, 134, 227, 263, 273, 283
Redi, Francesco, 15, 125n77, 203–5, 219, 279
Esperienze intorno alla generazione degl' insetti, 203, 215
Regius, Henricus, 40, 45n54, 63–4, 202, 209-10, 212, 225
Remak, Robert, 296
Rembrandt van Rijn, 72, 74n58
Reneri, Henricus, 61, 62n2
Rome, 21, 262
Römer, Ole, 26n111
Ronchi, Vasco, 8n17
Roos, Johannes Carolus, 261n9
Rösel von Rosenhof, Johann, 77n64, 261–2
Ross, Alexander, 226
rotifer, 261, 273, 276, 301n122
Roy, Henri de, *see* Regius, Henricus
Royal Society, 2, 12, 14, 47, 112, 27n114, 194, 199, 201, 203, 222n102, 237, 239;
and the microscope, 291–2
see also Leeuwenhoek, and the Royal Society
Rucellai, Giovanni, 8n17
Ruysch, Frederik, 46, 84, 86, 90, 108, 120, 120n64, 125n77, 170, 230
and injection, 85n24, 86–9, 94–8, 100
and the microscope, 85, 98, 100, 230
and the vascular physiology, 91, 94–7

scabies, 263n19
Schacht, Herman Oosterdijk, 84n18
Schelstrate, Emmanuel, 24n93
Schleiden, Matthias Jacob, 286, 294–7
Schon, Donald, 170n113
Schott, Gaspar, 22, 22n85, 24n93
Schrader, Justus, 47n71
Schriek, Otto Marseus van, 53
Schwann, Theodor, 286n38, 295–9, 300n116

Seghers, Daniel, 73n53
Sellius, Gottfried, 264–6, 277–8, 284
Severino, Marco Aurelio, 205
's Gravesande, Cornelis, 170
shipworm, 264–5, 277–8
Siebold, Theodor Ernst von, 296–7
silkworm, 48, 107, 112, 114, 116, 119n62, 121, 134n111, 167, 205, 253, 291
Sladus, Mattheus, 126
Sloane, Hans, 70n38, 263n19
Snel, Willebrord, 33n158
soldier fly, 135, 141, 145, 215, 248
Spallanzani, Lazzaro, 269, 292
species, visible (doctrine), 33–4
spermatozoa, 25–6, 260, 267, 281, 285–6
  discovery of, 23–5, 216
  see also Leeuwenhoek, and the spermatozoa
Spinoza, Baruch, 22n89
Stelluti, Francesco, 58
Steno, Nicolaus, 46–7, 82, 90, 101, 111, 120, 120n64, 122n69, 126, 205
  and the microscope, 85–6, 101
Stiles, F. H. Eyles, 280
Stratiomyia chamaeleon Linnaeus, 135n115; *see also* soldier fly
Stratiomyia furcata (Fabricius), *see* soldier fly
subtle anatomy, 41, 43, 45–6, 48, 61–2, 79, 81, 86, 88, 96, 293
  and the microscope, 79, 80–2, 85–6, 97, 99, 101, 103
Swammerdam, Jan, 22n89, 35, 44, 52n101, 53, 54, 64, 82, 105n4, 130, 146–7, 157, 167, 176n2, 239, 280, 287, 290, 293, 300–1, 304
  aesthetic response, 131, 134–6; *see also* Swammerdam, artistic sensibilities; ~ artistry in nature
  ambition, 119–24, 128, 130, 145
  and anatomy, 46–8, 105, 111–12, 117, 121–4, 129, 131–2, 134, 144; and the microscope, 85
  artistic sensibilities, 182, 301–2
  artistry in nature, 136–7, 140, 144, 182, 302
  and the bee, 109–10, 117, 120, 121, 144; anatomy, 110, 113–14, 116, 124, 126, 129, 135; dissection, 109; metamorphosis, 110
  and the butterfly, 117, 134, 136, 144; anatomy, 110, 135; dissection, 142; metamorphosis, 124, 131, 135
  collection (anatomy and natural history), 109, 110, 123, 150, 161
  and comparative anatomy, 108n14, 109, 111–12, 124n75, 132, 144, 180
  and competition, 29, 46–8, 86, 112, 119–24, 122, 128
  cultural legacy, its impact on microscopic research, 131, 134–5, 137–8, 141–3, 145
  and embryonic development, 241, 244–6, 249
  emotional crisis, 117, 118, 119, 120, 121, 125n77, 127
  and the erythrocyte, 188n51, 188, 189, 194
  and generation, 109n21, 124, 241, 245, 247; *see also* Swammerdam, and embryonic development; ~ and preexistence; ~ and preformation
  grace (spiritual) in science, 118, 143, 144, 302
  and injection, 85–9, 98, 110, 117, 120n64
  and insect anatomy, 29, 107, 110–11, 113–14, 116–17, 132, 135–7, 140–45, 206–7, 214, 302; and *see* Swammerdam, and insect dissection
  and insect dissection, 91, 105, 109–14, 116–17, 121, 126, 129–32, 135, 138, 142–3, 145, 167, 204–5, 214–15, 282–4
  insect life cycles, 117, 136, 212–14, 247–50
  and Leeuwenhoek, 30–1, 152, 163–4, 169–70
  and mechanism (doctrine), 144n176, 212, 244
  and metamorphosis, 53, 56, 117, 124, 131–2, 132n108, 135, 141, 213–15, 245–9
  microscopes, 15–16, 24n96, 24–5, 27, 29–30
  microscopic discovery, the potential for (attitude), 143–5, 302, 304
  and microscopic dissection, 282, 303; *see also* Swammerdam, and insect dissection
  microscopic techniques, 16, 98, 112–14, 129, 294
  and miniature painting, 132
  and the miraculous, viviparous, crystalline snail, 138–41, 144–5, 243, 244n90
  moth, 144; *see also* *Orgyia antiqua* (Linnaeus)
  and natural history, 53, 105, 109, 112, 123–4, 131, 136, 145, 212–14, 248–50
  and nature as revelation, 88, 105, 116–19, 137–44, 211–12, 246, 302
  and order in nature, 131–2, 135–7, 212–14, 220, 247–9
  personal relations, 46–8, 120, 125–8, 163, 169; *see also* Swammerdam, social relations
  and preexistence, 241–50, 252n123
  and preformation, 242–4, 246
  research: emotional returns of, 129–31, 214, 303; as piety, 116–19, 142, 145; as spiritually perilous, 105, 118–20

Swammerdam, Jan *(cont.)*
social relations 124–31, 145–7, 174, 300; *see also* Swammerdam, personal relations
and spontaneous generation, 53, 201, 203–4, 206–7, 211–15, 220, 247
Bybel der natuure, 136n130, 214n63, 246n102, 247n104, 282–3
Ephemeri vita, 56, 112–13, 118, 120–1, 123, 127, 129, 142, 282
Historia insectorum generalis, 106n4, 110–14, 120–1, 127, 131–2, 134, 136, 138, 140, 143, 213–14, 241–2, 245n90, 247, 249n111
Miraculum naturae, 122, 123, 242
Tractatus de respiratione, 244n90
Syen, Arnold, 24n96
Sylvius, Franciscus dele Boë, 33, 45–6, 143

tadpole, 95, 175, 176, 178, 184
telescope, 1–8, 12, 18, 27, 36–8
Teredo navalis Linnaeus, *see* shipworm
Thévenot, Melchisédech, 109nn22–3, 111n31, 120, 124–8, 139, 145, 204, 282
Tinaea granella Linnaeus, *see* grain moth
Torre, Giovanni Maria della, 27n120, 28, 30, 280
Torricelli, Evangelista, 21
trachea, 94, 114, 116–17, 135, 137, 140, 143, 145, 283
Trembley, Abraham, 10, 15, 272–3, 276, 280n2, 292
microscopic techniques, 16
Tschirnhaus, Ehrenfried Walter, Baron von, 168
Turner, Ruth D., 265n22
Tylenchus tritici, 273n54

Uffenbach, Zacharias von, 19, 21, 151–2, 155n40, 165, 169
Utrecht, University of, 30, 40, 45, 61
medical school, 225

Valentin, Gabriel, 297, 298n108
Valkenburg, Adrianus van, n38, 43
vascular physiology, 43–5, 89–93, 95–8, 101–2 192, 289, 293
Boerhaave and, 92–6, 98, 274, 289
and the microscope, 91, 93–5, 98–9, 101–3 289–90
Ruysch and, 91, 94–7
Velthuysen, Lambert van, 216, 219, 250
Verelst, Simon Pieterszoon, 73n53
Vermeer, Jan, 182n35
Vieussens, Raymond, 96n92
Virchow, Rudolf, 298
Viviparus viviparus (Linnaeus), *see* Swammerdam, and the miraculous . . . snail
Vogt, Carl, 296–7
Volder, Burchardus de, 67n25
Voltaire, 269n46
volvox, 261–2, 272
Vondel, Joost van den, 57, 73n53
Vossius, Isaac, 22, 34–5, 37, 209

Wagner, Rudolph, 297, 298n108
William III, Stadholder and Prince of Orange, 127n87, 160
Withoos, Mathias, 71n46
Withoos, Pieter, 71n46
Witsen, Nicolaas, 149n11
Wolff, Caspar Friedrich, 281–2, 288
wood anatomy, 150, 154, 158, 175, 198, 252n126
Wren, Christopher, 190
Wrisberg, Heinrich August, 261n6, 270–1, 275n62, 279n79
Würzburg, University of, 297

x-ray crystallography, 287n48, 300n118

yeast, 190–2, 197

Zuylen, J. van, 7n13, 12n29, 17n64, 31n142